AF531296

DPH MATHEMATICS SERIES

TEXT BOOK OF DIFFERENTIAL CALCULUS

(For B.A., B.Sc., B.Com., I.A.S., P.C.S.)

By

A.K. Sharma

DISCOVERY PUBLISHING HOUSE
NEW DELHI-110002

Published by:
Namit Wasan
DISCOVERY PUBLISHING HOUSE PVT. LTD.
4383/4B, Ansari Road, Darya Ganj
New Delhi-110 002 (India)
Phone : +91-11-23279245; 23253475; 43596065
E-mail : discoverybooksindia@gmail.com
discoverypublishinghouse@gmail.com
namitwasan9@gmail.com
web : www.discoverypublishinggroup.com

***Edition:* 2020**

ISBN: 978-81-7141-844-2

Text Book of Differential Calculus

Printed at:
Infinity Imaging Systems
Delhi

Preface

The book is written to meet the requirements of B.A., B.Sc., students of various Indian Universities. The subject matter is exhaustive and attempts are made to present things in an easy-to-understand style. In solving the questions, care has been taken to explain each step so that student can follow the subject matter them self without even consulting others. A large number and solved and self practice problems (with hint and answer) have been included in each chapter to make students familiar with the types of questions set in various examinations.

I believe that the book will serve specific need of the students appearing in the examination of B.A./B.Sc., and other competitive examinations.

I gratefully acknowledge the inspiration, encouragement and valuable suggestions received from the well wisher during the preparation of this book.

Suggestion and comments for the improvement of the book from reader will be thankfully received and duly incorporated in the subsequent editions.

A.K. Sharma

CONTENTS

Page

Preface

1. **Function of Real Variable, Limits, Continuity and Differentiability** 1

   Constant, Variable, Function, Types of Functions, Definition Limit, Limit of Inequalities, Algebra of Limits, Continuity, Continuity of Elementary Function, Continuity of Sum Products, Differentiability.

2. **Rolle's Theorem, Mean Value Theorems, Taylor's and Maclaurin's Theorems** 53

   Rolle's Theorem, Lagrange's Mean Value Theorem or First Mean Value Theorem, Some Important Deductions from Mean Value Theorem, Cauchy's Mean Value Theorem or Second Mean Value Theorem, Taylor's Theorem with Lagrange's Form of Remainder After n Terms, Taylor's Theorem with Cauchy's Form of Remainder.

3. **Differentiation** 83

   Definition, Some Standard Results, Differential Coefficient of A Function of A Function, Inverse Hyperbolic Functions, Hyperbolic Functions, Some More Methods of Differentiation, List of Standard Results to be Committed to Memory.

4. **Successive Differentiation** 110

   Definition, Standard Results, Leibnitz's Theorem.

5. **Expansions of Functions** 140

   Taylor's Series, Maclaurin's Series, Normal Expansions of Functions, Some Important Expansions.

6. **Partial Differentiation** 202

   Functions of Two or More Independent Variables, Homogeneous Functions, Euler's Theorem on Homogeneous

Functions, Homogeneous Functions, Euler's Theorem on Homogeneous Functions, Total Derivatives.

7. Indeterminate Forms **252**

Definition, The Form 0/0: L' Hospital's Rule, Method of Expansion (Algebraic Methods), Form III $0 \times \infty$, Form IV: $[\infty - \infty]$, The Forms 0^0, 1^∞, ∞^0.

8. Tangents and Normals **320**

Tangent, Tangents Parallel and Perpendicular to the X-Axis, Normal, Angle of Intersection of Two Curves, Length of Cartesian Tangent, Normal, Subtangent and Subnormal, Polar Coordinates, Angle Between Radius Vector and Tangent, Angle of Intersection of Two Polar Curves, Polar Subtangent and Polar Subnormal, Length of the Perpendicular From Pole to Tangent, Pedal Equation, Differential Coefficient of Arc Length Cartesian Co-Ordinates, Differential Coefficient of Arc Length. Polar Co-Ordnates.

9. Curvature **394**

Introduction, Definition, Radius of Curvature of Intrinsic Curves, Cartesian Formula For Radius of Curvature, Radius of Curvature for Parametric Curves, Radius of Curvature for Pedal Curves, Radius of Curvature for Polar Curves, Tangential Polar Formula For Radius of Curvature, Miscellaneous Formulae Radius of Curvature, When x And y Are Given as Functions of Arc Length, Radius of Curvature at the Origin, Co-Ordinates of Centre of Curvature, Chord of Curvature Through the Origin (Pole), Chord of Curvature Perpendicular to the Radius Vector, Chords of Curvature Parallel to the Co-Ordinate Axes.

10. Asymptotes **472**

Definition, Determination of Asymptotes, The Asymptotes of the general Rational Algebraic Curve, Non-Existence of Asymptotes, Case of Parallel Asymptotes, Asymptotes Parallel to the Co-Ordinate Axes, Total Number of Asymptotes of a Curve, Complete Working Rule for Finding the Asymptotes of Rational Algebraic Curves, Asymptotes by Expansion, Alternative Methods of Finding Asymptotes of Algebraic Curves, Asymptotes by Inspection, Intersection of a Curve and its Asymptotes, Asymptotes of the Polar Curves, Circular Asymptotes.

1

FUNCTION OF REAL VARIABLE, LIMITS, CONTINUITY AND DIFFERENTIABILITY

1.1 CONSTANT

It is a quantity which during any set of mathematical operations, retains the same value. The constant are generally denoted by earlier letter a, b, c, . . . , or by numbers 1, 2, 3, ...

1.2 VARIABLE

It is a quantity which during any set of mathematical operation, does not retain the same value but is capable of assuming different value. The variable are generally denoted by the last letters of the alphabet *i.e.*, u, v, w, x, y, z etc., the variable are of following kinds:

(a) Continuous and Discrete Variables

If a variable can take all values from one given numbers to another, then it is called a continuous variable, but if it cannot take all values from one given number to another, then it is called a discontinuous or discrete variable.

(b) Independent Variable

The variable which may take up any arbitrary value that may be assigned to it is called an independent variables.

(c) Dependent Variable

This is an variable which assumes its value in consequence of some second variable or a system of variables taking up any set of arbitrary value that may be assigned to them.

1.3 FUNCTION

If x is a variable then a symbol which has one definite value for every value which x can have, is called 0 function of x.

Notation: A function of x is generally denoted by a single letter y. More generally it is denoted by symbols like f(x), F(x), $\phi(x)$, $\psi(x)$, $f_1(x_2)$, $f_2(x)$ etc. If a is any particular value of x, the value of the function f(x) obtained by putting x = a in it, is called the functional value of f(x) and is denoted by f(a).

A function of x is generally written as

$$y = f(x)$$

Here x and y are called the related variable because the variation of y depends upon the variation of x.

To be more clear about variation, consider the function.

$$y = x^2 + 2x + 6 \qquad ...(1)$$

For different values of x, y (or the expression $x^2 + 2x + 6$) assumes different values as shown below:

x	1	2	3	4	5
y	9	14	21	20	41

Here y is a dependent variable and x is the independent variable.

If y = f(x) = c *i.e.* the value of function always remain 'c', whatever by the x, then such a function is called 0 constant function.

1.4 TYPES OF FUNCTIONS

(i) Algebraic Functions: A function consisting a finites number of term involving powers and roots of the variable x and the four basic mathematical operations (addition, subtraction, multiplication and division) is called an algebraic function. For example:

$2x^3 + 5x^2 + 6x + 1$, $(x + 8)^7$, $\dfrac{x}{(x^2 - 1)}$ etc. are all algebraic functions.

(ii) Transcendental Functions: The functions which are not algebraic are called transcendental functions. These functions include:

(a) Trigonometric function such as

$\sin x, \cos x, \tan x,$ etc.

(b) Inverse trigonometric functions such as

$\sin^{-1} x, \cos^{-1} x, \cot^{-1} x,$ etc.

(c) Logarithmic functions such as

$$\log_e x \ \log_a x \ \log\left(\sqrt{x} + \frac{1}{\sqrt{x}}\right) \text{ etc.}$$

(d) Exponential functions such as

$e^x, a^x, e^{\tan x}, x^{\cos x}, x^x$ etc.

(e) In commensurable power functions of x as

$x^6, a^{\sqrt{7}}, x^{5/2},$ etc.

(iii) Explicit Functions: A function which is directly expressed in terms of the independent variable, is called an explicit function.

For example in $y = x^2$, or $y = e^x + 2\cos x$, y is an explicit function of x.

(iv) Implicit Functions: A function in which the variable x and y both occur together in an equation, but y cannot be directly expressed in terms of x, then y is called and implicit functions of x.

For example $2 + x^4 + 6x^2y^2 + 4xy + y^2 = 0$ is a implicit function of x. The implicit functions are usually denoted by $f(x, y) = 0$.

(v) Even or Odd Functions: If a functions does not change its sign when the sign of its independent variable is changed, then it is said, to be on even functions such as x^6, cos x, etc., are even functions of x because if we put (–x) in place of x, their signs are unchanged. But if a function changed its sign when the sign of its variable is changed, then it is called an odd function such as x^7, sin x, cot x etc., are odd function because their signs are changed if we put (–x) in place of x in these.

(vi) Periodic Functions: If $f(x + T) = f(x)$, where T is a constant then f(x) is called a periodic function of x with period T. For example $\sin(2\pi + x) = \sin x$, and hence sin x is a periodic functions with period 2π.

(vii) Single-Valued and Many-Valued Functions: The function is a single-valued function when to every value of the independent variable there corresponds only one value of the function, otherwise it is a many-valued function. An implicit function is often a many-valued function.

1.5 DEFINITION LIMIT

A function f(x) is said to tend to the limit A as x tends to a, if corresponding to any arbitrary chosen positive number ε, however small (but not zero), there exists a positive number δ, such that

$$|f(x) - A| < \varepsilon,$$

for all values of x for which $0 < |x - a| < \delta$.

We write it as $\lim_{x \to a} f(x) = A$.

Right Hand Limit: A function f(x) is said to tend to the limit A as x → a from the right, if corresponding to any arbitrary chosen +ive number ε, however small (but not zero), there exists a +ive number δ, such that

$$|f(x) - A| < \varepsilon,$$

for all values of x for which $0 < x - a < \delta$.

The limit of f(x) as x → a from the right is called the *right hand limit* of f(x) and is denoted by $\lim_{x \to a+0} f(x)$ or by f(a + 0).

We calculate f(a + 0) by evaluating $\lim_{h \to 0} f(a + h)$, where h is +ive and sufficiently small.

Left Hand Limit: A function f(x) is said to tend to the limit A as x → a from the left, if corresponding to any arbitrary chosen +ive number ε, however small (but not zero), there exists a +ive number δ, such that

$$|f(x) - A| < \varepsilon,$$

for all values of x for which $0 < a - x < \delta$.

The limit of f(x) as x → a from the left is called the *left hand limit* of f(x) and is denoted by $\lim_{x \to a-0} f(x)$ or by f(a + 0).

We calculate f(a – 0) by evaluating $\lim_{h \to 0} f(a - h)$, where h is +ive and sufficiently small.

From the definition it is clear that the limit of f(x) as x → a has no necessary connection with the value of the function at a except that when the function is continuous at x = a. The two are equals; even in the case of such functions the limit is distinct in idea from the value.

Note 1: $\lim_{x \to a} f_1(x)$ exists only if, limit on the left = limit on the right.

Note 2: If $f(x) \to A$ and $f_2(x) \to B$, as $x \to a$, where A and B are both finite, we have

(i) $\lim_{x \to a} \{f_1(x) \pm f_2(x)\} = A \pm B$,

(ii) $\lim_{x \to a} \{f_1(x) \, . \, f_2(x)\} = A \, . \, B$,

(iii) $\lim_{x \to a} \left\{\frac{f_1(x)}{f_2(x)}\right\} = \frac{A}{B}$, provided $B \neq 0$,

(iv) $\lim_{x \to a} cf(x) = cA$ where c is constant.

1.6 LIMIT OF INEQUALITIES

It $\phi(x) + A$ and $\psi(x) \to B$ as $x \to a$ and if

$$\phi(x) < \psi(x), \; x \neq u$$

Then $\quad A \leq B$

Since $\phi(x) < \psi(x)$ for all values of a an which the limit depend one would naturally expect that $A < B$. But it is also possible that A may be equal to B. Let $\phi(x) = 1 + x^2$. and $\psi(x) = 1 + 2u^2$. We have $\lim_{x \to a} \phi(x) = 1$, $\lim_{x \to a} \psi(x) = 1$. Also if $x \neq 0$ we have $\phi(x) < \psi(x)$ for all values of x on which the limit depend.

1.7 ALGEBRA OF LIMITS

Example 1:

The limit of a sum is equal to the sum of the limits.

Solution:

Let $\lim_{x \to a} f_1(x) = A$ and $\lim_{x \to a} f_2(x) = B$.

Let ε be any arbitrary chosen positive number.

We can choose δ_1, δ_2 such that

$$|f_1(x) - A| < \frac{1}{2}\varepsilon, \text{ when } 0 < |x - a| < \delta_1,$$

and

$$|f_2(x) - B| < \frac{1}{2}\varepsilon, \text{ when } 0 < |x - a| < \delta_2,$$

Let δ be any positive number smaller than both δ_1 and δ_2. Then

$$|f_1(x) - A| < \frac{1}{2}\varepsilon \text{ and } |f_2(x) - B| < \frac{1}{2}\varepsilon, \text{ for } 0 < |x - a| < \delta$$

Now from algebra, we have

$$|\{f_1(x) + f_2(x)\} - (A + B)| = |\{f_1(x) - A\} + f_2(x)\} - B)\}|$$

$$\leq |\{f_1(x) - A| + |f_2(x)\} - B)\}| < \frac{\varepsilon}{2} + \frac{\varepsilon}{2} \text{ i.e., } < \varepsilon$$

$$\therefore \quad \lim_{x \to a} \{f_1(x) + f_2(x)\} = A + B.$$

Example 2:

The limit of a product is equal to the product of the limits.

(Meerut, 1975)

Solution:

Let $\lim_{x \to a} f_1(x) = A$ and $\lim_{x \to a} f_2(x) = B$.

Let $A \neq 0$ and $B \neq 0$

Again let, $f_1(x)f_2(x) - AB$

$$= [f_1(x) - A]\,[f_2(x)\} - B] + A\,[f_2(x) - B] + B[f_1(x) - A].$$

$\therefore$ by algebra, we have $|f_1(x) - f|\, f_2(x) - AB|$

$$\leq |f_1(x) - A| \,.\, |f_2(x) - B)| + |A| \,.\, |f_2(x) - B| + |B|\, f_1(x) - A|$$

Now, $\lim_{x \to a} f_1(x) = A$ and $\lim_{x \to a} f_2(x) = B$. Therefore, for a given ε we can take δ such that

$$|f_1(x) - A| < \frac{\varepsilon}{3|B|} \text{ and } |f_2(x) - B| < \frac{\varepsilon}{3|A|},$$

when $0 < |x - a| < \delta$. Then

$$|f_1(x)f_2(x) - AB| < \frac{\varepsilon}{3} \cdot \frac{\varepsilon}{3|A|\,|B|} + \frac{\varepsilon}{3} + \frac{\varepsilon}{3}.$$

Since only small values of ε need by considered, we can take ε such that $\frac{\varepsilon}{3|A|\,|B|} < 1$. Then

$$|f_1(x)f_2(x) - AB| < \frac{\varepsilon}{3} + \frac{\varepsilon}{3} + \frac{\varepsilon}{3} \text{ i.e, } < \varepsilon, \text{ when } 0 < |x - a| < \delta.$$

Hence $\lim_{x \to a} \{f_1(x) \,.\, f_2(x)\} = AB.$

SOME MORE SOLVED EXAMPLES

Example 1:

Define the limit of a function. Discuss the existence of the limit of the function $f(x) = 1$, if $x < 1$; $f(x) = 2 - x$, if $1 < x < 2$; $f(x) = 2$, if $x \geq 2$ at $x = 1$ and $x = 2$. **(Delhi, 1982)**

Solution:

For definition of limit (see definition)

R.H.L. At $x = 1$,

i.e., $f(1 + 0) = \lim_{h \to 0} f(1 + h)$,

where h is +ive and sufficiently small

$$= \lim_{h \to 0} [2 - (1 + h) = \lim_{h \to 0} (1 - h) = 1;$$

L.H.L. At $x = 1$,

i.e., $f(1 - 0) = \lim_{h \to 0} f(1 - h) = \lim_{h \to 0} (1) = 1$

since R.H.L. = L.H.L. = 1,

therefore $= \lim_{x \to 1} f(x)$ exists and is equal to 1.

R.H.L At $x = 2$,

i.e., $f(2 + 0) = \lim_{h \to 0} f(2 + h) = \lim_{h \to 0} (2) = 2;$

L.H.L. At $x = 2$,

i.e., $f(2 - 0) = \lim_{h \to 0} (2 - h) = \lim_{h \to 0} [2 - (2 - h)]$

$$= \lim_{h \to 0} h = 0.$$

$\because$ left hand limit $\neq$ right hand limit,

$\therefore$ $f = \lim_{x \to 2} f(x)$ does not exist.

Example 2:

Prove that $\lim_{h \to 0} (1 + x)^{1/x} = e$.

Solution:

Let $f(x) = (1 + x)^{1/x}$.

We have R.H.L.= $f(0 + 0) = \lim_{h \to 0} f(0 + h) = \lim_{h \to 0} f(h) = \lim_{h \to 0} (1 + h)^{1/h}$

$$= \lim_{h\to 0}\left[1+\frac{1}{h}.h+\frac{\frac{1}{h}\left(\frac{1}{h}-1\right)}{1.2}h^2+\frac{\frac{1}{h}\left(\frac{1}{h}-1\right)\left(\frac{1}{h}-2\right)}{1.2.3}h^3+\ldots\right]$$

$$= \lim_{h\to 0}\left[1+\frac{1}{1!}+\frac{1.(1-h)}{2!}+\frac{1.(1-h)(1-2h)}{3!}+\ldots\right]$$

$$= 1+\frac{1}{1!}+\frac{1}{2!}+\frac{1}{3!}+\ldots\infty = e.$$

We have, R.H.L. $f(0-0) = \lim_{h\to 0} f(0-h) = \lim_{h\to 0} f(-h)$

$$= \lim_{h\to 0} f(1-h)^{-1/h} = e.$$

Since both $f(0+0)$ and $f(0-0)$ exist and are equal to e, therefore $\lim_{x\to 0}(1+x)^{1/x} = e$.

Example 3:

Evaluate the following limits if they exist: $\lim_{x\to 2}\frac{x^2-4}{x-2}$.

(Delhi, 1982)

Solution:

Let $\quad f(x) = \dfrac{x^2-4}{x-2} = \dfrac{(x-2)(x+2)}{x-2}$

R.H.L. $\quad x = 2,\ f(2+0) = \lim_{h\to 0} f(2+h)$

$$= \lim_{h\to 0}\frac{(2+h-2)(2+h+2)}{2+h-2} = \lim_{h\to 0}\frac{h(h+4)}{h}$$

$$= \lim_{h\to 0}(h+4) = 4.$$

L.H.L. $x = 2$, *i.e.*, $f(2-0) = \lim_{h\to 0} f(2-h)$

$$= \lim_{h\to 0}\frac{(2-h-2)(2-h+2)}{2-h-2} = \lim_{h\to 0}\frac{-h(-h+4)}{-h}$$

$$= \lim_{h\to 0}(-h+4) = 4.$$

Since both $f(2+0)$ f = $f(2-0)$, therefore $\lim_{x\to 2} f(x)$ exists and is equal to 4.

Example 4:

Evaluate the limits: $\lim_{x\to 0}\frac{1+3x^2}{x}$.

Solution:

We have $\lim_{x\to 0}\frac{1+3x^2}{x}$

Let $f(x) = \frac{1+3x^2}{x}$

We get, R.H.L. *i.e.*, $f(0+0) = \lim_{h\to 0} f(0+h) = \lim_{h\to 0} f(h)$

$$= \lim_{h\to 0}\frac{1+3h^2}{h} = \lim_{h\to 0}\left(\frac{1}{h}+3h\right) = \infty$$

Again, L.H.L. *i.e.*, $f(0-0) = \lim_{h\to 0} f(0-h) = \lim_{h\to 0} f(-h)$

$$= \lim_{h\to 0}\frac{1+3h^2}{-h} = \lim_{h\to 0}\left(-\frac{1}{h}-3h\right) = -\infty$$

R.H.L. $\neq$ L.H.L

Since $f(0+0) \neq f(0-0)$, therefore $\lim_{x\to 0} f(x)$ does not exist.

Example 5:

Show that $\lim_{x\to 2}\frac{|x-2|}{x-2}$ *does not exist.*

Solution:

Let $f(x) = \frac{|x-2|}{(x-2)}$

Here, R.H.L., $f(2+0) = \lim_{h\to 0} f(2+h) = \lim_{h\to 0}\frac{|2+h-2|}{(2+h-2)} = 1$

L.H.L., $f(2-0) = \lim_{h\to 0} f(2-h) = \lim_{h\to 0}\frac{|2-h-2|}{(2-h-2)}$

$$= \lim_{h\to 0}\frac{|-h|}{-h} = \lim_{h\to 0}\frac{h}{-h} = -1.$$

Since both $f(2+0)$ and $f(2-0)$, therefore $\lim_{x\to 2}\frac{|x-2|}{x-2}$ does not exist.

Example 6:

Evaluate: $\lim\limits_{x\to 1}\dfrac{x^3-1}{x^2-1}$. **(Delhi 1981)**

Solution:

We have $\lim\limits_{x\to 1}\dfrac{x^3-1}{x^2-1} = \lim\limits_{x\to 1}\dfrac{(x-1)(x^2+x+1)}{(x-1)(x+1)}$

$\lim\limits_{x\to 1}\dfrac{x^2+x+1}{x+1} = \dfrac{3}{2}$ Ans.

Example 7:

Evaluate: $\lim\limits_{x\to 2}\dfrac{x^2+3x+2}{x-2}$

Solution:

Let $f(x) = \dfrac{x^2+3x+2}{x-2}$

We have, R.H.L. *i.e.*, $f(2+0) = \lim\limits_{h\to 0} f(2+h)$

$$= \lim_{h\to 0}\frac{(2+h)^2+3(2+h)+2}{2+h-2} = \lim_{h\to 0}\frac{12+7h+h^2}{h}$$

$$= \lim_{h\to 0}\left(\frac{12}{h}+7+h\right) = \infty$$

Also, L.H.L. *i.e.*, $f(2-0) = \lim\limits_{h\to 0} f(2-h)$

$$= \lim_{h\to 0}\frac{(2-h)^2+3(2-h)+2}{2-h-2} = \lim_{h\to 0}\frac{12-7h+h^2}{-h}$$

$$= \lim_{h\to 0}\left(-\frac{12}{h}+7-h\right) = -\infty$$

Since $f(2+0) \neq f(2-0)$, therefore $\lim\limits_{x\to 2} f(x)$ does not exist.

Example 8:

Evaluate: $\lim\limits_{x\to 0}\dfrac{(1+x)^{1/3}-(1+x)^{1/3}}{x}$ **(Delhi, 1980)**

Solution:

We have $\lim\limits_{x\to 0}\dfrac{(1+x)^{1/3}-(1-x)^{1/3}}{x} = \lim\limits_{h\to 0}\dfrac{(1+h)^{1/3}-(1-h)^{1/3}}{h}$

$= \lim\limits_{h\to 0}\dfrac{1+\frac{1}{3}h+\ldots-(1-\frac{1}{3}h+\ldots)}{h} = \lim\limits_{h\to 0}\dfrac{\frac{2}{3}h+\ldots}{h} = \dfrac{2}{3}$.

Example 9:

Evaluate: $\lim\limits_{x\to 0} e^{-1/x}$.

Solution:

We have $\lim\limits_{x\to 0} e^{-1/x}$ let $f(x) = e^{1/x}$.

R.H.L. $f(0 + 0) = \lim\limits_{h\to 0} f(0 + h) = \lim\limits_{h\to 0} f(h) = \lim\limits_{h\to 0} e^{-1/h} - \infty$

$= \lim\limits_{h\to 0}\dfrac{1}{e^{1/h}} = 0$.

Again, L.H.L. $f(0 - 0) = \lim\limits_{h\to 0} f(-h) = \lim\limits_{h\to 0} e^{1/h} - \infty$

Since $f(0 + 0) \neq f(0 - 0)$ therefore $\lim\limits_{x\to 0} e^{-1/x}$ does not exist.

Example 10:

Evaluate: $\lim\limits_{x\to 0}\dfrac{1}{1-e^{1/x}}$.

Solution:

We have $\lim\limits_{x\to 0}\dfrac{1}{1-e^{1/x}}$

Let $f(x) = \dfrac{1}{1-e^2x}$

We have $f(0 + 0) = \lim\limits_{h\to 0} f(h) = \lim\limits_{h\to 0}\dfrac{1}{1-e^{1/h}} = 0$

Again $f(0 - 0) = \lim\limits_{h\to 0} f(-h) = \lim\limits_{h\to 0}\dfrac{1}{1-e^{-1/h}}$

$= \lim\limits_{h\to 0}\dfrac{1}{1-(1/e^{1/h})} = 1$

L.H.L $\neq$ R.H.L

$\because$ $f(0+0) \neq f(0-0)$ $\therefore$ $\lim_{x \to 0} \frac{1}{1-e^{1/x}}$ does not exist.

Example 11:

Evaluate: $\lim_{x \to 0} \frac{\sin x}{x}$ **(Lucknow, 1982, 79).**

Solution:

Let $f(x) \frac{\sin x}{x}$

We have R.H.L. $= f(0+0) = \lim_{h \to 0} f(0+h) = \lim_{h \to 0} f(h) = \lim_{h \to 0} \frac{\sin h}{h}$

$$= \lim_{h \to 0} \frac{h - \frac{h^3}{3!} + \frac{h^5}{5!} - \ldots}{h} = \lim_{h \to 0} \left(1 - \frac{h^2}{3!} + \frac{h^4}{5!} - \ldots\right) = 1$$

L.H.L., $= f(0-0) = \lim_{h \to 0} f(0-h) = \lim_{h \to 0} f(-h)$

$$= \lim_{h \to 0} \frac{\sin(-h)}{h} = \lim_{h \to 0} \frac{\sin h}{h} = 1.$$

L.H.L. = R.H.L.

Since $f(0+0) \neq f(0-0) = 1$ therefore $\lim_{x \to 0} \frac{\sin x}{x} = 1.$

Example 12:

Evaluate: $\lim_{x \to \infty} \frac{\sin x}{x}$.

Solution:

Put $x = \frac{1}{y}$. Then $= \lim_{h \to \infty} = \frac{\sin x}{x} = \lim_{y \to 0} y \sin\left(\frac{1}{y}\right)$

Now let $f(y) = y \sin\left(\frac{1}{y}\right)$.

We have R.H.L. $f(0+0) = \lim_{h \to 0} f(0+h) = \lim_{h \to 0} f(h) = \lim_{h \to 0} h \sin\left(\frac{1}{h}\right) = 0$

Since $\sin\left(\frac{1}{h}\right)$ lies between -1 and 1.

Also L.H.L. $f(0-0) = \lim_{h \to 0} f(0-h) = \lim_{h \to 0} f(-h) = \lim_{h \to 0} (-h)$

$$= \lim_{h \to 0} f \sin\left(-\frac{1}{h}\right) = \lim_{h \to 0} h \sin\left(\frac{1}{h}\right) = 0$$

Since both the limits f(0 + 0) and f(0 – 0) exist and are equal to zero, therefore $\lim_{y \to 0} y \sin\left(\frac{1}{y}\right) = 0$, Hence $\lim_{x \to \infty} \frac{\sin x}{x} = 0$.

Example 13:

Evaluate: $\lim_{x \to 0} = \sin\frac{1}{x}$. **(Delhi, 1975)**

Solution:

Let $f(x) = \sin\frac{1}{x}$

We have $f(0 + 0) = \lim_{h \to 0} f(0 + h) = \lim_{h \to 0} f(h) = \lim_{h \to 0} \sin\left(\frac{1}{h}\right)$

As $h \to 0$, the value of $\sin\left(\frac{1}{h}\right)$ oscillates between + 1 and –1, passing through zero and intermediate values an infinite number of times. Thus, there is no definite number A to which $\sin\left(\frac{1}{h}\right)$ tends as h hends to zero. Therefore the right hand limit f(0 + 0) does not exist.

Similarly, the left hand limit f(0 – 0) also does not exist. Hence $\lim_{x \to 0} \sin\frac{1}{x}$ does not exist.

Example 14:

Evaluate: $\lim_{x \to a} = \frac{x^m - a^m}{x - a}$.

Solution:

$$\lim_{x \to a} \frac{x^m - a^m}{x - a}$$

Let $f(x) = \frac{x^m - a^m}{x - a}$

We have R.H.L. $f(a + 0) = \lim_{h \to 0} f(a + h) = \lim_{h \to 0} = \frac{(a - h)^m - a^m}{a + h - a}$

$$= \lim_{h \to 0} \frac{a^m\left[\left(1 + \frac{h}{a}\right)^m - 1\right]}{h}$$

$$= \lim_{h\to 0} \frac{a^m}{h}\left[1 + m.\frac{h}{a} + \frac{m(m-1)}{2!}\frac{h^2}{a^2} + \ldots - 1\right]$$

$$\lim_{h\to 0}\left[\frac{h}{a} + \frac{m(m-1)}{2}\frac{h}{a^2} + \ldots\right] = a^m.\frac{m}{a} = ma^{m-1}$$

L.H.L. $= f(a-0) = \lim_{h\to 0} f(a-h) = \lim_{h\to 0}\frac{(a-h)^m}{a-h-a} = ma^{m-1}$.

Left hand limit = Right hand limit

Thus, $f(a+0) = f(a-0) = ma^{m-1}$

Therefore $\lim_{x\to a} f(x) = ma^{m-1}$.

Example 15:

Evaluate: $\lim_{x\to 0}\frac{1}{x}e^{1/x}$.

Solution:

Let $f(x) = \frac{1}{x}e^{1/x}$

R.H.L. $= f(0+0) = \lim_{h\to 0} f(0+h) = \lim_{h\to 0} f(h) = \lim_{h\to 0}\frac{1}{h}e^{1/h} = \infty$

[Since both and $e^{1/h}$ tend to ∞ as $h\to 0$]

Again we have

L.H.L. $= f(0-0) = \lim_{h\to 0} f(0-h) = \lim_{h\to 0} f(-h) = \lim_{h\to 0} -\frac{1}{h}e^{-1/h}$

$$= \lim_{h\to 0}\frac{-1}{he^{1/h}} = \lim_{h\to 0}\frac{-1}{h\left(1 + \frac{1}{h} + \frac{1}{2!}\frac{1}{h^2} + \ldots\right)}$$

$$= \lim_{h\to 0}\frac{-1}{h + 1 + (1/2h) + \ldots} = 0.$$

L.H.L. $\neq$ R.H.L. Since both the limits $f(0+0)$ and $f(0-0)$ are unequal, therefore $\lim_{x\to 0}\frac{1}{x}e^{1/x}$ does not exit.

Example 16:

Here the right hand limit i.e., $f(0+0) = h \to 0\, f(0-h)$, where h is +ive and sufficiently small

Evaluate: $\lim\limits_{x \to 0} \dfrac{e^x - 1}{x}$.

Solution:

$$= \lim_{h \to 0} f(h) = \lim_{h \to 0} h, \qquad [\because h > 0 \text{ and } f(x) = x \text{ when } x > 0]$$

$$= 0.$$

Again the left hand limit i.e., $f(0 - 0) = = \lim\limits_{h \to 0} f(0 - h)$,

where h is +ive and sufficiently small

$$= \lim_{h \to 0} f(-h) = \lim_{h \to 0} -(-h), \qquad [\because -h < 0 \text{ and } f(x) = -x \text{ when } x < 0]$$

$$= \lim_{h \to 0} h = 0.$$

Since both the limits $f(0 + 0)$ *and* $f(0 - 0)$ *exist and are equal to zero, therefore* $\lim\limits_{x \to 0} f(x)$ *exists and is equal to zero.*

Solution:

Let $f(x) = \dfrac{e^x - 1}{x} = \dfrac{1 + x + \dfrac{x^2}{2!} + \ldots - 1}{x} = \dfrac{x\left(1 + \dfrac{x}{2!} + \ldots\right)}{x}$.

R.H.L. $f(0 + 0) = \lim\limits_{h \to 0} f(0 + h) = \lim\limits_{h \to 0} f(h)$

$$= \lim_{h \to 0} \frac{h\left[1 + (h/2!) + \ldots\right]}{h} = \lim_{h \to 0} \left(1 + \frac{h}{2!} + \ldots\right) = 1.$$

L.H.L. $f(0 - 0) = \lim\limits_{h \to 0} f(0 - h) = \lim\limits_{h \to 0} f(-h)$

$$= \lim_{h \to 0} \frac{-h\left[1 = (h/2!) + \ldots\right]}{-h} = 1.$$

Example 17:

Thus, $f(0 + 0) = f(0 - 0) = 1$. *Therefore* $\lim\limits_{x \to 0} (e^x - 1)/x = 1$.

Evaluate: $\lim\limits_{x \to 0} \dfrac{a^x - 1}{x}$.

Solution:

$$\lim_{x\to 0}\frac{a^x-1}{x}. \text{Let } f(x)=\frac{a^x-1}{x}$$

$$=\frac{1+x\log a+(x^2/2!)(\log a)^2+\ldots-1}{x}$$

$$\text{Now, } =\frac{x\left[\log a+\frac{1}{2}x(\log a)^2+\ldots\right]}{x}$$

$$\text{R.H.L.} = f(0+0)=\lim_{h\to 0} f(h)=\lim_{h\to 0}\frac{h\left[\log a+\frac{h}{2}(\log a)^2+\ldots\right]}{h}=\log a.$$

$$\text{L.H.L.} = f(0-0)=\lim_{h\to 0} f(0-h)=\lim_{h\to 0} f(-h)=\log a.$$

Thus, $f(0+0)=f(0-0)=\log a$. Therefore

$$\lim_{x\to 0}\frac{(a^x-1)}{x}=\log a.$$

Example 18:

Find the right hand and the left hand limits in the following cases and discuss the existence of the limit in each case:

(i) $\displaystyle \lim_{x\to 2}\frac{2x^2-8}{x-2}$;

(ii) $\displaystyle \lim_{x\to 0}\frac{e^{1/x}-1}{e^{1/x}+1}$; **(Delhi, 1993, 89)**

(iii) $\displaystyle \lim_{x\to 0} f(x)$, *where f(x) is defined as*

$f(x) = x$, *when* $x > 0$

$f(x) = 0$, *when* $x = 0$

$f(x) = -x$, *when* $x < 0$

Solution:

$$\text{Let } f(x)=\frac{2x^2-8}{x-2}.$$

We have L.H.L. = $f(2+0)=\lim_{h\to 0} f(2+h)=\lim_{h\to 0}\frac{2(2+h)^2-8}{2+h-2}$

$$=\lim_{h\to 0}\frac{2(4+4h+h^2)-8}{h}=\lim_{h\to 0}\frac{8h+2h^2}{h}=\lim_{h\to 0}\frac{h(8+2h)}{h}$$

$$=\lim_{h\to 0}(8+2h)=8.$$

Again L.H.L. *i.e.* = $f(2-0)=\lim_{h\to 0} f(2-h)=\lim_{h\to 0}\frac{2(2-h)^2-8}{2-h-2}$

$$=\lim_{h\to 0}\frac{2(4-4h+h^2)}{-h}=\lim_{h\to 0}\frac{-8h+2h^2}{-h}$$

$$=\lim_{h\to 0}\frac{-h(8-2h)}{-h}=\lim_{h\to 0}(8-2h)=8.$$

Since f(2 + 0) = f(2 – 0) = 8, therefore $\lim_{x\to 0}\frac{2x^2-8}{x-2}$ exists and is equal to 8.

(ii) Let f(x) $f(x)=\frac{e^{1/x}-1}{e^{1/x}+1}$

We have R.H.L.

$$f(0+0)=\lim_{h\to 0} f(0+h)=\lim_{h\to 0} f(h)=\lim_{h\to 0}\frac{e^{1/h}-1}{e^{1/h}+1}$$

$$=\lim_{h\to 0}\frac{e^{1/h}\left[1-\left(1/e^{1/h}\right)\right]}{e^{1/h}\left[1+\left(1/e^{1/h}\right)\right]}=\lim_{h\to 0}\frac{1-\left(1/e^{1/h}\right)}{1+\left(1/e^{1/h}\right)}=1.$$

Again L.H.L. = $f(0-0)=\lim_{h\to 0} f(0-h)=\lim_{h\to 0} f(-h)=\lim_{h\to 0}\frac{e^{-1/h}-1}{e^{-1/h}+1}$

$$=\lim_{h\to 0}\frac{\left(1/e^{1/h}\right)-1}{\left(1/e^{1/h}\right)+1}=\frac{0-1}{0+1}=-1.$$

L.H.L. ≠ R.H.L.

Since f(0 + 0) ≠ f(0 – 0), therefore $\lim_{x\to 0}\frac{e^{1/x}-1}{e^{1/x}+1}$ does not exist.

Example 19:

Evaluate: $\lim\limits_{x\to 0}\dfrac{(1+x)^n-1}{x}$.

Solution:

$\lim\limits_{x\to 0}\dfrac{(1+x)^n-1}{x}$. Let $f(x)=\dfrac{(1+x)^n-1}{x}$

R.H.L. $f(0+0) = \lim\limits_{x\to a} f(h) = \lim\limits_{x\to a}\dfrac{(1+h)^n-1}{h}$

$$=\lim_{h\to 0}\frac{1+nh+\frac{n(n-1)}{2!}h^2+\ldots-1}{h}$$

$$=\lim_{h\to 0}\frac{h\left[n+\frac{n(n-1)}{2!}h+\ldots\right]}{h}$$

$$=\lim_{h\to 0}\left[n+\frac{n(n-1)}{2!}h+\ldots\right]=n.$$

L.H.L. $= f(0-0)=\lim\limits_{h\to 0} f(-h)=\lim\limits_{h\to 0}\dfrac{(1-h)^n-1}{-h}$

$$=\lim_{h\to 0}\frac{1+n(-h)+\frac{n(n-1)}{2!}(-h)^2+\ldots-1}{-h}=n.$$

R.H.L. = L.H.L.

Thus, $f(0+0)=f(0-0)=n$. Therefore, $\lim\limits_{x\to 0} f(x)=n$.

EXERCISES

Evaluate:

1. $\lim\limits_{x\to 1}\dfrac{(x-1)\sqrt{2-x}}{x^2-1}$. **[Ans. $\frac{1}{2}$]**

2. $\lim\limits_{x\to 1/2}\dfrac{8x^3-1}{5x^2-5x+1}$. **[Ans. 6]**

3. $\lim\limits_{x\to 1}\dfrac{x^m-1}{x^n-1}$ (m and n are integers). **[Ans. $\frac{m}{n}$]**

4. $\lim\limits_{x\to 0} \dfrac{\sqrt{1+x^2} - 1}{x}$. [Ans. 0]

5. $\lim\limits_{x\to 0} \dfrac{\sqrt{1+x} - 1}{x^2}$. [Ans. 8]

6. $\lim\limits_{x\to 0} \dfrac{\sqrt{x-b} - \sqrt{a-b}}{x^2 - a^2}$ (a > b). $\left[\textbf{Ans. } \dfrac{1}{4a\sqrt{a-b}}\right]$

7. $\lim\limits_{x\to 0} \dfrac{3\sqrt{1+x^2} - 4\sqrt{1-2x}}{x+x^2}$. $\left[\textbf{Ans. } \dfrac{1}{2}\right]$

8. $\lim\limits_{x\to 1} \dfrac{3\sqrt{7+x^3} - \sqrt{3-x^2}}{x-1}$. $\left[\textbf{Ans. } -\dfrac{1}{4}\right]$

9. $\lim\limits_{x\to 0} \dfrac{1\_\sin x - \cos x}{1 - \sin x - \cos x}$. [Ans. – 1]

10. $\lim\limits_{y\to a} \left[\sin\dfrac{y-a}{2}.\tan\dfrac{\pi 4}{2a}\right]$. $\left[\textbf{Ans. } \dfrac{-a}{n}\right]$

1.8 CONTINUITY

Definition: *A function f(x) defined for x = a is said to be continuous at x = a if*

(i) f(a) i.e., the value of f(x) at x = a is a definite number.

(ii) the limit of the function f(x) as x → a exists and is equal to the value of f(x) at x = a. **(Meerut, 1993, 92, 91; Delhi, 93, 91, 90)**

Otherwise the function is **discontinuous** at x = a.

A function is said to be continuous in an interval (a, b) if it is continuous at every point of that interval.

Arithmetical Definition of Continuity: *A function f(x) is said to be continuous at x = a, if for any arbitrarily chosen positive number ε, however small (but not zero), we can find a corresponding number δ such that*

$$|f(x) - f(a)| < \varepsilon$$

for all values of x for which $|x - a| < \delta$.

This definition, as given by Cauchy, is known as Cauchy's definition of continuity.

Remark:

One comparing the definitions of limit and continuity, we find that a function f(x) is continuous at x = a if

$$\lim_{x \to a} f(x) = f(a).$$

Thus f(x) is continuous at x = a if we have L.H.L. = R.H.L. = f(a); otherwise it is discontinuous at x = a.

1.9 CONTINUITY OF ELEMENTARY FUNCTION

The student as supposed to be familiar with the elementary functions and to know that:

(i) x^n is continuous for all values of x when n is positive.

(ii) x^n is continuous for all values of x except x = 0 when n is negative. For when n is negative and equal to –m say, where m iis positive we can write x^n as $\frac{1}{x^m}$ and as $x \to 0$ $\frac{1}{x^m}$ either does not tends to a limit or $\to \infty$.

(iii) sin x and cos x are continuous for all values of x.

(iv) tan x is continuous for all values of x except $\pm\frac{\pi}{2}, \pm\frac{3\pi}{2}, \pm\frac{5\pi}{2}$ etc. as $x \to \frac{\pi}{2}$ from the right tan $x \to -\infty$ as $x \to \frac{\pi}{2}$ from the left tan $x \to +\infty$. Thus, $\lim_{x \to \pi/2}$ tan x does not exist.

(v) Similarly, statement can be made for cot x, sec x and cosec x.

(vi) $\sin^{-1}x$ and $\cos^{-1}x$ are not real for value of x out side the domain [– 1, 1] similarly, $\sec^{-1} x$ and $\text{cosec}^{-1} x$ are not real for values of x between + and.

(vii) All the inverse circular function are continuous at every point with in the domain in which they are real.

(viii) Log x is continuous for all values of x > 0. For values of x < 0 the function is not real and we shall not consider these values as x tends to zero from the right log x tends to $-\infty$.

1.10 CONTINUITY OF SUM, PRODUCIS etc.

(i) The sum of two continuous function is continuous.

(ii) The product of two continuous function is continuous.

(iii) The quotient of two continuous function is a continuous function except for value of x which make the denominator zero.

1.11 DIFFERENTIABILITY

Definition: *A function f(x) is said to be differentiable at x = a if both*

$$\lim_{h\to 0}\frac{f(a+h)-f(a)}{h},\ h>0 \text{ and } \lim_{h\to 0}\frac{f(a-h)-f(a)}{-h},\ (L.H.L. = R.H.L.)$$

exist and have a common value (finite or infinite). This common value is called the **derivative** of f(x) at the point x = a and is denoted by f'(a). Another name for derivative is known as differential co-efficient.

(Lucknow, 1970; Mysore, 1971; Delhi, 1972, 70)

Progressive and Regressive Derivatives: Progressive derivative or *the right hand differential coefficient of* f(x) at x = a is given by

$$\lim_{h\to 0}\frac{f(a+h)-f(a)}{h},\ h>0.$$

It is denoted by Rf'(a).

Regressive derivative or *the left hand differential coefficient* of f(x) at x = a is given by

$$\lim_{h\to 0}\frac{f(a-h)-f(a)}{-h},\ h>0.$$

It is denoted by Lf'(a).

A function f(x) is said to be differentiable at x = a if both Rf'(a) and Lf '(a) exist and are equal; otherwise it is said to be not differential.

Theorem:

Continuity is necessary but not a sufficient condition for the existence of a finite derivative. **(Delhi, 1992, 97, Meerut, 1996 S)**

Proof:

Let f(x) have a finite derivative f'(a) at x = a. Then

$$\lim_{h\to 0}\frac{f(a+h)-f(a)}{h} = f'(a). \quad ...(1)$$

Now $f(a + h) - f(a) = \dfrac{f(a-h)-f(a)}{h} \times h, h \neq 0.$

Taking limit of both sides when $h \to 0$, we get

$$\lim_{h\to 0}\left[f(a+h)-f(a)\right] = \lim_{h\to 0}\frac{f(a+h)-f(a)}{h} \times \lim_{h\to 0} h$$

$= f'(a) \times 0,$ [from (1)]

$= 0$, since $f'(a)$ is finite.

Now, $\lim_{h\to 0}$ $[f(a + h) - f(a)] = 0$

$\Rightarrow$ $\lim_{h\to 0}$ $f(a + h) = f(a).$ ...(2)

From the condition (2) we see that f(x) is necessarily continuous at x = a. Thus continuity is a necessary condition for the existence of a finite derivative.

In order to show that continuity is not a sufficient condition for differentiability, we should give an example of a function which is continuous at a point but is not differentiable at that point.

Example 1:

If $f(x) - x\tan^{-1}\left(\frac{1}{x}\right)$ *when* $x \neq 0$ *and* $f(0) = 0$ *show that f is continuous but not deriable for* $x = 0$.

Solution:

$$\lim_{x\to 0} f(x) = \lim_{x\to 0} x\tan^{-1}\left(\frac{1}{x}\right) = 0\left(-\frac{\pi}{2}\right) = 0$$

$$\lim_{x\to 0} f(x) = \lim_{x\to 0}+ x\tan^{-1}\left(\frac{1}{x}\right) = 0\left(\frac{1}{2}\pi\right) = 0$$

Also f(0) = 0 Hence f is continuous at x = 0

$$\text{Now } Lf'(0) = \lim_{x\to 0}-\frac{x\tan^{-1}\left(\frac{1}{x}\right)-0}{x-0} = \lim_{x\to 0}-\tan^{-1}\left(\frac{1}{x}\right) = \frac{-\pi}{2}$$

$$Rf'(0) = \lim_{x\to 0}+\tan^{-1}\left(\frac{1}{x}\right) = \frac{\pi}{2}$$

$\therefore$ $Lf'(0) \neq Rf'(0)$

Hence f is not deriable at x = 0.

Example 2:

Show that $f(x) = (x^2 - 1)/(x - 1)$ is continuous for all values of x except $x = 1$. **(Meerut, 1997)**

How may this function be defined to make it continuous at x = 1?

Solution:

At first we have to test continuity of f(x) at x = a where a ± 1 $f(a) = \frac{a^2-1}{a-1} =$ a definite real number because $a - 1 \neq 0$;

We have R.H.L. $= f(a+0) = \lim_{h\to 0} f(a+h) = \lim_{h\to 0} \frac{(a+h)^2-1}{a+h-1} = \frac{a^2-1}{a-1}$;

L.H.L. $f(a-0) = \lim_{h\to 0} f(a-h) = \lim_{h\to 0} \frac{(a-h)^2-1}{a-h-1} = \frac{a^2-1}{a-1}$.

L.H.L. = R.H.L.

Thus f(a + 0) = f(a – 0) = f(a). Therefore, f(x) is continuous at x = a where $a \neq 1$.

From the formula definiting f(x), we have $f(1) = \frac{1-1}{1-1} = \frac{0}{0}$ which is meaningless. Thus f(x) has no definite value at x = 1. Therefore, f(x) is not continuous at x = 1. To make f(x) continuous at x = 1, we shall find $\lim_{x\to 1} f(x)$. We have

$$f(1+0) = \lim_{h\to 0} f(1+h) = \lim_{h\to 0} \frac{(1+h)^2-1}{1+h-1} = \lim_{h\to 0} \frac{2h+h^2}{h}$$

$$= \lim_{h\to 0} \frac{h(2+h)}{h} = \lim_{h\to 0}(2+h) = 2;$$

and $$f(1-0) = \lim_{h\to 0} f(1-h) = \lim_{h\to 0} \frac{(1-h)^2-1}{1-h-1}$$

$$= \lim_{h\to 0} \frac{-2h+h^2}{-h}$$

$= \lim_{h \to 0} (2 - h) = 2$.

Thus, $\lim_{x \to 1} f(x) = 2$. So if we take $f(x) = \frac{(x^2 - 1)}{(x - 1)}$ when $x \neq 1$ and $f(x) = 2$ when $x = 1$, it will become continuous at $x = 1$.

Example 3:

Define continuity of function at a point and in an interval. Discuss the continuity of

$f(x) = 2, x \leq 0; f(x) = 3x + 2, 0 < x \leq 1; f(x) = x/(x - 1), 1 < x$ *at* $x = 0$ *and at* $x = 1$. **(Delhi, 1993, 91)**

Solution:

See definition of continuity.

(i) To test f(x) for continuity at x = 0. We have

$$f(0) = 2;\ \text{R.H.L.} = f(0 + 0) = \lim_{h \to 0} f(0 + h)$$

$$= \lim_{h \to 0} f(h) = \lim_{h \to 0} (3h + 2) = 2$$

$$\text{L.H.L.} = f(0 - 0) = \lim_{h \to 0} f(0 - h) = \lim_{h \to 0} f(-h)$$

$$= \lim_{h \to 0} (2) = 2.$$

Since L.H.L. = R.H.L. = f(0) = 2, the function is continuous at x = 0.

(ii) To test f(x) for continuity at x = 1. We have

$$f(1) = 3.1 + 2 = 5.$$

$$\text{R.H.L.} = f(1 + 0) = \lim_{h \to 0} f(1 + h)$$

$$= \lim_{h \to 0} \frac{1 + h}{(1 + h) - 1} = \lim_{h \to 0} \frac{1 + h}{h} = \infty,$$

$$\text{L.H.L.} = f(1 - 0) = \lim_{h \to 0} f(1 - h)$$

$$= \lim_{h \to 0} [3(1 - h) + 2] = \lim_{h \to 0} [5 - 3h] = 5.$$

Since R.H.L. ≠ L.H.L. the function is not continuous at x = 1.

Example 4:

Show that the function f(x) = |x| is continuous at x = 0, but not differentiable at x = 0, where |x| means the numerical value of x.

(Kashmir, 1993; Meerut, 1991; Delhi, 1991, 1994, 1993)

Solution:

To test the continuity of f(x) at x = 0.

We have f(0) = |x| = 0;

R.H.L. $f(0+h) = \lim_{h\to 0} f(0+h) = \lim_{h\to 0} f(h) = \lim_{h\to 0} |h|$

L.H.L. $f(0-0) = \lim_{h\to 0} f(0-h) = \lim_{h\to 0} f(-h) = \lim_{h\to 0} |-h|$

$= \lim_{h\to 0} h = 0$

L.H.L. = R.H.L.

Since f(0 + 0) = f(0 – 0) = f(0), therefore f(x) is continuous at x = 0.

To test the differentiability of f(x) at x = 0.

We have $Rf'(0) = \lim_{h\to 0} \frac{f(0+h)-f(0)}{h}$, where h is +ive and sufficiently small

$$= \lim_{h\to 0} \frac{f(h)-f(0)}{h} = \lim_{h\to 0} \frac{|h|-|0|}{h} = \lim_{h\to 0} \frac{h}{h}$$

$$= \lim_{h\to 0} 1 = 1;$$

and $Lf'(0) = \lim_{h\to 0} \frac{f(0-h)-f(0)}{-h}$. h being +ive and sufficiently small

$$= \lim_{h\to 0} \frac{f(-h)-f(0)}{-h} = \lim_{h\to 0} \frac{|-h|-|0|}{-h} = \lim_{h\to 0} \frac{h-0}{-h}$$

$$= \lim_{h\to 0} (-1)$$

$$= -1.$$

Since Rf'(0) ≠ Lf'(0); therefore the function f(x) is not differentiable at x = 0.

Example 5:

A function is defined by

$$f(x) = -x^2 \text{ if } x \leq 0$$
$$= 5x - 4 \text{ if } 0 < x < 1$$
$$= 4x^2 - 3x \text{ if } 1 < x < 2$$
$$= 3x + 4 \text{ if } x \geq 2.$$

Examine f for continuity at x = 0, 1, 2. Also discuss the kind of discontinuity if axis. **(D.U. B. Sec. (G), 2000]**

Solution:

At x = 0

$$\lim_{x\to 0} - f(x) = \lim_{x\to 0}\left(-x^2\right) = 0$$

$$\lim_{x\to 0} + f(x) = \lim_{x\to 0}(5x-4) = -4$$

Also f(0) = 0

Thus, $\lim_{x\to 0}$ + f(x) ± f(0) and so f has a does continuity of the first and from the right at x = 0

At x = 1

$$\lim_{x\to 1} - f(x) = \lim(5x-4) = 1$$

$$\lim_{x\to 1} + f(x) = \lim_{x\to 1}\left(4x^2 - 3x\right) = 1$$

Also f(1) = 5 × 1 – 4 = 1

Thus, 1 $\lim_{x\to 1-} f(x) = \lim_{x\to 1+} f(x)$ and so f(x) is continuous let x ≤ 1

At x = 2

$$\lim_{x\to 2-} f(x) = \lim_{x\to 2}\ (4x^2 - 3x) = 4 \times 4 - 3 \times 2 = 10$$

$$\lim_{x\to 2+} f(x) = \lim_{x\to 2}\ (34 + 4) = 10$$

Thus, $\lim_{x\to 2-} f(x) = \lim_{x\to 2+} f(x) = f(2)$

and so f(x) is continuous at x = 2.

Example 6:

Test the following function for continuity

f(x) = x cos (1/x) for x ' 0; f(0) = 0 at x = 0.

Solution:

We have

$$f(0+0)=\lim_{h\to 0} f(0+h)=\lim_{h\to 0} f(h)=\lim_{h\to 0} h\cos\frac{1}{h}=0.$$

Again $\quad f(0-0)=\lim_{h\to 0} f(0-h)=\lim_{h\to 0} f(-h)=\lim_{h\to 0} -h\cos\left(\frac{-1}{h}\right)=0.$

Also f(0) = 0. Thus f(0 + 0) = f(0 – 0) = f(0). Therefore f(x) is continuous at x = 0.

Example 7:

If a function f is continuous at x = a and if y is a function continuous at f(a) show that the function got is continuous at x = a.

(D.U. B. Sec. (G), 2000]

Solution:

If f: R = R and g: R → R then the product or composite of t and g. Written as fog: R → R is defined as

(fog) (x) = f[g(x)] for all x ∀ R.

Similarly (gof) (x) = g[f(x)] for all x ∀ R. Since g is continuous at f(a), so for any ε > 0. There exist some $\delta_1 > 0$, such that

|g[f(x)] – g(fog)] < ε when

$|f(x) - f(a)| < \delta_1$ ...(1)

Since f is continuous at a fog $\delta_1 > 0$ there exist some δ > 0 such that

|f(x) – f(a)| < δ when |x – a| < δ ...(2)

from (1) and (2) we obtain

|g(f(x)) – g(f(a))| < ε, when |x – a| < δ

|(g of) (x) – (gof) (a)| < ε when |x – a| < δ

If follows that gof is continuous at x = a.

Let y = [x] or E(x) denotes the integral part of x. Prove that this function is discontinuous where x has an integral value. Also draw the graph.

(Allahabad, 1993)

Solution:

From the definition of E(x), we have

E(x) = n – 1 for n – 1 ≤ x < n,

E(x) = n for n ≤ x < n + 1,

E(x) = n + 1 for n + 1 ≤ x < n + 2,

also so on where n is an integer.

We consider x = n. Then E(n) = n, E(n – 0) = n – 1 and E(n + 0) = n.

Since E(n + 0) ≠ E(n – 0), the function E(x) is discontinuous at x = n *i.e.*, when x has an integral value. Evidently it is continuous for all other values of x.

To draw the graph, we put

n = ..., – 4, – 3, – 2, – 1, 0, 1, 2, 3, 4, ..., so that

y = – 4, when – 4 ≤ x < – 3

y = – 3, when – 3 ≤ x < – 2

y = – 2, when – 2 ≤ x < – 1

y = – 1, when – 1 ≤ x < 0

y = 0, when 0 ≤ x < 1

y = 1 when 1 ≤ x < 2

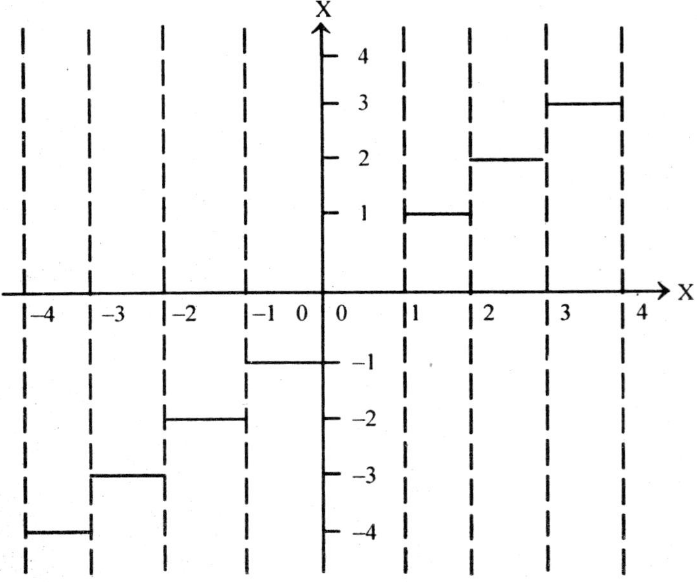

Fig. 1.1

y = 2, when 2 ≤ x < 3

y = 3, when 3 ≤ x < 4

y = 4, when 4 ≤ x < 5 and so on.

The graph is shown by thick lines.

Example 8:

The function for continuity

$$f(x) = 2^{1/x}, \text{ when } x \neq 0;\ f(0) = 0 \text{ at } x = 0.$$

Solution:

We have

$$f(0+0) = \lim_{h\to0} f(0+h) = \lim_{h\to0} f(h) = \lim_{h\to0} 2^{1/h} = \infty.$$

Again $f(0-0) = \lim_{h\to0} f(0-h) = \lim_{h\to0} f(-h) = \lim_{h\to0} 2^{1/h} = 2^{-\infty} = 0.$

Since $f(0+0) \neq f(0-0)$, the function f(x) is discontinuous at x = 0.

Note: A function f(x) is said to have an *infinite discontinuity* at a point x = a, if one or both of the limits f(a + 0) and f(a – 0) are infinite.

Test the function for continuity

Example 9:

$$f(x) = \frac{1}{1-e^{-1/x}};\ x = 0;\ f(0) = 0 \text{ at } x = 0.$$

Solution:

We have

$$f(0+0) = \lim_{h\to0} f(0+h) = \lim_{h\to0} f(h)$$

$$= \lim_{h\to0} \frac{1}{1-e^{-1/h}}$$

$$= \frac{1}{e^{-\infty}} = \frac{1}{1-0} = 1.$$

Again

$$f(0-0) = \lim_{h\to0} f(0-h) = \lim_{h\to0} f(-h) = \lim_{h\to0} \frac{1}{1-e^{1/h}}$$

$$= \frac{1}{1-e^{\infty}} = \frac{1}{1-\infty} = 0.$$

Since $f(0+0) \neq f(0-0)$, the function f(x) is discontinuous at x = 0.

Example 10:

Test the following functions for continuity:

$$f(x) = x^3 \text{ at } x = 2.$$

Solution:

We have $f(2) = 2^3 = 8;$

$$f(2+0) = \lim_{h\to 0} f(2+h) = \lim_{h\to 0} (2+h)^3 = 2^3 = 8;$$

and $f(2-0) = \lim_{h\to 0} f(2-h) = \lim_{h\to 0} (2-h)^3 = 2^3 = 8.$

Since $f(2) = f(2+0) = f(2-0)$, therefore $f(x)$ is continuous at $x = 2$.

Example 11:

Test two function for continuity

$f(x) = x \sin(1/x)$, $x \neq 0$ and $f(0) = 0$ at $x = 0$.

(Delhi, 1993, 90, Meerut, 1994; Allahabad, 1994)

Solution:

We have

$$f(0+0) = \lim_{h\to 0} f(0+h) = \lim_{h\to 0} f(h) = \lim_{h\to 0} h \sin\frac{1}{h} = 0.$$

Again $f(0-0) = \lim_{h\to 0} \lim_{h\to 0} f(0-h) = \lim_{h\to 0} f(-h)$

$$= \lim_{h\to 0}(-h)\sin\left(-\frac{1}{h}\right) = 0.$$

Also $f(0) = 0$. Thus $f(0+0) = f(0-0) = f(0)$. Therefore $f(x)$ is continuous at $x = 0$.

Note: If we are given that $f(0) = 2$, the above function becomes discontinuous at $x = 0$.

Example 12:

Give $\in - \delta$ definition of continuity of a function f at a point use this definition to show that if f is continuous at c, then so $|B|$ $|f|$ is the converse true 3 justify. **(D.U. B.Sc. (G), 2001)**

Solution:

A function f is said to be continuous at a point c_1 if for any $\in > 0$ three exist some $\delta > 0$ such that

$|f(x) - f(c)| < \in$ when $|x - c| < \delta$ ...(1)

Now we show that if f is continuous at c then $|f|$ is continuous at c.

We know

$||x| - |f|| \leq |x - y|$ for all $x, y \in R$

$||f(x)| - |f(c)|| \leq |f(x) - f(c)|$ ...(2)

from equation (1) and (2) we get

$\therefore \quad |f(x) - |f(c)| < \in$ when $|x - c| < \delta$

Hence $|f|$ is continuous at c

However, the converse is not true *i.e.* If $|t|$ is continuous at a point. Then f may not be continuous at that point

$$\text{Let } f(x) = \begin{cases} \frac{|x|}{x} \text{ if } x \neq 0 \\ 1 \quad \text{if } x = 0 \end{cases}$$

We have $\lim_{x \to 0} f(x) = \lim_{x \to 0} \frac{-x}{x} = -1$

$$\lim_{x \to 0} f(x) = \lim_{x \to 0} \frac{x}{n} = 1$$

$f(0) = 1$

Hence f is discontinuous at x = 0

We have $[\,|f|\,(x) = 1 \ \forall \ x \in n]$

i.e., $|f|$ is a constant function and so $|f|$ is continuous at ever point of R including x = 0

Example 13:

Discuss the points of discontinuity of the function given by:

$$f(x) = \begin{cases} -x, & \text{for} \quad x \leq 0 \\ x, & \text{for} \quad 0 < x \leq 1 \\ 2-x, & \text{for} \quad 1 < x \leq 2 \\ 1, & \text{for} \quad x > 2 \end{cases}$$ **(Meerut, 1993)**

Solution:

Here we shall test f(x) for continuity at the points x = 0, 1 and 2. Obviously at all other points the function is continuous.

(i) Continuity at x = 0. We have f(0) = 0, because f(x) = – x for x = 0.

L.H.L. of f(0) as $x \to 0$ *i.e.*,

$$f(0-0) = \lim_{h \to 0} f(0-h)\, f(-h) = \lim_{h \to 0} [-(-h)], \ [\because f(x) = -x \text{ or } x < 0]$$

$$= \lim_{h \to 0} (h) = 0\,;$$

R.H.L. of f(x) as $x \to 0$ *i.e.*,

$$f(0 + 0) = \lim_{h \to 0} f(0 + h) = \lim_{h \to 0} f(h) = \lim_{h \to 0} h,$$

$[\because f(x) = x \text{ for } 0 < x \leq 1]$

$$= 0.$$

R.H.L. = L.H.L.

Thus, $f(0 - 0) = f(0 + 0) = f(0)$ and so $f(x)$ is continuous at $x = 0$.

(ii) **Continuity at x = 1.** We have $f(1) = 1$, because $f(x) = x$ for $0 < x \leq 1$.

Now L.H.L. of $f(x)$ as $x \to 1$ *i.e.*,

$$f(1 - 0) = \lim_{h \to 0} f(1 - h) = \lim_{h \to 0} (1 - h),$$

$[\because f(x) = x \text{ for } 0 < x \leq 1]$

$$= 1;$$

and R.H.L. $f(1 + 0) = \lim_{h \to 0} f(1 + h) = \lim_{h \to 0} [2 - (1 + h)],$

$[\because f(x) = 2 - x \text{ for } 0 < x \leq 2]$

$$\lim_{h \to 0} (1 - h) = 1.$$

L.H.L. = R.H.L.

Since $f(1 - 0) = f(1 + 0) = f(1)$, therefore $f(x)$ is continuous at $x = 1$.

(iii) **Continuity at x = 2.** We have $f(2) = 2 - 2 = 0$, because $f(x)\ 2 - x$, for $1 < x \leq 2$.

L.H.L. of $f(x)$ as $x \to 2$ *i.e.*,

$$f(2 - 0) = \lim_{h \to 0} f(2 - h) = \lim_{h \to 0} [2 - (2 - h)],$$

$[\because f(x) = 2 - x, \text{ for } 1 < x \leq 2]$

$$= \lim_{h \to 0} (h) = 0;$$

and R.H.L. $= f(2 + 0) = \lim_{h \to 0} f(2 + h) = \lim_{h \to 0} (1),$

$$= 1$$

$[\because f(x) = 1 \text{ for } x > 2]$

L.H.L. = R.H.L.

Since $f(2 + 0) \neq f(2 - 0)$, the function $f(x)$ is discontinuous at $x = 2$.

Example 14:

Discuss the continuity of the function f(x) at $x = \frac{1}{2}$ *where* $f(x) = x,\ 0 \le x < \frac{1}{2};\ f(x) = 1,\ x = \frac{1}{2};\ f(x) = 1 - x,\ \frac{1}{2} < x \le 1.$ **(Delhi, 1999)**

Solution:

We have f(x) = 1, when $x = \frac{1}{2}$.

Now R.H.L. $= \lim_{h\to 0} f\left(\frac{1}{2} + h\right)$, where h is +ive and sufficiently small

$$= \lim_{h\to 0}\left[1 - \left(\frac{1}{2} + h\right)\right], \qquad \left[\because f(x) = 1 - x \text{ when } \frac{1}{2} < x \le 1\right]$$

$$= \lim_{h\to 0}\left(\frac{1}{2} - h\right) = \frac{1}{2},$$

$$\text{L.H.L.} = \lim_{h\to 0} f\left(\frac{1}{2} - h\right), \qquad \left[\because f(x) = x \text{ when } 0 \le x < \frac{1}{2}\right]$$

Since R.H.L. = L.H.L. ≠ the value of f(x) at $x = \frac{1}{2}$, therefore the function is discontinuous at $x = \frac{1}{2}$.

Example 15:

Define continuity of a function f(x) at x = a. Discuss the continuity of the function f(x) defined as:

$f(0) = 0;\ f9x) = 1 - x,\ 0 < x < 1;\ f(1) = 1;$

$f(x) = 2 - x,\ 1 < x < 2;\ f(2) = 0$

at the points x = 0, 1, 2,. **(Delhi, 1990)**

Solution:

For definition of continuity

(i) **To test f(x) for continuity at x = 0.** We have f(0) = 0.

Now $f(0 + 0) = \lim_{h\to 0} f(1 + h) = \lim_{h\to 0} f(h) = \lim_{h\to 0} (1 - h)$

$[\because f(x) = 1 - x \text{ for } 0 < x < 1]$

= 1.

As $f(0 + 0) \neq f(0)$, therefore f(x) is discontinuous at x = 0.

(ii) To test f(x) for continuity at x = 1. We have f(1) = 1.

Now $f(1 - 0) = \lim_{h\to 0} f(1 - h) = \lim_{h\to 0} [1 - (1 - h)]$

$[\because f(x) = 1 - x \text{ for } 0 < x \leq 1]$

$= 0,$

and $f(1 + 0) = \lim_{h\to 0} f(1 + h) = \lim_{h\to 0} [2 - (1 + h)],$

$[\because f(x) = 2 - x, \text{ for } 1 < x < 2]$

$= 1.$

$\because$ $f(1 - 0) \neq f(1 + 0)$, therefore f(x) is discontinuous at x = 1.

(iii) To test f(x) for continuity at x = 2. We have f(2) = 0.

Now, $f(2 - 0) = \lim_{h\to 0} f(2 - h) = \lim_{h\to 0} [2 - (2 - h)],$

$[\because f(x) = 2 - x \text{ for } 1 < x < 2]$

$= 0.$

Since f(2) = f(2 – 0), f(x) is continuous at x = 2.

Note that here f(x) is not defined for x < 2. So the question of finding f(2 + 0) does not arise. Here f(x) will be continuous at x = 2 if f(2) = f(2 – 0).

Example 16:

Determine the continuity of the function f(x) defined as follows:

$$f(0) = 0;\ f(x) = \frac{1}{2} - x,\ 0 < x < \frac{1}{2};\ f\left(\frac{1}{2}\right) = 0\ ;$$

$$f(x) = 4x^2 - 1,\ \frac{1}{2} < x < 3/4;\ f(x) = 1 - x^2,\ 3/4 \leq x \leq 1.$$

(D.U. B.Sc. (G), 1998)

Solution:

The function is continuous at x = 0, 3/4.

Example 17:

A function f(x) is defined as follows:

$$f(x) = \begin{cases} \left(x^2/a\right) - a, & \text{when } x < 0 \\ 0, & \text{when } x = a \\ a - \left(a^2/x\right), & \text{when } x > 0 \end{cases}$$

Prove that the function f(x) is continuous at x = a. **(Delhi, 1993, 98)**

Solution:

We have f(a) = 0.

Also the right hand limit of f(x) at x = a *i.e.*,

$$f(a+0) = \lim_{h\to 0} f(a+h) = \lim_{h\to 0}\left[a - \frac{a^2}{(a+h)}\right],$$

[∵ $f(x) = a - (a^2/x)$ for $x > a$]

$= a - (a^2/a) = a - a = 0;$

and the left hand limit of f(x) at x = a *i.e.*,

$$f(a-0) = \lim_{h\to 0} f(a-h) = \lim_{h\to 0}\left[\frac{(a-h)^2}{a} - a\right],$$

[∵ $f(x) = (x^2/a) - a$ for $x < a$]

$= (a^2/a) - a = a - a = 0.$

∵ f (a + 0) = f(a – 0) = f(a), therefore f(x) is continuous at x = a.

Example 18:

Give ∈ – δ definition of continuity of a function f at a point and use two definition to show that if f is continuous at c then so is | f | is the converse true? Justify. **(D.U. B.Sc. (G), 2001)**

Solution:

A function f is said to be continuous at a point c. If for any ∈ > 0. There exist some δ > 0 such that

$|f(x) - f(c)| < \in$ when $|x - c| < \delta$ ...(1)

Now we show that if f is continuous at c then |f| f is continuous at c

We know $||x| - |y|| \le |x - y|$, for all x, y ∈ R

$\therefore \quad ||f(x)| - f(c)|| \le f(x) - f(c)|$ ...(2)

from (1) and (2) we get

$||f(x) - |f(c)|| < \in$ when $|x - c| < \delta$

Hence δ is continuous at c

However, the converse is not true *i.e.*, if |f| is continuous at a point. Then f may not be continuous at that point

$$\text{Let } f(x) = \begin{cases} \dfrac{|x|}{x} & \text{if} \quad x \neq 0 \\ 1 & \text{if} \quad x = 0 \end{cases}$$

we have $\lim\limits_{x \to 0-} f(x) = \lim\limits_{x \to 0} \dfrac{-x}{x} = -1$

$$\lim_{x \to 0+} f(x) = \lim_{x \to 0} \frac{x}{x} = 1$$

$$f(0) = 1$$

Hence f is continuous at x = 0

We have (| f |) (x) = 1 ∀ x ∈ R

i.e. | f | is a constant function and so | f | is continuous at every point of R including x = 0.

Example 19:

Examine the function

f(x) = (x – a) sin {1/(x – a)}, when x ≠ a and f(a) = 0 for continuity at x = a.

Solution:

We have R.H.L. $= f(a + 0) = \lim\limits_{h \to 0} f(a + h)$

$= \lim\limits_{h \to 0} (a + h - a) \sin \{1/(a + h - a)\} = \lim\limits_{h \to 0} h \sin (1/h) = 0.$

Again

L.H.L. $= f(a - 0) = \lim\limits_{h \to 0} f(a - h)$

$= \lim\limits_{h \to 0} (a - h - a) \sin \{1/(a - h - a)\}$

$= \lim\limits_{h \to 0} - h \sin (-1/h) = 0.$

Also f(a) = 0. Thus, f(a + 0) = f(a – 0) = f(a). Therefore f(x) is continuous at x = a.

Example 20:

Investigate the points of continuity and discontinuity of the following function:

$$f(x) = \frac{1}{x-a} \operatorname{cosec}\left(\frac{1}{x-a}\right), \; x \neq a;$$

$f(x) = 0, \qquad x = a.$ **(Lucknow, 1990)**

Solution:

We have L.H.L. f(a + 0) = $\lim_{h \to 0} f(a+h)$

$$= \lim_{h \to 0} \frac{1}{a+h-a} \operatorname{cosec}\left(\frac{1}{a+h-a}\right) = \lim_{h \to 0} \frac{1}{h \sin(1/h)}$$

$= +\infty$, since h sin (1/h) → 0 as h → 0.

Again,

$$\text{R.H.L.} = f(a-0) = \lim_{h \to 0} f(a-h) = \lim_{h \to 0} \frac{1}{a-h-a} \operatorname{cosec}\left(\frac{1}{a-h-a}\right)$$

$$= \lim_{h \to 0} -\frac{1}{h} \cdot \frac{1}{\sin\{-(1/h)\}} = \lim_{h \to 0} \frac{1}{h \sin(1/h)}$$

$= +\infty$, since h sin (1/h) → 0 as h → 0.

Also as given f(a) = 0.

Since f(a + 0) = f(a – 0) → f(a), the function f(x) is discontinuous at x = a, having an infinite discontinuity of second kind.

Example 21:

Examine the function defined below for continuity at x = 0:

$$f(x) = \frac{\sin^2 ax}{x^2} \text{ for } x \neq 0,\ f(x) = 1 \text{ for } x = 0.$$

Solution:

We have f(0) = 1;

$$\text{R.H.L.} = f(0+0) = \lim_{h \to 0} f(0+h) = \lim_{h \to 0} f(h) = \lim_{h \to 0} \frac{\sin^2 ah}{h^2}$$

$$= \lim_{h \to 0} \left(\frac{\sin ah}{ah}\right)^2 . a^2 = 1.a^2 = a^2;$$

$$\text{and L.H.L. } f(0-0) = \lim_{h \to 0} f(0-h) = \lim_{h \to 0} f(-h) = \lim_{h \to 0} \frac{\sin^2(-ah)}{(-h)^2}$$

$$= \lim_{h \to 0} \frac{\sin^2 ah}{h^2} = a^2 .$$

Now f(x) is continuous at x = 0 if and only if

f(0) = f(0 + 0) = f(0 – 0).

Therefore f(x) is discontinuous at x = 0 unless a = ± 1.

Example 22:

Discuss the continuity at $x = 0, 1, 2$ of function $f(x)$ defined as follows:

$f(x) = -x^2$ *for* $x \le 0$, $f(x) = 5x - 4$ *for* $0 < x \le 1$.

$f(x) = 4x^2 - 3x$ *for* $1 < x < 2$ *and* $f(x) = 3x + 4$ *for* $x \ge 2$.

(Delhi, 1992; Meerut, 1991)

Solution:

(i) Continuity at x = 0. We have $f(0) = -0^2 = 0$;

We have

L.H.L. $= f(0-0) = \lim_{h\to 0} f(0-h) = \lim_{h\to 0} f(-h) = \lim_{h\to 0} -(-h)^2$

$= \lim_{h\to 0} -h^2 = 0$;

and R.H.L. $f(0+0) = \lim_{h\to 0} f(0+h) = \lim_{h\to 0} f(h) = \lim_{h\to 0} (5h-4)$.

$[\because 0 < h < 1]$

$= 4$.

Since R.H.L. $\neq$ L.H.L.

$\Rightarrow f(0+0) \neq f(0-0)$, the function $f(x)$ is discontinuous at $x = 0$.

(ii) Continuity at x = 1. We have $f(1) = 5 \times 1 - 4 = 1$;

R.H.L $= f(1+0) = \lim_{h\to 0} f(1+h) = \lim_{h\to 0} [4(1+h)^2 - 3(1+h)]$,

$[\because 1 < 1 + h < 2]$

$= \lim_{h\to 0} [4h^2 + 5h + 1] = 1$;

and L.H.L. $f(1-0) = \lim_{h\to 0} f(1-h) = \lim_{h\to 0} [5(1-h) - 4]$,

$[\because 0 \le 1 - h < 1]$

$= \lim_{h\to 0} (1 - 5h) = 1$.

Thus, $f(1+0) = f(1-0) = f(1)$. Therefore $f(x)$ is continuous at $x = 1$.

(iii) Continuity at x = 2. We have $f(2) = 3 \times 2 + 4 = 10$;

L.H.L. $= f(2-0) = \lim_{h\to 0} f(2-h) = \lim_{h\to 0} [4(2-h)^2 - 3(2-h)]$;

$[\because 1 < 2 - h < 2]$

$= \lim_{h\to 0} [4h^2 - 13h + 10] = 10$;

and R.H.L.$= f(2+0) = \lim_{h\to 0} f(2+h) = \lim_{h\to 0} [3(2+h) + 4]$

$= \lim_{h\to 0} (3h + 10) = 10$

$f(2) = f(2 - 0) = f(2 + 0)$ therefore f(x) continuous at x = 2.

Example 23:

Let $f(x) = x\dfrac{e^{1/x} - e^{-1/x}}{e^{1/x} + e^{-1/x}}, x \neq 0; f(0) = 0.$

Show that f(x) is continuous but not derivable at x = 0.

Solution:

We have f(0) = 0; **(D.U. B.Sc. (G), 1984)**

$$\text{R.H.L.} = f(0+0) = \lim_{h\to 0} f(0+h) = \lim_{h\to 0} f(h)$$

$$= \lim_{h\to 0} h\frac{e^{1/h} - e^{-1/h}}{e^{1/h} + e^{-1/h}} = \lim_{h\to 0} h\frac{1-e^{-2h}}{1+e^{-2h}}, \text{ dividing Nr. and Dr. by } e^{1/h}$$

$$= 0 \times \frac{1-0}{1+0} = 0 \times 1 = 0;$$

and $\text{L.H.L.} \; f(0-0) = \lim_{h\to 0} f(0-h) = \lim_{h\to 0} f(-h)$

$$= \lim_{h\to 0} -h.\frac{e^{-1/h} - e^{1/h}}{e^{-1/h} + e^{1/h}}$$

$$= \lim_{h\to 0} -h\frac{e^{-2/h} - 1}{e^{-2/h} + 1} = 0 \times \frac{0-1}{0+1} = 0.$$

R.H.L. = L.H.L.

$\Rightarrow f(0 + 0) = f(0 - 0) = f(0)$, the function is continuous at x = 0

Now to test f(x) for differentiability at x = 0, e have

$$Rf'(0) = \lim_{h\to 0} \frac{f(0+h) - f(0)}{h} = \lim_{h\to 0} \frac{f(h) - f(0)}{h}$$

$$= \lim_{h\to 0} \left[h\frac{e^{1/h} - e^{-1/h}}{e^{1/h} - e^{-1/h}} - 0 \right] / h = \lim_{h\to 0} \frac{1-e^{-2/h}}{1+e^{-2/h}} = \frac{1-0}{1+0} = 1,$$

and $Lf'(0) = \lim_{h\to 0} \dfrac{f(0-h) - f(0)}{-h} = \lim_{h\to 0} \dfrac{f(-h) - f(0)}{-h}$

$$= \lim_{h\to 0}\left[(-h).\frac{e^{-1/h}-e^{1/h}}{e^{-1/h}+e^{1/h}}-0\right]/(-h)$$

$$= \lim_{h\to 0}\frac{e^{-2/h}-1}{e^{-2/h}+1}=\frac{0-1}{0+1}=-1.$$

Since Rf '(0) ≠ Lf '(0), the function is not differentiable at x = 0.

Example 24:

Consider the function f(x) = x – [x], where x is a positive variable, and [x] denotes the integral part of x and show that it is discontinuous for all integral values of x, and continuous for all others. Draw the graph.

(Meerut, 1993, 90)

Solution:

From the definition of the function f(x), we have

$$f(x) = x - (n-1) \qquad \text{for } n-1 < x < n,$$

$$f(x) = 0 \qquad \text{for } x = n,$$

$$f(x) = x - n \qquad \text{for } n < x < n+1$$

and so on, where n is an integer.

We shall test the function f(x) for continuity at x = n. We have f(n) = 0;

$$f(n+0) = \lim_{h\to 0} f(n+h) = \lim_{h\to 0}\left[(n+h)-n\right] = \lim_{h\to 0} h = 0;$$

$$[\because n < n+h < n+1]$$

and L.H.L. $= f(n-0) = \lim_{h\to 0} f(n-h) = \lim_{h\to 0}\left[(n-h)-(n-1)\right],$

$$[\because n-1 < n-h < n]$$

$$= \lim_{h\to 0}(1-h) = 1.$$

Since f(n + 0) ≠ f(n – 0), the function f(x) is discontinuous at x = n. Thus, f(x) is discontinuous for all integral values of x. It is obviously continuous for all other values of x.

Since x is a positive variable, putting n = 1, 2, 3, 4, 5, ..., we observe that the graph of f(x) consists of the following:

$$y = x \qquad \text{when } 0 < x < 1, y = 0 \text{ when } x = 1$$

$$y = x - 1 \qquad \text{when } 1 < x < 2, y = 0 \text{ when } x = 2$$

$$y = x - 2 \qquad \text{when } 2 < x < 3, y = 0 \text{ when } x = 3$$

$$y = x - 3 \qquad \text{when } 3 < x < 4, y = 0 \text{ when } x = 4$$

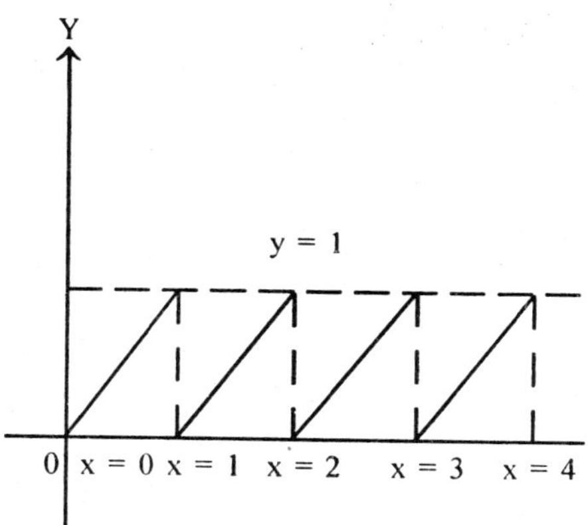

Fig. 1.2

and so on.

The graph of the function thus obtained is shown by thick lines from x = 0 to x = 4.

Example 25:

If $f(x) = \dfrac{xe^{1/x}}{1+e^{1/x}}$ *for* $x \neq 0$ *and* $f(0) = 0$*, show that f(x) is continuous at x = 0, but f '(0) does not exist.* **(Delhi, 1992; Meerut, 95S, 97)**

Solution:

Here f(0) = 0. Also right hand limit *i.e.*,

R.H.L. $f(0+0) = \lim_{h\to 0} f(0+h) = \lim_{h\to 0} f(h)$

$$= \lim_{h\to 0} \frac{he^{1/h}}{1+e^{1/h}} = \lim_{h\to 0} \frac{h}{e^{-1/h}+1} \text{ (Note)} = \frac{0}{0+1} = 0,$$

and L.H.L. $f(0-0) = \lim_{h\to 0} f(0-h) = \lim_{h\to 0} f(-h)$

$$= \lim_{h\to 0}\left[\frac{-he^{-1/h}}{1+e^{-1/h}}\right] = \frac{0}{1+0} = 0.$$

Since f(0 + 0) = f(0 − 0) = f(0), the function is continuous at x = 0.

Now we have to the derivative of f(x) at x = 0.

We have,

$$Rf'(0) = \lim_{h\to0} \frac{f(0+h)-f(0)}{h} = \lim_{h\to0} \frac{f(h)-f(0)}{h}$$

$$= \lim_{h\to0} \frac{\left(\dfrac{he^{1/h}}{1+e^{1/h}}\right)}{h} = \lim_{h\to0} \frac{e^{1/h}}{1+e^{1/h}} = \lim_{h\to0} \frac{1}{e^{1/h}+1}$$

$$= \frac{1}{0+1} = 0$$

and
$$Lf'(0) = \lim_{h\to0} = \frac{f(0-h)-f(0)}{-h} = \lim_{h\to0} \frac{f(-h)-f(0)}{-h}$$

$$= \lim_{h\to0} \frac{\dfrac{-he^{-1/h}}{1+e^{-1/h}}}{-h} = \lim_{h\to0} \frac{e^{-1/h}}{1+e^{-1/h}} = \frac{0}{1+0} = 0\cdot$$

Since $Rf'(0) \neq Lf'(0)$, the derivative of f(x) at x = 0 does not exist.

Example 26:

Let $f(x) = \dfrac{x^2}{1+x^2} + \dfrac{x^2}{(1+x^2)^2} + \dfrac{x^2}{(1+x^2)^3} + \dfrac{x^2}{(1+x^2)^4} = \ldots \infty.$

Is f(x) continuous at the origin? Give reasons for your answer.

Solution:

Here the function f(x) forms an infinite G.P whose first terms a is $\dfrac{x^2}{1+x^2}$ and common ratio r is $\dfrac{1}{1+x^2}$ which is less than one when $x \neq 0$.

$$\therefore \quad f(x) = \frac{a}{1-r} = \frac{x^2/(1+x^2)}{1-1/(1+x^2)} = \frac{x^2/(1+x^2)}{x^2/(1+x^2)} = 1, \qquad \text{provided } x \neq 0.$$

Thus, f(x) = 1 for $x \neq 0$ and f(x) = 0 for x = 0.

We have

$$f(0+0) = \lim_{h\to0} f(0+h) = \lim_{h\to0} f(h) = \lim_{h\to0} 1 = 1.$$

Thus $f(0+0) \neq f(0)$ and so f(x) is not continuous at x = 0.

Example 27:

A function f(x) is defined as follows:

f(x) = 0   for   x = 0

$$\phi(x) = \frac{1}{2} - x \quad for \quad 0 < x < \frac{1}{2}$$
$$\phi(x) = \frac{1}{2} \quad for \quad x = \frac{1}{2}$$
$$\phi(x) = \frac{3}{2} - x \quad for \quad \frac{1}{2} < x < 1$$
$$\phi(x) = 1 \quad for \quad x = 1.$$

Find the points of discontinuity. Also draw the graph of the function.

(Meerut, 1990; Lucknow, 99; Delhi, 97)

Solution:

We shall test the function for continuity at $x = 0, \frac{1}{2}$ and 1.

(i) For $x = 0$, we have $\phi(0) = 0$, $\phi(0 + 0) = \lim_{h \to 0} \phi(0 + h)$

$= \lim_{h \to 0} \phi(h) = \lim_{h \to 0} \left(\frac{1}{2} - h\right) = \frac{1}{2}$. Since $\phi(0) \neq \phi(0 + 0)$, the function $\phi(0)$ is discontinuous at $x = 0$.

(ii) For $x = \frac{1}{2}$, we have $\phi\left(\frac{1}{2}\right) = \frac{1}{2}$, $\phi\left(\frac{1}{2} - 0\right) = \lim_{h \to 0} \phi\left(\frac{1}{2} - h\right)$

$= \lim_{h \to 0} \left[\frac{1}{2} - \left(\frac{1}{2} - h\right)\right]$, $\left(\text{Note that } 0 < \frac{1}{2} - h < \frac{1}{2}\right)$

$= \lim_{h \to 0} h = 0$.

Since $\phi\left(\frac{1}{2} - 0\right) \neq \phi\left(\frac{1}{2}\right)$, the function $\phi(x)$ is discontinuous at $x = \frac{1}{2}$.

(iii) For $x = 1$, we have $\phi(1) = 1$, $\phi(1 - 0) = \lim_{h \to 0} \phi(1 - h)$

$= \lim_{h \to 0} [3/2 - (1 - h)]$, $\left[\text{Note that } \frac{1}{2} < 1 - h < 1\right]$

$= \lim_{h \to 0} \left(\frac{1}{2} + h\right) = \frac{1}{2}$.

Since $\phi(1) \neq \phi(1 - 0)$, the function $\phi(x)$ is discontinuous at $x = 1$.

Hence the function $\phi(x)$ has three points of discontinuity at $x = 0, \frac{1}{2}$ and 1.

The graph of the function consists of the point (0, 0); the segment of the line $y = \frac{1}{2}, 0 < x < \frac{1}{2}$;

the point $\left(\frac{1}{2}, \frac{1}{2}\right)$; the segment of the line

$y = (3/2) - x, \frac{1}{2} < x < 1$; and the point (1, 1).

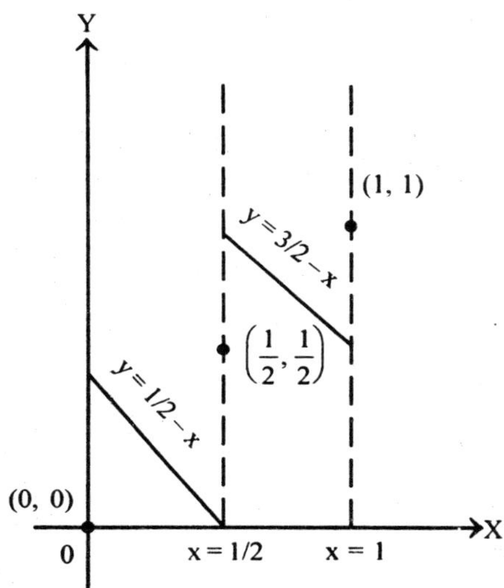

Fig. 1.3

Thus, the graph is as shown in the figure. From the graph we observe that the function is discontinuous at $x = 0, \frac{1}{2}$ and 1.

Example 28:

If $f(x) = -x$ for $x \leq 0$ and $f(x) = +x$ for $x \geq 0$, prove that $f(x)$ is continuous but not differentiable at $x = 0$.

Solution:

(i) To test f(x) for continuity at x = 0. We have f(0) = 0; f(0 + 0)

$= \lim_{h \to 0} f(0+h) = \lim_{h \to 0} f(h) = \lim_{h \to 0} -(-h) = \lim_{h \to 0} h = 0$.

Since f(0 + 0) = f(0 − 0) = f(0), the function f(x) is continuous at x = 0.

(ii) To test f(x) for differentiability at x = 0.

We have $Rf'(0) = \lim_{h \to 0} \frac{f(0+h) - f(0)}{h} = \lim_{h \to 0} \frac{f(h) - f(0)}{h}$

$$= \lim_{h\to 0} \frac{h-0}{h} = \lim_{h\to 0} 1 = 1.$$

Again $Lf'(0) = \lim_{h\to 0} \frac{f(0-h)-f(0)}{-h} = \lim_{h\to 0} \frac{f(-h)-f(0)}{-h}$

$$= \lim_{h\to 0} \frac{-(-h)-0}{-h} = \lim_{h\to 0} \frac{h}{-h} = \lim_{h\to 0} -1 = 1.$$

Since $Lf'(0) \neq Lf'(0)$ the function f(x) is not differentiable at x = 0.

Example 29:

Discuss the continuity of the function f(x) defined as follows:

$f(x) = x^2$ for $x < -2$, $f(x) = 4$ for $-2 \le x \le 2$, $f(x) = x^2$ for $x > 2$.

Solution:

We shall test f(x) for continuity only at the points x = – 2 and 2. It is obviously continuous at all other points.

(i) Continuity of f(x) at x = – 2. We have f(– 2) = 4;

$$f(-2+0) = \lim_{h\to 0} f(-2+h) = \lim_{h\to 0} 4 = 4;$$

$$f(-2-0) = \lim_{h\to 0} f(-2-h) = \lim_{h\to 0} (-2-h)^2 \qquad [\because -2-h < -2]$$

= 4.

Thus f(– 2 + 0) = f(– 2 – 0) = f(– 2). Therefore f(x) is continuous at x = – 2.

(ii) Continuity of f(x) at x = 2. We have f(2) = 4;

$$f(2+0) = \lim_{h\to 0} f(2+h) = \lim_{h\to 0} (2+h)^2 = 4.$$

and $f(2-0) = \lim_{h\to 0} f(2-h) = \lim_{h\to 0} (2-h)^2 = 4 = 4.$

Thus, f(x) is also continuous at x = 2.

Example 30:

Discuss the continuity and differentiability of the function f(x) defined as follows:

$f(x) = 1$ for $-\infty < x\ 0$, $f(x) = 1 + \sin x$ for $0 \le x < \pi/2$,

$$f(x) = 2 + \left(x - \frac{\pi}{2}\right)^2 \text{ for } \frac{\pi}{2} \le x < \infty.$$

Solution:

We shall test f(x) for continuity and differentiability at x = 0 and π/2. It is obviously continuous as well as differentiable at all other points.

(i) Continuity and differentiability of f(x) at x = 0. We have

$$f(0) = 1 + \sin 0 = 1;\ f(0+0) = \lim_{h\to 0} f(0+h)$$

$$= \lim_{h\to 0} f(h) = \lim_{h\to 0} (1+\sin h) = 1;\ \text{and}$$

$$f(0-0) = \lim_{h\to 0} f(0-h)$$

$$= \lim_{h\to 0} f(-h) = \lim_{h\to 0} 1 = 1.$$ Thus, f(0) = f(0 + 0) = f(0 – 0) and therefore f(x) is continuous at x = 0.

$$\text{Now } Rf'(0) = \lim_{h\to 0} \frac{f(0+h)-f(0)}{h} = \lim_{h\to 0} \frac{f(h)-f(0)}{h}$$

$$= \lim_{h\to 0} \frac{(1+\sin h)-(1+\sin 0)}{h} = \lim_{h\to 0} \frac{\sin h}{h} = 1;$$

$$\text{and } Lf'(0) = \lim_{h\to 0} \frac{f(0-h)-f(0)}{-h} = \lim_{h\to 0} \frac{f(-h)-f(0)}{-h}$$

$$= \lim_{h\to 0} \frac{1-(1+\sin 0)}{-h} = \lim_{h\to 0} \frac{0}{-h} = \lim_{h\to 0} 0 = 0.$$

Since Rf '(0) ≠ Lf '(0), f(x) is not differentiable at x = 0.

(ii) Continuity and differentiability of f(x) at x = π/2. We have

$$f\left(\frac{\pi}{2}\right) = 2 + \left(\frac{\pi}{2} - \frac{\pi}{2}\right)^2 = 2;$$

$$f\left(\frac{\pi}{2}+0\right) = \lim_{h\to 0} f\left(\frac{\pi}{2}+h\right) = \lim_{h\to 0}\left[2 + \frac{\pi}{2}\left\{\left(\frac{\pi}{2}+h\right) - \frac{\pi}{2}\right\}^2\right]$$

$$= \lim_{h\to 0} (2+h^2) = 2;$$

$$\text{and } f\left(\frac{\pi}{2}-0\right) = \lim_{h\to 0} f\left(\frac{\pi}{2}-h\right) = \lim_{h\to 0}\left[1+\sin\left(\frac{\pi}{2}-h\right)\right]$$

$$= \lim_{h\to 0} (1 + \cos h) = 1 + 1 = 2.$$

Thus, $f\left(\frac{\pi}{2}+0\right)=f\left(\frac{\pi}{2}-0\right)=f\left(\frac{\pi}{2}\right)$

and therefore f(x) is continuous at $x = \frac{\pi}{2}$.

Now $Rf'\left(\frac{\pi}{2}\right)=\lim_{h\to 0}\frac{f(\pi/2+h)-f(\phi/2)}{h}$

$$=\lim_{h\to 0}\frac{[2+\{\pi/2+h-\pi/2\}]-[2+(\pi/2-\pi/2)^2]}{h}$$

$$=\lim_{h\to 0}\frac{2+h^2-2}{h}=\lim_{h\to 0}h=0;$$

and $Lf'\left(\frac{\pi}{2}\right)=\lim_{h\to 0}\frac{f(\pi/2-h)-f(\pi/2)}{-h}$

$$=\lim_{h\to 0}\frac{1+\sin(\pi/2-h)-2}{-h}$$

$$=\lim_{h\to 0}\frac{-1+\cos h}{-h}=\lim_{h\to 0}\frac{1-\cos h}{h}=\lim_{h\to 0}\frac{2\sin^2(h/2)}{h}$$

$$=\lim_{h\to 0}\left[\frac{\sin h/2}{h/2}\sin h/2\right]=1\times 0=0.$$

Thus, Rf '(0) = Lf '(0) and so f(x) is differentiable at $x = \pi/2$.

Example 31:

A function f(x) is defined as follows:

f(x) = 1 + x if x ≤ 2 and f(x) = 5 – x if x ≥ 2.

Is the function continuous at x = 2?

(Lucknow, 1982; Meerut, 96, 98)

Solution:

We have f(2) = 1 + 2 or 5 – 2 = 3;

$f(2 + 0) = \lim_{h\to 0} f(2+h)$, where h is +ive and sufficiently small

$=\lim_{h\to 0}[5-(2+h)]$, [∵ 2 + h > 2 and f(x) = 5 – x if x > 2]

$=\lim_{h\to 0}(3-h)=3$

and f(2 – 0) = $\lim_{h\to 0} f(2-h)$, where h is +ive and sufficiently small

$$= \lim_{h\to 0}\left[1+(2-h)\right], \qquad [\because 2 - h < 2 \text{ and } f(x) = 1 + x \text{ if } x < 2]$$

$$= \lim_{h\to 0}(2-h) = 3.$$

Since f(2 + 0) = f(2 – 0) = f(2), the function f(x) is continuous at x = 2.

Example 32:

If f(x) = x^2 sin(1/x) for x ' 0 and f(0) = 0, show that f(x) is continuous and differentiable at x = 0. **(Delhi, 1992, 90; Meerut, 1997)**

Solution:

(i) To test f(x) for continuity at x = 0. We have f(0) = 0;

$$f(0+0) = \lim_{h\to 0} f(0+h) = \lim_{h\to 0} f(h) = \lim_{h\to 0} h^2 \sin\frac{1}{h} = 0;$$

and $$f(0-0) = \lim_{h\to 0} f(0-h) = \lim_{h\to 0} f(-h)$$

$$= \lim_{h\to 0} (-h)^2 \sin(-1/h)$$

$$= \lim_{h\to 0} h^2 \sin\frac{1}{h} = 0.$$

Now f(0 + 0) = f(0 – 0) = f(0) implies that f(x) is continuous at x = 0.

(ii) To test f(x) for differentiability at x = 0. We have

$$Rf'(0) = \lim_{h\to 0}\frac{f(0+h)-f(0)}{h} = \lim_{h\to 0}\frac{f(-h)-f(0)}{-h}$$

$$= \lim_{h\to 0}\frac{h^2\sin(1/h)-0}{h} = \lim_{h\to 0} h \sin\frac{1}{h} = 0.$$

Again $$Lf'(0) = \lim_{h\to 0}\frac{f(0-h)-f(0)}{-h} = \lim_{h\to 0}\frac{f(-h)-f(0)}{-h}$$

$$= \lim_{h\to 0}\frac{(-h)^2\sin(-1/h)-0}{-h} = \lim_{h\to 0}\frac{-h^2\sin(1/h)}{-h}$$

$$= \lim_{h\to 0} h\sin\frac{1}{h} = 0.$$

Now, Rf(0) = Lf '(0) implies that f(x) is differentiable at x = 0. The derivative of f(x) at x = 0 has the value zero.

Example 33:

Examine the following curve for continuity and differentiability:

$y = x^2$ *for* $x \le 0$, $y = 1$ *for* $0 < x < 1$, $y = 1/x$ *for* $x > 1$.

(Meerut, 1996 P)

Solution:

Let y = f(x). We need to test f(x) for continuity and differentiability at the points x = 0 and 1. It is obviously continuous and differentiable at all other points.

At x = 0. We have $f(0) = 0^2 = 0$; $f(0 + 0) = \lim_{h \to 0} f(0+h)$

$= \lim_{h \to 0} f(h) = \lim_{h \to 0} 1 = 1$. Since $f(0 + 0) \neq f(0)$, the function f(x) is not continuous at x = 0. Consequently it is also not differentiable at x = 0.

At x = 1. The function f(x) is not defined at x = 1. Therefore it is discontinuous as well as non-differentiable at x = 1.

Example 34:

Give an example to show that a continuous function need not be differentiable. **(Delhi, 1990; Meerut, 1992)**

Solution:

Here we shall give one more such example. Consider the function f(x) defined as follows:

$$f(x) = x \sin (1/x) \text{ for } x \neq 0,\ f(0) = 0.$$

As already proved in part (b) this function f(x) is continuous at x = 0. Now we shall show that the function f(x) is not differentiable at x = 0. We have

$$Rf'(0) = \lim_{h \to 0} \frac{f(0+h) - f(0)}{h} = \lim_{h \to 0} \frac{f(h) - f(0)}{h}$$

$$= \lim_{h \to 0} \frac{h \sin(1/h) - 0}{h}$$

$$= \lim_{h \to 0} \sin \frac{1}{h}, \text{ which does not exist.}$$

Similarly, Lf (0) does not exist. Hence f(x) is not differentiable at x = 0.

Example 35:

Examine the following curve for continuity and differentiability at the points $x = \pm 1$:

$$y = x - 1 \text{ for } x \geq 1,\ y = \frac{3}{2} + x \text{ for } x \leq -1,$$

$$y = 1 + x + x^2 + x^3 + \ldots \infty \text{ for } -1 < x < 1.$$

Solution:

Obviously $y = \dfrac{1}{1-x}$ for $-1 < x < 1$. Let $y = f(x)$.

(i) At x = 1. We have $f(1) = 1 - 1 = 0$;

$$f(1+0) = \lim_{h \to 0} f(1+h) = \lim_{h \to 0} (1+h-1) = 0;$$

and $f(1-0) = \lim_{h \to 0} f(1-h) = \lim_{h \to 0} \dfrac{1}{1-(1-h)} = \lim_{h \to 0} \dfrac{1}{h} = \infty$.

Since $f(1 + x)$ ¹ $f(1 - 0)$, the function $f(x)$ is not continuous at $x = 1$. Consequently $f(x)$ is also not differentiable at $x = 1$.

(ii) At x = − 1. We have $f(-1) = \dfrac{3}{2} - 1 = \dfrac{1}{2}$;

$$f(-1+0) = \lim_{h \to 0} f(-1+h) = \lim_{h \to 0} \frac{1}{1-(-1+h)}$$

$$= \lim_{h \to 0} \frac{1}{2-h} = \frac{1}{2};$$

and $f(-1-0) = \lim_{h \to 0} f(-1-h) = \lim_{h \to 0} \left[\dfrac{3}{2} + (-1-h)\right]$

$$= \lim_{h \to 0} \left(\frac{1}{2} - h\right) = \frac{1}{2}.$$

Since $f(-1+0) = f(-1-0) = f(-1)$, the function $f(x)$ is continuous at $x = -1$.

Now $Rf'(-) = \lim_{h \to 0} \dfrac{f(-1+h) - f(-1)}{h}$

$$= \lim_{h \to 0} \frac{\dfrac{1}{1-(-1+h)} - \left(\dfrac{3}{2} - 1\right)}{h} = \lim_{h \to 0} \frac{\dfrac{1}{2-h} - \dfrac{1}{2}}{h}$$

$$= \lim_{h \to 0} \frac{2-2+h}{2h(2-h)} = \lim_{h \to 0} \frac{1}{2(2-h)} = \frac{1}{4}.$$

Again $Lf'(-1) = \lim_{h \to 0} \dfrac{f(-1-h) - f(-1)}{-h}$

$$= \lim_{h \to 0} \frac{\left[\frac{3}{2} + (-1-h)\right] - \left(\frac{3}{2} - 1\right)}{-h}$$

$$= \lim_{h \to 0} \frac{\frac{1}{2} - h - \frac{1}{2}}{-h} = \lim_{h \to 0} 1 = 1 \cdot$$

Since Rf ' (– 1) ≠ Lf ' (– 1), the function f(x) is not differentiable at x = – 1.

Example 36:

If $f(x) = \dfrac{xe^{1/x}}{1+e^{1/x}}$ *for* $x \neq 0$ *and* $f(0) = 0$, *show that f(x) is continuous at x = 0, but f '(0) does not exist.* **(Delhi, 1992; Meerut, 95 S, 97)**

Solution:

Here f(0) = 0. Also right hand limit *i.e.*,

$$\text{R.H.L.} = f(0 + 0) = \lim_{h \to 0} f(0+h) = \lim_{h \to 0} f(h)$$

$$= \lim_{h \to 0} \frac{he^{1/e}}{1+e^{1/h}} = \lim_{h \to 0} \frac{h}{e^{-1/h}+1} \text{ (Note)} = \frac{0}{0+1} = 0,$$

$$\text{and L.H.L.} = f(0 - 0) = \lim_{h \to 0} f(0-h) = \lim_{h \to 0} f(-h)$$

$$= \lim_{h \to 0} \left[\frac{-he^{1/h}}{1+e^{-1/h}}\right] = \frac{0}{1+0} = 0$$

Since f(0 + 0) = f(0 – 0) = f(0), the function is continuous at x = 0.

Now we have to the derivatives of f(x) at x = 0.

We have,

$$Rf'(0) = \lim_{h \to 0} \frac{f(0+h) - f(0)}{h} = \lim_{h \to 0} \frac{f(h) - f(0)}{h}$$

$$= \lim_{h \to 0} \frac{\left(\frac{he^{1/h}}{1+e^{1/h}}\right)}{h} = \lim_{h \to 0} \frac{e^{1/h}}{1+e^{1/h}} = \lim_{h \to 0} \frac{1}{e^{1/h}+1}$$

$$= \frac{1}{0+1} = 0$$

$$\text{and } Lf'(0) = \lim_{h \to 0} \frac{f(0-h) - f(0)}{-h} = \lim_{h \to 0} \frac{f(-h) - f(0)}{-h}$$

$$= \lim_{h\to 0} \frac{\dfrac{-he^{-1/h}}{1+e^{-1/h}} - 0}{-h} = \lim_{h\to 0} \frac{e^{-1/h}}{1+e^{-1/h}} = \frac{0}{1+0} = 0\cdot$$

Since $Rf'(0) \neq Lf'(0)$, the derivative of $f(x)$ at $x = 0$ does not exist.

EXERCISES

1. Show that x^r, where x is a positive integer, its continuous at $x = a$.
2. Show that $\frac{1}{x}$ is continuous at $x = a$, $(a \neq 0)$.
3. If $f(x) = x$ for $x \leq 1$ and $f(x) = x^2$ for $x < 1$ show that f is continuous at $x = 1$ but not derivable at $x = 1$.
4. Show that $f(x) = x|x|$ is derivable at the origin.
5. Show that the function $f(x) = (x - a) \sin \frac{1}{x-a}$.
6. Show that the function $f(x) = |x - 1| + |x + 1|$ is not derivable at $x = -1, 2$ and is deriable at every other point.

2

ROLLE'S THEOREM, MEAN VALUE THEOREMS, TAYLOR'S AND MACLAURIN'S THEOREMS

2.1 ROLLE'S THEOREM

If a function f(x) which is continuous in the closed interval [a, b] and differentiable in the open interval (a, b), be zero at x = a and also at x = b. Then f(x) is zero for at least one value of x between a and b.

(Meerut, 1985, 91; Agra, 82, 77, 73; Indore, 70; Gorakhpur, 78)

OR

If function f(x) is such that

(i) f(x) is continuous in the closed interval $a \le x \le b$,

(ii) f '(x) exists for every point in the open interval $a < x < b$,

(iii) f '(a) = f(b), then there exits at least one value of x, say c, where $a < c < b$, such that f '(c) = 0.

Proof:

Since f(a) = f(b), unless the function f(x) is a constant in which case the theorem is at once established, f(x) should either increase or decrease when x takes values greater than a. Suppose it increases; then since it again takes a value f(b) = f(a), it must cease to increase and begin to decrease at some point c, such that $a < c < b$.

At this point c the function f(x) has a maximum value and so $f(c + h) - f(c)$ and $f(c - h) - f(c)$ are both negative, h being small and positive.

$$\therefore \quad \frac{f(c+h) - f(c)}{h} < 0 \text{ and } \frac{f(c-h) - f(c)}{-h} > 0.$$

Obviously as $h \to 0$, the above expressions tend to being –ive and +ive respectively unless each of them has the limit zero.

If they have different limits, then $Rf'(c) \neq Lf'(c)$ and therefore $f'(c)$ does not exist, contradicting the hypothesis.

Hence each of the above limits must be zero,

i.e., $f'(c) = 0$ where $a < c < b$.

Note 1 : There may be more than one point like c at which $f'(x)$ vansihes.

Note 2 : Rolle's theorem will not hold good

(i) if $f(x)$ is discontinuous at some point in the interval

$a \leq x \leq b$,

or (ii) if $f'(x)$ does not exist at some point in the interval

$a \leq x \leq b$,

or (iii) if $f(a) \neq f(b)$.

Geometrical Interpretation of Rolle's Theorem

Suppose the function $f(x)$ satisfies the conditions of Rolle's theorem in the interval $[a, b]$. Then its geometrical interpretation is that on the curve $y = f(x)$ there is at least one point lying in the open interval (a, b) the tangent at which is parallel to the axis of x.

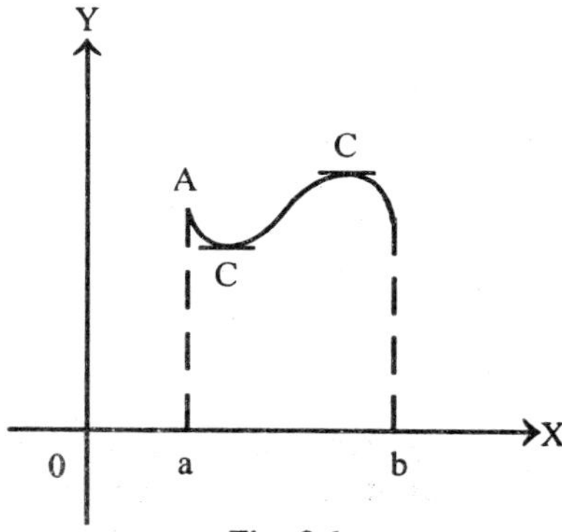

Fig. 2.1

Example 1:

Verify Rolle's Theorem for $f(x) = \sqrt{1 - x^2}$ *in* $[-1, 1]$

Solution:

The given function being an algebraic function of x is continuous in $[-1, 1]$

Now $f'(x) = \frac{-x}{\sqrt{1-x^2}}$ for $-1 < x < x < 1$ so f(x) is derivable in]–1, 1[such that $f'(c) = 0$

Thus, f(x) satisfies all the conditions of Rolle's Theorem.

So there exist a point $c \in]-1, 1[$ such that $f'(c) = 0$

Clearly $f'(c) = 0$ for $c \in]-1, 1[$

Hence, the given function satisfies the hypotheses and the conclusion of Rolle's Theorem.

Example 2:

Verify Rolle's Theorem in the case of the functions

(Agra, 1982, 80)

(i) $f(x) = 2x^3 + x^2 - 4x - 2$,

(ii) $f(x) = \sin x$ in $[0, \pi]$,

(iii) $f(x) = (x - a)^m (x - b)^n$, *where m and n are +ive integers, and x lies in the interval [a, b].* **(Agra, 1981)**

Solution:

(i) Here f(x) is a rational integral function of x. We know that rational functions are continuous and differentiable for all real values of x. Thus, the first two conditions of Rolle's Theorem are satisfied in any interval.

Again let $f(x) = 0$. Then $2x^3 + x^2 - 4x - 2 = 0$

or $(x^2 - 2)(2x - 1) = 0$ *i.e.* $x = \pm \sqrt{2}, -\frac{1}{2}$.

Thus $f(\sqrt{2}) = f(-\sqrt{2}) = f\left(-\frac{1}{2}\right) = 0$.

Now, let us consider the interval $[-\sqrt{2}, \sqrt{2}]$. In this interval all the conditions of Rolle's Theorem are satisfied. Therefore there is at least one value of x in the open interval $(-\sqrt{2}, \sqrt{2})$ where $f'(x) = 0$.

Now $f'(x) = 0$ where $6x^2 + 2x - 4 = 0$

or $3x^2 + x - 2 = 0$ or $(3x - 2)(x + 1) = 0$

or $x = -1, 2/3$.

Thus, $f'(x)\,(-1) = f'\left(\frac{2}{3}\right) = 0$

Since both the points $x = -1$ and $x = \frac{2}{3}$ lie in the open interval $(-\sqrt{2}, \sqrt{2})$, Rolle's Theorem is verified.

(ii) Here $f(0) = \sin 0 = 0$ and $f(\pi) = \sin \pi = 0$. Thus

$$f(0) = f(\pi) = 0.$$

We know that sin x is continuous and differentiable in $[0, \pi]$. Hence all the three conditions of Roll's Theorem are satisfied. Therefore $f'(x) = 0$ for at least one value of x in the open interval $(0, \pi)$.

Now $f'(x) = 0$ gives $\cos x = 0$ or $x = \pm \frac{\pi}{2}, \pm \frac{3\pi}{2}, \pm \frac{5\pi}{2}, \ldots$ Since $x = \pi/2$ lies in the open interval $(0, \pi)$, the Rolle's Theorem is verified.

(iii) Here $f(x) = (x - a)^m (x - b)^n$.

As m and n are positive integers, $(x - a)^m$ and $(x - b)^n$ are polynomials in x on being expanded by binomial theorem. Hence f(x) is also a polynomial in x. Consequently f(x) is continuous and differentiable in the closed interval [a, b]. Also $f(x) = f(b) = 0$. Thus all the three conditions of Rolle's Theorem are satisfied. So $f'(x) = 0$ for at least one value of x lying in the open interval (a, b).

Now, $f'(x) = (x - a)^m \cdot n\,(x - b)^{n-1} + m\,(x - a)^{m-1}(x - b)^n$.

The equation $f'(x) = 0$, on being solved, gives

$$x = a, b, \frac{na + mb}{m + n}.$$

Out of these values the value $\frac{na + mb}{m + n}$ is a point lying in the open interval (a, b) as it divides the interval (a, b) internally in the ratio m : n. Thus, the Rolle's Theorem is verified.

Example 3:

(a) Discuss the applicability of Rolle's Theorem for $f(x) = 2 + (x-1)^{2/3}$ in the interval [0, 2]. **(G.N.U., 1977)**

Solution:

Given $f(x) = 2 + (x - 1)^{2/3}$. Obviously $f(0) = 3 = f(2)$, showing that the third condition of Rolle's Theorem is satisfied.

The function f(x), being an algebraic function of x, is continuous in the closed interval [0, 2]. Thus the first condition of Rolle's Theorem is satisfied.

Now $f'(x) = \frac{2}{3} \cdot \left[\frac{1}{(x-1)^{1/3}}\right]$. We observe that at x = 1, f '(x) is not finite while x = 1 is a point of the open interval $0 < x < 2$. Thus, the second condition for Rolle's Theorem is not satisfied.

Hence the Rolle's Theorem is not applicable for the function $2 + (x - 1)^{3/2}$ in the given interval [0, 2].

Example 4:

Discuss the applicability of Rolle's Theorem in the interval [–1, 1] to function f(x) = | x |. **(Meerut 1994)**

Solution:

Here $f(-1) = |-1| = 1$ and $f(1) = |1| = 1$, so that $f(-1) = f(1)$.

Also the function f(x) is continuous throughout the closed interval [–1, 1] but it is not differentiable at x = 0 which is a point of the open interval (–1, 1). Therefore the second condition for Rolle's Theorem is not satisfied, *i.e.*, the Rolle's Theorem is not applicable here.

Example 5:

Are the conditions of Rolle's Theorem satisfied in the case of the following functions?

(i) $f(x) = x^2$ *in* $2 \le x \le 3$,

(ii) $f(x) = \cos(1/x)$ *in* $-1 \le x \le 1$,

(iii) $f(x) = \tan x$ *in* $0 \le x \le \pi$,

Solution:

(i) The function $f(x) = x^2$ is continuous and differentiable in the interval [2, 3]. Thus, the first two conditions of Rolle's Theorem are satisfied.

Also f(2) = 4 and f(3) = 9, so that $f(2) \neq f(3)$. Hence the third condition is not satisfied.

(ii) Here $f(-1) = \cos(-1) = \cos 1$ and $f(1) = \cos 1$. Thus, $f(-1) = f(1)$ *i.e.*, the third condition is satisfied.

But the first two conditions of Rolle's Theorem are not satisfied as the function is discontinuous at x = 0 and consequently is not differentiable there.

(iii) Here f(0) = tan 0 = 0 and f(π) = tan π = 0. Thus, f(0) = f(π) *i.e.*, the third condition is satisfied.

But the first two conditions of Rolle's Theorem are not satisfied here as the function is not continuous at $x = \pi/2$ and consequently is non-differentiable there.

Example 6:

If f(x), φ(x), ψ(x) have derivatives when a ≤ x ≤ b, show that there is a value c of x lying between a and b such that

$$\begin{vmatrix} f(a) & \phi(a) & \psi(a) \\ f(b) & \phi(b) & \psi(b) \\ f'(c) & \phi'(c) & \psi'(c) \end{vmatrix} = 0$$

(Agra, 1993)

Solution:

Consider the following function

$$F(x) = \begin{vmatrix} f(a) & \phi(a) & \psi(a) \\ f(b) & \phi(b) & \psi(b) \\ f(x) & \phi(x) & \psi(x) \end{vmatrix}$$

On expanding the determinant, we observe that the function F(x) is of the form A f(x) + B φ(x) + C ψ(x), where A, B, C are some real numbers.

Since the functions f(x), φ(x) and ψ(x) have derivatives when $a \le x \le b$, therefore, the function F(x) also possesses derivatives when $a \le x \le b$. Consequently F(x) is also continuous when $a \le x \le b$. Further F(a) = F(b) = 0 because then the two rows of the determinant become identical. Thus, F(x) satisfies all the three conditions of Rolle's Theorem. Hence F '(x) = 0 for at least one value of x, say x = c, lying between a and b. Thus there is a value c of x lying between a and b such that

$$\begin{vmatrix} f(a) & \phi(a) & \psi(a) \\ f(b) & \phi(b) & \psi(b) \\ f'(c) & \phi'(c) & \psi'(c) \end{vmatrix} = 0.$$

Example 7:

Discuss the applicability of Rolle's Theorem to $f(x) = \log\left[\frac{x^2 + ab}{(a+b)x}\right]$, *in the interval [a, b].* **(Meerut, 1990)**

Solution:

We have $f(a) = \log\left[\dfrac{a^2 + ab}{(a+b)a}\right] = \log 1 = 0,$

and $f(b) = \log\left[\dfrac{b^2 + ab}{(a+b)b}\right] = \log 1 = 0.$

Hence f(a) = f(b) = 0.

Also $Rf'(x) = \lim_{h \to 0} \dfrac{f(x+h) - f(x)}{h}$

$$= \lim_{h \to 0} \frac{1}{h}\left[\log\left\{\frac{(x+h)^2 + ab}{(a+b)(x-b)}\right\} - \log\left\{\frac{x^2 + ab}{(a+b)x}\right\}\right]$$

$$= \lim_{h \to 0} \frac{1}{h}\left[\log \frac{(x^2 + 2xh + h^2 + ab)(a+b)x}{(a+b)(x-h)(x^2+ab)}\right]$$

$$= \lim_{h \to 0} \frac{1}{h}\left[\log\left\{\frac{(x^2 + 2xh + h^2 + ab)}{(x^2+ab)} \times \frac{x}{x+h}\right\}\right]$$

$$= \lim_{h \to 0} \frac{1}{h}\left[\log\left\{1 + \frac{xh + h^2}{x^2 + ab}\right\} - \log\left\{1 + \frac{h}{x}\right\}\right]$$

$$= \lim_{h \to 0} \frac{1}{h}\left[\frac{2hx + h^2}{x^2 + ab} - \frac{h}{x} + \ldots\ldots\right] \qquad \ldots(1)$$

[$\because \log(1+y) = y - 1/2\, y^2 + \ldots.$]

$$= \frac{2x}{x^2 + ab} - \frac{1}{x}.$$

Again $Lf'(x) = \lim_{h \to 0}\left[\dfrac{f(x-h) - f(x)}{-h}\right]$

$$= \lim_{h \to 0} \frac{1}{(-h)}\left[\frac{-2hx + h^2}{x^2 + ab} - \frac{(-h)x}{x} + \ldots\right],$$

[replacing h by – h in (1)]

$$= \frac{2x}{x^2 + ab} - \frac{1}{x}.$$

Thus, Rf '(x) = Lf '(x), showing that f(x) is differentiable for all values of x in [a, b]. Consequently f(x) is also continuous for all values of x in [a, b]. Hence all the three conditions of Rolle's Theorem are satisfied.

$\therefore$ f '(x) = 0 for at least one value of x in the open interval a < x < b.

Now $\quad f'(x) = 0$ where $\dfrac{2x}{x^2 + ab} = \dfrac{1}{x} = 0$

or $\quad 2x^2 = (x^2 + ab) = 0$ or $\quad x^2 = ab$

or $\quad x = \sqrt{(ab)}$, which being the geometric mean of a and b lies in the open interval (a, b). Hence Rolle's Theorem is verified.

Example 8:

Verify Rolle's Theorem for

(i) $f(x) = x^3 - 4x$ *in* $[-2, 2]$, **(G.N.U., 1975)**

(ii) $f(x) = x(x + 3)\, e^{-x/2}$ *in* $[-3, 0]$, **(Gorakhpur, 1990)**

(iii) $f(x) = e^x (\sin x - \cos x)$ *in* $[\pi/4, 5\pi/4]$,

Solution:

(i) Here $f(x) = x^3 - 4x$. Since f(x) is a polynomial in x, therefore it is continuous and differentiable for every real value of x; Also $f(-2) = 0 = f(2)$.

$\therefore$ f(x) satisfies all the three conditions of Rolle's Theorem.

$\therefore$ in the open interval

(–2, 2) for which f '(c) = 0.

Now $\quad f'(x) = 0$ gives $3x^2 - 4 = 0$

or $\quad x = \pm \dfrac{2}{\sqrt{3}} = \pm 1.155$ (aprox).

Both these values lie in the open interval (–2, 2). Thus the theorem is verified.

(ii) Here $f(x) = x(x + 3)\, e^{-x/2} = (x^2 + 3x)\, e^{-x/2}$.

We have $f'(x) = (2x + 3)\, e^{-x/2} + (x^2 + 3x) \cdot e^{-x/2} \cdot \left(-\dfrac{1}{2}\right)$

$= e^{-x/2}$ [2x + 3 – 1/2 $(x^2 + 3x) = -1/2\ (x^2 - x - 6)\ e^{-x/2}$,

which exists for every value of x in the interval [–3, 0]. Therefore f(x) is differentiable and also continuous in the interval [–3, 0].

Also f(–3) = 0 = f(0). Therefore all the three conditions of Rolle's Theorem are satisfied.

∴ There must exist at least one number, say c, in the open interval (–3, 0) for which f '(c) = 0 *i.e.*, $-1/2\ (c^2 - c - 6)\ e^{-c/2} = 0$ or $c^2 - c - 6 = 0$ or (c – 3) (c + 2) = 0 or c = 3, –2.

The value c = – 2 lies in the open interval (–3, 0). Hence the theorem is verified.

(iii) Here f(x) = e^x (sinx – cosx).

We have f(π/4) = $e^{\pi/4}$ {sin (π/4) = cos (π/4)}

$$= e\pi/4\ [(1/\sqrt{2}) - (1/\sqrt{2})] = 0$$

and $$f\left(\frac{5\pi}{4}\right) = e^{5\pi/4}\left[\sin\frac{5\pi}{4} - \cos\frac{5\pi}{4}\right] = e^{5\pi/4}\left[-\frac{1}{\sqrt{2}} + \frac{1}{\sqrt{2}}\right] = 0.$$

∴ f(π/4 = f(5π/4) = 0.

Further the function f(x) is continuous and differentiable in [π/4, 5π/4]. Therefore all the three conditions of Rolle's Theorem are satisfied.

∴ There must exist at least one number, say c, in the open interval (π/4, 5π/4) for which f '(c) = 0.

Now f '(x) = e^x (cosx + sinx) + e^x (sinx – cosx) = 2 e^x sin x.

From f '(x) = 0 we get $2e^x$ sinx = 0

or sinx = 0, [∵ $e^x \neq 0$]

or x = 0, ± π, ± 2π, ± 3π, ...

Out of these values x = π lies in the open interval (π/4, 5π/4). Thus the Rolle's Theorem is verified.

2.2 LAGRANGE'S MEAN VALUE THEOREM OR FIRST MEAN VALUE THEOREM

(Lucknow, 1983, 81; Meerut, 81, 84P, 86, 91)

If a function f(x) is

(i) continuous in the closed interval a ≤ x ≤ b, and

(ii) differentiable in the open interval (a, b) *i.e.*, $a < x < b$, then there exists at least one value 'c' of x lying in the open interval $a < x < b$ such that

$$\frac{\mathbf{f(b) - f(a)}}{\mathbf{b - a}} = \mathbf{f'(c)}.$$

Proof:

Consider the function $\phi(x)$ defined by

$$\phi(x) = f(x) + Ax, \qquad ...(1)$$

where A is a constant to be determined such that $\phi(a) = \phi(b)$ *i.e.*,

$$f(a) + Aa = f(b) + Ab$$

or
$$-A = \frac{f(b) - f(a)}{b - a} \qquad ...(2)$$

Now f(x) is given to be continuous in $a \le x \le b$ and differential in $a < x < b$.

Again, A being a constant, Ax is also continuous in $a \le x \le b$ and differentiable in $a \le x \le b$.

$\therefore$ $\phi(x) = f(x) + Ax$ is continuous in $a \le x \le b$ and differentiable in $a < x < b$. Also by our choice of A, we have $\phi(a) = \phi(b)$. Thus, $\phi(x)$ satisfies all the conditions of Rolle's Theorem in the interval [a, b]. Hence there exists at least one point, say $x = c$, of the open interval $a < x < b$, such that $\phi'(c) = 0$.

But $\phi'(x) = \phi'(x) + A$, from (1).

$\therefore$ $\phi'(c) = 0$ gives $f'(c) + A = 0$

or
$$f'(c) = -A = \frac{f(b) - f(a)}{b - a}, \text{ from (2).}$$

This proves the theorem.

Another Form of Lagrange's Mean Value Theorem

If a function f(x) is

(i) continuous in the closed interval [a, a + h], and

(ii) differentiable in the open interval (a, a + h), then there exists at least one number θ lying between 0 and 1 such that

$$f(a + h) = f(a) + hf'(a + \theta h). \qquad \textbf{(K.U. 1994)}$$

Proof:

Let $a + h = b$. Then $b - a = h$ = the length of the interval.

Now give the complete proof of Lagrange's mean value theorem.

Since c lies between a and a + h, therefore it is greater than a by a fraction of h and may be written as $c = a + \theta h$, where $0 < \theta < 1$.

Hence the result of Lagrange's mean value theorem can be written as

$$f(a + h) - f(a) = hf'(a + \theta h), \qquad [0 < \theta < 1].$$

Geometrical Interpretation of the Mean Value Theorem

(Meerut, 1997, 85 P; Lucknow, 80)

In the figure let ACB be the graph of f(x) in (a, b) and let the chord AB make an angle α with the x-axis so that

$$\tan \alpha = \frac{f(b) - f(a)}{b - a}$$

$= f'(c)$, by the Mean Value Theorem where $a < c < b$.

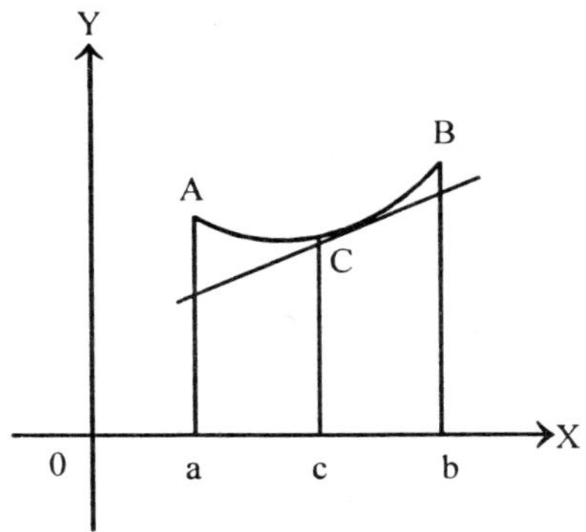

Fig. 2.2

Thus, there is some point c within (a, b) such that the tangent to the curve at the point [c, f(c)] is parallel to the chord AB.

2.3 SOME IMPORTANT DEDUCTIONS FROM MEAN VALUE THEOREM

Theorem 1:

If f(x) is

(i) continuous in the closed interval [a, b],

(ii) differentiable in the open interval (a, b), and

(iii) $f'(x)$ is –ive in $a < x < b$, then $f(x)$ is a monotonically decreasing function in the closed interval $[a, b]$.

Proof:

Let x_1, x_2 be any two points belonging to the closed interval [a, b] such that $x_2 > x_1$.

Applying Lagrange's mean value theorem to f(x) in the interval $[x_1, x_2]$, we have

$$f(x_2) - f(x_1) = (x_2 - x_1)\, f'(c), \text{ where } x_1 < c < x_2. \qquad ...(1)$$

Now $x_2 - x_1 > 0$. Since by hypothesis $f'(x)$ is negative for every x in (a, b), therefore $f'(c) < 0$. Hence from (1), we have

$$f(x_2) - f(x_1) < 0, \text{ i.e., } f(x_2) < f'(x_1).$$

Thus, f(x) is a decreasing function of x in [a, b].

Similarly, we can prove that *a function having a positive derivative for every value of x in an interval is a monotonically increasing function in that interval.* **(Mysore 1991)**

Corollary: *The function f(x) is strictly decreasing or increasing in [a, b] if $f'(x) < 0$ or ($f'(x) > 0$) for every x in (a, b) except for a finite number of points where the derivative is zero.*

Theorem 2:

If a function f(x) be such that $f'(x)$ is zero throughout the interval (a, b), then f(x) must be constant throughout the interval.

Proof:

Let x_1, x_2 be any two points in the interval (a, b), such that $x_2 > x_1$. Since $f'(x)$ exists through out the interval (a, b), therefore f(x) satisfies the conditions of Lagrange's mean value theorem in the interval $[x_1, x_2]$. So applying this theorem for f(x) in the interval $[x_1, x_2]$, we get

$$\frac{f(x_2) - f(x_1)}{x_2 - x_1} = f'(c), \text{ where } x_1 < c < x_2.$$

But by hypothesis $f'(x) = 0$ throughout the interval (a, b).

$\therefore \quad f'(c) = 0$ or $f(x_2) - f(x_1) = 0$

or $\quad f(x_2) = f(x_1)$.

Thus, the values of f(x) at every two points of (a, b) are equal.

Hence f(x) must be constant throughout (a, b)

Theorem 3:

If f(x) and $\phi(x)$ be two functions such that $f'(x) = \phi'(x)$ throughout the interval (a, b), then f(x) and $\phi(x)$ differ only by a constant.

Proof:

Consider the function $F(x) = f(x) - \phi(x)$.

Throughout the interval (a, b), we have

$$F'(x) = f'(x) - f'\phi(x) = 0, \qquad [\because f'(x) = \phi'(x)].$$

Therefore, from theorem 1, we have

$$F(x) = \text{const, or } f(x) - \phi(x) = \text{const.}$$

2.4 CAUCHY'S MEAN VALUE THEOREM OR SECOND MEAN VALUE THEOREM

(Meerut 1991; Gorakhpur 82; Allahabad 82)

If two function f(x) and g(x) are

(i) continuous in the closed interval [a, b]

(ii) differentiable in the open interval (a, b), and

(vii) $g'(x) \neq 0$ for any point of the open interval (a, b), then there exists at least one value c of x in the open interval (a, b), such that

$$\frac{f(b) - f(a)}{g(b) - g(a)} = \frac{f'(c)}{g'(c)}, \ a < c < b.$$

Proof:

First we note that $g(b) - g(a) \neq 0$. For if g(b) – g(a0 = 0 *i.e.*, g(b) = g(a), then the function g(x) satisfies the conditions of Rolle's Theorêm and so its derivative g '(x) should vanish for at least one value of x lying in the open interval (a, b). But this is contrary to our hypothesis.

Now consider the function F(x) defined by

$$F(x) = f(x) + Ag(x), \qquad ...(1)$$

where A is a constant to be determined such that F(a) = F(b) *i.e.*,

$$f(a) + Ag(a) = f(b) + Ag(b)$$

or
$$-A = \frac{f(b) - f(a)}{g(b) - g(a)}. \qquad ...(2)$$

Since $g(b) - g(a) \neq 0$, therefore A is a definite real number.

Now, the function F(x) obviously satisfies the conditions of Rolle's Theorem in the interval [a, b]. Therefore, there exists, at least one value, say c, of x in the open interval (a, b) such that $F'(c) = 0$.

But $\quad F'(x) = f'(x) + Ag'(x)$, from (1).

$\therefore \quad F'(c) = 0$ gives $f'(c) + Ag'(c) = 0$

or $$-A = \frac{f'(c)}{g'(c)}. \qquad ...(3)$$

From (2) and (3), we get

$$\frac{f(b) - f(a)}{g(b) - g(a)} = \frac{f'(c)}{g'(c)}.$$

Another Form: Let $b = a + h$. Then $a + \theta h = a$ when $\theta = 0$ and $a + \theta h = b$ when $\theta = 1$. Therefore $a + \theta h$, where $0 < \theta < 1$, means some value between a and b. So putting $b = a + h$ and $c = a + \theta h$, the result of the above theorem can be written as

$$\frac{f(a+h) - f(a)}{g(a+h) - g(a)} = \frac{f'(a+\theta h)}{g'(a+\theta h)}, \; 0 < \theta < 1.$$

Note: Lagrange's mean value theorem is a particular case of Cauchy's mean value theorem.

Let us set $g(x) = x$ in Cauchy's mean value theorem which is justified because $g(x) = x$ means $g(b) = b$, $g(a) = a$, $g'(x) = 1$ and so $g'(c) = 1$. Putting these values in Cauchy's mean value theorem, we get which is nothing but the result of Lagrange's mean value theorem.

$$\frac{f(b) - f(a)}{b-a} = f'(c), \; (a<c<b)$$

Example 1:

State the conditions for the validity for the formula and investigate how far these conditions are satisfied and whether the result is true, when $f(x) = x \sin (1/x)$ (being defined to be zero at $x = 0$) and $x < 0 < x + h$.

Solution:

The conditions for the validity of the given formula are :

(i) The function f(x) must be continuous in the closed interval [x, x + h].

(ii) The function f(x) must be differentiable in the open interval (x, x + h).

(iii) θ is a real number such that $0 < \theta < 1$.

Now consider the function f(x) defined as:

$$f(x) = x \sin (1/x) \text{ for } x \neq 0, f(0) = 0.$$

The first condition is satisfied because f(x) is continuous in the closed interval [x, x + h] for x < 0 < x + h. [The students should show here that f(x) is continuous at x = 0].

But the second condition is the satisfied because f(x) is not differentiable at x = 0 which is a point lying in the open interval (x, x + h) for x < 0 < x + h. [Show here that f(x) is not differentiable at x = 0].

Hence the result of the given formula is not true for this function f(x).

Example 2:

A function f(x) is continuous in the closed interval $0 \leq x \leq 1$ and differentiable in the open interval $0 < x < 1$, prove that $f'(x_1) = f(1) - f(0)$, where $0 < x_1 < 1$.

Solution:

Here a = a, b = 1. Therefore

$$\frac{f(b) - f(a)}{b - a} = \frac{f(1) - f(0)}{1 - 0} = f(1) - f(0).$$

If we take $c = x_1$ and substitute these values in the result of Lagrange's mean value theorem, we get

$$f(1) - f(0) = f'(x_1) \text{ where } 0 < x_1 < 1.$$

Example 3:

Separate the intervals in which the polynomial $2x^3-15x^2+36x+1$ is increasing or decreasing.

Solution:

Let $f(x) = 2x3 = 15x^2 + 36x + 1.$

Then $f'(x) = 6x^2 = 30x + 36 = 6(x - 2)(x - 3).$

Now $f'(x) > 0$ for $x < 2$; $f'(x) < 0$ for $2 < x < 3$; $f'(x) > 0$ for $x > 3$; $f'(x) = 0$ for x = 2 and 3.

Thus, f(x) is positive in the intervals $(-\infty, 2)$ and $(3, \infty)$ and negative in the interval (2, 3).

Hence f(x) is monotonically increasing in the intervals $(-\infty, 2]$, $[3, \infty)$ and monotonically decreasing in the interval [2, 3].

Example 4:

Find 'c' so that $f'(c) = \dfrac{f(b)-f(a)}{b-a}$ *in the following cases :*

(i) $f(x) = x^2 - 3x - 1;\ a = -11/7,\ b = 13/7$

(ii) $f(x) = x^x;\ a = 0,\ b = 1.$

Solution :

(i) Here $f(a) = f\left(-\dfrac{11}{7}\right) = \dfrac{121}{49} + \dfrac{33}{7} - 1 = \dfrac{303}{49}$

and $f(b) = f\left(-\dfrac{13}{7}\right) = \dfrac{169}{49} + \dfrac{39}{7} - 1 = \dfrac{153}{49}$

$\therefore \quad \dfrac{f(b)-f(a)}{b-a} = \dfrac{-456/49}{24/7} - \dfrac{19}{7}$

Now $f'(x) = 2x - 3;\ \therefore\ f'(c) = 2c - 3.$

From Lagrange's mean value theorem, we have

$$2c - 3 = -19/7 \text{ or } c = 1/7.$$

(ii) Here $f(a) = f(0) = e^0 = 1$, and $f(b) = f(1)\ e^1 = e$. Also $f'(x) = e^x$, so that $f'(c) = e^c$

$\therefore$ using Lagrange's mean value theorem, we have

$$\frac{e-1}{1-0} = e^c \text{ or } e^c = e - 1$$

or $c \log_e (e - 1)$

Example 5:

Compute the value of θ in the first mean value theorem

$$f(x + h) = f(x) + hf'(x + \theta h),$$

if $f(x)\ ax^2 + bx + c.$

Solution :

We have $f(x) = ax^2 + bx + c.$

$\therefore \quad f(x + h) = a(x + h)^2 + b(x + h) + c,$

$f'(x) = 2ax + b,\ f'(x + \theta h) = 2a(x + \theta h) + b.$

Putting all these values in the Lagrange's mean value theorem, we have

$$a(x + h)^2 + b(x + h) + c = ax^2 + bx + c + h\,[2a(x + \theta h) + b] \quad ...(1)$$

The relation (1) is identically true for all values of x. So when $x \to 0$, we get

$$ah^2 + bh + c = c + h[2a\theta h + b]$$

or $\quad ah^2 = 2a\theta h^2$

or $\quad \theta = 1/2$

Example 6:

Verify Cauchy's mean value theorem for the functions x^2 and x^3 in the interval [1, 2].

Solution :

Let $f(x) = x^2$ and $g(x) = x^3$. Both f(x) and g(x) are continuous n the closed interval [1, 2] and differentiable in the open interval (1, 2). Also $g'(x) = 3x^2 \neq 0$ for any point in the open interval (q, 2). Therefore by Cauchy's mean value theorem there exists at least one real number c in the open interval (1, 2), such that

$$\frac{f(2)-f(1)}{g(2)-g(1)} = \frac{f'(c)}{g'(c)} \qquad ...(1)$$

Now $\quad \dfrac{f(2)-f(1)}{g(2)-g(1)} = \dfrac{4-1}{8-1} = \dfrac{3}{7}$

Also $\quad f'(x) = 2x,\ g'(x) = 3x^2.$

Therefore $\quad \dfrac{f'(c)}{g'(c)} = \dfrac{2c}{3c^2} = \dfrac{2}{3c}$

Substituting these values in (1), we get $\dfrac{3}{7} = \dfrac{2}{3c}$ or $c = \dfrac{14}{9}$ which lies in the open interval (1, 2). This verifies the theorem.

Example 7:

If $f(x) = (x-1)(x-2)(x-3)$ and $a = 0$, $b = 4$, find 'c' using Lagrange's mean value theorem.

Solution :

We have

$$f(x) = (x-1)(x-2)(x-3) = x^3 - 6x^2 + 11x - 6$$

$\therefore \quad f(a) = f(0) = -6$, and $f(b) = f(4) = 6$.

$$\therefore \quad \frac{f(b)-f(a)}{b-a} = \frac{6-(-6)}{4-0} - \frac{12}{4} = 3$$

Also $f'(x) = 3x^2 - 12x + 11$, so that $f'(c) = 3c^2 - 12c + 11$. Substituting these values in Lagrange's mean value theorem.

$$\frac{f(b)-f(a)}{b-a} = f'(c),\ (a < c < b), \text{ we have}$$

$$3 = 3c^2 - 12c + 11$$

or $\quad 3c^2 - 12c + 8 = 0$

or $\quad c = \dfrac{12 \pm \sqrt{(144-96)}}{6} = 2 \pm \dfrac{2\sqrt{3}}{3}$

Both of these values of c lie in the interval (0, 4). Hence both of these are the required values of c.

Example 8:

Find 'c' of the mean value theorem, if

$$f(x) = x(x-1)(x-2);\ a = 0,\ b = 1/2.$$

Solution :

Here $f(a) = f(0) = 0$ and

$$f(b)\ f\left(\frac{1}{2}\right) = \frac{1}{2}\left(\frac{1}{2}-1\right)\left(\frac{1}{2}-2\right) = \frac{3}{8}.$$

$$\therefore \quad \frac{f(b)-f(a)}{b-a} = \frac{\frac{3}{8}-0}{\frac{1}{2}-0} = \frac{3}{4}.$$

Now $f(x) = x^3 - 3x^2 + 2x$.

$\therefore \quad f'(x) = 3x^2 - 6x + 2$, so that $f'(c) = 3c^2 - 6c + 2$.

Substituting these values in Lagrange's mean value theorem

$$\frac{f(b)-f(a)}{b-a} = f'(c),\ (a < c < b), \text{ we have}$$

$$\frac{3}{4} = 3c^2 - 6c + 2$$

or $\quad 12c^2 - 24c + 5 = 0.$

$$c = \frac{24 \pm \sqrt{(24\times 24 - 4 - \times 12 \times 5}}{24}$$

$$= \frac{24 \pm 4\sqrt{36-15}}{24} = 1 \pm \frac{\sqrt{21}}{6}$$

Out of these two values of c only $1 - \frac{\sqrt{21}}{6}$ lies in the open interval $\left(0, \frac{1}{2}\right)$ which is therefore the required value of c.

Example 9:

If in the Cauchy's mean value theorem, we write f(x) = ex and g(x) = e^{-x}, show that 'c' is the arithmetic mean between a and b.

Solution :

Here $$\frac{f(b)-f(a)}{g(b)-g(a)} = \frac{e^b - e^a}{e^{-b} - e^{-a}} = -e^a e^b = -e^{a+b}.$$

Also $$\frac{f'(x)}{g'(x)} = \frac{e^x}{-x^{-x}}$$

so that $$\frac{f'(c)}{g'(c)} = \frac{e^c}{-x^{-c}} = e^{2c}$$

Substituting these values in Cauchy's mean value theorem, we get

$$-e^{a+b} = e^{2c} \text{ or } 2c = a + b$$

or $\frac{1}{2}(a+b)$.

Hence c is the arithmetic mean between a and b.

Example 10:

If, in the Cauchy's mean value theorem, we write

(i) f(x) = $\sqrt{x}$ and g(x) = 1/$\sqrt{x}$, then c is the geometric mean between a and b, and if

(ii) f(x) = $1/x^2$ and g(x) = 1/x, then c is the harmonic mean between a and b.

Solution :

Here $$\frac{f(b)-f(a)}{g(b)-g(a)} = \frac{\sqrt{b}-\sqrt{a}}{(1/\sqrt{b}) - (1/\sqrt{2})} = -\sqrt{(ab)}.$$

Also $$\frac{f'(x)}{g'(x)}=\frac{\frac{1}{2}x^{-1/2}}{-\frac{1}{2}x^{-3/2}},\text{ so that }\frac{f'(c)}{g'(c)}=-\frac{c^{-1/2}}{c^{-3/2}}=-c.$$

Substituting these values in Cauchy's mean value theorem, we get $-\sqrt{(ab)}=-c$ or $c=-\sqrt{(ab)}$ *i.e.*, c is the geometric mean between a and b.

(ii) From the Cauchy's mean value theorem, we have

$$\frac{f(b)-f(a)}{g(b)-g(a)}=\frac{f'(c)}{g'(c)}.$$

Putting f(x) $1/x^2$ and g(x) = 1/x, we get

$$\frac{(1/b^2)-(1/a^2)}{(1/b)-(1/a)}=\frac{-2c^{-3}}{-c^{-2}}\text{ or }\frac{a+b}{ab}=\frac{2}{c}$$

or $$c=\frac{2ab}{a+b}$$

i.e., c is the harmonic mean between a and b.

Example 11:

Show that x^3-3x^2+2 is monotonically increasing in every interval.

Solution :

Let $f(x)=x^3-3x^2+3x+2.$

Then $f'(x)=3x^2-6x+3=3(x-1)^2.$

We see that $f'(x)>0$ for every real value of x except 1 where its value is zero. Hence f(x) is monotonically increasing in every interval.

Example 12:

Show that $\frac{x}{1+x}<\log(1+x)<x$ for $x>0$. **(Delhi, 1993)**

Solution :

Let $$f(x)=\log(1+x)-\frac{x}{1+x}$$

Then $$f'(x)=\frac{1}{1+x}-\frac{1.(1+x)-x.1}{(1+x)^2}$$
$$=\frac{1}{1+x}-\frac{1}{(1+x)^2}=\frac{x}{(1+x)^2}$$

We see that $f'(x) > 0$ for $x > 0$. Therefore, $f(x)$ is monotonically increasing in the interval $[0, \infty)$. But $f(x) = 0$. Therefore, $f(x) > f(0) = 0$ for $x > 0$ *i.e.*, $\left[\log(1+x) - \frac{x}{1+x}\right] > 0$ for $x > 0$

Hence $\log(1+x) > \frac{x}{1+x}$ for $x > 0$. ...(1)

Again let $\phi(x) = x - \log(1+x)$; then

$$\phi'(x) = 1 - \frac{1}{1+x} = \frac{x}{1+x}$$

We see that $\phi'(x) > 0$ for $x > 0$. Therefore, $\phi(x)$ is monotonically increasing in the interval $[0, \infty)$. But $\phi(0) = 0$. Therefore $\phi(0) > \phi(0) = 0$ for $x > 0$ *i.e.*, $[x - \log(1+x)] >$ for $x > 0$.

Hence $x > \log(1+x)$ for $x > 0$. ...(2)

From (1) and (2), we have

$$\frac{x}{1+x} < \log(1+x) < x \text{ when } x > 0.$$

2.5 TAYLOR'S THEOREM WITH LAGRANGE'S FORM OF REMAINDER AFTER n TERMS

(Delhi, 1971; K.U. 73; Meerut, 90)

If $f(x)$ is a single valued function of x such that

(i) all the derivatives of $f(x)$ upto $(n-1)^{th}$ are continuous in $a \le x \le a+h$.

(ii) $f^{(n)}(x)$ exists in $a < x < a+h$, then

$$f(a+h) = f(a) + hf'(a) + \frac{h^2}{2!}f''(a) + \ldots + \frac{h^{n-1}}{(n-1)!}f^{(n-1)}(a) + \frac{h^n}{n!}(a+\theta h), \text{ where } 0 < \theta < 1.$$

Proof:

Consider the function $\phi(x)$ defined by

$$\phi(x) = f(x) + (a+h-x)f'(x) + \frac{(a+h-x)^2}{2!}f''(x) + \ldots + \frac{(a+h-x)^{n-1}}{(n-1)!}f^{(n-1)}(x) + \frac{A}{n!}(a+h=x)^n,$$

where A is a constant to be determined such that

$$\phi(a) = \phi(a + h).$$

Now $$\phi(a) = \phi(a) + hf'(a) + \frac{h^2}{2!}f''(a) + \ldots + \frac{h^{n-1}}{(n-1)!}f^{(n-1)}(a) + \frac{A}{n!}h^n,$$

and $$\phi(a + h) = f(a + h).$$

Therefore A is given by

$$f(a + h) = f(a) + hf'(a) + \frac{h^2}{2!}f''(a) + \ldots + \frac{h^{n-1}}{(n-1)!}f^{(n-1)}(a) + \frac{h^n}{n!}A \qquad \ldots(1)$$

Now, by hypothesis, all the functions

$$f(x),\ f'(x),\ f''(x), \ldots, f^{(n-1)}(x)$$

are continuous in the closed interval [a, a + h]

and differentiable n the open interval (a, a + h).

Also $(a + h - x)$, $(a + h - x)2/2!$, ..., $(a + h - x)^n/n!$, all being polynomial, are continuous in the closed interval [a, a + h] and differentiable in the open interval (a, a + h). Further, A is a constant.

∴ $\phi(x)$ is continuous in the closed interval [a, a + h] and differentiable in the open interval (a, a + h). Also by our choice of A, $\phi(a) = \phi(a + h)$. Thus $\phi(x)$ satisfies all the conditions of Rolle's Theorem.

$$\therefore \quad \phi'(a + \theta h) = 0, \text{ where } 0 < \theta < 1.$$

Now $$\phi'(x) = f'(x) - f'(x) + (a + h - x)f''(x) - (a + h - x)f''(x) + \ldots + \frac{(a+h-x)^{n-1}}{(n-1)!}f^{(n)}(x) - \frac{A}{(n-1)!}(a+h=x)^{n-1}$$

$$= \frac{(a+h-x)^{n-1}}{(n-1)!}[f^{(n)}(x) - A], \text{ since other terms cancel in pairs.}$$

∴ $\phi'(a + \theta h) = 0$ gives

$$\frac{[a+h-(a+\theta h)]^{n-1}}{(n-1)!}[f^{(n)}(a + \theta h) - A] = 0$$

or $$\frac{h^{n-1}(1+\theta)]^{n-1}}{(n-1)!}[f^{(n)}(a + \theta h) - A] = 0.$$

Now $h \neq 0$. Also $(1 - \theta) \neq 0$ because $0 < \theta < 1$.

$$\therefore \qquad f^{(n)}(a + \theta h) - A + 0$$

$$\text{or} \qquad A = f^{(n)}(a + \theta h).$$

Substituting this value of A in (a), we get

$$\mathbf{f(a + h) = f(a) + hf'(a) + \frac{h^2}{2!} f''(a) \ldots}$$

$$\mathbf{+ \frac{h^{n-1}}{(n-1)!} f^{(n-1)}(a) + \frac{h^n}{n!} f^{(n)}(a + \theta h).}$$

This is **Taylor's Development** of $f(a + h)$ in ascending integral powers of h. The $(n + 1)^{th}$ term $\mathbf{\frac{h^2}{n!} f^{(n)}(a + \theta h)}$ is called **Lagrange's Form of Remainder** after n terms in Taylor's expansion of $f(a + h)$.

Note: If we take $n = 1$, we observe that Lagrange's mean value theorem is a particular case of Taylor's theorem.

Corollary. (Maclaurin's Development): Instead of considering the interval $[a, a + h]$, let us take the interval $[0, x]$. Then changing a to 0 and h to x in Taylor's theorem, we get

$$\mathbf{f(x) = f(0) + xf'(0) + \frac{x^2}{2!} f''(0) + \ldots + \frac{x^{n-1}}{(n-1)!} f^{(n-1)}(0) + \frac{x^n}{2!} f^{(n)}(\theta x).}$$

which is known as **Maclaurin's Theorem** or **Maclaurin's Development** of $f(x)$ in the interval $[0, x]$ with **Lagrange's Form of Remainder** $\mathbf{\frac{x^n}{n!} f^{(n)}(\theta x)}$.

2.6 TAYLOR'S THEOREM WITH CAUCHY'S FORM OF REMAINDER

(G.N.U., 1195; Meerut, 91)

If $f(x)$ is a single valued function of x such that

(i) all the derivatives of $f(x)$ upto $(n - 1)^{th}$ are continuous in $a \leq x \leq a + h$, and

(ii) $f^{(n)}(x)$ exists in $a < x < a + h$, then

$$f(a + h) = f(a) + hf'(a) + \frac{h^2}{2!} f''(a) + \ldots$$

$$+ \frac{h^{n-1}}{(n-1)!} f^{(n-1)}(a) + \frac{h^n}{(n-1)!} (1 - \theta)^{n-1}(a + \theta h),$$

where $0 < \theta < 1$.

Proof:

Consider the function $\phi(x)$ defined by

$$\phi(x) = f(x) + (a + h - x)\, f'(x) + \frac{(a+h-x)^2}{2!} f''(x) + \ldots$$
$$+ \frac{(a+h-x)^{n-1}}{(n-1)!} f^{(n-1)}(x) + (a+h-x)A,$$

where A is a constant to be determined such that

$$\phi(a) = \phi(a + h).$$

Now $$\phi(a) = \phi(a) + hf'(a) + \frac{h^2}{2!} f''(a) + \ldots$$
$$+ \frac{h^{n-1}}{(n-1)!} f^{(n-1)}(a) + \frac{A}{n!} hA,$$

and $$\phi(a + h) = f(a + h).$$

Therefore A is given by

$$f(a + h) = f(a) + hf'(a) + \frac{h^2}{2!} f''(a) + \ldots$$
$$+ \frac{h^{n-1}}{(n-1)!} f^{(n-1)}(a) + hA. \quad \ldots(1)$$

As shown in (2.5), we can easily show that $\phi(x)$ satisfies all the conditions of Rolle's Theorem.

$\therefore \quad \phi'(a + \theta h) = 0$, where $0 < \theta < 1$.

Now

$$\phi'(x) = \frac{(a+h-x)^{n-1}}{(n-1)!} f^{(n)}(x) - A,$$ since other terms cancel in pairs.

$\therefore \quad \phi'(a + \theta h) = 0$ gives

$$\frac{[a+h-(a+\theta h)]^{n-1}}{(n-1)!} f^{(n)}(a + \theta h) - A = 0$$

or $$A = \frac{h^{n-1}(1+\theta)]^{n-1}}{(n-1)!} (1 - \theta)^{n-1} f^{(n)}(a + \theta h).$$

Substituting this value of A in (a), we get

$$\mathbf{f(a + h) = f(a) + hf'(a) + \frac{h^2}{2!} f''(a) \ldots + \frac{h^{n-1}}{(n-1)!} f^{(n-1)}(a)}$$
$$\mathbf{+ \frac{h^n}{(n-1)!} (1-\theta)^{n-1} f^{(n)}(a + \theta h).}$$

The $(n + 1)^{th}$ term $\frac{\mathbf{h^n}}{\mathbf{(n-1)!}}(\mathbf{1-\theta})^{\mathbf{n-1}}\ \mathbf{f^{(n)}(a + \theta h)}$ is called **Cauchy's Form of Remainder** after n terms in Taylor's expansion of f(a + h) in ascending integral powers of h.

Corollary. (Maclaurin's Development with Cauchy's Form of Remainder): Changing a to 0 and h to x in above result, we get

$$\mathbf{f(x) = f(0) + xf'(0) + \frac{x^2}{2!} f''(0) + ... + \frac{x^{n-1}}{(n-1)!} f^{(n-1)}(0)}$$

$$\mathbf{+ \frac{x^{n-1}}{(n-1)!}(1 - \theta)^{n-1} f^{(n)}(\theta x),}$$

which is known as **Maclaurin's Development** of f(x) in the interval [0, x] with **Cauchy's Form of Remainder** after n terms.

Example 1:

Find 'θ' if $f(x + h) = f(x) + hf'(x) + \frac{h^2}{2!} f''(x + \theta h),\ 0 < \theta < 1$ *and*

(i) $f(x) = ax^3 + bx^2 = cx + d,$ **(Lucknow, 1981)**

(ii) $f(x) = x^3.$ **(Lucknow, 1983, 84)**

Solution :

(i) Here $f(x) = ax^3 + bx^2 = cx + d$

$\therefore\quad f(x + h) = a(x + h)^3 + b(x + h)^2 + c(x + h) + d,$

$f'(x) = 3ax^2 + 2bx + c,\ f''(x) = 6ax + 2b,$

$f''(x + \theta h) = 6a(x + \theta h) + 2b.$

The given relation

$$f(x + h) = f(x) + hf'(x) + \left(\frac{h^2}{2!}\right) f''(x + \theta h), \text{ we have}$$

$$a(x + h)^3 + b(x + h)^2 + c(x + h) + d,$$

$$= ax^3 + bx^2 + cx + d + h(3ax^2 + 2bx + c)$$

$$+ \left(\frac{h^2}{2!}\right) \{6a(x + \theta h) + 2b\} \quad ...(1)$$

The relation (1) is an identify in x. Letting $x \to 0$ on both sides of (1), we have

$$ah^3 + bh^2 + ch + d = d + ch + \left(\frac{h^2}{2}\right)(6a\theta h) + 2b)$$

or $$ah^3 + bh^2 + ch + d = d + ch + 3a\theta h^3 + bh^2$$

or $\quad ah^3 = 3a\theta h^3$

or $\quad \theta = \dfrac{1}{3}$ $\qquad [\because ah^3 \neq 0]$

(ii) Proceed as in part (i) of this question. The required value of θ is 1/3.

Example 2:

Expand the following by Maclaurin's theorem with Lagrange's form of remainder after n terms:

(i) a^x, *(ii)* e^x.

Solution :

(i) Here $\quad f(x) = a^x$. ...(1)

$\therefore \quad f^{(n)}(x) = a^x(\log a)^n$. ...(2)

Putting $x = 0$ in (1) and (2), we get

$$f(0) = a^0 = 1,\ f^{(n)}(0) = a^0(\log a)^n = (\log a)^n.$$

$\therefore \quad f'(0) = \log a,\ f''(0) = (\log a)^2, \ldots, f^{(n-1)}(0) = (\log a)^{n-1}$.

Also changing x to θx in (2), we get

$$f^{(n)}(\theta x) = a^{\theta x}(\log a)^n.$$

Now by Maclaurin's theorem with Lagrange's form of remainder after n terms, we have

$$f(x) = f(0) + xf'(0) + \frac{x^2}{2!}f''(0) + \ldots + \frac{x^{n-1}}{(n-1)!}f^{(n-1)}(0) + \frac{x^n}{n!}f^{(n)}(\theta x), \text{ where } 0 < \theta < 1. \quad \ldots(3)$$

Substituting values found above in (3), we get

$$a^x = 1 + x \log a \ \frac{x^2}{2!}(\log a)^2 + \ldots + \frac{x^{n-1}}{(n-1)!}(\log a)^{(n-1)} + \frac{x^n}{n!}a^{\theta x}(\log a)^n$$

Here Lagrange's form of remainder after n terms

$$+\frac{x^n}{n!}a^{\theta x}(\log a)^n, \quad \text{where } 0 < \theta < 1.$$

(ii) Here $\quad f(x) = e^x$. Therefore $f^{(n)}(x) = e^x$.

Putting $x = 0$, in these, we get

$f(0) = e^0 = 1,\ f^{(n)}(0) = e^0 = 1$. Also $f^{(n)} = (\theta x)\ e^{\theta x}$.

Substituting these values in Maclaurin's theorem with Lagrange's form of remainder after n terms, we get

$$e^x = 1 + x + \frac{x^2}{2!} + \frac{x^3}{2!} + \ldots + \frac{x^{n-1}}{(n-1)!} + \frac{x^n}{n!} e^{\theta x}$$

Example 3:

Show that

(i) $\sin x = x - \frac{x^3}{3!} + \frac{x^5}{5!} - \ldots + (-1)^{n-1} \frac{x^{2n-1}}{(2n-1)!} + (-1)^n \frac{x^{2n}}{(2n)!} \sin \theta x,$

for every real value of x. **(K.U., 1973)**

(i) $\log(1 + x) = x - \frac{x^2}{2} + \frac{x^3}{3} - \ldots + (-1)^{n-2} \frac{x^{n-1}}{n-1}$

$$+ (-1)^{n-1} \frac{x^n}{n(1 + \theta x)^n},\ \textit{for } x > -1.$$

Solution:

(i) Here $f(x) = \sin x$.

We know that sin x possesses derivatives of every order for every real number x and

$$f^{(n)}(x) = \sin x\ (x + 1/2n\pi). \qquad \ldots(1)$$

Putting $x = 0$ in (1) and (2), we get

$f(0) = \sin 0 = 0,\ f^{(n)}(0) = \sin(1/2n\pi)$.

$\therefore \quad f'(0) = \sin(1/2\pi) = 1,\ f''(0) = \sin \pi = 0$.

$f'''(0) = \sin(3\pi/2) = -1$.

$f^{iv}(0) = \sin 2\pi = 0,$

$f^{v}(0) = \sin(5\pi/2) = \sin(2\pi + 1/2\pi) = \sin 1/2\ \pi = 1, \ldots,$

$f^{(2n-1)}(0) = \sin(n - 1)\ \pi = 0,$

$f^{(2n-1)}(0) = \sin\{1/2\ (2n - 1)\ \pi\} = \sin(n\pi = 1/2\pi)$

$= (-1)^n \sin(-\pi/2),$

$[\because \sin(n\pi + \theta) = (-1)^n \sin\theta]$

$= (-1)^n (-1) = (-1)^{n+1} = (-1)^{n-1}$.

Also changing n to 2n and x to θx in (2), we get

$f^{(2n)}(\theta x) = \sin(\theta x + n\pi) = (-1)^n \sin \theta x$.

Substituting these values in Maclaurin's theorem with Lagrange's form of remainder after 2n terms *i.e.*,

$$f(x) = f(0) + sf'(0) + \frac{x^2}{2!} f''(0) + \ldots$$

$$+ \frac{x^{2n-1}}{(2n-1)!} f^{(2n-1)}(0) + \frac{x^{2n}}{(2n)!} f^{(2n)}(\theta x),$$

we get $\sin x = x - \frac{x^3}{3!} + \frac{x^5}{5!} - \ldots$

$$+ (-1)^{n-1} \frac{x^{2n-1}}{(2n-1)!} + (-1)^n \frac{x^{2n}}{(2n)!} \sin \theta x.$$

(ii) Here $f(x) = \log (1 + x)$. ...(1)

We know that $\log (1 + x)$ possesses derivatives of every order when $(1 + x) > 0$ *i.e.*, $x > -1$.

Also, $f^{(n)}(x) = (-1)^{n-1} (n - 1)!\,(1 + x)^{-n}$ ...(2)

Putting x = 0 in (1) and (2), we get

$$f(0) = \log 1 = 0,\ f^{(n)}(0) = (-1)^{n-1} (n - 1)!.$$

Also changing x to θx in (2), we get

$$f^{(n)}(\theta x) = (-1)^{n-1} (n - 1)!\,(1 + \theta x)^{-n}.$$

Substituting these values in Maclaurin's theorem with Lagrange's form of remainder after n terms *i.e.*,

$$f(x) = f(0) \frac{x}{1!} f'(0) + \frac{x^2}{2!} f''(0) + \ldots$$

$$+ \frac{x^{n-1}}{(n-1)!} f^{(n-1)}(0) + \frac{x^n}{n!} f^{(n)}(\theta x),$$

we get

$$\log (1 + x) = 0 + \frac{x}{1!}.1 + \frac{x^2}{2!}.(-1).1 + \frac{x^3}{3!}(-1)^2.2! + \ldots$$

$$+ \frac{x^{n-1}}{(n-1)!}(-1)^{n-2}(n-2)! + \frac{x^n}{n!}(-1)^{n-1}(n-1)!(1+\theta x)^{-n}$$

$$= x \frac{x^2}{2} + \frac{x^3}{3} - \ldots + (-1)^{n-2} \frac{x^{n-1}}{n-1} + (-1)^{n-1} \frac{x^n}{n(1+\theta x)}$$

Example 4:

Show that $f(x) = |x - 1|$, $0 \le x \le 2$ is not derivable at $x = 1$. Is it continuous in [0, 2]? **(Delhi, 1999, 97)**

Solution:

(i) To test for continuity at x = 1. We have f(1) = 0.

Also R.H.L. = f(1 + 0) = $\lim_{h \to 0}$ f(1 + h) = $\lim_{h \to 0}$ |1 + h − 1) = 0;

and L.H.L. = f(1 − 0) = f(1 − 0) = f(1), therefore f(x) is continuous at x = 1.

In fast f(x) is continuous at every point in the interval [0, 2].

(ii) To test f(x) for differentiability at x = 1.

We have

$$Rf'(1) = \lim_{h \to 0} \frac{f(1+h) - f(1)}{h} = \lim_{h \to 0} \frac{|1+h-1| - 0}{h}$$

$$= \lim_{h \to 0} \frac{h}{h} = 1;$$

$$\text{and } Lf'(1) = \lim_{h \to 0} \frac{f(1-h) - f(1)}{-h} = \lim_{h \to 0} \frac{|1-h-1| - 0}{-h}$$

$$= \lim_{h \to 0} \frac{h}{-h} = -1.$$

Since Rf'(1) ≠ Lf'(1), the function f(x) is not differentiable at x = 1.

Example 5:

Show that the function f(x), where

f(x) = 2 + x if x ≥ 0; f(x) = 2 − x if x < 0 is not derivable at the point x = 0. **(Delhi, 1981)**

Solution:

We have

$$Rf'(0) = \lim_{h \to 0} \frac{f(0+h) - f(0)}{h} = \lim_{h \to 0} \frac{f(h) - f(0)}{h}$$

$$= \lim_{h \to 0} \frac{2+h-2}{h} = 1$$

$$\text{and } Lf'(0) = \lim_{h \to 0} \frac{f(0-h) - f(0)}{-h} = \lim_{h \to 0} \frac{f(-h) - f(0)}{-h}$$

$$= \lim_{h \to 0} \frac{\{2-(-h)\} - 2}{-h} = -1.$$

Since Rf' (0) ≠ Lf'(0), the function f(x) is not differentiable at x = 0.

EXERCISES

1. Show that $f(x) = |x - a|$ is continuous at $x = a$ but not derivable at $x = a$.

2. Prove that the function f(x) defined by

$$f(x) = \begin{cases} 2-x & \text{if } x < 2 \\ -2+3x-x^2 & \text{if } x > 2 \end{cases}$$

is derivable at x = z **(D.U. B. Sc.(G), 1995)**

3. Construct a function which is continuous in [1, 5] but not derivable at 2, 3, 4.

4. Determine whether f(x) is continuous and has a derivative at two origin,

While $f(x) = 2 + x, \text{ if } x \geq 0$
$\quad\quad = 2 + x, \text{ if } x \geq 0$

5. Show that function $f(x) = |x - 1| + |x + 1|$ is not derivable at $x = -1, 1$; and is derivable at every other point.

6. Show that $f(x) = \dfrac{xe^{1/x}}{1+e^{1/x}}$, $x \neq 0$ and $f(x) = 0$ is continuous but not derivable at $x = 0$.

3

DIFFERENTIATION

3.1 DEFINITION

If f is a function of x

$\lim_{h\to 0} \frac{f(a+h) - f(a)}{h}$ is called the differential coefficient of f (x) at x = a. It is generally denoted by f '(a).

The differential coefficient is also called the derivative, or the derived function. We shall use the latter term only for the set of ordered pairs (x, f '(x) and denoted by f '.

The process of finding the differential coefficient is called differentiation. We are said to differential f (x).

It should be carefully noted that in taking the limit {f (a + h) – f (a)}/h is regarded as a function of h, h is regarded as a variable; and a is regarded as a constant. The sequal will show that $\lim_{h\to 0}$ [f (a + h) – f (a)] h is independent of h.

Generally it is more convenient to write x at self for a. It being understood that in the process of finding the limit when h → 0, x is to be kept constant with this understanding we can write :

The differential coefficient of f (x) = $\lim_{h\to 0} \frac{f(x+h) - f(x)}{h}$

The differential coefficient of f (x) is written as $\frac{df(x)}{dx}$ or $\frac{d}{dx}$[f (x)] or f '(x) or D[f(x)].

The differential coefficient of x = a is generally reperesented by f '(x) or $\left[\frac{df(x)}{dx}\right]$x = a.

3.2 SOME STANDARD RESULTS

To find the differential coefficients of some standard functions from the first principle.

3.2.1 Differential Coefficient of a Constant

Let f (x) = c, where c is a constant. Then f (x + δx) = c, because f (x) takes the same value c for every value of x.

Now by the definition of a diff. coeff., we have

$$\frac{d}{dx} f(x) = \lim_{\delta x \to 0} \frac{f(x+\delta x)-f(x)}{\delta x} = \lim_{\delta x \to 0} \frac{c-c}{\delta x}$$

$$= \lim_{\delta x \to 0} \frac{0}{\delta x} = \lim_{\delta x \to 0} 0 = 0$$

Thus, the diff. coeff. of a constant is zero. *i.e.* $\frac{d}{dx}(c) = 0$

3.2.2 Differential Coefficient of the Product of Two Functions

(Delhi, 1982)

Let y = Then y + δy = $f_1(x + \delta x) . f_2(x + \delta x)$.

$$\therefore \quad \frac{dy}{dx} = \lim_{\delta \to 0} \frac{f_1(x+\delta x)\, f_2(x+\delta x) - f_1(x) . f_2(x)}{\delta x}$$

[by the def. of a diff. coeff.]

$$= \lim_{\delta x \to 0} \left[f_1(x+\delta x) . \frac{f_2(x+\delta x)-f_2(x)}{\delta x} + f_2(x) \frac{f_1(x+\delta x)-f_1(x)}{\delta x} \right]$$

$$= \lim_{\delta x \to 0} f_1(x+\delta x) \frac{f_2(x+\delta x)-f_2(x)}{\delta x} + \lim_{\delta x \to 0} f_2(x) . \frac{f_1(x+\delta x)-f_1(x)}{\delta x}$$

$$\frac{d}{dx}[f_1(x) . f_2(x)] = f_1(x) . \frac{d}{dx} f_2(x) + f_2(x) \frac{d}{dx} f_1(x).$$

This can also be written as.

∴ the diff. coeff. of the product of two functions = (first function) × (diff. coeff. of the second function) + (second function) × (diff. coeff. of the first function).

3.2.3 Differential Coefficient of the Quotient of Two Functions

Let y = $f_1(x)/f_2(x)$. Then y + δy = $f_1(x + \delta x)/ f_2(x + \delta x)$.

$$\therefore \quad \frac{dy}{dx} = \lim_{\delta x \to 0} \frac{\left[\frac{f_1(x+\delta x)}{f_2(x+\delta x)} - \frac{f_1(x)}{f_2(x)}\right]}{\delta x}$$

$$= \lim_{\delta x \to 0}\left[\frac{f_1(x+\delta x).f_2(x) - f_1(x).f_2(x+\delta x)}{\delta x.f_2(x+\delta x).f_2(x)}\right]$$

$$= \lim_{\delta x \to 0}\left[\frac{f_1(x+\delta x)\,f_2(x) - f_1(x)f_2(x) - f_2(x)\,f_2(x+\delta x) + f_1(x).f_2(x)}{f_2\,(x+\delta x)\,.f_2(x)}\right]$$

$$= \lim_{\delta x \to 0}\left[f_2(x).\frac{\left\{\frac{f_1(x+\delta x) - f_1(x)}{\delta x}\right\} - f_2(x).\left\{\frac{f_2(x+\delta x) - f_2(x)}{\delta x}\right\}}{f_2(x+\delta x)\,f_2(x)}\right]$$

$$\frac{d}{dx}\left[\frac{f_1(x)}{f_2(x)}\right] = \frac{f_2(x).\frac{d}{dx}.f_1(x) - f_1(x).\frac{d}{dx}f_2(x)}{[f_2(x)]^2}.$$

This can also be written as.

$\therefore$ **the diff. coeff. of the quotient of two functions**

$$= \frac{\textbf{(diff. coeff. of Nr.) (Dr.) – (Nr.) (diff. coeff. of Dr.)}}{\textbf{square of the denominator}}$$

3.2.4 $\left(\frac{d}{dx}\right) x^n = nx^{n-1}$

Let $y = x^n$. Then $y + \delta y = (x + \delta x)^n$

$$\therefore \quad \frac{dy}{dx} = \lim_{\delta x \to 0} \frac{(x+\delta x)^n - x^n}{\delta x} = \lim_{\delta x \to 0} x^n \left[\frac{\left(1 + \frac{\delta x}{x}\right)^n - 1}{\delta x}\right]$$

$$= \lim_{\delta x \to 0} x^n \left[\frac{1 + n\frac{\delta x}{x} + \frac{n(n-1)}{2!}\frac{(\delta x)^2}{x^2} + \ldots - 1}{\delta x}\right]$$

$$= \lim_{\delta x \to 0} x^n \left[\frac{n}{x} + \frac{n(n-1)}{2!}\frac{\delta x}{x^2} + \text{terms containing higher powers of } \delta x\right]$$

$$= x^n \left(\frac{n}{x}\right) = nx^{n-1}$$

$$\Rightarrow \left[\frac{d}{dx}(x^n) = nx^{n-1}\right].$$

3.2.5 $\left(\frac{d}{dx}\right) e^x = e^x$

Let $y = e^x$. Then $y + \delta y = e^{x+\delta x}$

$$\therefore \quad \frac{dy}{dx} = \lim_{\delta x \to 0} \frac{e^{x+\delta x} - e^x}{\delta x}$$

$$= \lim_{\delta x \to 0} e^x \left[\frac{e^{\delta x} - 1}{\delta x}\right] = \lim_{\delta x \to 0} e^x \left[\frac{1 + \delta x + (\delta x)^2/2! + \ldots - 1}{\delta x}\right]$$

$$= \lim_{\delta x \to 0} e^x \left[1 + \frac{\delta x}{2!} + \text{terms containing higher power of } \delta x\right] = e^x.$$

$$\Rightarrow \frac{d}{dx}(e^x) = e^x.$$

3.2.6 $\left(\frac{d}{dx}\right) \tan x = \sec^2 x$

Let y = log x. Then y + δy = log (x + δx)

$$\therefore \quad \frac{dy}{dx} = \lim_{\delta x \to 0} \frac{\tan(x + \delta x) - \tan x}{\delta x}$$

$$= \lim_{\delta x \to 0} \left[\frac{\dfrac{\sin(x+\delta x)}{\cos(x+\delta x)} - \dfrac{\sin x}{\cos x}}{\delta x}\right]$$

$$= \lim_{\delta x \to 0} \frac{\sin(x+\delta x)\cos x - \sin x \cos(x+\delta x)}{\delta x \cos(x+\delta x).\cos x}$$

$$= \lim_{\delta x \to 0} \frac{\sin\{(x+\delta x) - x\}}{\delta x.\cos(x+\delta x)\cos x}$$

$$= \lim_{\delta x \to 0} \frac{\sin \delta x}{\delta x} \cdot \left[\frac{1}{\cos(x+\delta x).\cos x}\right] = \frac{1}{\cos^2 x} = \sec^2 x .$$

$$= \left[\frac{d}{dx}(\tan x) = \sec^2 x)\right].$$

3.2.7 $\left(\frac{d}{dx}\right)\log_e x = \frac{1}{x}$

Let y = log x. Then y + δy = log (x + δx)

$$\therefore \quad \frac{d}{dx} = \lim_{\delta x \to 0} \frac{\log(x + \delta x) - \log x}{\delta x} = \lim_{\delta x \to 0} \frac{\log(x + \delta x)/x}{\delta x}$$

$$= \lim_{\delta x \to 0} \frac{\log\left(1+\frac{\delta x}{x}\right)}{\delta x} = \lim_{\delta x \to 0} \frac{\frac{\delta x}{x} - \frac{(\delta x)^2}{2x^2} + \ldots}{\delta x}$$

$$= \lim_{\delta x \to 0}\left[\frac{1}{x} = \frac{\delta x}{2x^2} + \ldots\right] = \frac{1}{x}.$$

$$\frac{d}{dx}[\log x] = \frac{1}{x}.$$

3.2.8 $\left(\frac{d}{dx}\right)a^x = a^x \log_e a$

Let y = a^x. Then y + δy = a^x + δx

$$\therefore \quad \frac{dy}{dx} = \lim_{\delta x \to 0} \frac{a^x + \delta x - a^x}{\delta x}$$

$$= \lim_{\delta x \to 0} a^x \left[\frac{a^{\delta x} - 1}{\delta x}\right]$$

$$= \lim_{\delta x \to 0} a^x \left[\frac{1 + \delta x \log a + \frac{(\delta x)^2 (\log a)^2}{2!} + \ldots - 1}{\delta x}\right]$$

$= \lim_{\delta x \to 0} a^x$[log a + (δx/2!) (log a)^2 + terms containing higher power of δx]

$$\frac{d}{dx} = (a^x) = a^x \log a$$

3.2.9 Differential Coefficient of the Product of a Constant and a Function

We have, $\frac{d}{dx}\{cf(x)\} = \lim_{\delta x \to 0} \frac{cf(x+\delta x) - cf(x)}{\delta x}$

$$c \lim_{\delta x \to 0} \frac{f(x+\delta x) - f(x)}{\delta x} \; c\frac{d}{dx} f(x)$$

$$\frac{d}{dx}[cf(x)] = c - \frac{d}{dx}[f(x)].$$

3.2.10 Differential of a Sum or a Difference of Two Functions

Let $f(x) = f_1(x) \pm f_2(x)$.

Then $f(x + \delta x) = f_1(x + \delta x) \pm f_2(x + \delta x)$.

$$\therefore \quad \frac{dy}{dx} f(x) = \lim_{\delta x \to 0} \frac{\{f_1(x+\delta x) \pm f_2(x+\delta x) - f_1(x) \pm f_2(x)\}}{\delta x},$$

$$= \lim_{\delta x \to 0} \frac{f_1(x+\delta x)}{\delta x} \pm \lim_{\delta x \to 0} \frac{f_2(x+\delta x) - f_2(x)}{\delta x}$$

$$= \frac{d}{dx} f_1(x) \pm \frac{d}{dx} f_2(x)$$

$$\Rightarrow \quad \frac{d}{dx}[f_1(x) \pm f_2(x)] = \frac{d}{dx}.[f_1(x)] + \frac{d}{dx}[f_2(x)].$$

3.2.11 $\left(\frac{d}{dx}\right) \log a \; x = \frac{1}{x \log_e a}$

We have $\log_a x = \log_e x \,.\, \log_a e = \log_a e \,.\, \log_e x$, where $\log_a e$ is simply a constant.

$$\therefore \quad \frac{dy}{dx} = (\log_a x) = (\log_a e)\frac{d}{dx}(\log_e x) = (\log_a e)\frac{1}{x} = \frac{1}{x}.\log_a e$$

$$= \frac{1}{x \log_e a}, \text{ since } \log_a e \,.\, \log_e a = 1.$$

$$\Rightarrow \quad \frac{d}{dx} \log_a x = \frac{1}{x \log_e a}.$$

3.2.12 $\left(\frac{d}{dx}\right)\sin x = \cos x$

Let y = sin x. Then y + δy = sin (x + δx).

$$\therefore \quad \frac{dy}{dx} = \lim_{\delta x \to 0} \frac{\sin(x+\delta x) - \sin x}{\delta x}$$

$$= \lim_{\delta x \to 0} \frac{2\cos\{x + (\delta x/2)\}\sin(\delta x/2)}{\delta x}$$

$$= \lim_{\delta x \to 0} \frac{\cos\{x + (\delta x/2)\}\sin(\delta x/2)}{(\delta x/2)}$$

$$= \lim_{\delta x \to 0} \cos\left(x + \frac{\delta x}{2}\right) . \lim_{\delta x \to 0} \frac{\sin(\delta x/2)}{(\delta x/2)}$$

$$= (\cos x) . 1 = \cos x.$$

$$\Rightarrow \frac{d}{dx}(\sin x) = \cos x .$$

3.2.13 $\left(\frac{d}{dx}\right)\cos x = -\sin x$

Let y = cos x. Then y + δy = sin (x + δx).

$$\therefore \quad \frac{dy}{dx} = \lim_{\delta x \to 0} \frac{\cos(x+\delta x) - \cos x}{\delta x}$$

$$= \lim_{\delta x \to 0} \frac{-2\sin\{x + (\delta x/2)\}\sin(\delta x/2)}{\delta x}$$

$$= \lim_{\delta x \to 0} -\sin\left(x - \frac{\delta x}{2}\right) . \frac{\sin(\delta x/2)}{(\delta x/2)} = (-\sin x).1 = -\sin x.$$

$$\Rightarrow \frac{d}{dx}(\cos x) = -\sin x$$

3.2.14 $\left(\frac{d}{dx}\right)\sec x = \sec x \tan x$

Let y = sec x. Then y + δy = sec (x + δx).

$$\therefore \quad \frac{dy}{dx} = \lim_{\delta x \to 0} \frac{\sec(x+\delta x) - \sec x}{\delta x}$$

$$= \lim_{\delta x \to 0} \frac{\dfrac{1}{\cos(x+\delta x)} - \dfrac{1}{\cos x}}{\delta x}$$

$$= \lim_{\delta x \to 0} \frac{\cos x - \cos(x+\delta x)}{\delta x \, . \, \cos(x+\delta x)\cos x}$$

$$= \lim_{\delta x \to 0} \frac{2\sin\left(s + \frac{1}{2}\delta x\right)\sin\left(\frac{\delta x}{2}\right)}{\delta x \, . \, \cos(x+\delta x)\cos x}$$

$$= \lim_{\delta x \to 0} \left[\frac{\sin\left(x + \frac{1}{2}\delta x\right)}{\cos(x+\delta x)\cos x} \cdot \frac{\sin(\delta x/2)}{(\delta x/2)}\right]$$

$$= \frac{\sin x}{\cos x \cos x} = \sec x \tan x.$$

3.2.15 $\left(\dfrac{d}{dx}\right)\mathbf{cosec\,x = -\,cosec\,x\;cot\,x}$

Proof do your self

3.2.16 $\left(\dfrac{d}{dx}\right)(\sin^{-1} x) = \dfrac{1}{\sqrt{(1-x^2)}}$

Let $y = \sin^{-1} x$. Then $x = \sin y$ and so $x + \delta x = \sin(y + \delta y)$. As $\delta x \to 0$. δy also $\to 0$.

Now $\delta x = \sin(y + \delta y) - \sin y$.

$$\therefore \quad 1 = \frac{\sin(y+\delta x) - \sin y}{\delta x}, \qquad \text{[on dividing both sides by } \delta x]$$

$$\text{or} \quad 1 = \frac{\sin(y+\delta y) - \sin y}{\delta y} \cdot \frac{\delta y}{\delta x}.$$

Taking limits of both sides when $\delta x \to 0$, we get

$$1 = \lim_{\delta y \to 0} \frac{\sin(y+\delta y) - \sin v}{\delta y} \cdot \lim_{\delta y \to 0} \frac{\delta y}{\delta x},$$

$$[\because \delta y \to 0 \text{ when } \delta x \to 0]$$

$$= \left[\lim_{\delta y \to 0} \frac{2\cos(y + 1/2\,\delta y)\sin(1/2\,\delta y)}{\delta y}\right] \cdot \frac{dy}{dx}$$

$$= \left[\lim_{\delta y \to 0} \cos\left(y + \frac{1}{2}\,\delta y\right) . \frac{\sin\,(1/2\,\delta y)}{\delta y/2} \right] . \frac{dy}{dx}$$

$= (\cos\, y)\, .\, (dy/dx).$

$$\therefore \quad \frac{dy}{dx} = \frac{1}{\cos y} = \frac{1}{\sqrt{(1-\sin^2 y)}} = \frac{1}{\sqrt{(1-x^2)}}.$$

$$\Rightarrow \quad \left[\frac{d}{dx}(\sin^{-1} x) = \frac{1}{\sqrt{1-x^2}} \right]$$

3.2.17 $\mathbf{\dfrac{d}{dx}(sec^{-1}x) = \dfrac{1}{x\sqrt{x^2-1}}}$

Let $y = \sec^{-1} x$. Then $x = \sec y$ and so $x + \delta x = \sec\,(y + \delta y)$.

As $\delta x \to 0$, δy also $\to 0$.

Now $\delta x = \sec\,(y + \delta y) - \sec\, y$.

$$\therefore \quad 1 = \frac{\sec(y+\delta y) - \sec y}{\delta y} . \frac{\delta y}{\delta x}.$$

Taking limits of both sides when $\delta x \to 0$, we get

$$1 = \lim_{\delta y \to 0} \frac{\sec(y+\delta y) - \sec y}{\delta y} . \lim_{\delta x \to 0} \frac{\delta y}{\delta x}$$

$$= \frac{dy}{dx} . \lim_{\delta y \to 0} \left[\frac{1}{\cos\,(y+\delta y)} - \frac{1}{\cos y} \right] / \delta y$$

$$= \frac{dy}{dx} . \lim_{\delta y \to 0} \frac{\cos y - \cos(y+\delta y)}{\delta y . \cos y \cos(y+\delta y)}$$

$$= \frac{dy}{dx} . \lim_{\delta y \to 0} \frac{2 \sin\left(y + \frac{1}{2}\delta y\right) \sin\left(\frac{1}{2}\delta y\right)}{\delta y . \cos y \cos(y+\delta y)}$$

$$= \frac{dy}{dx} . \lim_{\delta y \to 0} \left[\frac{\sin\left(y + \frac{1}{2}\delta y\right)}{\cos y \cos(y+\delta y)} . \frac{\sin\left(\frac{1}{2}\delta y\right)}{\frac{1}{2}\,\delta y} \right]$$

$$= \frac{dy}{dx} \cdot \frac{\sin y}{\cos y \cos y} = \frac{dy}{dx}.\sec y \tan y.$$

$$\therefore \quad = \frac{dy}{dx} = \frac{1}{\sec y \tan y} = \frac{1}{\sec y \sqrt{(\sec^2 y - 1)}} = \frac{1}{x\sqrt{(x^2-1)}}$$

$$\left[\frac{d}{dx}(\sec^{-1} x) = \frac{1}{x\sqrt{x^1 - 1}}\right].$$

3.2.18 $\frac{d}{dx}\left(\text{cosec}^{-1} x\right) = -\frac{1}{x\sqrt{(x^2-1)}}$

3.2.19 $\frac{d}{dx}\left(\cos^{-1} x\right) = -\frac{1}{\sqrt{(1-x^2)}}$

For proof do your self

3.2.20 $\frac{d}{dx}\left(\tan^{-1} x\right) = -\frac{1}{1+x^2}$

Let $y = \tan^{-1} x$. Then $x = \tan y$ and so $x + \delta x = \tan (y + \delta y)$.

As $\delta x \to 0$, δy also $\to 0$.

Now $\delta x = \tan (y + \delta y) - \tan y$.

Taking limits of both sides when $\delta x \to 0$, we get

$$\therefore \quad 1 = \frac{\tan(y+\delta y) - \tan y}{\delta y} \frac{\delta y}{\delta x}$$

$$1 = \lim_{\delta y \to 0} \frac{\tan(y+\delta y) - \tan y}{\delta y} . \lim_{\delta x \to 0} \frac{\delta y}{\delta x}$$

$$= \left[\lim_{\delta y \to 0} \left\{\frac{\sin(y+\delta y)}{\cos(y+\delta y)} - \frac{\sin y}{\cos y}\right\} \Big/ \delta y\right] . \frac{dy}{dx}$$

$$= \frac{dy}{dx} . \lim_{\delta y \to 0} \frac{\sin(y-\delta y)\cos y - \cos(y+\delta y)\sin y}{\delta y \cos(y+\delta y)\cos y}$$

$$= \frac{dy}{dx} . \lim_{\delta y \to 0} \frac{\sin(y+\delta y - y)}{\delta y . \cos(y+\delta y)\cos y}$$

$$= \frac{dy}{dx} \cdot \lim_{\delta y \to 0}\left[\frac{\sin \delta y}{\delta y} \cdot \frac{1}{\cos(y+\delta y)\cos y}\right]$$

$$= \frac{dy}{dx} \cdot \frac{1}{\cos^2 y} = \frac{dy}{dx} \cdot \sec^2 y.$$

$$\therefore \quad = \frac{dy}{dx} \cdot \frac{1}{\sec^2 y} = \frac{1}{1+\tan^2 y} = \frac{1}{1+x^2}$$

$$= \frac{d}{dx}(\tan^{-1} x) = \frac{1}{1+x^2}$$

3.2.21 $\frac{d}{dx}\left(\cot^{-1} x\right) = -\frac{1}{1+x^2}$

Proof do your self.

3.3 DIFFERENTIAL COEFFICIENT OF A FUNCTION OF A FUNCTION

Consider the function log sin x. Here log (sin x) is a function of sin x whereas sin x is itself a function of x. Thus we have the case of a function of a function.

Let $y = f\{\phi(x)\}$.

Put $t = \phi(x)$. Then $t + \delta t = \phi(x + \delta x)$. As $\delta x \to 0$, δt also $\to 0$.

We have

$$\frac{dy}{dx} = \lim_{\delta x \to 0} \frac{\delta y}{\delta x} = \lim_{\delta x \to 0} \frac{\delta y}{\delta t} \cdot \frac{\delta t}{\delta x} = \left(\lim_{\delta x \to 0} \frac{\delta y}{\delta t}\right) \cdot \left(\lim_{\delta x \to 0} \frac{\delta t}{\delta x}\right)$$

$$= \left(\lim_{\delta t \to 0} \frac{\delta y}{\delta t}\right) \cdot \left(\lim_{\delta x \to 0} \frac{\delta y}{\delta x}\right), \text{ since } \delta t \to 0 \text{ when } \delta x \to 0$$

$$= \frac{dy}{dt} \cdot \frac{dt}{dx}.$$

Thus, *if y is a function of t and t is a function of x, then y is a function of x and we have*

$$\frac{dy}{dx} = \frac{dy}{dt} \cdot \frac{dt}{dx}.$$

Similarly, *if y is a function of u, u is a function of v and v is a function of x, then y is also a function of x, and we have*

$$\frac{dy}{dx} = \frac{dy}{du} \cdot \frac{du}{dv} \cdot \frac{dv}{dx}.$$

This is known as the **chain rule of differentiation.** We can extend it still further.

Cor. To prove that $\frac{dy}{dx} \times \frac{dx}{dy} = 1$

Let $y = f(x)$. Then $y + \delta y = f(x + \delta x)$. As $\delta x \to 0$, δy also $\to 0$.

We have

$$\frac{\delta y}{\delta x} \cdot \frac{\delta x}{\delta y} = 1$$

Taking limits of both sides when $\delta x \to 0$, we get

$$\lim_{\delta x \to 0} \left(\frac{\delta y}{\delta x} \cdot \frac{\delta x}{\delta y} \right) = 1 \text{ or } \left(\lim_{\delta x \to 0} \frac{\delta y}{\delta x} \right) \left(\lim_{\delta x \to 0} \frac{\delta x}{\delta y} \right) = 1$$

or $\left(\lim_{\delta x \to 0} \frac{\delta y}{\delta x} \right) \cdot \left(\lim_{\delta y \to 0} \frac{\delta x}{\delta y} \right) = 1$, since $\delta y \to 0$ as $\delta x \to 0$

or $\frac{dy}{dx} \cdot \frac{dx}{dy} = 1$. Thus $\frac{dy}{dx} = \frac{1}{dx/dy}$.

3.4 INVERSE HYPERBOLIC FUNCTIONS

Let $y = \sin h^{-1} x$. Then $\sin h\, y = x$. Therefore

$\cosh y = \sqrt{(\sin h^2 y + 1)} = \sqrt{(x^2 + 1)}$. Adding these, we get $\sin h\, y + \cosh y = x + \sqrt{(x^2 + 1)}$

or $\quad 1/2(e^y - e^{-y}) + 1/2(e^y - e^{-y}) = x + (x^2 + 1)$

or $\quad e^y = x + \sqrt{(x^2 + 1)}$.

$\therefore \quad y = \log [x + \sqrt{(x^2 + 1)}]$

Thus $\sin h^{-1} x = \log [x + \sqrt{(x^2 + 1)}]$

Similarly, $\cos h^{-1} x = \log [x + \sqrt{(x^2 - 1)}]$;

$$\tan h^{-1} x = \frac{1}{2} \log \frac{1+x}{1-x}; \coth^{-1} x = \frac{1}{2} \log \frac{x+1}{x-1};$$

$$\sec h^{-1} x = \log \frac{1+\sqrt{(1-x^2)}}{x}; \operatorname{cosech}^{-1} x = \log \frac{1+\sqrt{(1+x^2)}}{x}.$$

Derivatives of Inverse Hyperbolic Functions

Let $\sin h^{-1} x = y$, Then $x = \sin h\, y$

$$\therefore \quad \frac{d}{dx}(x) = \frac{d}{dx}(\sin h\, y) \text{ or } 1 = \left[\frac{d}{dy}(\sin h\, y)\right]\frac{dy}{dx} = \cos h\, y \frac{dy}{dx}.$$

$$\therefore \quad \frac{dy}{dx} = \frac{1}{\cosh y} = \frac{1}{\sqrt{(\text{siin } h^2\, y + 1)}} = \frac{1}{\sqrt{(x^2 + 1)}}.$$

$$\therefore \quad \frac{d}{dx}\sin h^{-1} x = \frac{1}{\sqrt{(x^2+1)}}.$$

Similarly, $\frac{d}{dx}\cosh^{-1} x = \frac{1}{(x^2 - 1)}$

$$\frac{d}{dx}\tan h^{-1} x = \frac{1}{1 - x^2}, \quad \frac{d}{dx}\cot h^{-1} x = 1\, \frac{1}{x^2 - 1}$$

$$\frac{d}{dx}\sec h^{-1} x = \frac{1}{x\sqrt{(1-x^2)}}; \quad \frac{d}{dx}\operatorname{cosec} h^{-1} x = \frac{1}{x\sqrt{(1+x^2)}}.$$

3.5 HYPERBOLIC FUNCTIONS

These are defined as follows:

$\sin h\, x = 1/2\, (e^x - e^{-x})$; $\cos h\, x = 1/2\, (e^x + e^{-x})$;

$$\tan h\, x\ \frac{\sin h\, x}{\cos h\, x} = \frac{e^x - e^{-x}}{e^x + e^{-x}}; \quad \cot h\, x = \frac{\cos h\, x}{\sin h\, x} = \frac{e^x + e^{-x}}{e^x - e^{-x}};$$

$$\sec h\, x\ \frac{1}{\cos h\, x}; \text{ and } \operatorname{cosec} h\, x = \frac{1}{\sin h\, x}.$$

Remember the Following Eelations Between Hyperbolic Functions

(i) $\cos h^2 x - \sin h^2 x = 1$;

(ii) $\sin h\, 2x\, x = 2 \sin h\, x \cos h\, x$;

(iii) $1 - \tan h^2 x = \sec h^2 x$;

(iv) $\cot h^2 x - 1 = \operatorname{cosec} h^2 x$;

(v) $\cos h\, 2x = \cos h^2 x \sin h^2 x$,

(v) $\tan h\, 2x = \frac{2 \tan h\, x}{1 + \tan h^2 x}$.

To get any of these relations first write the corresponding relation between circular functions. Then convert each circular function into the corresponding hyperbolic function and also change the sign of the term which contains the product of two sines.

Derivatives of Hyperbolic Functions

We have, $\frac{d}{dx}$ sin h x = $\frac{d}{dx}\left\{\frac{1}{2}(e^x - e^{-x})\right\} = \frac{1}{2}(e^x + e^{-x}) = \cos h\, x$

Similarly, we can find the derivatives of other hyperbolic function. Remember the following results.

$$\frac{d}{dx}\sin h\, x = \cos h\, x, \frac{d}{dx}\cos h\, x = \sin h\, x, \frac{d}{dx}\tan h\, x = \sec h^2\, x,$$

$$\frac{d}{dx}\operatorname{cost} h\, x = \operatorname{cosec} h^2\, x, \frac{d}{dx}\sec h\, x = \sec h\, x\ \tan h\, x,$$

$$\frac{d}{dx}\operatorname{cosec} h\, x, = -\operatorname{cosec} h\, x\ \cot h\, x.$$

We not that these results are similar to the corresponding results on circular functions. The only change is that here the derivatives of sin h x, cos h x and tan h x have +ive sign placed before them and the derivatives of the remaining three functions have –ive sign placed before them.

3.6 SOME MORE METHODS OF DIFFERENTIATION

(a) **Logarithmic Differentiation :** When a function consists of the product or the quotient of a number of functions, we take the logarithm and differentiate. This process, called *logarithmic differentiation,* is also useful when a function of x is raised to a power which is itself a function of x.

(b) **Implict Function:** If the relation between y and x is represented by an equation from which y cannot be easily expressed in terms of x, then y is called an *implicit function* of x. But if x is given directly in terms of x, then y is called an *explicit function* of x.

To find dy/dx in the case of implicit functions, differentiate each term of the given equation w.r.t. 'x' and solve the resulting equation for dy/dx.

Note: Here the value of dy/dx shall usually contain both x and y.

(c) **Parametric Equations:** Sometimes x and y are both expressed in terms of a third variable, say, t. This variable is called a parameter and the equations thus given are known as parametric equations. In the case of parametric equations $x = f_1(t)$ and $y = f_2(t)$, we find dy/dx in the following manner:

$$\frac{dy}{dx} = \frac{dy}{dt} \cdot \frac{dt}{dx} = \frac{dy/dt}{dx/dt}.$$

(d) **Trigonometrical Transformations:** Sometimes it is easy to differentiate after making some trigonometrical transformations.

Following formulae of trigonometry are of frequent use:

(i) $\cos x = 2\cos^2(x/2) - 1 = 1 - 2\sin^2(x/2)$;

(ii) $\sin x = 2\tan\frac{1}{2}x \Big/ \left(1+\tan^2\frac{1}{2}x\right)$;

(iii) $\cos x = \left(1-\tan^2\frac{1}{2}x\right) \Big/ \left(1+\tan^2\frac{1}{2}x\right)$;

(iv) $\tan x = 2\tan\frac{1}{2}x \Big/ \left(1-\tan^2\frac{1}{2}x\right)$;

(v) $\tan^{-1}x + \tan^{-1}y = \tan^{-1}\frac{x+y}{1-xy}$;

(vi) $\tan^{-1}x - \tan^{-1}y = \tan^{-1}\frac{x-y}{1+xy}$;

(vii) $2\tan^{-1}x = \tan^{-1}2x/(1-x^2)$;

(viii) $3\tan^{-1}x = \tan^{-1}\frac{3x-x^3}{1+3x^2}$;

(ix) $\sin 3x = 3\sin x - 4\sin^3 x$;

(x) $\cos 3x = 4\cos^3 x - 3\cos x$.

(e) **Differentiation of a Function w.r.t. a Function**

Suppose we have to differentiate f (x) w.r.t. $\phi(x)$. Put $\phi(x) = t$.

Then $$\frac{df(x)}{d\phi(x)} = \frac{df(x)}{dt} = \frac{df(x)}{dx}\cdot\frac{dx}{dt} = \frac{df(x)}{dx}\Big/\frac{dt}{dx}.$$

Hence $$\frac{df(x)}{d\phi(x)} = \frac{df(x)}{dx}\Big/\frac{d\phi(x)}{dx}.$$

3.7 LIST OF STANDARD RESULTS TO BE COMMITTED TO MEMORY

$\frac{d}{dx}x^n = nx^{n-1}$

$\frac{d}{dx}\tan x = \sec^2 x$

$\frac{d}{dx}e^x = e^x$

$\frac{d}{dx}\cot x = -\text{cosec}^2 x$

$\frac{d}{dx}\log_e x = \frac{1}{x}$

$\frac{d}{dx}\sec x = \sec x \tan x$

$\frac{d}{dx}\log_a x = \frac{1}{x}\log_a e$

$\frac{d}{dx}\text{cosec}\, x = -\text{cosec}\, x \cot x$

$\frac{d}{dx}a^x = a^x \log_e a$

$\frac{d}{dx}\sin h\, x = \cos h\, x$

$\frac{d}{dx}\sin x = \cos x$

$\frac{d}{dx}\cos h\, x = \sin h\, x$

$\frac{d}{dx}\cos x = -\sin x$

$\frac{d}{dx}\tan h\, x = \sec h^2 x$

$\frac{d}{dx}\cot h\, x = -\text{cosec}\, h^2 x$

$\frac{d}{dx}\cot^{-1} x = -\frac{1}{1+x^2}$

$\frac{d}{dx}\sec h\, x = -\sec h\, x \tan h\, x$

$\frac{d}{dx}\sec^{-1} x = \frac{1}{x\sqrt{(x^2-1)}}$

$\frac{d}{dx}\text{cosec}\, h\, x = -\text{cosec}\, h\, x \cot h\, x$

$\frac{d}{dx}\text{cosec}^{-1} x = -\frac{1}{x\sqrt{(x^1-1)}}$

$\frac{d}{dx}\sin^{-1} x = \frac{1}{\sqrt{(1-x^2)}}$

$\frac{d}{dx}\sin h^{-1} x = \frac{1}{\sqrt{(x^2+1)}}$

$\frac{d}{dx}\cos^{-1} x = -\frac{1}{\sqrt{(1+x^2)}}$

$\frac{d}{dx}\cos h^1 x = \frac{1}{\sqrt{(x^2-1)}}$

$\frac{d}{dx}\tan^{-1} x = \frac{1}{1+x^2}$

$\frac{d}{dx}\tan h^{-1} x = \frac{1}{1-x^2}$.

SOME MORE SOLVED EXAMPLES

Example 1:

Find the differential coefficients of the following w.r.t. x.

(i) $7 \sin x + 2 \log x = e^x + (x^2 - 7x + 4)$

(ii) $(\cos x) . (\log x)$

(iii) $e^{\sin^{-1} x}$

(iv) $e^{ax} \cos (bx + c)$

(v) $\log \sec x$

(vi) $a^{2x} \sin h\, 2x$

(vii) $\log \frac{+\sqrt{x}}{1-\sqrt{x}}$

(viii) $\dfrac{e^x}{1+x}$ *(ix)* $\dfrac{\tan x}{x+e^x}$

(x) $\log [\sqrt{(1 + \log x)} - \sin x]$ *(xii)* $\tan (\log \tan^{-1} \sqrt{x})$.

Solution:

(i) Let $y = 7 \sin x + 2 \log x + e^x + (x^2 - 7x + 4)$.

Then $\dfrac{dy}{dx} = 7. \dfrac{d}{dx} (\sin x) + 2 \dfrac{d}{dx} (\log x) - \dfrac{d}{dx}(e^x) + \dfrac{d}{dx}(x^2 - 7x + 4)$

$= 7 \cos x + 2/x - e^x + 2x - 7.$ **Ans.**

(ii) Let $y = (\cos x) . (\log x)$.

Then $\dfrac{dy}{dx} = (\cos x) \dfrac{d}{dx} (\log x) + (\log x) \dfrac{d}{dx} (\cos x)$

$= (\cos x) (1/x) + (\log x) (-\sin x)$

$= (1/x) \cos x - (\sin x) \log x.$ **Ans.**

(iii) Let $y = e^{\sin^{-1} x}$. Then $\dfrac{dy}{dx} = e^{\sin^{-1} x} \dfrac{d}{dx} \sin^{-1} x$

$= e^{\sin^{-1} x} \dfrac{1}{\sqrt{(1-x^2)}}.$ **Ans.**

(iv) Let $y = e^{ax} \cos (bx + c)$.

Then $\dfrac{dy}{dx} = e^{ax} . \dfrac{d}{dx} \cos (bx + c) + \cos (bx + c) \dfrac{d}{dx} e^{ax}$

$= e^{ax} [\{-\sin (bx + c)\} b] + \cos (bx + c) . e^{ax} . a$

$= e^{ax} [a \cos (bx + c) - b \sin (bx + c)].$ **Ans.**

(v) Let $y = \log \sec x$. Then $\dfrac{dy}{dx} = \dfrac{1}{\sec x} . \dfrac{d}{dx} (\sec x)$

$= \dfrac{1}{\sec x} \sec x \tan x = \tan x.$ **Ans.**

(vi) Let $y = a^{2x} \sin h\, 2x$. Then $\dfrac{dy}{dx} = a^{2x} . \dfrac{d}{dx} \sin h\, 2x + \sin h\, 2x . \dfrac{d}{dx} a^{2x}$

$= 2a^{2x} \cos h\, 2x + (\sin h\, 2x) . (a^2x \log a) . 2$

$= 2a^{2x} [\cos h\, 2x + (\log a) . \sin h\, 2x].$ **Ans.**

(vii) Let $y = \log \dfrac{1+\sqrt{x}}{1-\sqrt{x}} = \log (1 + \sqrt{x}) - \log (1 - \sqrt{x})$

Then $\frac{dy}{dx} = \frac{1}{1+\sqrt{x}} \cdot \frac{d}{dx}(1+\sqrt{x}) - \frac{1}{1-\sqrt{x}} \cdot \frac{d}{dx}(1-\sqrt{x})$

$$= \frac{1}{2\sqrt{x}}\left[\frac{1}{1+\sqrt{x}} + \frac{1}{1-\sqrt{x}}\right]$$

$$= \frac{1}{2\sqrt{x}} \cdot \frac{2}{1-x} = \frac{1}{(1-x)\sqrt{x}}$$ **Ans.**

(viii) Let $y = e^x/(1+x)$

Then $\frac{dy}{dx} = \frac{(1+x) \cdot \frac{d}{dx} e^x - e^x \cdot \frac{d}{dx}(1+x)}{(1+x)^2}$

$$= \frac{(1+x)e^x - e^x .1}{(1+x)^2} = \frac{x e^x}{(1+x)^2}$$ **Ans.**

(ix) Let $y = \tan x/(x+e^x)$

Then $\frac{dy}{dx} = \frac{(x+e^x)\sec^2 x - (\tan x)(1+e^x)}{(1+e^x)^2}$

$$= \frac{x\sec^2 x + e^x(\sec^2 x - \tan x) - \tan x}{(x+e^x)^2}$$

(x) Let $y = \log[\sqrt{(1+\log x)} - \sin x]$.

Then $\frac{dy}{dx} = \frac{1}{\sqrt{(1+\log x)} - \sin x} \cdot \left[\frac{1}{2x\sqrt{(1+\log x)}} - \cos x\right]$ **Ans.**

(xi) Let $y = \tan(\log \tan^{-1} \sqrt{x})$.

Then $\frac{dy}{dx} = \sec^2(\log \tan^{-1}\sqrt{x})\left[\frac{1}{\tan^{-1}(\sqrt{x})} \cdot \frac{1}{1+x} \cdot \frac{1}{2\sqrt{x}}\right]$

$$= \frac{\sec^2(\log \tan^{-1}\sqrt{x})}{(2\sqrt{x})(1+x)\tan^{-1}(\sqrt{x})}$$ **Ans.**

Example 2:

Find the differential coefficients of the following w.r.t. x.

(i) $\sqrt{(2x - x^{-2})}$ *(ii)* $\dfrac{\sqrt{(5-2x)}}{2x+1}$

(iii) $\log \dfrac{2}{x}$ *(iv)* $\sqrt{\{(1 + \sin x)/(1 - \sin x)\}}$

(v) $\dfrac{ax^2+b}{ax^2-b} + \dfrac{ax^2-b}{ax^2+b}$ *(vi)* $\log(\log x)$.

Solution:

(i) $\dfrac{d}{dx}(2x - x^{-2})^{1/2} = \dfrac{1}{2}(2x - x^{-2})^{-1/2}(2 + 2x^{-3})$

$= (2x - x^{-2})^{-1/2}(1 + x^{-3}) = = \dfrac{1+x^{-3}}{\sqrt{(2x-x^{-2})}}$ **Ans.**

(ii) $\dfrac{d}{dx}\left\{\dfrac{\sqrt{(5-2x)}}{2x+1}\right\} = \dfrac{d}{dx}\{(5-2x)^{1/2} \,.\, (2x+1)^{-1}\}$

$= \dfrac{1}{2}(5 - 2x)^{-1/2}(-2)(2x + 1)^{-1} = (2x + 1)^{-2} \,.\, 2 \,.\, (5 - x)^{1/2}$

$= \dfrac{-1}{(2x+1)\sqrt{(5-2x)}} - \dfrac{2\sqrt{(5-2x)}}{(2x+1)^2}$

$= \dfrac{(2x+1) - 2(5-2x)}{(2x+1)^2\sqrt{(5-2x)}} - \dfrac{2x-11}{(2x+1)^2\sqrt{(5-2x)}}$ **Ans.**

(iii) $\dfrac{d}{dx}\log\dfrac{2}{x} = \dfrac{d}{dx}(\log 2 - \log x) = -\dfrac{1}{x}$ **Ans.**

(iv) We have $\sqrt{\left(\dfrac{1-\sin x}{1-\sin x}\right)} = \sqrt{\left\{\dfrac{(1+\sin x)(1+\sin x)}{(1-\sin x)(1+\sin x)}\right\}}$

$= \sqrt{\left\{\dfrac{(1+\sin x)^2}{1-\sin^2 x}\right\}} = \dfrac{1+\sin x}{\cos x} = \sec x + \tan x$

$\therefore \dfrac{d}{dx}\sqrt{\left(\dfrac{1+\sin x}{1-\sin x}\right)} = \dfrac{d}{dx}(\sec x + \tan x)$

$= \sec x \tan x + \sec^2 x.$ **Ans.**

(v) $\frac{d}{dx}\left(\frac{ax^2+b}{ax^2-b}+\frac{ax^2-b}{ax^2+b}\right)=\frac{d}{dx}\left(\frac{2a^2x^4+2b^2}{a^2x^4-b^2}\right)$

$$=\left(\frac{8a^2x^3(a^2x^4-b^2)-4a^2x^3(2a^2x^4+2b^2)}{(a^2x^4-b^2)^2}\right)$$

$$=-\frac{16a^2b^2x^3}{(a^2x^4-b^2)^2}$$ **Ans.**

(vi) $=\frac{d}{dx}\{\log(\log x)\}=\frac{1}{\log x}\frac{d}{dx}(\log x)=\frac{1}{x\log x}$. **Ans.**

Example 3:

Find (dy/dx) if sin y = $log_{sin\,x}$ cos x. **(Delhi, 1980)**

Solution:

We have

$$\sin y=\log_{\sin x}\cos x=\frac{\log_e\cos x}{\log_e\sin x}$$ **[Note]**

Differentiating both sides w.r.t. x, we have

$$(\cos y)\,.\,\frac{dy}{dx}=\frac{(-\tan x)\,.\,\log\sin x-(\cot x)\,.\,\log\cos x}{(\log_e\sin x)^2}$$

$$\therefore\quad\frac{dy}{dx}=-\frac{(\tan x)\log\sin x+(\cot x)\log\cos x}{(\cos y)(\log\sin x)^2}$$ **Ans.**

Example 4:

Find (dy/dx), when **(Delhi, 1983, 79)**

(i) $y=x^{(x^x)}$

(ii) $y=(\tan x)^{\log x}+(\cot x)^{\sin x}$

(iii) $y=(\cot x)^{\cot x}+(\cosh x)^{\cos x}$

(iv) $y=x^x+x^{\sin x}$.

Solution:

(i) Taking logarithm of both sides, we get $\log y=(x^x)\log x$

Now differentiating both sides w.r.t. x, we get

$$\frac{1}{y}\frac{dy}{dx} = x^x \cdot \frac{1}{x} + \log x \cdot \frac{d}{dx}(x^x). \quad ...(1)$$

Let $z = x^x$. Then $\log z = x \log x$.

Differentiating w.r.t. x, we have

$$\frac{1}{z}\frac{dz}{dx} = x\frac{1}{x} + \log x \qquad \therefore \quad \frac{dz}{dx} = x^x(1+\log x)$$

Now equation (1) gives, $\frac{dy}{dx} = x^{(x^x)} \cdot [x^{x-1} + x^x (\log x).(1+\log x)]$.

(ii) We have $y = (\tan x)^{\log x} + (\cot x)^{\sin x}$.

Let $(\tan x)^{\log x} = u$ and $(\cot x)^{\sin x} = v$.

Then $y = u + v$. $\quad \therefore \quad \frac{dy}{dx} = \frac{du}{dx} + \frac{dv}{dx}$

Now $\log u = (\log x)(\log \tan x)$.

$$\therefore \quad \frac{1}{u} \cdot \frac{du}{dx} = \frac{\sec^2 x}{\tan x} + \frac{1}{x}\log \tan x$$

or $\quad \frac{du}{dx} = (\tan x)^{\log x} . [\cot x \sec^2 \log x + (1/x) \log \tan x]$

Again $\quad \log v = (\sin x)(\log \cot x)$.

$$\therefore \quad \frac{1}{v}\frac{dv}{dx} = \sin x \cdot \frac{1}{\cot x}(-\operatorname{cosec}^2 x) + \cos x \log \cot x$$

or $\quad \frac{dv}{dx} = (\cot x)^{\sin x}[\cos x \log \cot x - \sec x]$.

Hence $\quad \frac{dy}{dx} = (\tan x)^{\log x}[\cot x \sec^2 x \log x + (1/x) \log \tan x]$

$\qquad + (\cot x)^{\sin x}[\cos x \log f\cot x - \sec x]$

(iii) Here $y = (\cot x)^{\cot x} + (\cos h\, x)^{\cos h\, x}$

Let $(\cot x)^{\cot x} = u$ and $(\cos h\, x)^{\cos h\, x} = v$

Then $y = u + v$. Therefore $dy/dx = du/dx + dv/dx$

Now $\log u = \cot x. \log \cot x$

$$\therefore \quad \frac{1}{u}\frac{du}{dx} = \cot x \frac{1}{\cot x}(-\operatorname{cosec}^2 x) - \operatorname{cosec}^2 x \cdot \log \cot x$$

or $\quad \frac{du}{dx} = -\operatorname{cosec}^2 x\,(1 + \log \cot x)$

Again log v = cos h x log cos h x

$$\therefore \quad \frac{1}{v}\frac{dv}{dx} = \cos h\, x \cdot \frac{1}{\cos h\, x} \cdot \sin h\, x + \sin h\, x \log \cos h\, x$$

or $\dfrac{dv}{dx} = v \sin h\, x\,(1 + \log \cos h\, x)$

Hence

$$\frac{dy}{dx} = -\,u \operatorname{cosec}^2 x\,(1 + \log \cot x) + v \sin h\, x\,(1 + \log \cos h\, x)$$

$$= -\,(\cot x)^{\cot x} \operatorname{cosec}^2 x\,(1 + \log \cot x) + (\cos h\, x)^{\cos h\, x} \sin h\,(1 + \log \cos h\, x)$$

(iv) We have $y = x^x + (x)^{\sin x}$

Let $x^x = u$ and $(x)^{\sin x}$

Then $\quad y = u + v$ and $\dfrac{dy}{dx} = \dfrac{du}{dx} + \dfrac{dv}{dx}$

Now $\quad \log u = x \log x$

$$\therefore \quad \frac{1}{u}\frac{du}{dx} = x\,\frac{1}{x} + \log x \quad \text{or} \quad \frac{du}{dx} = x^x\,(1 + \log x)$$

Again log v = (sin x) log x

$$\therefore \quad \frac{1}{v}\cdot\frac{dv}{dx} = (\sin x)\cdot\frac{1}{x} + (\cos x)\log x$$

$$\text{or} \quad \frac{dv}{dx} = x^{\sin x}\,.\,(1 + \log x)\left[\frac{1}{x}\sin x + (\cos x)\log x\right]$$

$$\text{Hence} \quad \frac{dy}{dx} = x^x\,.\,(1 + \log x) + (x)^{\sin x}\left[\frac{1}{x}\sin x + (\cos x)\log x\right]$$

Example 5:

(a) If sin y = x sin 9a + y), prove that

$$\frac{dy}{dx} = \frac{\sin^2 (a + y)}{\sin a}$$ **(Delhi 1982, 95, 92)**

(b) If xy = ex–y prove that

$$\frac{dy}{dx} = \frac{\log x}{(1 + \log x)^2}$$ **(Delhi, 1981, 96; Meerut, 81S)**

(c) If $x = e^{\tan^{-1}\{(y-x^2)/x^2\}}$, *find (dy/dx), expressing it as a function of x only.*

(d) If x/(x − y) = log {a/(x − y)}, prove that (dy/dx) = 2 − (x/y).

(Delhi, 1983, 94)

Solution:

(a) We have x = sin y/sin (a + y)

Differentiating both sides w.r.t. x, we get

$$1 = \frac{\cos y \sin(a+y) - \sin y \cos(a+y)}{\sin^2(a+y)} \cdot \frac{dy}{dx}$$

or $$1 = \frac{\sin\{(a+y) - y\}}{\sin^2(a+y)} \cdot \frac{dy}{dx}$$

Hence $$\frac{dy}{dx} = \frac{\sin^2(a+y)}{\sin a}.$$

(b) We have, $x^y = e^{x-y}$.

∴ y log x = (x − y) log e = x − y

or y + y log x = x

or y = x/(1 + log x).

Hence $$\frac{dy}{dx} = \frac{1(1+\log x) - x.(1/x)}{(1+\log x)^2}$$

$$= \log x/(1 + \log x)^2.$$

(c) We have $x = t\ e^{\tan^{-1}\{(y-x^2)/x^2\}}$,

Therefore $$\log x = \tan^{-1} \frac{y-x^2}{x^2} \log e$$

or $$\frac{y-x^2}{x^2} = \tan \log x$$

or $y = x^2 + x^2 \tan \log x$.

∴ $$\frac{dy}{dx} = 2x + x^2 (\sec^2 \log x) \,.\, 1/x + 2x \,.\, \tan \log x$$

$$= x\,[2 + \sec^2 \log x + 2 \tan \log x].$$

(d) We have x/(x − y) = log {a/(x − y)}.

Differentiating w.r.t. x, we have

$$\frac{d}{dx}\left(\frac{x}{x-y}\right)=\frac{d}{dx}\left(\log\frac{a}{x-y}\right)=\frac{d}{dx}[\log a-\log(x-y)]$$

or $$\frac{1.(x-y)-\{1-(dy/dx)\}.x}{(x-y)^2}=\left[0-\frac{1}{(x-y)}\left(1-\frac{dy}{dx}\right)\right]$$

or $(x - y) - x + x\,(dy/dx) = -(x - y) + (x = y)\,(dy/dx)$

or $y\,(dy/dx) = 2y - x$ *i.e.*, $dy/dx = 2 - (x/y)$. **Ans.**

Example 6:

Find $\frac{dy}{dx}$ *if*

(i) $y = (x)^{\tan x} + (\sin x)^{\cos x}$ **(Delhi, 1980)**

(ii) $y = (\sin x)^{\cos x} + (\cos x)^{\sin x}$

Solution:

(i) Let $u = (x)^{\tan x}$ and $v = (\sin x)^{\cos x}$

$\therefore\ y = u + v \Rightarrow (dy/dx) = (du/dx) + (dv/dx)$ ...(1)

Now $u = (x)^{\tan x}$ $\log u = \tan x \log x$

$\therefore$ Differentiating w.r.t. x, we have

$(1/u)\,(du/dx) = (\sec^2 x)\,.\,\log x + (\tan x)/x$

or $$\frac{du}{dx}=(x)^{\tan x}\left[\sec^2 x\log x+\frac{\tan x}{x}\right]$$

Similarly, $\frac{dv}{dx} = (\sin x)^{\cos x}\,[(-\sin x)\,.\,\log\sin x + (\cos x)\,.\,\cot x]$.

$\therefore$ from (1), we get

$$\left(\frac{dy}{dx}\right)=\left(\frac{du}{dx}\right)+\left(\frac{dv}{dx}\right)$$

$$= (x)^{\tan x}\,[\sec^2 x \log x + (\tan x)/x]$$
$$+ (\sin x)^{\cos x}\,[\cos x \cot x - \sin x\,.\,\log \sin x].$$

(ii) Proceed exactly as in part (i).

Example 7:

Find (dy/dx) *if* $y = \tan^{-1}\{(ax - b)/(bx + a)\}$. **(Delhi, 1980)**

Solution:

We have $y = \tan^{-1}\left(\dfrac{ax-b}{bx+a}\right)$

$$\therefore \quad \frac{dy}{dx} = \frac{1}{1+(ax-b)^2/(bx+a)^2} \cdot \frac{d}{dx}\left(\frac{ax-b}{bx+a}\right)$$

$$\text{or} \quad \frac{dy}{dx} = \frac{(bx+a)^2}{(bx-a)^2+(ax-b)^2} \cdot \frac{a(bx+a)-b(ax-b)}{(bx+a)^2}$$

$$= \frac{a^2+b^2}{(a^2+b^2)x^2 + (a^2+b^2)} = \frac{1}{1+x^2}$$

Example 8:

Find (dy/dx), when

(i) $x = a(t - \sin t),\ y = a(1 - \cos t)$

(ii) $x = a(\cos t + \log \tan t/2),\ y = a \sin t$

(iii) $y = \tan^{-1}\dfrac{2t}{1-t^2},\ x = \sin^{-1}\dfrac{2t}{1+t^2}$

(iv) $x = a\sqrt{\left(\dfrac{t^2-1}{t^2+1}\right)},\ y = at\sqrt{\left(\dfrac{t^2-1}{t^2+1}\right)}$

Solution:

(i) We have $x = a(t - \sin t),\ y = a(1 - \cos t)$

$$\therefore \quad \frac{dx}{dt} = a(1-\cos t) \text{ and } \frac{dy}{dt} = a \sin t$$

$$\text{Now } \frac{dy}{dx} = \frac{dy/dt}{dx/dt} = \frac{a \sin t}{a(1-\cos t)} = \frac{2 \sin t/2 \cos t/2}{2 \sin^2 t/2} = \cot t/2$$

(ii) Here $x = a(\cos t + \log \tan t/2),\ y = a \sin t$

$$\therefore \quad \frac{dx}{dt} = a\left[-\sin t + \frac{1}{\tan t/2}(\sec^2 t/2) \cdot \frac{1}{2}\right]$$

$$= a\left(-\sin t + \frac{1}{2 \sin t/2 \cos t/2}\right)$$

$$= a\left[\frac{1}{\sin t} - \sin t\right] = a\left[\frac{1-\sin^2 t}{\sin t}\right] = \frac{a\cos^2 t}{\sin t}$$

Again $(dy/dt) = a \cos t$.

Now $\dfrac{dy}{dx} = \dfrac{dy/dt}{dx/dt} = a \cos t \cdot \dfrac{\sin t}{a\cos^2 t} = \tan t$

(iii) Let $t = \tan\theta$. Then

$$y = \tan^{-1}\frac{2\tan\theta}{1-\tan^2\theta} = \tan^{-1}\tan 2\theta = 2\theta = 2\tan^{-1} t$$

$\therefore \quad (dy/dt) = 2/(1 + t^2)$

Again $x = \sin^{-1}\dfrac{2t}{1+t^2} = \sin^{-1}\dfrac{2\tan\theta}{1+\tan^2\theta}$

$$= \sin^{-1}(\sin 2\theta) = 2\theta = 2\tan^{-1} t$$

$\therefore \quad (dx/dt) = 2/(1 + t^2)$

Now $\dfrac{dy}{dx} = \dfrac{dy/dt}{dx/dt} = \dfrac{2/(1+t^2)}{2/(1+t^2)} = 1$

(iv) Here $x = a\sqrt{\left\{\dfrac{t^2-1}{t^2+1}\right\}}$ and $y = at\sqrt{\left\{\dfrac{t^2-1}{t^2+1}\right\}}$

We have, $\log y = \log a + \log t + \dfrac{1}{2}\log(t^2-1) = \dfrac{1}{2}\log(t^2+1)$

$$\therefore \quad \frac{1}{y}\frac{dy}{dt} = \frac{1}{t} + \frac{t}{t^2-1} - \frac{t}{t^2+1} = \frac{t^4+2t^2-1}{t(t^4-1)}$$

or $\quad \sqrt{\left\{\dfrac{t^2-1}{t^2+1}\right\}} \cdot \dfrac{t^4+2t^2-1}{(t^4-1)}$

Again $\log x = \log a + 1/2 \log(t^2-1) - 1/2\log(t^2+1)$

$$\therefore \quad \frac{1}{x}\frac{dx}{dt} = \frac{t}{t^2-1} - \frac{1}{t^2+1} = \frac{2t}{t^4-1}$$

or $\quad \dfrac{dx}{dt} = a\sqrt{\left(\dfrac{t^2-1}{t^2+1}\right)} \cdot \dfrac{2t}{t^4-1}$

Now $\frac{dx}{dx} = \frac{dy/dt}{dx/dt} = \frac{t^4 + 2t^2 - 1}{2t}$.

Example 9:

Find $\frac{dy}{dx}$, *when* $x^y + (\sin x)^{\log x} = a.$ **(D.U. B.Sc. (G), 1996)**

Solution:

Let $u = x^y$, $v = (\sin x)^{\log x}$ so that $u + v = a$

$$\therefore \quad \frac{du}{dx} + \frac{dv}{dx} = 0 \qquad ...(1)$$

We have $\log x = y \log x$ and $\log v = \log x \log \sin x$

Differentiating w.r.t. 'x' we get

$$\frac{1}{u} \cdot \frac{dv}{dx} = \frac{y}{x} + \log x \frac{dy}{dx}$$

$$\frac{du}{dx} = x^y \left[\frac{y}{x} + \log x \frac{dy}{dx}\right] \qquad ...(2)$$

Again $\frac{1}{v}\frac{dv}{dx} = \frac{1}{x} \log \sin x + \frac{\cos x}{\sin x} \log x$

$$\frac{dv}{dx} = (\sin x)^{\log x} \left[\frac{1}{x} \log \sin x + \cot x \log x\right] \qquad ...(3)$$

Putting equations (2) and (3) in (1) we get

$$y\, x^{y-1} + x^y \log x \frac{dy}{dx} + (\sin x)^{\log x} \left[\frac{1}{x} \log \sin x + \cot x \log x\right] = 0$$

Hence $\frac{dy}{dx} = \frac{(\sin x)^{\log x} \left[\frac{1}{x} \log \sin x + \cot x \log x\right] + yx^{y-1}}{x^y \log x}$.

4

SUCCESSIVE DIFFERENTIATION

4.1 DEFINITION

It y be a function of x, etc., differential coefficient $\frac{dy}{dx}$, will be in general a function of x which can be differentiated. The differential co-efficient of y. Similarly, the differential co-efficient of the second differential co-efficient is called the third differential co-efficient, and so on the successive differential co-efficients of y are denoted by

$$\frac{dy}{dx}, \frac{d^2y}{dx^2}, \frac{d^3y}{dx^3} \cdots$$

The nth differential co-efficient of y being $\frac{d^2y}{dx^n}$.

Alternative methods of writing the nth differential co-efficient are

$$\left(\frac{d}{dx}\right)^n y, D^n y, y_n, \frac{d^2y}{dx^n}, y(x).$$

In the last case the first second etc. differential co-efficients would be written as y', y", etc.

The value of a differential co-efficient at x = a is usually indicated by adding a suffix. Thus, $(y_n)x \to a$ or $(y_n)a$. It y = f(x). The same thing can also be indicated by $f^{(n)}(a)$.

4.2. STANDARD RESULTS

(i) If $y = (ax + b)^m$, then $y_1 = ma\,(ax + b)^{m-1}$;

$y_2 = m(m - 1)\,a^2(ax + b)^{m-2}$, $y_3 = m(m - 1)(m - 2)\,a^3(ax+b)^{m-3}$, and so on.

In general,

$$y_n = m(m-1)(m-2) \ldots \{m-(n-1)\} a^n (ax+b)^{m-n}.$$

Thus,

$$D^n (ax+b)^m = m(m-1)(m-2) \ldots (m-n+1) a^n (ax+b)^{m-n}.$$

If m is a positive integer, the above result can be written in a compact form by using the factorial notation. Thus in this case $D^n (ax+b)^m$

$$= \frac{m(m-1)\ldots(m-n+1)(m-n)(m-n-1)\ldots 1}{(m-n)(m-n-1)\ldots 2.1} a^n (ax+b)^{m-n}$$

$$= \frac{m!}{(m-n)!} a^n (ax+b)^{m-n}.$$

Note 1: If m is a + ive integer, then $D^m (ax+b)^m$

$= (m!/0!) \cdot a^m (ax+b)^0 = m!\, a^m$. In particular, $D^m x^m$ - m!.

Note 2: If m is a + ive integer, the mth diff. co-eff. of $(ax+b)^m$ is constant. Therefore the $(m+1)^{th}$ and all the higher differential co-efficients of $(ax+b)^m$ are zero.

Note 3: If m is a negative integer, say $m = -p$, where p is a positive integer, then

$$D^n(ax+b)^{-p} = (-p)(-p-1)(-p-2) \ldots \{-p-(n-1)\} a^n (ax+b)^{-p-n}$$

$$= (-1)^n p(p+1) \ldots (p+n-1) a^n (ax+b)^{-p-n}$$

$$= (-1)^n \frac{(p+n-1)!}{(p-1)!} a^n (ax+b)^{-p-n}.$$

(ii) If $y = e^{ax} \sin(bx+c)$, then

$$y_1 = ae^{ax} \sin(bx+c) + be^{ax} \cos(bx+c)$$
$$= e^{ax} [a \sin(bx+c) + b \cos(bx+c)].$$

Putting $a = r\cos\phi$ and $b = r\sin\phi$, we get

$$y_1 = re^{ax} \sin(bx+c+\phi),$$

where $r^2 = a^2 + b^2$ and $\phi = \tan^{-1}(b/a)$.

Similarly, $y_2 = r^2 e^{ax} \sin(bx+c+2f)$, and so on.

Thus, $D^n\{e^{ax} \sin(bx+c)\} = r^n e^{ax} \sin(bx+c+n\phi)$, where

$r = (a^2+b^2)^{1/2}$ and $\phi = \tan^{-1}(b/a)$.

Similarly, $D^n\{e^{ax} \cos bx + c)\} = r^n e^{ax} \cos(bx+c+n\phi)$, where

$r = (a^2+b^2)^{1/2}$ and $\phi = \tan^{-1}(b/a)$.

(iii) If $y = (ax + b)^{-1}$, then $y_1 = (-1)\, a\, (ax + b)^{-2}$,
$y^2 = (-1)(-2)\, a^2 (ax + b)^{-3}$, $y_3 = (-1)(-2)(-3)\, a^3 (ax + b)^{-4}$, and so on.

In general, $y_n = (-1)(-2)(-3) \ldots (-n)\, a^n (ax + b)^{-(n+1)}$.

Thus, $D^n (ax + b)^{-1} = (-1)^n n!\, a^n (ax + b)^{-n-1}$.

(iv) If $y = e^{ax+b}$, then $y_1 = a\, e^{ax+b}$, $y_2 = a^2 e^{ax+b}$, $y_3 = a^3 e^{ax+b}$, and so on.

Thus, $D^n e^{ax+b} = a^n e^{ax+b}$. In particular, $D^n e^{ax} = a^n e^{ax}$.

(Meerut, 1992)

(v) If $y = a^x$, then $y_1 = a^x \log a$, $y_2 = a^x (\log a)^2$, $y_3 = a^x (\log a)^3$, and so on. Thus, $D^m a^x = a^x (\log)^n$

(vi) If $y = \log (ax + b)$, then $y_1 = a/(ax + b) = a\,(ax + b)^{-1}$,.
$y_2 = (-1)\, a^2 (ax + b)^{-2}$, $y_3 = (-1)(-2)\, a^3 (ax + b)^{-3}$, and so on.
In general, $y_n = (-1)(-2) \ldots \{-(n-1)\}\, a^n (ax + b)^{-n}$.

Thus, $D^n \log (ax + b) = \dfrac{(-1)^{n-1}(n-1)!\, a^n}{(ax+b)^n}$. **(Meerut, 1991)**

In particular, $D^n \log x = \dfrac{(-1)^{n-1}(n-1)!}{x^n}$.

(vii) If $y = \sin (ax + b)$, then

$y_1 = a \cos (ax + b) = a \sin \left(ax + b + \frac{1}{2}\pi\right)$. Comparing y with y_1, we find that

$$y_2 = a^2 \sin \left(ax + b + \frac{1}{2}\pi + \frac{1}{2}\pi\right) = a^2 \sin \left(ax + b + \frac{2}{2}\pi\right),$$

$$y_3 = a^3 \sin \left(ax + b + \frac{3}{2}\pi\right), \text{ and so on.}$$

Thus $D^n \sin (ax + b) = a^n \sin \left(ax + b + \frac{1}{2}n\pi\right)$. **(Meerut, 1996)**

(viii) Similarly, $D^n \cos (ax + b) = a^n \cos \left(ax + b + \frac{1}{2}n\pi\right)$.

(Meerut, 1996 BP)

Example 1:

(i) $e^x \sin^2 x$,

(ii) $e^{ax} \cos^2 x \sin x$,

(iii) $\cos^2 x \sin^3 x$, **(Kashmir, 1993)**

(iv) $\sin^2 x \sin 2x$,

(v) $\sin^5 x \cos^3 x$,

(vi) $\sin x \cos 3x$,

(vii) $e^{2x} \sin^3 x$. **(Meerut, 1998 S)**

Solution:

(i) Let $y = e^x \sin^2 x = \frac{1}{2}e^x\left(2\sin^2 x\right) = \frac{1}{2}e^x\left(1 - \cos 2x\right)$

$= \frac{1}{2}\left[e^x - e^x \cos 2x\right]$.

Now $D^n e^{ax} = a^n e^{ax}$.

Also $D^n e^{ax} \cos(bx + c) = r^n e^{ax} \cos(bx + c + n\phi)$, where

$r = (a^2 + b^2)^{1/2}$ and $\phi = \tan^{-1}(b/a)$.

$\therefore\ y_n = \frac{1}{2}\left[e^x - r^n e^x \cos(2x + n\phi)\right]$,

where $r = \sqrt{\left(1^2 + 2^2\right)} = \sqrt{5}$ and $\phi = \tan^{-1}\frac{2}{1} = \tan^{-1} 2$.

(ii) Let $y = e^{ax} \cos^2 x \sin x = \frac{1}{2}e^{ax}\sin(1 + \cos 2x) \sin x$

$= \frac{1}{2}e^{ax}\sin x + \frac{1}{2}e^{ax}\cos 2x \sin x$

$= \frac{1}{2}e^{ax}\sin x + \frac{1}{4}e^{ax}(2\cos 2x \sin x)$

$= \frac{1}{2}e^{ax}\sin x + \frac{1}{4}e^{ax}(\sin 3x - \sin x) = \frac{1}{4}\left[e^{ax}\sin x + e^{ax}\sin 3x\right]$.

Now using the formula for $D^n\{e^{ax}\sin(bx + c)\}$, we get

$y_n = \frac{1}{4}\left[(a^2 + 1)^{n/2} e^{ax} \sin\{x + n\tan^{-1}(1/a)\}\right.$

$\left. + (a^2 + 9)^{n/2} e^{ax} \sin\{3x + n\tan^{-1}(3/a)\right]$.

(iii) Let $y = \cos^2 x \sin^3 x = \frac{1}{2}(1 + \cos 2x)\frac{1}{4}(3\sin x - \sin 3x)$

$= \frac{1}{8}(3\sin x - \sin 3x + 3\sin x \cos 2x - \sin 3x \cos 2x)$

$= \frac{1}{8}\left[3\sin x - \sin 3x + \frac{3}{2}(2\sin x \cos 2x) - \frac{1}{2}(2\sin 3x \cos 2x)\right]$

$$= \frac{1}{8}\left[3\sin x - \sin 3x + \frac{3}{2}(\sin 3x - \sin x) - \frac{1}{2}(\sin 5x + \sin x)\right]$$

$$= \frac{1}{16}(6\sin x - 2\sin 3x + 3\sin 3x - 3\sin x - \sin 5x - \sin x)$$

$$= \frac{1}{16}(2 \sin x + \sin 3x - \sin 5x).$$

Now using the standard formula $D^n \sin(ax + b)$, we get

$$y_n = \frac{1}{16}\left[2\sin\left(x + \frac{1}{2}n\pi\right) + 3^n \sin\left(3x + \frac{1}{2}n\pi\right) - 5^n \sin\left(5x + \frac{1}{2}n\pi\right)\right].$$

(iv) Let $y = \sin^2 x \sin 2x = \frac{1}{2}(1 - \cos 2x)\sin 2x$

$$= \frac{1}{2}(\sin 2x - \sin 2x \cos 2x)$$

$$= \frac{1}{2}\left(\sin 2x - \frac{1}{2}.2\sin 3x \cos 2x\right) = \frac{1}{2}\left(\sin 2x - \frac{1}{2}n\pi\right)$$

$$= \frac{1}{4}(2 \sin 2x - \sin 4x).$$

Then $y_n = \frac{1}{4}\left[2\cdot 2^n \sin\left(2x + \frac{1}{2}n\pi\right) - 4^n \sin\left(4x + \frac{1}{2}n\pi\right)\right]$

$$= 2n^{-1} \sin\left(2x + \frac{1}{2}n\pi\right) - 4^{n-1} \sin\left(4x + \frac{1}{2}n\pi\right).$$

(v) Let $z = \cos x + i \sin x$.

Then $z^{-1} = (\cos x + i \sin x)^{-1} = \cos x - i \sin x$.

Therefore $z + z^{-1} = 2 \cos x$ and $z - z^{-1} = 2i \sin x$.

Also by De-Moivre's Theorem, $z^m = (\cos x + i \sin x)^m$

$= \cos mx + i \sin mx$ and $z^{-m} = \cos mx - i \sin mx$.

$\therefore\ z^m + z^{-m} = 2 \cos mx$ and $z^{-m} - z^{-m} = 2i \sin mx$.

Now $(2i \sin x)^5 (2 \cos x)^3 = (z - z^{-1})^5.(z + z^{-1})^3$

$= (z^8 - z^{-8}) - 2(z^6 - z^{-6}) - 2(z^4 - z^{-4}) + 6(z^2 - z^{-2})$

$= 2i \sin 8x - 2(2i \sin 6x) - 2(2i \sin 4x) + 6(2i \sin 2x)$.

$\therefore\ \sin^5 x \cos^3 x = 2^{-7}[\sin 8x - 2 \sin 6x - 2 \sin 4x + 6 \sin 2x]$.

Now let $y = \sin^5 x \cos^3 x$. Then using the standard result for $D^n \sin(ax + b)$, we get

$$y_n = 2^{-7}\left[8^n \sin\left(8x+\frac{1}{2}n\pi\right)-2.6^n \sin\left(6x+\frac{1}{2}n\pi\right)\right.$$

$$\left.-2.4^n \sin\left(4x+\frac{1}{2}n\pi\right)+6.2^n \sin\left(2x+\frac{1}{3}n\pi\right)\right].$$

(vi) Let $y = \sin x \cos 3x = \frac{1}{2}(\sin 4x - \sin 2x)$.

Then $y_n = \frac{1}{2}\left\{4^n \sin\left(4x+\frac{1}{2}n\pi\right)-2^n \sin\left(2x+\frac{1}{2}n\pi\right)\right\}$.

(vii) Let $y = e^{2x} \sin^3 x$.

We know that $\sin 3x = 3 \sin x - 4 \sin^3 x$.

$\therefore\ 4 \sin^3 x = 3 \sin x - \sin 3x$ $\quad [\because \sin 3x = 3 \sin x - 4 \sin^3 x]$

Now $D^n \sin (ax + b) = a^n \sin\left(ax+b+\frac{1}{2}n\pi\right)$.

$$\therefore\ y_n = \frac{3}{4}\sin\left(x+\frac{n\pi}{2}\right)-\frac{1}{4}.3^n \sin\left(3x+\frac{n\pi}{2}\right).$$

(viii) Let $y = \cos^4 x = (\cos^2 x)^2 = \left[\frac{1}{2}(1+\cos 2x)\right]^2$

$$=\frac{1}{4}\left(1+2\cos 2x+\cos^2 2x\right)=\frac{1}{4}\left[1+2\cos 2x+\frac{1}{2}(1+\cos 4x)\right]$$

$$=\frac{1}{4}\left(\frac{3}{2}+2\cos 2x+\frac{1}{2}\cos 4x\right)=\frac{3}{8}+\frac{1}{2}\cos 2x+\frac{1}{8}\cos 4x.$$

Now $D^n \cos (ax + b) = a^n \cos\left(ax+b+\frac{1}{2}n\pi\right)$.

$$\therefore\ y_n = 0+\frac{1}{2}.2^n \cos\left(2x+\frac{1}{2}n\pi\right)+\frac{1}{8}.4^n \cos\left(4x+\frac{1}{2}n\pi\right)$$

$$=2^{n-1} \cos\left(2x+\frac{1}{2}n\pi\right)+2^{2n-3} \cos\left(4x+\frac{1}{2}n\pi\right).$$

(ix) Let $y = \cos x \cos 2x \cos 3x = \frac{1}{2}\cos x\,(2 \cos 2x \cos 3x)$

$$=\frac{1}{2}\cos x\,(\cos 5x + \cos x) = \frac{1}{4}(2 \cos x \cos 5x + 2 \cos^2 x)$$

$$=\frac{1}{4}(\cos 6x + \cos 4x + \cos 2x + 1).$$

$$\therefore\ y_n = \frac{1}{4}\left\{6^n \cos\left(6x+\frac{1}{2}n\pi\right)+4^n \cos\left(4x+\frac{1}{2}n\pi\right)+2^n \cos\left(2x+\frac{1}{2}n\pi\right)\right\}.$$

(x) Let $y = e^{ax} \sin bx \cos cx = \frac{1}{2} e^{ax} (2 \sin bx \cos cx)$

$$= \frac{1}{2} e^{ax} [\sin (bx + cx) + \sin (bx - cx)]$$

$$= \frac{1}{2} [e^{ax} \sin (b + c) x + e^{ax} \sin (b - c)x].$$

Now $D^n\{e^{ax} \sin (bx + c)\}$

$= (a^2 + b^2)^{n/2} e^{ax} \sin \{bx + c + n \tan^{-1} (b/a)\}.$

$$\therefore\ y_n = \frac{1}{2} [\{a^2 + (b + c)^2\}^{n/2} e^{ax} \sin \{(b + c)x + n \tan^{-1} (b + c)/a\}$$

$$+ \{a^2 + (b - c)^2\}^{n/2} e^{ax} \sin \{(b - c) x + n \tan^{-1} (b - c)/a\}].$$

or $\sin^3 x = (1/4) (3 \sin x - \sin 3x)$.

$$\therefore\ y = \frac{1}{4} e^{2x} [3 \sin x - \sin 3x]$$

$$= \frac{3}{4} e^{2x} \sin x - \frac{1}{4} e^{2x} \sin 3x.$$

$$\therefore\ y_n = \frac{3}{4} [(2^2 + 1^2)^{1/2}]^n e^{2x} \sin[x + n \tan^{-1} (1/2)]$$

$$- \frac{1}{4} [(2^2 + 3^2)^{1/2}]^n e^{2x} \sin [2x + n \tan^{-1} (3/2)].$$

Example 2:

Find the nth differential coefficients of

(i) *log [(ax + b) (cx + d)],*

(ii) *sin ax cos bx,*

(iii) *$\sin^3 x$,*

(iv) *$\cos^4 x$,*

(v) *cos x cos 2x cos 3x,*

(vi) *e^{ax} sin bx cos cx.* **(Meerut, 1998 S)**

Solution:

(i) Let y log [(ax + b) (cx + d)]

$= \log (ax + b) + \log (cx + d).$

We know that $D^n \log (ax + b) = (-1)^{n-1} (n - 1)!\ a^n (ax + b)^{-n}$

$\therefore\ y_n = (-1)^{n-1}(n-1)!\ a^n (ax+b)^{-n}$
$+ (-1)^{n-1}(n-1)!\ c^n (cx+d)^{-n}$

$$= (-1)^{n-1}(n-1)!\left[\frac{a^n}{(ax+b)^n}+\frac{c^n}{(cx+d)^n}\right].$$

(ii) Let $y = \sin ax \cos x\ bx = \frac{1}{2}[2 \sin ax \cos bx]$

$= \frac{1}{2}[\sin (a+b) x + \sin (a-b) x]$.

We know that $D^n \sin (ax+b) = a^n \sin\left(ax+b+\frac{1}{2}n\pi\right)$.

$$\therefore\ y_n = \frac{1}{2}\left[(a+b)^n \sin\left\{(a+b)x+\frac{1}{2}n\pi\right\}\right.$$
$$\left.+(a-b)^n \sin\left\{(a-b)x+\frac{1}{2}n\pi\right\}\right].$$

(iii) Let $y = \sin^3 x = \frac{1}{4}(3 \sin x - \sin 3x)$,

Example 3:

(i) $\dfrac{1}{(x-1)^3(x-2)}$

(ii) $\dfrac{x^4}{(x-1)(x-2)}$ **(Gorakhpur, 1999; Meerut, 1997)**

(iii) $\dfrac{1}{x^2-a^2}$

(iv) $\dfrac{1}{a^2-x^2}$ **(Agra, 1993)**

(v) $\dfrac{x^2}{(x+2)(2x+3)}$ **(Meerut, 1998)**

Solution:

(i) Let $y = \dfrac{1}{(x-1)^3(x-2)}$. To resolve y into partial fractions we put $x - 1 = z$. Then $y = \dfrac{1}{z^3}.\dfrac{1}{z-1} = \dfrac{1}{z^3}.\dfrac{1}{-1+z}$, arranging the numerator and the denominator both in ascending powers of z.

Now dividing the numerator 1 by the denominator $-1+z$ till z^3 is a common factor in the remainder, we get

$$y=\frac{1}{z^3}\left[-1-z-z^2+\frac{z^3}{-1+z}\right]=-\frac{1}{z^3}-\frac{1}{z^2}-\frac{1}{z}+\frac{1}{-1+z}$$

$$=-\frac{1}{(x-1)^3}-\frac{1}{(x-1)^2}-\frac{1}{x-1}+\frac{1}{x-2}$$

$= -(x-1)^{-3}-(x-1)^{-2}-(x-1)^{-1}+(x-2)^{-1}$.

Now $D^n (ax+b)^{-p} = (-p)(-p-1)(-p-2)\ldots$

$\ldots \{(-p-(n-1)\}\, a^n (ax+b)^{-p-n}$

$$=\frac{(-1)^n (p+n-1)!}{(p-1)!}a^n(ax+b)^{-p-n}.$$

$$\therefore\ y_n=-\frac{(-1)^n(3+n-1)!}{(3-1)!}(x-1)^{-3-n}-\frac{(-1)^n(2+n-1)!}{(2-1)!}.(x-1)^{-2-n}$$

$-(-1)^n n!\,(x-1)^{-n-1}+(-1)^n n!\,(x-2)^{-n-1}$

$$=(-1)^n\left[-\frac{(n+2)!}{2!}(x-1)^{-3-n}-\frac{(n+1)!}{1!}(x-1)^{-2-n}\right.$$

$$\left.-n!(x-1)^{-n-1}+n!+n(x-2)^{-n-1}\right]$$

$$=(-1)^{n+1}n!\left[\frac{(n+2)(n+1)}{2(x-1)^{n+3}}+\frac{n+1}{(x-1)^{n+2}}+\frac{1}{(x-1)^{n+1}}-\frac{1}{(x-2)^{n+1}}\right].$$

(ii) Let $y=\dfrac{x^4}{(x-1)(x-2)}=\left[x^2+3x+7+\dfrac{15x-14}{(x-1)(x-2)}\right]$,

dividing N^r by the D^r

$=\left[x^2+3x+7+\dfrac{16}{x-2}-\dfrac{1}{x-1}\right]$, on resolving into partial fractions.

Then $y_n = 16(-1)^n.n!\,(x-2)^{-1-n} - (-1)^n n!\,(x-1)^{-1-n}$, if $n>2$

$= (-1)^n \,.\, n!\,[16(x-2)^{-n-1}-(x-1)^{-n-1}]$.

(iii) Let $y=\dfrac{1}{x^2-a^2}=\dfrac{1}{(x-a)(x+a)}=\dfrac{1}{2a}\left[\dfrac{1}{x-a}-\dfrac{1}{x+a}\right]$

$$= \frac{1}{2a}\left[(x-a)^{-1} - (x+a)^{-1}\right].$$

Then $y_n = \frac{1}{2a}$ $(-1)^n$ $n!$ $\{(x-a)^{-n-1} - (x+a)^{-n-1}\}$.

(iv) Let $y = \frac{1}{a^2 - x^2} = \frac{1}{(a-x)(a+x)} = \frac{1}{2a}\left[\frac{1}{a-x} + \frac{1}{a+x}\right]$

$$= \frac{1}{2a}\left[\frac{1}{x+a} - \frac{1}{x-a}\right] = \frac{1}{2a}\left[(x+a)^{-1} - (x-a)^{-1}\right]$$

Then $y_n = \frac{1}{2a}(-1)^n n!$ $\{(x+a)^{-n-1} - (x-a)^{-n-1}\}$.

(v) Let $y = x^2/[(x+2)(2x+3)]$.

The given fraction is not a proper one. If we divide the numerator by the denominator, we observe orally that the quotient will be 1/2. So let

$$\frac{x^2}{(x+2)(2x+3)} \equiv \frac{1}{2} + \frac{A}{x+2} + \frac{B}{2x+3}.$$

Then $A = -4$ and $B = 9/2$.

Hence $y = \frac{1}{2} - \frac{4}{x+2} + \frac{9}{2(2x+3)}$

$$= \frac{1}{2} - 4(x+2)^{-1} + \frac{9}{2}(2x+3)^{-1}.$$

$\therefore y_n = -4(-1)^n$ $n!$ $(x+2)^{-n-1}$

$$+ \frac{9}{2} \cdot (-1)^n\, n! \,.\, 2^n (2x+3)^{-n-1}$$

$$= (-1)^n n!\left[\frac{9.2^{n-1}}{(2x+3)^{n+1}} - \frac{4}{(x+2)^{n+1}}\right].$$

Example 4:

Find the nth derivatives of

(i) $\frac{x}{1+3x+2x^2}$

(ii) $\dfrac{1}{1-5x-6x^2}$ **(Kanpur, 1990; Meerut, 90)**

(iii) $\dfrac{17x^2+26x-42}{6x^3-25x^2-29x+20}$

(iv) $\dfrac{x^2}{(x-a)(a-b)}$.

Solution:

(i) Let $y=\dfrac{x}{1+3x+2x^2}=\dfrac{x}{(1+x)(1+2x)}$

Resolving into partial fractions, we get

$$y=\frac{-1}{(1-2)(1+x)}+\frac{-\frac{1}{2}}{\left(1-\frac{1}{2}\right)(1+2x)}=\frac{1}{x+1}-\frac{1}{2x+1}.$$

Now using the standard result for $D^n(ax+b)^{-1}$, we get

$$y=\frac{(-1)^n n!}{(x+1)^{n+1}}-\frac{(-1)^n n!\,2^n}{(2x+1)^{n+1}}$$

$$=(-1)^n n!\left[\frac{1}{(x+1)^{n+1}}-\frac{2^n}{(2x+1)^{n+1}}\right].$$

(ii) Let $y=\dfrac{1}{6x^2-5x+1}=\dfrac{1}{(3x-1)(2x-1)}$

$=\dfrac{2}{2x-1}-\dfrac{3}{3x-1}$, (on resolving into partial fractions)

$=2(2x-1)^{-1}-3(3x-1)^{-1}$.

Now $D^n(ax+b)^{-1}=(-1)^{-n}\,n!\,a^n\,(ax+b)^{-n-1}$.

$\therefore\ y_n=2\,(-1)^n\,n!\,2^n\,(2x-1)^{-n-1}$

$-3\,(-1)^n\,n!\,3^n\,(3x-1)^{-n-1}$

$=(-1)^n\,n!\,[2^{n+1}(2x-1)^{-n-1}-3^{n+1}(3x-1)^{-n-1}]$.

(iii) Let y = the given traction. Resolving into partial fractions, we get

$$y=\frac{1}{2x-1}-\frac{2}{3x+4}+\frac{3}{x-5}.$$

$$\therefore\ y_n = \frac{(-1)^n n!\, 2^n}{(2x-1)^{n+1}} - \frac{2(-1)^n n!\, 3^n}{(3x+4)^{n+1}} + \frac{3(-1)^n .n!}{(x-5)^{n+1}}$$

$$= (-1)^n .n! \left\{ \frac{2n}{(2x-1)^{n+1}} - \frac{2.3^n}{(3x+4)^{n+1}} + \frac{3}{(x-5)^{n+1}} \right\}.$$

(iv) Let $y = \frac{x^2}{(x-a)(x-b)}$. Since the given fraction is not a proper one, therefore we should first divide the numerator by the denominator before resolving it into partial fractions. Here we observe orally that the quotient will be 1. So let

$$\frac{x^2}{(x-a)(x-b)} \equiv 1 + \frac{A}{x-a} + \frac{B}{x-b}.$$

Then $A = \frac{a^2}{a-b}$ and $B = \frac{b^2}{b-a}$

Hence $y = 1 + \frac{a^2}{(a-b)(x-a)} + \frac{b^2}{(b-a)(x-b)}$.

$$= 1 + \frac{a^2}{(a-b)}(x-a)^{-1} - \frac{b^2}{(a-b)}(x-b)^{-1}.$$

Now differentiating both sides n times, we get

$$y_n = \frac{a^2}{(a-b)}(-1)^n n!\,(x-a)^{-n-1} - \frac{b^2}{(a-b)}(-1)^n n!\,(x-b)^{-n-1}$$

$$= \frac{(-1)^n n!}{(a-b)} \left[\frac{a^2}{(x-a)^{n+1}} - \frac{b^2}{(x-b)^{n+1}} \right].$$

Example 5:

(a) If $y = \sin^{-1} \{2x/(1 + x^2)\}$, prove that
$y_n = 2(-1)^{n-1}(n-1)!\, \sin^n \sin n\theta$, where $\theta = \cot^{-1} x$.

(Rohilkhand, 1990)

(b) If $y = \tan^{-1} \left\{ \frac{\sqrt{(1+x^2)} - 1}{x} \right\}$, show that

$$y_n = \frac{1}{2}(-1)^{n-1}\ (n-1)!\ \sin^n\ \theta \sin n\theta,$$

where $\theta = \cot^{-1} x$. **(Meerut, 1992, 95)**

Solution:

(a) Put $x = \tan \phi$. Then

$$y \sin^{-1}\left(\frac{2\tan\phi}{1+\tan^2\phi}\right) = \sin^{-1} \sin 2\phi = 2\phi$$

$= 2 \tan^{-1} x$.

(b) Let $y = \tan^{-1}\left\{\frac{\sqrt{(1+x^2)}-1}{x}\right\}$.

Put $x = \tan \phi$. Then

$$y = \tan^{-1}\left[\frac{\sqrt{(1+\tan^2\phi)}-1}{\tan\phi}\right] = \tan^{-1}\frac{\sec\phi - 1}{\tan\phi}$$

$$= \tan^{-1}\frac{1-\cos\phi}{\sin\phi} = \tan^{-1}\frac{2\sin^2(\phi/2)}{2\sin(\phi/2)\cos(\phi/2)}$$

$= \tan^{-1} \tan (\phi/2) = \phi/2 = 1/2\tan^{-1} x$.

$\therefore\ y_1 = 1/\{2(1 + x^2)\}$.

We get

$$y_n = \frac{1}{2}(-1)^{n-1}(n-1)!\ \sin^n\theta \sin n\theta, \text{ where } \theta = \cot^{-1} x.$$

Example 6:

Prove that the value of the nth differential coefficient of $x^3/(x^2 - 1)$ *for* $x = 0$, *is zero if n is even, and is* $-n!$ *if n is odd and greater than 1.*

Solution:

$$\text{Let } y = \frac{x^3}{x^2-1} = x + \frac{x}{x^2-1} = x + \frac{x}{(x-1)(x+1)}$$

$$= x + \frac{1}{(1+1)(x-1)} + \frac{-1}{(-1-1)(x+1)}$$

$$= x + \frac{1}{2(x-1)} + \frac{1}{2(x+1)}.$$

$\therefore$ When n > 1, we have

$$y_n = \frac{(-1)^n n!}{2}\left[\frac{1}{(x-1)^{n+1}} + \frac{1}{(x+1)^{n+1}}\right].$$

Putting x = 0 in the expression for y_n, we get

$$(y_n)_0 = \frac{(-1)^n n!}{2}\left[\frac{1}{(-1)^{n+1}} + \frac{1}{1}\right] = \frac{(-1)^n n!}{2}\left[\frac{1}{(-1)^{n+1}} + 1\right].$$

When n is even, $(y_n)_0 = \frac{n!}{2}\left[\frac{1}{(-1)} + 1\right] = \frac{n!}{2}.0 = 0.$

When n is odd, $(y_n)_0 = -\frac{n!}{2}[1+1] = -n!.$

Example 7:

Find the nth *derivative of* $\tan^{-1}\left(\frac{1+x}{1-x}\right)$.

Solution:

Let

$$y = \tan^{-1}\left(\frac{1+x}{1-x}\right) = \tan^{-1}\left\{\frac{1+x}{1-1.x}\right\} = \tan^{-1} 1 + \tan^{-1} x.$$

Then $y_1 = 1/(1 + x^2)$. We have

$y_n = (-1)^{n-1} (n-1)! \sin^n \phi \sin n\phi$, where $\phi = \tan^{-1}(1/x)$.

Example 7(a):

If $y = x(a^2 + x^2)^{-1}$, *prove that*

$y_n = (-1)^n n! a^{-n-1} \sin^{n+1} \phi \cos (n+1) f$, *where* $\phi = \tan^{-1}(a/x)$.

(Meerut, 1992 S, 93 S; Rohilkhand, 98, Gorakhpur, 92)

Solution:

We have $y = \frac{x}{a^2 + x^2} = \frac{x}{(x-ia)(x+ia)}$

$$= \frac{1}{2}\left[\frac{1}{(x-ia)} + \frac{1}{(x+ia)}\right],$$ on resolving into partial fractions.

Differentiating both sides n times w.r.t 'x', we get

$$y_n = \frac{1}{2}[(-1)^n\, n!\, (x - ia)^{-n-1} + (-1)^n\, n!\, (x + ia)^{-n-1}]$$

$$= \frac{1}{2}(-1)^n\, n!\, [(x - ia)^{-n-1} + (x + ia)^{-n-1}].$$

Putting $x = r\cos\phi$ and $a = r\sin\phi$, we get

$$y_n = \frac{1}{2}(-1)^n\, n!\, [r^{-n-1}(\cos\phi - i\sin\phi)^{-n-1} + r^{-n-1}(\cos\phi + i\sin\phi)^{-n-1}]$$

$$= \frac{1}{2}(-1)^n\, n!\, r^{-n-1}\, [\{\cos(n+1)\phi + i\sin(n+1)\phi\} + \{\cos(n+1)\phi - i\sin(n+1)\phi\}]$$

$$= \frac{1}{2}(-1)^n\, n!\, r^{-n-1}\, 2\cos(n+1)\phi$$

$= (-1)^n\, n!\, (a/\sin\phi)^{-n-1}\cos(n+1)\phi$, since $r = a/\sin\phi$

$= (-1)^n.n!\, a^{-n-1}\sin^{n+1}\phi\cos(n+1)\phi$,

where $\phi = \tan^{-1}(a/x)$.

Example 8:

Find the nth differential coefficient of $1/(x^2 + a^2)$.

(Lucknow, 1993; Gorakhpur, 92; Magadh, 94)

Solution:

We have $y = \dfrac{1}{x^2 + a^2} = \dfrac{1}{(x + ia)(x - ia)}$

$$= \frac{1}{2ia}\left\{\frac{1}{x - ia} - \frac{1}{x + ia}\right\}.$$

$$\therefore\ y_n = \frac{1}{2ia}.(-1)^n n!\left\{\frac{1}{(x - ia)^{n+1}} - \frac{1}{(x + ia)^{n+1}}\right\}.$$

Let $x = r\cos\phi$ and $a = r\sin\phi$, so that $\phi = \tan^{-1}(a/x)$. Then

$$y_n = \frac{(-1)^n n!}{2ia}\left\{\frac{1}{(r\cos\phi - ir\sin\phi)^{n+1}} - \frac{1}{(r\cos\phi + ir\sin\phi)^{n+1}}\right\}$$

$$= \frac{(-1)^n n!}{2ia\, r^{n+1}}\{(\cos\phi - i\sin\phi)^{-(n+1)} - (\cos\phi + i\sin\phi)^{-(n+1)}\}$$

$$= \frac{(-1)^n n!}{2ia\, r^{n+1}} [\{\cos (n+1)\phi + i \sin (n+1)\phi\}$$

$$- \{\cos (n+1)\phi - i \sin (n+1)\phi\}],$$

(by De-Moivre's Theorem)

$$= \frac{(-1)^n n!}{2ia\, r^{n+1}} 2i \sin (n+1)\phi = \frac{(-1)^n n!}{a.a^{n+1}} \sin (n+1)\phi \sin^{n+1} \phi,$$

since $r = a/\sin \phi$

$= (-1)^n n!\, a^{-(n+2)} \sin (n+1)\phi \sin^{n+1}\phi,$

where $\phi = \tan^{-1} (a/x)$.

Example 9:

(a) If $y = \sin^{-1} x$, prove that

$$(1 - x^2)(d^2y/dx^2) = x.(dy/dx).$$

(b) Find d^2y/dx^2 for the cycloid whose equation is

$$x = a(\theta - \sin \theta),\ y = a(1 - \cos \theta).$$

(c) If $x = a(t - \sin t)$ and $y = a(1 + \cos t)$, prove that

$$\frac{d^2y}{dx^2} = \frac{1}{4a} \operatorname{cosec}^4\left(\frac{t}{2}\right).$$ **(Meerut, 1995)**

(d) If $x = a(\cos \theta + \theta \sin \theta)$, $y = a(\sin \theta - \theta \cos \theta)$, find d^2y/dx^2. **(Meerut, 1991 P)**

(e) If $x = 2 \cos t - \cos 2t$, $y = 2 \sin t - \sin 2t$, find the value of (d^2y/dx^2) at $t = \pi/2$.

(f) If $y = \sin (\sin x)$, prove that

$$(d^2y/dx^2) + (dy/dx)(\tan x + y \cos^2 x = 0.$$ **(Kashmir, 1993)**

Solution:

(a) We have $y = \sin^{-1} x$; $\therefore dy/dx = 1/\sqrt{(1-x^2)}$

or $(1 - x^2)(dy/dx)^2 = 1.$

Differentiating again with respect to x, we get

$$(1-x^2).2\frac{dy}{dx}.\frac{d^2y}{dx^2} - 2x\left(\frac{dy}{dx}\right)^2 = 0$$

or $$(1-x^2)\frac{d^2y}{dx^2} = x\frac{dy}{dx}.$$ $\left[\because \frac{dy}{dx} \neq 0\right]$

(b) We have $x = a(\theta - \sin \theta)$; $\therefore (dy/d\theta) = a(1 - \cos \theta)$.

Also $y = a(1 - \cos\theta)$; $\therefore\ (dy/d\theta) = a \sin\theta$.

Now $\dfrac{dy}{dx} = \dfrac{dy/d\theta}{dx/d\theta} = \dfrac{a \sin\theta}{a(1-\cos\theta)} = \dfrac{2 \sin\theta/2 \cos\theta/2}{2 \sin^2\theta/2} = \cot\dfrac{\theta}{2}$.

$$\therefore\ \frac{d^2y}{dx^2} = \frac{d}{dx}\left(\frac{dy}{dx}\right) = \frac{d}{dx}(\cot\theta/2) = \left[\frac{d}{d\theta}(\cot\theta/2)\right]\frac{d\theta}{d}$$

$$= -\frac{1}{2}\operatorname{cosec}^2\frac{\theta}{2}.\frac{1}{a(1-\cos\theta)}$$

$$= -\frac{1}{2}\operatorname{cosec}^2\frac{\theta}{2}.\frac{1}{2a\sin^2\theta/2} = -\frac{1}{4a}\operatorname{cosec}^4\frac{\theta}{2}.$$

(c) Proceed as in (b).

(d) We have $x = a(\cos\theta + \theta\sin\theta)$.

$\therefore\ dx/d\theta = a(-\sin\theta + \sin\theta + \theta\cos\theta) = a\theta\cos\theta$.

Also $y = a(\sin\theta - \theta\cos\theta)$.

$\therefore\ dy/d\theta = a(\cos\theta - \cos\theta + \theta\sin\theta) = a\theta\cos\theta$.

Now $\dfrac{dy}{dx} = \dfrac{dy/d\theta}{dx/d\theta} = \dfrac{a\theta\sin\theta}{a\theta\cos\theta} = \tan\theta$.

$$\therefore\ \frac{d^2y}{dx^2} = \frac{d}{dx}\left(\frac{dy}{dx}\right) = \frac{d}{dx}(\tan\theta) = \left(\frac{d}{d\theta}(\tan\theta)\right)\frac{d\theta}{dx}$$

$$= \sec^2\theta.\frac{1}{a\theta\cos\theta} = \frac{1}{a}\frac{\sec^3\theta}{\theta},$$

(e) We have $x = 2\cos t - \cos 2t$; $\therefore\ \dfrac{dx}{dt} = -2\sin t + 2\sin 2t$.

Also $y = 2\sin t - \sin 2t$; $\therefore\ \dfrac{dy}{dt} = 2\cos t - 2\cos 2t$.

Now $\dfrac{dy}{dx} = \dfrac{dy/dt}{dx/dt} = \dfrac{2\cos t - 2\cos 2t}{-2\sin t + 2\sin 2t} = \dfrac{\cos t - \cos 2t}{\sin 2t - \sin t}$

$$= \frac{2\sin\frac{3t}{2}.\sin\frac{t}{2}}{2\cos\frac{3t}{2}.\sin\frac{t}{2}} = \tan\frac{3t}{2}.$$

$$\therefore \frac{d^2y}{dx^2} = \frac{d}{dx}\left(\frac{dy}{dx}\right) = \frac{d}{dx}\left(\tan\frac{3t}{2}\right) = \left[\frac{d}{dt}\left(\tan\frac{3t}{2}\right)\right]\frac{dt}{dx}$$

$$= \frac{3}{2}\sec^2\frac{3t}{2}\cdot\frac{1}{2\sin 2t - 2\sin t}.$$

When $t = \frac{\pi}{2}, \frac{d^2y}{dx^2} = \frac{3}{2}\sec^2\frac{3\pi}{4}\cdot\frac{1}{2\sin\pi - 2\sin\frac{\pi}{2}}$

$$\therefore \left(\frac{d^2y}{dx^2}\right)_{at\, t=\pi/2} = \frac{3}{2}.2.\frac{1}{(0-2)} = -\frac{3}{2}.$$

(f) We have $y = \sin(\sin x)$. ...(1)

$$\therefore \frac{dy}{dx} = [\cos(\sin x)]\cos x.$$

Differentiating again, we have

$$(d^2y/dx^2) = [-\sin(\sin x)]\cos^2 x - [\cos(\sin x)]\sin x$$

$$= -y\cos^2 x - [\cos(\sin x)]\cos x.(\sin x/\cos x),$$

[$\because$ from (1), $y = \sin(\sin x)$]

$$= -y\cos^2 x - \left(\frac{dy}{dx}\right).\tan x.$$

$\left[\because \text{from (2)}, \frac{dy}{dx} = \{\cos(\sin x)\}\cos x\right]$

$$\therefore \frac{d^2y}{dx^2} + \frac{dy}{dx}\tan x + y\cos^2 x = 0.$$

Example 10:

Find the nth derivative of $\tan^{-1}\{2x/(1-x^2)\}$.

(Lucknow, 1992)

Solution:

Let $y = \tan^{-1}\{2x/(1-x^2)\} = 2\tan^{-1} x.$

Then $y_1 = \frac{2}{1+x^2} = \frac{2}{(x-i)(x+i)} = \frac{1}{i}\left[\frac{1}{x-i} - \frac{1}{x+i}\right]$, on resolving into partial fractions.

Now differentiating both sides (n – 1) times w.r.t. 'x', we have

$$y_n = (-1)^{n-1}(n-1)!\,[(x-i)^{-n} - (x+i)^{-n}],$$

Putting x = r cos ϕ and 1 = r sin ϕ we have

$$y_n = \frac{(-1)^{n-1}(n-1)!}{i}\left[r^{-n}(\cos\phi - i\sin\phi)^{-n} - r^{-n}(\cos\phi + i\sin\phi)^{-n}\right]$$

$$= \frac{(-1)^{m-1}(n-1)!}{i}(r^{n}[(\cos n\phi + i\sin n\phi) - (\cos n\phi - i\sin n\phi)]$$

$= 2(-1)^{n-1}(n-1)!\, r^{-n} \sin n\phi$

$= 2(-1)^{n-1}(n-1)!\, (1/\sin\phi)^{-n} \sin n\phi$, since $r = 1/\sin\phi$

$= 2(-1)^{n-1}(n-1)!\, \sin^n\phi \sin n\phi$, where $\phi = \tan^{-1}(1/x)$.

Example 11:

(i) *If* $y = A \sin mx + B \cos mx$, *prove that*

$y_2 + m^2 y = 0.$ **(Indore, 1994)**

(ii) *If* $y = e^{ax} \sin bx$, *prove that*

$y_2 - 2ay_1 + (a^2 + b^2)\, y = 0.$ **(Gorakhpur, 1991)**

(iii) *If* $y = \sin nx + \cos nx$, *show that*

$y_r = n^r\{1 + (-1)^r \sin 2nx\}^{1\,2}.$

(Meerut, 1993, 93 P, 98; Lucknow, 93; Gorakhpur, 94)

Solution:

(i) We have, $y = A \sin mx + B \cos mx.$ ...(1)

Differentiating both sides w.r.t. 'x' we get

$y_1 = Am \cos mx - Bm \sin mx.$

Again $y_2 = -Am^2 \sin mx - Bm^2 \cos mx$

$= -m^2(A \sin mx + B \cos mx) = -m^2 y$, from (1).

$\therefore \quad y_2 = m^2 y = 0$

(ii) Here $y = e^{ax} \sin bx.$ ...(1)

$\therefore \quad y_1 = e^{ax}\, b \cos bx + a \,.\, e^{ax} \sin bx.$

Replacing $e^{ax} \sin bx$ by y, we get

$y_1 = be^{ax} \cos bx + ay.$ ...(2)

Differentiating both sides (2) w.r.t. 'x', we get

$y_2 = abe^{ax} \text{cox } bx - b^2 e^{ax} \sin bx + ay_1$

$= a(y_1 - ay) - b^2 y + ay_1$, [$\because$ from (2), $be^{ax} \cos bx = y_1 - ay$ and from (1), $e^{ax} \sin bx = y$]

$= 2ay_1 - (a^2 + b^2)y.$

$\therefore\ y_2 - 2ay_1 + (a^2 + b^2)y = 0.$

(iii) Here $y = \sin nx + \cos nx.$

$$\therefore\ y_r = n^r \sin\left(nx + \frac{1}{2}r\pi\right) + n^r \cos\left(nx + \frac{1}{2}r\pi\right)$$

$$= n^r\left[\left\{\sin\left(nx + \frac{1}{2}r\pi\right) + \cos\left(nx + \frac{1}{2}r\pi\right)\right\}^2\right]^{1/2}$$

$$= n^r\left[\sin^2\left(nx + \frac{1}{2}r\pi\right) + \cos^2\left(nx + \frac{1}{2}r\pi\right)\right.$$

$$\left. + 2\sin\left(nx + \frac{1}{2}r\pi\right)\cos\left(nx + \frac{1}{2}r\pi\right)\right]^{1/2}$$

$$= n^r\left[1 + \sin 2\left(nx + \frac{1}{2}r\pi\right)\right]^{1/2} = n^r\left[1 + \sin\left(2nx + r\pi\right)\right]^{1/2}$$

$= n^r[1 + (-1)^r \sin 2nx]$, since $(r\pi + \theta) = (-1)^r \sin \theta$.

Example 12:

Find the nth differential coefficient of $\tan^{-1}(x/a)$.

(Meerut, 1998, 96P, 93, 92, 91; Jhansi, 1998; Avadh, 1998)

Solution:

Let $y = \tan^{-1}\left(\frac{x}{a}\right)$. Then $y_1 = \frac{a}{x^2 + a^2} = \frac{a}{(x + ia)(x - ia)}$

$$= \frac{a}{(-ia - ia)(x + ia)} + \frac{a}{(ia + ia)(x - ia)}$$

$$= -\frac{1}{2i(x + ia)} + \frac{1}{2i(x - ia)} = \frac{1}{2i}\left[(x - ia)^{-1} - (x + ia)^{-1}\right].$$

Now differentiating both sides (n – 1) times w.r.t. 'x', we get

$$y_n = \frac{1}{2i}[(-1)^{n-1}(n-1)!\,(x - ia)^{-n} - (-1)^n (n-1)!\,(x + ia)^{-n}]$$

$$= \frac{(-1)^{n-1}(n-1)!}{2i}[(x - ia)^{-n} - (x + ia)^{-n}].$$

Put $x = r\cos\phi$ and $a = r\sin\phi$. Then

$$y_n = \frac{(-1)^{n-1}(n-1)!}{2i}[(r\cos\phi - ir\sin\phi)^{-n} - (r\cos\phi + ir\sin\phi)^{-n}]$$

$$= \frac{(-1)^{n-1}(n-1)!}{2i} r^{-n}[(\cos\phi - i\sin\phi)^{-n} - (\cos\phi + i\sin\phi)^{-n}]$$

$$= \frac{(-1)^{n-1}(n-1)!}{2i} r^{-n} [(\cos n\phi + i \sin n\phi) - (\cos n\phi - i \sin n\phi)]$$

$$= \frac{(-1)^{n-1}(n-1)!}{2i} r^{-n} 2i \sin n\phi = (-1)^{n-1} (n-1)! \, r^{-n} \sin n\phi$$

$= (-1)^{n-1} (n-1)! \, (a/\sin \phi)^{-n} \sin n\phi$, since $r = a/\sin \phi$

$= (-1)^{n-1} (n-1)! \, a^{-n} \sin^n \phi \sin n\phi$, where $\phi = \tan^{-1} (a/x)$.

Example 13:

Prove that the value when $x = 0$ of $D^n (\tan^{-1} x)$ is 0; $(n-1)!$ or $-(n-1)!$ according as n is of the form 2p, 4p + 1 or 4p + 3 respectively.

(Kanpur, 1993)

Solution:

Let $y = \tan^{-1} x$. Then

$$y_1 = \frac{1}{1+x^2} = \frac{1}{(x-i)(x+i)} = \frac{1}{2i}\left[\frac{1}{x-i} - \frac{1}{x+i}\right]$$

Differentiating both sides (n – 1) times w.r.t. 'x', we get

$$y_n = \frac{(-1)^{n-1}(n-1)!}{2i} [(x-i)^{-n} - (x+i)^{-n}].$$

Putting x = 0 in the expression for y_n, we get

$$(y_n)_{x=0} = \frac{(-1)^{n-1}(n-1)!}{2i}\left[(-i)^{-n} - i^{-n}\right]$$

$$= \frac{(-1)^{n-1}(n-1)!}{2i} [(-1)^{-n} i^{-n} - i^{-n}]$$

$$= \frac{(-1)^{n-1}(n-1)!}{2i} i^{-n} [(-1)^{-n} - 1].$$

Now we shall consider the three cases.

(i) If n is of the form 2p, we have

$$(y_n)_0 = \frac{(-1)^{2p-1}(n-1)!}{2i} i^{-2p}\left[(-1)^{-2p} - 1\right]$$

$$= -\frac{(n-1)!}{2i} i^{-2p}[1-1] = -\frac{(n-1)!}{2i} i^{-2p}.0 = 0.$$

(ii) If n is of the form 4p + 1, we have

$$(y_n)_0 = \frac{(-1)^{4p}(n-1)!}{2i} i^{-(4p+1)}\left[(-1)^{-(4p+1)} - 1\right]$$

$$= \frac{(n-1)!}{2.i^{4p+2}}[-1-1],$$

$[\because (-1)^{-(4p+1)} = 1/(-1)^{4p+1} = 1/(-1) = -1]$

$$= \frac{(n-1)!}{2.i^{4p+2}}(-2) = -\frac{(n-1)!}{i^{4p}.i^2} = -\frac{(n-1)!}{(i^4)^p i^2} = -\frac{(n-1)!}{4p(-1)}$$

$[\because i^4 = 1 \text{ and } i^2 = -1]$

$= (n-1)!.$

(iii) If n is of the form 4p + 3, we have

$$(y_n)_0 = \frac{(-1)^{4p+2}(n-1)!\, i^{-(4p+3)}}{2i} [(-1)^{-(4p+3)} - 1]$$

$$= \frac{(n-1)!}{2.i^{4p+4}}[-1-1] = \frac{(n-1)!}{2.i^{4p+4}}(-2)$$

$$= -\frac{(n-1)!}{(i^4)^{p+1}} = -\frac{(n-1)!}{1^{p+1}} = -(n-1)!.$$

Example 14:

If $p^2 = a^2 \cos^2 \theta + b^2 \sin^2 q$, prove that

$$p + \frac{d^2p}{d\theta^2} = \frac{a^2b^2}{p^3}.$$ **(Meerut, 1993, 93S, 92P; Delhi, 91)**

Solution:

We have $p^2 = a^2 \cos^2 \theta + b^2 \sin^2 \theta$. ...(1)

Differentiating both sides of (1) w.r.t. θ, we have

$$2p\left(\frac{dp}{d\theta}\right) = -2a^2 \cos\theta \sin\theta + 2b^2 \sin\theta \cos\theta$$

or $$p\left(\frac{dp}{d\theta}\right) = (b^2 - a^2) \sin\theta \cos\theta.$$ ...(2)

Now differentiating both sides of (2) w.r.t. θ, we have

$$p\frac{d^2p}{d\theta^2} + \left(\frac{dp}{d\theta}\right)^2 = (b^2 - a^2)(\cos^2\theta - \sin^2\theta)$$

$$= (b^2 \cos^2\theta + a^2 \sin^2\theta) - (a^2 \cos^2\theta + b^2 \sin^2\theta)$$

$= (b^2 \cos^2\theta + a^2 \sin^2\theta) - p^2.$ ...[from (1)]

$$\therefore\ p\frac{d^2p}{d\theta^2} + p^2 = \left(b^2 \cos^2\theta + a^2 \sin^2\theta\right) - \left(\frac{dp}{d\theta}\right)^2$$

$$= \left(b^2 \cos^2\theta + a^2 \sin^2\theta\right) - \frac{\left(b^2 - a^2\right)^2 \sin^2\theta \cos^2\theta}{p^2}$$

substituting for $\frac{dp}{d\theta}$ from (2)

$$= \frac{1}{p^2}\ [p^2\ (b^2 \cos^2\theta + a^2 \sin^2\theta) - (b^2 - a^2) \sin^2\theta \cos^2\theta]$$

$$= \frac{1}{p^2}\ [(a^2 \cos^2\theta + b^2 \sin^2\theta)(b^2 \cos^2\theta + a^2 \sin^2\theta) - (b^2 - a^2)^2 \sin^2\theta \cos^2\theta]$$

$$= \frac{1}{p^2}[a^2b^2 \cos^4\theta + a^2b^2 \sin^4\theta + 2a^2b^2 \sin^2\theta \cos^2\theta]$$

$$= \frac{a^2b^2}{p^2}\left(\cos^2\theta + \sin^2\theta\right)^2 = \frac{a^2b^2}{p^2}.$$

Thus, $$p\frac{d^2p}{d\theta^2} + p^2 = \frac{a^2b^2}{p^2}.$$

Dividing both sides by p, we have

$$\frac{d^2p}{d\theta^2} + p = \frac{a^2b^2}{p^3}.$$

4.3 LEIBNITZ'S THEOREM

This theorem helps us to find the nth differential coefficient of the product of two functions. The statement of the theorem is as follows:

If u and v are any two functions of x such that all their desired differential coefficients exist, then the nth differential coefficient of their product is given by

$$D^n(uv) = (D^nu).v + {}^nC_1\ D^{n-1}\ u\ Dv + {}^nC_2\ D^{n-2}\ u.D^2v + \ldots + {}^nC_r\ D^{n-r}\ u.\ D^rv + \ldots + u\ D^nv.$$

(Kanpur, 1998; Allahabad, 97; Delhi, 91, 90; Bihar, 95; Gorakhpur, 99, 96; Lucknow, 93, 91; Meerut, 93; Ranchi, 95)

Proof:

We shall prove the theorem by mathematical induction. By actual differentiation, we have

$$D(uv) = (Du).v + u.Dv. \quad \ldots(1)$$

From (1) we see that the theorem is true for n = 1.

Now assume that the theorem is true for a particular value of n. Then we have

$$D^n (uv) = (D^n u).v + {}^nC_1 D^{n-1}\, u\, Dv = {}^nC_2\, D^{n-2}\, u.D^{2v} + \ldots$$
$$+ {}^nC_r D^{n-r}\, u\, D^r v + {}^nC_{r+1}\, D^{n-r-1}\, u\, D^{r+1}\, v + \ldots + u\, D^n v. \quad \ldots(2)$$

Differentiating both sides of (2) with respect to x, we get

$$D^{n+1} (uv) = \{(D^{n+1}u).v + D^n u\, Dv\}$$
$$+ \{{}^nC_1\, D^n u.Dv + {}^nC_1\, D^{n-1}\, u.D^2 v\} + \{{}^nC_2 D^{n-1}\, u.D^2 v$$
$$+ {}^nC_2 D^{n-2} u.D^3 v\} + \ldots + \{{}^nC_r D^{n-r+1} u.D^r v + {}^nC_r D^{n-r} u.D^{r+1} v\}$$
$$+ \{{}^nC_{r+1}\, D^{n-r} u.D^{r+1}\, v + {}^nC_{r+1}\, D^{n-r-1} u.D^{r+2} v\} + \ldots$$
$$\ldots + \{Du\, D^n v + u\, D^{n+1}\, v\}.$$

Rearranging the terms, we get

$$D^{n+1} (uv) = (D^{n+1}\, u).v + (1 + {}^nC_1)\, (D^n u\, Dv)$$
$$+ ({}^nC_1 + {}^nC_2)\, D^{n-1}\, u.D^2 v$$
$$+ \ldots + ({}^nC_r + {}^nC_{r+1}\, (D^{n-r}\, u\, D^{r+1} v) + \ldots + u\, D^{n+1}\, v. \quad \ldots(3)$$

But from algebra, ${}^nC_r + {}^nC_{r+1} = {}^{n+1}C_{r+1}$. Therefore

$1 + {}^nC_1 = {}^nC_0 + {}^nC_1 = {}^{n+1}C_1$, ${}^nC_1 + {}^nC_2 = {}^{n+1}C_2$, and so on.

Hence equation (3) gives

$$D^{n+1}(uv) = (D^{n+1}\, u).v + {}^{n+1}C_1\, (D^n\, u).Dv$$
$$+ {}^{n+1}C_2 (D^{n-1}\, u).(D^2 v) + \ldots$$
$$+ {}^{n+1}C_{r+1} D^{n-r}\, u.D^{r+1} v + \ldots + u.D^{n+1}\, v. \quad \ldots(4)$$

From (4) we see that if the theorem is true for any value of n it is also true for the next value of n. But we have already seen that the theorem is true for n = 1. Hence it must be true for n = 2 and so for n = 3, and so on. Thus the theorem is true for all positive integral values of n.

Important Note: While applying Leibnitz's theorem if we observe that one of the two functions is such that all its differential coefficients after a certain stage become zero, then we should take that function as the second function.

Example 1:

If $y = x^2 \tan^{-1} x$, *find* y_n.

Solution:

Let $u = \tan^{-1} x$ and $v = x^2$. Then $y = uv$. Differentiating n times by Leibnitz's theorem, we get

$y_n = D^n(uv) = (D^n u).v + {}^nC_1(D^{n-1} u).Dv + {}^nC_2(D^{n-2}u).(D^2v) + \ldots$

Now $\quad D^n u = D^n (\tan^{-1} x) = (-1)^{n-1} (n-1)! \sin^n \theta \sin n\theta,$

where $\quad \theta = \tan^{-1} (1/x).$

Also $Dv = Dx^2 = 2x$, $D^2v = 2$ = constant and so the third and the higher derivatives of v all vanish.

Hence $y_n = [(-1)^{n-1} (n-1)! \sin^n \theta \sin n\theta].x^2$

$$+ {}^nC_1[(-1)^{n-2} (n-2)! \sin^{n-1} \theta \sin (n-1) \theta].2x$$

$$+ {}^nC_2[(-1)^{n-3} (n-3)! \sin^{n-2} \theta \sin (n-2)\theta].2$$

$$= (-1)^{n-1} (n-3)! [(n-1)(n-2) x^2 \sin^n \theta \sin n\theta$$

$$- 2nx (n-2) \sin^{n-1} \theta \sin (n-1) \theta$$

$$+ n (n-1) \sin^{n-2} \theta \sin (n-2) \theta], \text{ where } \theta = \tan^{-1} (1/x).$$

Example 2:

If $y = (\sin^{-1} x)\sqrt{(1-x^2)}$, *prove that*

$(1-x^2)y_{n+1} - (2n+1)xy_n - n^2y_{n-1} = 0.$

(Delhi, 1993; Gorakhpur, 91)

Solution:

We have $y\sqrt{(1-x^2)} = \sin^{-1} x.$

Differentiating both sides, we get

$$y_1\sqrt{(1-x^2)} + y.\frac{1}{2}\frac{-2x}{\sqrt{(1-x^2)}} = \frac{1}{\sqrt{(1-x^2)}}$$

or $\quad y_1(1-x^2) - yx - 1 = 0. \quad \ldots(1)$

Differentiating (1) n times by Leibnitz's theorem, we get

$$y_{n+1} (1-x^2) + {}^nC_1y_n (-2x) + {}^nC_2y_{n-1} (-2)$$

$$- y_n.x - {}^nC_1 y_{n-1}.1 = 0$$

or $(1-x^2)y_{n+1} - (2n+1)xy_n - n^2y_{n-1} = 0.$

Example 3:

If $y = (\sin^{-1} x)^2$, *prove that*

$$\left(1-x^2\right)\frac{d^2y}{dx^2} - x\frac{dy}{dx} - 2 = 0. \quad \textbf{(Kumaun, 1993)}$$

Differentiate the above equation n times w.r.t. 'x'.

(Delhi, 1990; Ranchi, 96; Garhwal, 93; Lucknow, 92; Meerut, 95)

Solution:

We have $y = (\sin^{-1} x)^2$.

Therefore $y_1 = 2(\sin^{-1} x)/\sqrt{(1-x^2)}$,

or $$y_1^2 = \frac{4(\sin^{-1} x)^2}{1-x^2} = \frac{4y}{1-x^2}, \qquad [\because (\sin^{-1} x)^2 = y]$$

or $y_1^2(1 - x^2) - 4y = 0$. Differentiating again, we get

$$2y_1y_2(1 - x^2) - 2xy_1^2 - 4y_1 = 0$$

or $$2y_1[y_2(1 - x^2) - xy_1 - 2] = 0.$$

Cancelling $2y_1$, since $2y_1 \neq 0$, we get

$$y_2(1 - x^2) - y_1x - 2 = 0$$

proving the first result

Defferentiating (1) n times by Leibnitz's theorem, we get

$$y_{n+2}(1 - x^2) + {}^nC_1y_{n+1}.(-2x) + {}^nC_2y_n.(-2) - y_{n+1}.x - {}^nC_1y_n.1 = 0$$

or $(1 - x^2)y_{n+2} - (2n + 1)xy_{n+1} - n^2y_n = 0$.

Example 4:

Find the nth differential coefficient of x^3 cos x. **(Meerut, 1991S)**

Solution:

Since the fourth and higher derivatives of x^3 are all zero, therefore for the sake of convenience we shall take x^3 as the second function. Applying Leibnitz's Theorem, we have

$$D^n[(\cos x)x^3] = (D^n \cos x).x^3 + {}^nC_1(D^{n-1} \cos x).(Dx^3)$$
$$+ {}^nC_2(D^{n-2} \cos x).(D^2x^3) + {}^nC_3 (D^{n-3} \cos x).(D^3x^3),$$

since all other terms become zero

$$= \left[\cos\left(x + \frac{1}{2}n\pi\right).x^3 + n\left[\cos\left\{x + \frac{1}{2}(n-1)\pi\right\}\right].3x^2\right.$$
$$+ \frac{n(n-1)}{1.2}\left[\cos\left\{x + \frac{1}{2}(n-2)\right\}\pi\right].6x$$
$$\left.+ \frac{n(n-1)(n-2)}{1.2.3}\left[\cos\left\{x + \frac{1}{2}(n-3)\pi\right\}\right].6\right.$$

$$= x^3 \cos\left(x + \frac{1}{2}n\pi\right) + 3nx^2 \sin\left(x + \frac{1}{2}n\pi\right)$$

$$- 3n\,(n-1)\,x\cos\left(x+\frac{1}{2}n\pi\right) - n\,(n-1)\,(n-2)\sin\left(x+\frac{1}{2}n\pi\right)$$

$$= [x^3 - 3n(n-1)x]\cos\left(x+\frac{1}{2}n\pi\right)$$

$$+\left[3x^2n - n(n-1)(n-2)\right]\sin\left(x+\frac{1}{2}n\pi\right).$$

Example 5:

If $y = x^2e^x \cos x$, find y_n.

Solution:

Let $u = e^x \cos x$ and $v = x^2$. Then $y = uv$. Differentiating n times by Leibnitz's Theorem, we get

$y_n = D^n(uv) = (D^nu)v + {}^nC_1\,(D^{n-1}u).(Dv) + {}^nC_2\,(D^{n-2}u).(D^2v) + \ldots$

Now $D^nu = D^n(e^x \cos x) = r^ne^x \cos(x + n\phi)$, where

$r = \sqrt{(1+1)} = \sqrt{2}$ and $\phi = \tan^{-1}(1/1) = \tan^{-1} 1 = \frac{\pi}{4}$.

Thus, $D^nu = 2^{n/2}e^x \cos\left(x+\frac{1}{4}n\pi\right)$. Also $Dv = Dx^2 = 2x$,

$D^2v = 2$ = constant and so the third and the higher derivatives of v all vanish. Hence

$$y_n = \left[2^{n/2}e^x\cos\left(x+\frac{1}{4}n\pi\right)\right].x^2 + n\left[2^{(n-1)/2}e^x\cos\left\{x+\frac{1}{4}(n-1)\pi\right\}\right].2x$$

$$+\frac{1}{2}n(n-1)\left[2^{(n-2)/2}e^x\cos\left\{x+\frac{1}{4}(n-2)\pi\right\}\right].2$$

$$= e^x\left[2^{n/2}x^2\cos\left(x+\frac{1}{4}n\pi\right) + 2nx\,2^{(n-1)/2}\cos\left(x+\frac{1}{4}(n-1)\pi\right)\right.$$

$$\left.+n(n-1)\,2^{(n-2)/2}\cos\left\{x+\frac{1}{4}(n-2)\pi\right\}\right].$$

Example 6:

Find the nth differential coefficient of $x^{n-1}\log x$.

(Agra, 1994; Kanpur, 95)

Solution:

Let $y = x^{n-1}\log x$ ...(1)

Then $y_1 = x^{n-1}(1/x) + (n-1).x^{n-2}.\log x$.

Multiplying both sides by x, we have

$$xy_1 = x^{n-1} + (n-1)\, x^{n-1} \log x$$

or $$xy_1 = x^{n-1} + (n-1)\, y. \quad ...(2)$$

$[\because$ from (1), $y = x^{n-1} \log x]$

Differentiating both sides of (2) (n – 1) times, we have

$$D^{n-1}(y_1 x) = D^{n-1} x^{n-1} + (n-1)\, D^{n-1} y$$

or $$(D^{n-1} y_1).x + {}^{n-1}C_1 (D^{n-2} y_1).1 = (n-1)! + (n-1)\, y_{n-1}$$

or $$xy_n + (n-1)y_{n-1} = (n-1)!\ (n-1)y_{n-1}$$

or $$xy_n = (n-1)! \text{ or } y_n = (n-1)!/x.$$

Example 7:

Find the nth derivative of $1/(1 + x + x^2 + x^3)$ and show that for $x = 0$ it is zero when n is of the form $4p + 2$ or $4p + 3$ and is n! or – n! according as n is of the form $4p + 1$.

Solution:

Let $$y = \frac{1}{1+x+x^2+x^3} = \frac{1}{(1+x)\left(x^2+1\right)}$$

$$= \frac{1}{2(x+1)} - \frac{x-1}{2\left(x^2+1\right)}, \text{ (on resolving into partial fractions)}$$

$$= \frac{1}{2(x+1)} - \frac{x-1}{2(x+i)(x-i)}$$

$$= \frac{1}{2(x+1)} - \frac{1}{2}\left[\frac{-i-1}{(-i-i)(x+i)} + \frac{i-1}{(i+i)(x-i)}\right]$$

$$= \frac{1}{2(x+1)} - \frac{1+i}{4i(x+i)} + \frac{1-i}{4i(x-i)}.$$

Differentiating n times w.r.t. 'x', we get

$$y_n = (-1)^n\, n! \left[\frac{1}{2(x+1)^{n+1}} - \frac{1+i}{4i(x+i)^{n+1}} + \frac{1-i}{4i(x-i)^{n+1}}\right]$$

Putting x = 0 in the expression for y_n, we get

$$(y_n)_0 = (-1)^n\, n! \left[\frac{1}{2} - \frac{1+i}{4.i^{n+2}} + \frac{1-i}{4(-1)^{n+1}\, i^{n+2}}\right].$$

Now we shall discuss the **different cases**.

(i) When $n = 4p + 2$, we have $(-1)^n = (-1)^{4p+2} = 1$,
$(-1)^{n+1} = (-1)^{4p+3} = -1$,
$i^{n+2} = i^{4p+4} = i^{4(p+1)} = 1^{p+1} = 1$.

$$\therefore (y_n)_0 = n!\left[\frac{1}{2}-\frac{1+i}{4}-\frac{1-i}{4}\right] = (n!).0 = 0.$$

(ii) When $n = 4p + 3$, we have $(-1)^n = -1$, $(-1)^{n+1} = 1$,
$i^{n+2} = i^{4p+5} = i^{4p+4}\, i = i$.

$$\therefore (y_n)_0 = -n!\left[\frac{1}{2}-\frac{1+i}{4i}+\frac{1-i}{4i}\right] = -(n!).0 = 0.$$

(iii) When $n = 4p$, we have $(-1)^n = 1$, $(-1)^{n+1} = -1$,
$i^{n+2} = i^{4p+2} = i^{4p}\, i^2 = -1$.

$$\therefore \quad (y_n)_0 = n!\left[\frac{1}{2}+\frac{1+i}{4}+\frac{1-i}{4}\right] = n!.$$

(iv) When $n = 4p + 1$, we have $(-1)^n = -1$, $(-1)^{n+1} = 1$,
$i^{n+2} = i^{4p+3} = i^{4p}\, i^3 = i^3 = -i$.

$$\therefore (y_n)_0 = -n!\left[\frac{1}{2}+\frac{1+i}{4i}-\frac{1-i}{4i}\right] = -n!.$$

Example 8:

Find the nth derivative of $x^2 e^x \cos^3 x$.

Solution:

Let $y = x^2 e^x \cos^3 x$. Since $\cos 3x = 4\cos^3 x - 3\cos x$,

therefore $\cos^3 x = \frac{1}{4}(3\cos x + \cos 3x)$.

Hence $y = \frac{1}{4}x^2 e^x (3\cos x + \cos 3x)$

$$= \frac{3}{4}\left(e^x \cos x\right)x^2 + \frac{1}{4}\left(e^x \cos 3x\right).x^2.$$

Now differentiating n times by Leibnitz's theorem, we get

$$y_n = \frac{3}{4}[\{r^n e^x \cos(x + n\phi)\}.x^2 + {}^nC_1\{r^{n-1} e^x \cos(x + n - 1\phi)\}.2x$$

$$+ {}^nC_2\{r^{n-2} e^x \cos(x + n - 2\phi)\}.2] + \frac{1}{4}\left[r_1^n e^x \cos\left(3x + n\phi_1\right).x^2\right.$$

$$+{}^nC_1\left\{r_1^{n-1}e^x\cos\left(3x+n-1\phi_1\right)\right\}.2x+{}^nC_2\left\{r_1^{n-2}e^x\cos\left(3x+n-2\phi_1\right)\right\}.2\Big],$$

where $r = \sqrt{2}$, $\phi = \tan^{-1} = \frac{\pi}{4}$; $r_1 = \sqrt{10}$, $\phi_1 = \tan^{-1} 3$.

EXERCISES

1. If $(x + y) = 1$ prove that

$$\frac{d^n}{dx^n}(x^n y^n) = n!\ [y^n - ({}^nC_1)^2 y^{n-1}x + ({}^nC_2)^2 y^{n-2} x^2 + \ldots + (-1)^x x^n].$$

2. If $y = \frac{b+cx}{a+2bx+cx_2}$, where $ac > b^2$ prove that $y_n = (-1)^n (n)!$

$$\left(\frac{C}{a+2bx+cx^2}\right)^{(n+1)/2}\cos\left[(n+1)\tan^{-1}\frac{\sqrt{(ac-b^2)}}{b+cx}\right].$$

3. Show that the 2nd differential co-efficient of $\cos^{2m}x.-(1)^n2 - 2^{m+1}$ $[(2m)^{2n}$ co cos 2m + $(2m-2)^{2n}$ $C_1.\cos(2m-2)^x + \ldots 2^{2n}(m-2)$ cos 2x] where C_r is the co-efficient of x^r in the expansion of $(1 + x)^{2m}$ in powers of x.

4. If $p^2 = a^2 \cos 2\theta + b^2 \sin^2 \theta$, prove that

$$p+\frac{d^2p}{d\theta^2}=\frac{a^2b^2}{p^2}.$$

5. If $y^{1/m} + y^{-1/m} = 2x$ prove that $(x^2 - 1)\, y_{n+2} + (2n + 1)\, xy_{n+1} + (x^2 - m^2)y_n = 0$ where y_n denotes the nth derivative of y.

6. If $u = \sin nx + \cos nx$ show that

$$u_r = n^r[1 + (-1)^r \sin 2m]^{1/3}$$

where u_r denotes the rth differential co-efficient of u with respect to x.

7. If $y = (x^2 - 1)^n$ prove that

$$(x^2 - 1)y_{n+2} + 2xy_{n+1} - n(x + 1)\, y_n = 0$$

Hence, if $P_n = \frac{d^n}{dx^n}(x^2 - 1)^n$, show that

$$\frac{d}{dx}\left[\left(1-x^2\right)\frac{dP_n}{dx}\right]+n(n+1)\,P_n=0.$$

5

EXPANSIONS OF FUNCTIONS

5.1 TAYLOR'S SERIES

Suppose f(x) possesses continuous derivatives of all orders in the interval [a, a + h]. Then for every positive integral value of n, we have

$$f(a + h) = f(a) + hf'(a) + \frac{h^2}{2!}f''(a) + ... + \frac{h^{n-1}}{(n-1)!}f(n-1)(a) + R_n, \quad ...(1)$$

where $R_n = \frac{h^n}{n!}$ f(n) (a + θh), (0 < θ < 1).

Suppose $R_n \to 0$,a s $n \to \infty$. Then taking limits of both sides of equation (1) when $n \to \infty$, we get

$$f(a + h) = f(a) = hf'(a) = \frac{h^2}{2!}f''(a) + ... + \frac{h^n}{n!}f^{(n)}(a) + ... \quad ...(2)$$

The series given in (2) is known as **Taylor's Infinite Series** for the expansion of f(a + h) as a **Power Series** in h.

A more useful form is obtain from the above on replacing ah by (x – a) Thus

$$f(x) = f(a) + (x - a) f'(a) + \frac{(x-a)^2}{2!} f''(a) + \frac{(x-a)^3}{3!} f'''(a)$$

$+ ... + \frac{(x-a)^n}{n!} f^n(a) + ...$ which is an expansion in powers of (x – a).

5.2 MACLAURIN'S SERIES

Let f(x) be a function of x which can be expanded in power of x and let the expansion be differentiable term by term any number of times.

$$f(x) = f(0) + xf' + \frac{x^2}{2!}f''(0) \ldots + \frac{x^{n-1}}{(n-1)!}f^{(n-1)}(0) + R_n, \qquad \ldots(1)$$

where $R_n = \frac{x^n}{n!}f^{(n)}(\theta x), (0 < \theta < 1)$.

Suppose $R_n \to 0$, as $n \to \infty$. Then taking limits of both sides of (1) when $n \to \infty$, we get

$$f(x) = f(0) + xf'(0) + \frac{x^2}{2!}f''(0) + \ldots + \frac{x^n}{n!}f^{(n)}(0) + \ldots \qquad \ldots(2)$$

The series given in (2) is known as **Maclaurin's Infinite Series** for the expansion of f(x) as a **Power Series** in x. Maclaurin's series is a particular case of Taylor's series. If in Taylor's series we put a = 0 and h = x we get Maclaurin's series.

Note: Maclaurin's Expansion of f(x) fails if any of the functions f(x), f'(x), f''(x), ... becomes infinite or discontinuous at any point of the interval [0, x] or if R_n does not tend to zero as $n \to \infty$.

Example 1:

Explained by Maclaurin's Theorem $e^x/(1 + e^x)$ as far as the term x^3.

Solution:

Let $y = \frac{e^x}{1+e^x} = \frac{1+e^x-1}{1+e^x} = 1 - \frac{1}{1+e^x}$.

Then $(y)_0 = \frac{e^0}{1+e^0} = \frac{1}{2}$,

$y_1 = 0 + \frac{e^x}{(1+e^x)^2} = \frac{e^x}{1+e^x}\,\frac{1}{1+e^x} = y(1-y) = y - y^2$ so that

$(y_1)_0 = \frac{1}{2} - \frac{1}{4} = \frac{1}{4}$,

$y_2 = y_1 - 2yy_1$ so that $(y_2)_0 = \frac{1}{4} - 2.\frac{1}{2}.\frac{1}{4} = 0$,

$y_3 = y_2 - 2y_1^2 - 2yy_2$ so that $(y_3)_0 = 0 - 2.(1/4)^2 - 0 = -1/8$ and so on.

Substituting these values in Maclaurin's theorem, we get

$$\frac{e^x}{1+e^x} = \frac{1}{2} + x.\frac{1}{4} + \frac{x^2}{2!}.0 + \frac{x^3}{3!}.(-1/8) + \ldots$$

$$= \frac{1}{2} + \frac{x}{4} - \frac{1}{48}x^3 + \ldots$$

Example 2:

Expand the following functions by Maclaurin's theorem:

(i) $e^{\sin x}$ **(Meerut, 1990S; Gorakhpur, 96; G.N.U., 91; K.U., 99)**

(ii) $e^{x \cos x}$

(iii) $e^x \log (1 + x)$

(iv) $\log (1 + \sin x)$ **(Gorakhpur, 1989; Kanpur, 96, Meerut, 92)**

Solution:

(i) Let $y = e^{\sin x}$. Then $(y)_0 = e^{\sin 0} = e^0 = 1$,

$y_1 = e^{\sin x} \cos x = y \cos x$ so that $(y_1)_0 = (y)_0 \cos 0 = 1 \times 1 = 1$,

$y_2 = y_1 \cos x - y \sin x$ so that $(y_2)_0 = 1 \times 1 - 1 \times 0 = 1$,

$y_3 = y_2 \cos x - y_1 \sin x - y_1 \sin x - y \cos x$

$= y_2 \cos x - 2y_1 \sin x - y \cos x$ so that $(y_3)_0 = 1 - 0 - 1 = 0$,

$y_4 = y_3 \cos x - 3y_2 \sin x - 3y_1 \cos x + y \sin x$ so that $(y_4)_0 = -3$,

and so on.

Now by Maclaurin's Theorem, we have

$$y = (y)_0 + x(y_1)_0 + \frac{x^2}{2!}(y_2)_0 + \frac{x^3}{3!} + \frac{x^4}{4!}(y_4)_0 + ...$$

$$\therefore\ e^{\sin x} = 1 + x \,.\, 1 + \frac{x^2}{2!}.1 + \frac{x^3}{3!}.0 + \frac{x^4}{4!}.(-3) + ...$$

$$= 1 + x + \frac{x^2}{2} - \frac{x^4}{8} + ...$$

(ii) Let $y_1 = e^{x \cos x}$. Then $(y)_0 = e^0 = 1$,

so that $(y_1)_0 = 1.(1 - 0) = 1$,

$y_2 = y_1(\cos x - x \sin x) + (-2 \sin x - x \cos x)$ giving

$(y_2)_0 = 1.(1 - 0) + 1.0 = 1$,

$y_3 = y_2 (\cos x - x \sin x) + y_1(-2 \sin x - x \cos x)$

$+ y_1(-2 \sin x - x \cos x) + y(-3 \cos x + x \sin x)$

$= y_2(\cos x - \sin x) + 2y_1 (-2 \sin x - x \cos x)$

$+ y(-3 \cos x + x \sin x)$

so that $(v_3)_0 = 1.(1 - 0) + 0 + 1.(-3) = -2$,

$y_4 = y_3 (\cos x - x \sin x) + 3y_2 (-2 \sin x - x \cos x)$

$+ 3y_1 (-3 \cos x + x \sin x) + y(4 \sin x + x \cos x)$

giving $(y_4)_0 = -2.(1 - 0) + 0 + 3.1.(-3 + 0) + 0 - 11$,

$y_5 = y_4 (\cos x - x \sin x) + 4y_3 (-2 \sin x - x \cos x)$

$+ 6y_2 (-3 \cos x + x \sin x) + 4y_1 (4 \sin x + x \cos x)$

$+ y (5 \cos x - x \sin x)$

so that $(y_5)_0 = -11 + 6.1.(-3) + 5 = -24$, and so on.

Substituting these values in Maclaurin's theorem, we get

$$e^{x\cos x} = 1 + x.1 + \frac{x^2}{2!}.1 + \frac{x^3}{3!}.(-2) + \frac{x^4}{4!}.(-11) + \frac{x^5}{5!}.(-24) + \ldots$$

$$= 1 + x + \frac{x^2}{2} - \frac{x^3}{3} - \frac{11x^4}{24} - \frac{x^5}{5} + \ldots$$

(iii) Let $y = e^x \log(1 + x)$. Then $(y)_0 = e^0 \log 1 = 1 \times 0 = 0$,

$y_1 = e^x \log(1 + x)^{-1} = y + e^x(1 + x)^{-1}$ so that $(y_1)_0 = 0 + 1 = 1$,

$y_2 = y_1 + e^x(1 + x)^{-1} - e^x(1 + x)^{-2}$ so that $(y_2)_0 = 1 + 1 - 1 = 1$,

$y_3 = y_2 + e^x(1 + x)^{-1} - e^x(1 + x)^{-2} - e^x(1 + x)^{-2} + 2e^x(1 + x)^{-3}$

$= y_2 + e^x(1 + x)^{-1} - 2e^x(1 + x)^{-2} + 2e^x(1 + x)^{-3}$

so that $(y_3)_0 = 1 + 1 - 2 + 2 = 2$,

$y_4 = y_3 + e^x(1 + x)^{-1} - 3e^x(1 + x)^{-2} + 6e^x(1 + x)^{-3}$

$- 6e^x(1 + x)^{-4}$ so that $(y_4)_0 = 2 + 1 - 3 + 6 - 6 = 0$,

$y_5 = y_4 + e^x(1 + x)^{-1} - 4e^x(1 + x)^{-2} + 12e^x(1 + x)^{-3}$

$- 24e^x(1 + x)^{-4} + 24e^x(1 + x)^{-5}$ so that

$(y_5)_0 = 0 + 1 - 4 + 12 - 24 + 24 = 9$, and so on.

Substituting these values in Maclaurin's Theorem, we get

$$e^x \log(1 + x) = 0 + x.1 + \frac{x^2}{2!}.1 + \frac{x^3}{3!}.2 + \frac{x^4}{4!}.0 + \frac{x^5}{5!}.9 + \ldots$$

$$= x + \frac{x^2}{2!} + \frac{2x^3}{3!} + \frac{9x^5}{5!} + \ldots$$

(iv) Let $y = \log(1 + \sin x)$. Then $(y)_0 = 0$,

$$y_1 = \frac{\cos x}{1 + \sin x} \text{ so that } (y_1)_0 = 1,$$

$$y_2 = \frac{-\sin x(1 + \sin x) - \cos^2 x}{(1 + \sin x)^2} = \frac{-(1 + \sin x)}{(1 + \sin x)^2}$$

$$= -\frac{1}{1 + \sin x} \text{ so that } (y_2)_0 = -1,$$

$$y_3 = \frac{\cos x}{(1 + \sin x)^2} = \frac{\cos x}{1 + \sin x}.\frac{1}{1 + \sin x} = -y_1 y_2$$

so that $(y_3)_0 = -1.(-1) = 1$,

$y_4 = -y_1 y_3 - y_2^2$ so that $(y_4)_0 = -1.1 - (-1)^2 = -1 - 1 = -2$,

$y_5 = -y_1y_4 - y_2y_3 - 2y_2y_3 = -y_1y_4 - 3y_2y_3$ so that

$(y_5)_0 = -1.(-2) - 3.(-1).1 = 2 + 3 = 5$, and so on.

Substituting these values in Maclaurin's theorem, we get

$$\log(1 + \sin x) = 0 + x.1 + (x^2/2\,!).(-1) + (x^3/3\,!).1 + (x^4/4\,!).(-2) + (x^5/5\,!).5 + \ldots$$

$$= x - \frac{x^2}{2} + \frac{x^3}{6} - \frac{x^4}{12} + \frac{x^5}{24} + \ldots$$

Example 3:

Use Maclaurin's formula to show that

$$e^x \sec x = 1 + x + \frac{2x^2}{2!} + \frac{4x^3}{3!} + \ldots$$

(Garhwal, 1993; Kanpur, 90; Gorakhpur, 93)

Solution:

Let $y = e^x \sec x$. Then $(y)_0 = 1$,

$y_1 = e^x \sec x\ e^x \sec x \tan x = y + y \tan x$ so that $(y_1)_0 = 1$,

$y_2 = y_1 + y_1 \tan x + y \sec^2 x$ so that $(y_2)_0 = 1 + 0 + 1 = 2$,

$y_3 = y_2 + y_2 \tan x + 2y_1 \sec^2 x + 2y \sec^2 x \tan x$ so that

$(y_3)_0 = 2 + 2 = 4$, and so on.

Substituting these values in Maclaurin's Theorem, we get

$$e^x \sec x = 1 + x + \frac{2x^2}{2!} + \frac{4x^3}{3!} + \ldots$$

Example 4:

Expand the following by Maclaurin's Theorem:

(i) a^x

(ii) $\tan x$ **(Meerut, 1991; Vikram, 97; Agra, 91; Kashmir, 94)**

(iii) $\log \sec x$ **(Rohilkhand, 1991; Meerut, 92; 96; 98; 95)**

(iv) $\log \cos x$

(v) $\sec x$

(vi) $\log(\sec x + \tan x)$.

Solution:

(i) Let $f(x) = a^x$. Then $f(0) = a^0 = 1$

$f'(x) = a^x \log a$ so that $f'(0) = \log a$

$f''(x) = a^x (\log a)^2$ so that $f''(0) = (\log a)^2$

$f'''(x) = a^x (\log a)^3$ so that $f'''(0) = (\log a)^3$, and so on.

In general, $f^n(x) = a^x(\log a)^n$ so that $f^n(0) = (\log a)^n$.

Now by Maclaurin's Theorem, we have

$$f(x) = f(0) + xf'(0) + \frac{x^2}{2!}f''(0) + \frac{x^3}{3!}f'''(0) + \ldots + \frac{x^n}{n!}f^n(0) + \ldots$$

$$\therefore a^x = 1 + x\log a + \frac{x^2}{2!}(\log a)^2 + \frac{x^3}{3!}(\log a)^3 + \ldots + \frac{x^n}{n!}(\log a)^n + \ldots$$

(ii) Let $y = \tan x$. Then $(y)_0 = \tan 0 = 0$,

$y_1 = \sec^2 x = 1 + \tan^2 x = 1 + y^2$ so that

$$(y_1)_0 = 1 + (y)_0^2 = 1 + 0 = 1,$$

$y_2 = 2yy_1$ so that $(y_2) = 2(y)_0 (y_1)_0 = 2 \times 0 \times 1 = 0$,

$y_3 = 2y_1y_1 + 2yy_2 = 2y_1^2 + 2yy_2$ so that $(y_3)_0 = 2 \times 1^2 + 0 = 2$,

$y_4 = 4y_1y_2 + 2y_1y_2 + 2yy_3 = 6y_1y_2 + 2yy_3$ so that

$$(y_4)_0 = 6 \times 1 \times 0 \times 2 \times 0 \times 2 = 0,$$

$y^5 = 6y_2^2 + 6y_1y_3 + 2y_1y_3 + 2yy_4 = 6y_2^2 + 8y_1y_3 + 2yy_4$ so that

$$(y_5)_0 = 0 + 8 \times 1 \times 2 + 0 = 16, \text{ and so on.}$$

Now by Maclaurin's Theorem, we have

$$y = (y)_0 + x(y_1)_0 + \frac{x^2}{2!}(y_2)_0 + \frac{x^3}{3!}(y_3)_0 + \frac{x^4}{4!}(y_4)_0 + \frac{x^5}{5!}(y_5)_0 + \ldots$$

$$\therefore \tan x + x.1 + \frac{x^2}{2!}.0 + \frac{x^3}{3!}.0 + \frac{x^3}{3!}.2 + \frac{x^4}{4!}.0 + \frac{x^5}{5!}.16 + \ldots$$

$$= x + \frac{x^3}{3} + \frac{2}{15}x^5 + \ldots$$

(iii) Let $y = \log \sec x$. Then $(y)_0 = \log \sec 0 = \log 1 = 0$,

$y_1 = \frac{1}{\sec x}.\sec x \tan x + \tan x$ so that $(y_2)_0 = 0$,

$y_2 = \sec^2 x = 1 + \tan^2 x = 1 + y_1^2$ so that $(y_2)_0 = 1 + (y_1)_0^2 = 1$,

$y_3 = 2y_1y_2$ so that $(y_3)_0 = 2(y_1)_0 (y_2)_0 = 0$,

$y_4 = 2y_2^2 + 2y_1y_3$ so that $(y_4)_0 = 2 \times 1^2 + 0 = 2$,

$y_5 = 4y_2y_3 + 2y_2y_3 + 2y_1y_4 = 6y_2y_3 + 2y_1y_4$ so that

$$(y_5)_0 = 6 \times 1 \times 0 + 2 \times 0 \times 2 = 0,$$

$y_6 = 6y_3^2 + 6y_2y_4 + 2y_2y_4 + 2y_1y_5 + 6y_3^2 + 8y_2y_4 + 2y_1y_5$ so that

$(y_6)_0 = 0 + 8 \times 1 \times 2 + 0 = 16$, and so on.

Now by Maclaurin's Theorem, we have

$$y = (y)_0 + x(y_1)_0 + \frac{x^2}{2!}(y_2)_0 + \frac{x^3}{3!}(y_3)_0 + \frac{x^4}{4!}(y_4)_0 + ...$$

$$\therefore \log \sec x = 0 + x.0 + \frac{x^2}{2!}.1 + \frac{x^3}{3!}.0 + \frac{x^4}{4!}.2 + \frac{x^5}{5!}.0 + \frac{x^5}{6!}.16 + ...$$

$$= \frac{x^2}{2} + \frac{x^4}{12} + \frac{x^6}{45} + ...,$$

(iv) Let $y = \log \cos x$. Then $(y)_0 = 0$,

$y_1 = \frac{-\sin x}{\cos x} = -\tan x$ so that $(y_1)_0 = 0$,

$y_2 = -\text{sex}^2 x = -(1 + \tan^2 x) = -(1 + y_1^2) = -1 - y_1^2$

so that $(y_2)_0 = -1 - 0 = -1$,

$y_3 = -2y_1y_2$ so that $(y_3)_0 = 0$,

$y_4 = -2y_2^2 - 2y_1y_3$ so that $(y_4)_0 = -2(-1)^2 - 0 = -2$,

$(y_5)_0 = 0$, $(y_6)_0 = -16$, and so on.

[For calculation of $(y_5)_0$ and $(y_6)_0$ proceed as in part (iii)]

Now substituting these values in Maclaurin's Theorem, we get

$\log \cos x = 0 + x.0 +$

$$\frac{x^2}{2!}.(-1) + \frac{x^3}{3!}.0 + \frac{x^4}{4!}.(-2) + \frac{x^5}{5!}.0 + \left(\frac{x^6}{6!}\right).(-16) + ...$$

$$= -\left(\frac{x^2}{2}\right) - \left(\frac{x^4}{12}\right) - \left(\frac{x^6}{45}\right) + ...$$

(v) Let $y = \sec x$. Then $(y)_0 = \sec 0 = 1$,

$y_1 = \sec x \tan x$ so that $(y_1)_0 = 1 \times 0 = 0$,

$y_2 = \sec x \sec^2 x + \sec y \tan x \tan x = \sec^3 x + \sec x \tan^2 x$

$= \sec^3 x + \sec x (\sec^2 x - 1) = 2 \sec^3 x - \sec x = 2y^3 - y$

so that $(y_2)_0 = 2 \times 1^3 - 1 = 1$,

$y_3 = 6y^2y_1 - y_1$ so that $(y_3)_0 = 0 - 0 = 0$,

$y_4 = 6y^2y_2 + 12yy_1^2 - y_2$ so that $(y_4)_0 = 6 \times 1^2 \times 1 + 0 - 1 = 5$, and so on.

Now by Maclaurin's Theorem, we have

$$y = (y)_0 + x(y_1)_0 + \frac{x^2}{2!}(y_2)_0 + \frac{x^3}{3!}(y_3)_0 + \frac{x^4}{4!}(y_4)_0 + \ldots$$

$$\therefore \sec x = 1 + x.0 + \frac{x^2}{2!}.1 + \frac{x^3}{3!}.0 + \frac{x^4}{4!}.5 + \ldots$$

$$= 1 + \left(\frac{x^2}{2!}\right) + \left(\frac{5x^4}{4!}\right) + \ldots$$

(vi) Let y = log (sec x + tan x). Then $(y)_0 = \log(1 + 0) = 0$,

$$y_1 = \frac{\sec x \tan x + \sec^2 x}{\sec x + \tan x} = \sec x \text{ so that } (y_1)_0 = \sec 0 = 1,$$

$y_2 = \sec x \tan x$ so that $(y_2)_0 = 1 \times 0 = 0$,

$y_3 = \sec x \sec^2 x + \sec x \tan^2 x = \sec^3 x + \sec x (\sec^2 x - 1)$

$= 2\sec^3 x - \sec x = 2y_1^3 - y_1$ so that $(y_3)_0 = 2 \times 1^3 - 1 = 1$,

$y_4 = 6y_1^2 y_2 - y_2$ so that $(y_4)_0 = 0$,

$y_5 = 6y_1^2 y_3 + 12y_1 y_2^2 - y_3$ so that $(y_5)_0 = 6 \times 1^2 \times 1 + 0 - 1 = 5$, and so on.

Substituting these values in Maclaurin's theorem, we get

$$\log(\sec x + \tan x) = 0 + x.1 + \frac{x^2}{2!}.0 + \frac{x^3}{3!}.1 + \frac{x^4}{4!}.0 + \frac{x^5}{5!}.5 + \ldots$$

$$= x + \left(\frac{x^3}{6}\right) + \left(\frac{x^5}{24}\right) + \ldots$$

Example 5:

Apply Maclaurin's Theorem to find the expansion in ascending powers of x of $\log_e(1 + e^x)$ to the term containing x^4.

(Gorakhpur, 1991; Kanpur, 98; Ranchi, 94)

Solution:

Let $y = \log_e(1 + e^x)$. Then $(y)_0 = \log_e(1 + e^0) = \log_e 2$,

$$y_1 = \frac{e^x}{1+e^x} = \frac{(1+e^x) - 1}{1+e^x} = 1 - \frac{1}{1+e^x} \text{ so that } (y_1)_0 = 1 - \frac{1}{2} = \frac{1}{2},$$

$$y_2 = 0 + \frac{e^x}{(1+e^x)^2} = \frac{e^x}{(1+e^x)} \cdot \frac{1}{1+e^x} = y_1(1 - y_1) = y_1 - y_1^2$$

$$\text{so that } (y_2)_0 = \frac{1}{2} - \left(\frac{1}{2}\right)^2 = \frac{1}{4},$$

$y_3 = y_2 - 2y_1y_2$ so that $(y_3)_0 = \frac{1}{4} - 2.\frac{1}{2}.\frac{1}{4} = 0,$

$y_4 = y_3 - 2y_2^2 - 2y_1y_3$ so that $(y_4)_0 = 0 - 2.\left(\frac{1}{4}\right)^2 - 0 = -\frac{1}{8}$, and so on.

Substituting these values in Maclaurin's Expansion.

$$y = (y)_0 + x(y_1)_0 + \frac{x^2}{2!}(y_2)_0 + \frac{x^3}{3!}(y_3)_0 + \frac{x^4}{4!}(y_4)_0 + ...$$

$\therefore$ log $(2 + e^x)$

$$= \log 2 + x.\frac{1}{2} + \frac{x^2}{2!}.\frac{1}{4} + \frac{x^3}{3!}.0 + \frac{x^4}{4!}.\left(-\frac{1}{8}\right) + ...$$

$$= \log 2 + \frac{x}{2} + \frac{x^2}{8} - \frac{x^4}{192} + ...$$

Example 6:

Expand $e^{a \sin^{-1} x}$ *by Maclaurin's theorem and find the general term. Hence show that*

$$e^{\theta} = 1 + \sin\theta + \frac{1}{2!}\sin^2\theta + \frac{2}{3!}\sin^3\theta + ...$$

(Meerut, 1992, 94; Allahabad, 97; Rohilkhand, 98; Agra, 97; Magadh, 97; Ranchi, 95; Indore, 93; Lucknow, 99; Kanpur, 97; 99)

Solution:

Let $y = e^{a \sin^{-1} x}$. We have

$y(0) = 1$, $y_1(0) = a$, $y_2(0) = a^2$,

and $y_{n+2}(0) = (n^2 + a^2)y_n(0)$. ...(1)

Putting n = 1, 2, 3, 4, ... in (1), we get

$y_3(0) = (1^2 + a^2)\,y_1(0) = (1^2 + a^2)\,a$, $y_4(0) = (2^2 + a^2)\,y_2(0)$

$= (2^2 + a^2)\,a^2$, $y_5(0) = (3^2 + a^2)\,y_3(0) = (3^2 + a^2)(1^2 + a^2)a$

$y_6(0) = (4^2 + a^2)\,y_4(0) = (4^2 + a^2)(2^2 + a^2)\,a^2$, etc.

In general,

$$y_n(0) = \begin{cases} a\left(1^2 + a^2\right)\left(3^2 + a^2\right)...\left[(n-2)^2 + a^2\right] \text{ if n is odd} \\ a^2\left(2^2 + a^2\right)\left(4^2 + a^2\right)...\left[(n-2)^2 + a^2\right] \text{ if n is even.} \end{cases}$$

Substituting these values in Maclaurin's Expansion

$$y = y(0) + xy_1(0) + \frac{x^2}{2!}y_2(0) + ... + \frac{x^n}{n!}y_n(0) + ...,$$

we get

$$e^a \sin^{-1} x + 1 + ax + \frac{a^2}{2!}x^2 + \frac{a(1^2 + a^2)}{3!}x^3 + \frac{a^2(2^2 + a^2)}{4!}x^4 + ...$$

...(2)

The general term is $(x^n/n!)\ y_n(0)$, where $y_n(0)$ is as given above.

Now putting $x = \sin\theta$ and $a = 1$ in (2), we get

$$e^\theta = 1 + \sin\theta + \frac{1}{2!}\sin^2\theta + \frac{2}{3!}\sin^3\theta + ...$$

Example 7:

Applying Maclaurin's theorem to obtain terms into x^4 in the expansion of $\log(1 + \sin^2 x)$. **(Rohilkhand, 1979)**

Solution:

Let $y = \log(1 + \sin^2 x)$. Then $(y)_0 = 0$.

Now $e^y = 1 + \sin^2 x$.

Differentiating, we get

$$e^y y_1 = 2 \sin x \cos x = \sin 2x \qquad ...(1)$$

Putting $x = 0$ in (1), we get

$$e^0 (y_1)_0 = 0 \text{ or } (y_1)_0 = 0.$$

Differentiating (1), we get

$$e^y(y_1^2 + y_2) = 2\cos 2x \qquad ...(2)$$

Putting $x = 0$ in (2), we get

$$(y_2)_0 = 2.$$

Differentiating (2), we get

$$e^y\left[(y_1^2 + y_2)y_1 + 2y_1y_2 + y_3\right] = -4\sin 2x$$

or $$e^y(y_1^3 + 2y_1y_2 + y_3) = -4\sin 2x. \qquad ...(3)$$

Putting $x = 0$ in (3), we get

$$(y_3)_0 = 0.$$

Differentiating (3), we get

$$e^y y_1\left(y_1^3+3y_1y_2+y_3\right)+e^y\left(3y_1^2y_2+3y_2^2+3y_1y_3+y_4\right)=-8\cos 2x$$

or $e^y\left(y_1^4+6y_1^2y_2+4y_1y_3+3y_2^2+y_4\right)=-8\cos 2x.$ ...(4)

Putting x = 0 in (4), we get

$$3.2^2+(y_4)_0=-8 \text{ or } (y_4)_0=-20.$$

Now substituting these values in Maclaurin's theorem, we get

$$\log\left(1+\sin^2 x\right)=0+x.0+\frac{x^2}{2!}.2+\frac{x^3}{3!}.0+\frac{x^4}{4!}(-20)+\ldots$$

$$=x^2-\frac{5}{6}x^4+\ldots$$

Example 8:

Expand log $\left\{x+\sqrt{\left(1+x^2\right)}\right\}$ *in ascending powers of x and find the general term.* **(Meerut, 1991; K.U., 95; Bihar, 94)**

Solution:

Let y log = $\left\{x+\sqrt{\left(1-x^2\right)}\right\}$, then

$$y_1+\frac{1}{x+\sqrt{\left(1+x^2\right)}}.\left\{1+\frac{2x}{2\sqrt{\left(x^2+1\right)}}\right\}=\frac{1}{\sqrt{\left(x^2+1\right)}}$$

Therefore $y_1^2\left(x^2+1\right)-1=0$.

Differentiating again, we get

$$(x^2+1)\,2y_1\,y_2+2xy_1^2=0$$

or $\quad (x^2+1)y_2+xy_1=0$, since $2y_1\neq 0$.

Now differentiating n times by Leibnitz's theorem, we get

$$(x^2+1)y_{n+2}+(2n+1)xy_{n+1}+n^2y_n=0.$$

Putting x = 0 in the above relations, we get

$(y)_0=0$, $(y_1)_0=1$, $(y_2)_0=0$, and

$(y_{n+2})_0=-n^2(y_n)_0$. ...(1)

Now putting n = 1, 3, 5, ... in (1), we get

$(y_3)_0=-1^2(y_1)_0=-1^2$, $(y_5)_0=(-3^2)(y_3)_0=(-3^2)(-1^2)=3^2.1^2$,

$(y_7)_0=(-5^2)(y_5)_0=(-5^2)(-3^2)(-1^2)=-5^2.3^2.1^2$, etc.

In general, if n is odd, we have

$(y_n)_0 = \{-(n-2)^2\}\{-(n-4)^2 \ldots (-5^2)(-3^2)(-1^2)$

$= (-1)^{(n-1)/2}(n-2)^2(n-4)^2 \ldots 5^2.3^2.1^2.$ ...(2)

Again, putting n = 2, 4, 6 ... in (1), we get

$(y_4)_0 = -2^2\ (y_2)_0 = 0,\ (y_6)_0 = -4^2\ (y_4)_0 = 0$, etc.

Thus, if n is even, we have $(y_n)_0 = 0$.

Now by Maclaurin's theorem, we have

$$\log\left[x+\sqrt{(1+x^2)}\right] = (y)_0 + x(y_1)_0 + \frac{x^2}{2!}(y_2)_0 + \frac{x^3}{3!}(y_3)_0 + \ldots$$

$$= 0 + x.1 + \frac{x^2}{2!}.0 + \frac{x^3}{3!}(-1^2) + \frac{x^4}{4!}.0 + \frac{x^5}{5!}(3^2.1^2)$$

$$= x - \frac{x^3}{3!}1^2 + \frac{x^5}{5!}(3^2.1^2) - \frac{x^7}{7!}(5^2.3^2.1^2) + \ldots$$

The general term = $(x^n/n!)\ (y_n)_0$, where $(y_n)_0$ is given by (2) when n is odd and $(y_n)_0 = 0$, when n is even.

Putting 2n – 1 in place of n in (2), we find that

$(x_{2n-1})_0 = (-1)^{n-1}\ (2n-3)^2 \ldots 5^2.3^2.1^2.$

$$\log\left[x+\sqrt{(1+x^2)}\right]$$

$$= x.1^2.\frac{x^3}{3!} + 1^2.3^2\frac{x^5}{5!} - 1^2.3^2.5^2.\frac{x^7}{7!} + \ldots$$

$$+(-1)^{(n-1)}1^2.3^2.5^2.(2n-3)^2.\frac{x^{2n-1}}{(2n-1)^2}$$

Example 10:

Find the first four terms in the expansion of log (1 + tan x) in powers of x. **(Magadh, 1996)**

Solution:

Let y = log (1 + tan x). Then $(y)_0$ = log (1 + tan 0) = 0.

Now $e^y = 1 + \tan x$. Differentiating both sides w.r.t. x, we get

$e^y\ y_1 = \sec^2 x.$ ...(1)

Putting x = 0 on both sides of (1), we get

$e^0\ (y_1)_0 = 1$ or $(y_1)_0 = 1.$

Differentiating (1), we get

$e^y y_1^2 + e^y y_2 = 2 \sec^2 x \tan x$

or $$e^y\left(y_1^2+y_2\right)=2\sec^2 x \tan x. \qquad ...(2)$$

Putting x = 0 in (2), we get

$$1+(y_2)_0=0 \text{ or } (y_2)_0=1.$$

Differentiating (2), we get

$$e^y y_1\left(y_1^2+y_2\right)+e^y\left(2y_1y_2+y_3\right)=2\sec^4 x+4\sec^2 x\tan^2 x$$

or $$e^y\left(y_1^3+3y_1y_2+y_3\right)=2\sec^4 x+4\sec^2 x\tan^2 x. \qquad ...(3)$$

Putting x = 0 in (3), we get

$$1+3.1.(-1)+(y_3)_0=2 \text{ or } (y_3)_0=4.$$

Differentiating (3), we get

$$e^y y_1\left(y_1^3+3y_1y_2+y_3\right)+e^y\left(3y_1^2y_2+3y_2^2+3y_1y_3+y_4\right)$$
$$=8\sec^4 x\tan x+8\sec^2 x\tan^3 x+8\sec^4 x\tan x$$

or $$e^y\left(y_1^4+6y_1^2y_2+4y_1y_3+3y_2^2+y_4\right)$$
$$=16\sec^4 x\tan x+8\sec^2 x\tan^3 x. \qquad ...(4)$$

Putting x = 0 in (4), we get

$$1+6.1.(-1)+4.1.4+3.(-1)^2+(y_4)_0=0$$

or $$(y_4)_0+14=0 \text{ or } (y_4)_0=-14.$$

Now substituting these values in Maclaurin's theorem, we get

$$\log(1+\tan x)=0+x.1+\frac{x^2}{2!}.(-1)+\frac{x^3}{3!}.4+\frac{x^4}{4!}.(-14)+...$$
$$=x-\frac{1}{2}x^2+\frac{2}{3}x^3-\frac{7}{12}x^4+...$$

Example 10:

Show that

(i) $$e^x os\, x=1+x-\frac{2x^3}{3!}-\frac{2^2x^4}{4!}-\frac{2^2x^5}{5!}+\frac{2^3x^7}{7!}+...$$
$$+2^{n/2}\cos\frac{n\pi}{4}.\frac{x^n}{n!}+...$$

(Meerut, 1994, 96; Kanpur, 95; Rohilkhand, 98, 99; Jhansi, 98)

(ii) $e^x \sin x = x + x^2 - \frac{2}{3!}x^3 - \frac{2^2}{5!}x^5 - ... + \sin\left(\frac{1}{4}n\pi\right)\frac{2^{n/2}}{n!}x^n + ...$

(Meerut, 1993, 95 96 BP; Avadh, 97; Lucknow, 92; Gorakhpur, 91; Agra, 90, 94, 99)

Solution:

(i) Let $y = e^x \cos x$. Then $(y)_0 = e^0 \cos 0 = 1$,

$y_1 = e^x \cos x - e^x \sin x = e^x (\cos x - \sin x)$

so that $(y_1)_0 = 1\ (1 - 0) = 1$,

$y_2 = e^x (\cos x - \sin x) + e^x (-\sin x - \cos x) = -2e^x \sin x$

so that $(y_2)_0 = 0$,

$y_3 = -2e^x \sin x - 2e^x \cos x = -2e^x (\sin x + \cos x)$

so that $(y_3)_0 = -2$,

$y_4 = -2e^x(\sin x + \cos x) - 2e^x (\cos x - \sin x)$

$= -4e^x \cos x = -2^2 y$ so that $(y_4)_0 = -2^2$,

$y_5 = -2^2 y_1$ so that $(y_5)_0 = -2^2$,

$y_6 = -2^2 y_2$ so that $(y_6)_0 = 0$,

$y_7 = -2^2 y_3$ so that $(y_7)_0 = 2^3$, and so on.

In general

$y_n = (1 + 1)^{n/2} \cos (x + n \tan^{-1} 1) = 2^{n/2} \cos (x + n\pi/4)$

so that $(y_n)_0 = 2^{n/2} \cos\left(\frac{1}{4}n\pi\right)$.

Now by Maclaurin's theorem, we have

$$y = (y)_0 + x(y_1)_0 + \frac{x^2}{2!}(y_2)_0 + ... + \frac{x^n}{n!}(y_n)_0 + ...$$

$$= 1 + x.1 + \frac{x^2}{2!}.0 + \frac{x^3}{3!}.(-2) + \frac{x^4}{4!}.(-2^2) + \frac{x^5}{5!}.(-2^2)$$

$$+ \frac{x^6}{6!}.0 + \frac{x^7}{7!}.2^3 + ... + \frac{x^n}{n!}2^{n/2}\cos\left(\frac{1}{4}n\pi\right) + ...$$

$$= 1 + x - \frac{2x^3}{3!} - \frac{2^2 x^4}{4!} - \frac{2^2 x^5}{5!} + \frac{2^3 x^7}{7!} + ... + 2^{n/2}\cos\left(\frac{1}{4}n\pi\right)\frac{x^n}{n!} + ...$$

(ii) Proceed as in part (i).

Example 11:

Apply Maclaurin's theorem to prove that

(i) $$e^{ax}\sin bx = bx + abx^2 + \frac{3a^2b-b^3}{3!}x^3 + \ldots + \frac{(a^2+b^2)^{n/2}}{n!}x^n \sin\left(n\tan^{-1}(b/a)\right)\Big\} + \ldots$$ **(Meerut, 1984)**

(ii) $$e^{ax}\cos bx = 1 + ax + \frac{a^2-b^2}{2}x^2 + \frac{a(a^2-3b^2)}{3!}x^3 + \ldots + \frac{(a^2+b^2)^{n/2}}{n!}x^n\cos\left\{n\tan^{-1}(b/a)\right\} + \ldots$$

(Meerut, 1998; 94; 93S, 97; Gorakhpur, 99; Allahabad, 94; Kurukshetra, 92; Magadh, 93; Delhi, 98; G.N.U., 95)

Hence deduce that

$$e^{x\cos\alpha}\cos(x\sin\alpha) = 1 + x\cos a + \frac{x^2}{2!}\cos 2\alpha + \frac{x^3}{3!}\cos 3\alpha + \ldots$$

(Kanpur 1992)

Solution:

(i) Let $y = e^{ax}\sin bx$. Then $(y)_0 = e^0 \sin 0 = 0$,

$y_1 = ae^{ax}\sin bx + be^{ax}\cos bx = ay + be^{ax}\cos bx$

so that $(y_1)_0 = b$,

$y_2 = ay_1 + abe^{ax}\cos bx - b^2e^{ax}\sin bx$

$= ay_1 - b^2y + abe^{ax}\cos bx$ so that $(y_2)_0 = ab - 0 + ab = 2ab$,

$y_3 = ay_2 - b^2y_1 + a^2be^{ax}\cos bx - ab^2e^{ax}\sin bx$

$= ay_2 - b^2y_1 - ab^2y + a^2be^{ax}\cos bx$

so that $(y_3)_0 = 2a^2b - b^3 + a^2b = 3a^2b - b^3$, and so on.

In general

$$y_n = (a^2+b^2)^{n/2}\sin\{bx + n\tan^{-1}(b/a)\}$$

so that $(y_n)_0 = (a^2+b^2)^{n/2}\sin\{n\tan^{-1}(b/a)\}$.

Now by Maclaurin's theorem, we have

$$y = (y)_0 + \frac{x}{1!}.(y_1)_0 + \frac{x^2}{2!}(y_2)_0 + \frac{x^3}{3!}(y_3)_0 + \ldots + \frac{x^n}{n!}(y_n)_0 + \ldots$$

$$= 0 + \frac{x}{1!}.b + \frac{x^2}{2!}.(2ab) + \frac{x^3}{3!}\left(3a^2b - b^3\right) + \ldots$$

$$+ \frac{x^n}{n!}\left(a^2 + b^2\right)^{n/2} \sin\left\{n \tan^{-1}(b/a)\right\} + \ldots$$

$$= bx + abx^2 + \frac{3a^2b - b^3}{3!}x^3 + \ldots$$

$$+ \frac{\left(a^2 + b^2\right)^{n/2}}{n!} x^n \sin\left\{n \tan^{-1}(b/a)\right\} + \ldots$$

(ii) Let $y = e^{ax} \cos bx$. Then $(y)_0 = e^0 \cos 0 = 1$,

$y_1 = ae^{ax} \cos bx - be^{ax} \sin bx = ay - be^{ax} \sin bx$ so that $(y_1)_0 = a$,

$y_2 = ay_1 - abe^{ax} \sin bx - b^2e^{ax} \cos bx = ay_1 - b^2y - abe^{ax} \sin bx$

so that $(y_2)_0 = a^2 - b^2$,

$y_3 = ay_2 - b^2y_1 - a^2be^{ax} \sin bx - ab^2e^{ax} \cos bx$

$= ay_2 - b^2y_1 - ab^2y - a^2be^{ax} \sin bx$

so that $(y_3)_0 = a(a^2 - b^2) - b^2a - ab^2 = a(a^2 - 3b^2)$, and so on.

In general, $y_n = (a^2 + b^2)^{n/2} \cos \{bx + n \tan^{-1} (b/a)\}$ so that

$(y_n)_0 = (a^2 + b^2)^{n/2} \cos \{n \tan^{-1} (b/a)\}$.

Substituting these values in Maclaurin's theorem, we get

$$e^{ax} \cos bx = 1 + ax + \frac{a^2 - b^2}{2!}x^2 + \frac{a\left(a^2 - 3b^2\right)}{3!}x^3 + \ldots$$

$$+ \frac{\left(a^2 + b^2\right)^{n/2}}{n!} x^n \cos\left\{n \tan^{-1}(b/a)\right\} + \ldots$$

Deduction: Putting $a = \cos \alpha$ and $b = \sin \alpha$, we get

$(y_n)_0 = (\cos^2 \alpha + \sin^2 \alpha)^{n/2} \cos (n \tan^{-1} \tan \alpha) = \cos n\alpha$ so that

$(y_1)_0 = \cos \alpha$, $(y_2)_0 = \cos 2\alpha$, $(y_3)_0 = \cos 3\alpha$, etc.

$\therefore e^{x \cos \alpha} \cos (x \sin \alpha)$

$= 1 + x \cos \alpha + (x^2/2\,!) \cos 2\alpha + (x^3/3\,!) \cos 3\alpha + \ldots$

Example 12:

Expand log sin x in powers of (x – a).

(Vikram, 1998; Gurunanak, 94)

Solution:

Let $f(x) = \log \sin x$. We can write $f(x) = f[a + (x - a)]$. Expanding $f[a + (x - a)]$ by Taylor's theorem in powers of $(x - a)$, we get

$$f(x) = f(a) + (x - a) f'(a) + (1/2 !) (x - a)^2 f''(a) + (1/3 !) (x - a)^3 f'''(a) + \ldots \qquad \ldots(1)$$

Now $f(x) = \log \sin x$. Therefore $f(a) = \log \sin a$,

$f'(a) = (1/\sin x).\cos x = \cot x$, giving $f'(c) = \cot a$

$f''(x) = - \text{cosec}^2 x$ so that $f''(a) = - \text{cosec}^2 a$,

$f'''(x) = 2 \text{ cosec}^2 x \cot x$ so that $f'''(a) = 2 \text{ cosec}^2 a \cot a$, and so on.

Substituting these values in (1), we get

$$\log \sin x = \log \sin a + (x - a) \cot a - \frac{(x-a)^2}{2!} \text{cosec}^2 a + \frac{(x-a)^3}{3!} 2 \text{cosec}^2 a \cot a + \ldots$$

Example 13:

Expand $\tan^{-1} x$ by Maclaurin's theorem. Write also the general term.

(Delhi, 1991; Bundelkhand, 92; Kashmir, 97)

Solution:

Let $y = \tan^{-1} x$.

$(y)_0 = 0, (y_1)_0 = 1, (y_2)_0 = 0,$

and $(y_{n+2}) = - \{(n + 1) n\} (y_n)_0$. ...(1)

Putting $n = 1, 2, 3, \ldots$ in (1), we get

$(y_3)_0 = - (2.1) (y_1)_0 = - 2!, (y_4)_0 = - (3.2) (y_2)_0 = 0,$

$(y_5)_0 = - (4.3) (y_3)_0 = - (4.3) . (- 2 !) = 4 !$, etc.

Now by Maclaurin's theorem, we have

$$y = (y)_0 + x(y_1)_0 + \frac{x^2}{2!}(y_2)_0 + \frac{x^3}{3!}(y_3)_0 + \ldots$$

$$\therefore \tan^{-1} x = 0 + x.1 + \frac{x^2}{2!}.0 + \frac{x^3}{3!}.(-2!) + \frac{x^4}{4!}.0 + \frac{x^5}{5!}.(4!) + \ldots$$

$$= x - \left(\frac{x^3}{3}\right) + \left(\frac{x^5}{5}\right) - \ldots$$

The general term in this expansion is $(x^n/n !) (y_n)_0$.

So we need the value of $(y_n)_0$. Putting $(n - 2)$ in place of n in (1), we get

$$(y_n)_0 = - \{(n - 1)\} (n - 2)\}(y_{n-2})_0$$

$= [-\{(n-1)(n-2)\}][-\{(n-3)(n-4)](y_{n-4})_0.$

Now there arise two cases:

Case I: When n is even, we have

$(y_n)_0 = [-\{(n-1)(n-2)\}][-\{(n-3)(n-4)] \ldots [-(3)(2)](y_2)_0$

$= 0$, since $(y_2)_0 = 0$.

Case II: When n is odd, we have

$(y_n)_0 = [-\{(n-1)(n-2)\}][-\{(n-3)(n-4)\}] \ldots$

$[-(4)(3)][-(2)(1)](y_1)_0$

$= (-1)^{(n-1)/2}(n-1)!$, since $(y_1)_0 = 1$.

Thus, in the expansion of $\tan^{-1} x$, the coefficient of x^n is 0 if n is even and is $\dfrac{(-1)^{(n-1)}(n-1)!}{n!}$ i.e., $\dfrac{(-1)^{(n-1)/2}}{n}$ if n is odd.

$$\therefore \tan^{-1} x = x - \frac{x^3}{3} + \frac{x^5}{5} - \frac{x^7}{7} + \ldots + \frac{(-1)^{[(2n-1)-1]/2}}{2n-1} x^{2n-1} + \ldots$$

$$= x - \frac{x^3}{3} + \ldots + (-1)^{n-1}\frac{x^{2n-1}}{2n-1} + \ldots$$

Example 14:

Prove that

$$\sin^{-1} x = x + x + \frac{1}{2}.\frac{x^3}{3} + \frac{1}{2}.\frac{3}{4}.\frac{x^5}{5} + \frac{1}{2}.\frac{3}{4}.\frac{5}{6}.\frac{x^7}{7} + \ldots$$

(Rohilkhand, 1990, 87S)

Solution:

Let $y = \sin^{-1} x$. Then $y_1 = 1/\sqrt{(1-x^2)}$

or $\quad (1-x^2)y_1^2 - 1 = 0$.

Differentiating again, we get

$$(1-x^2)2y_1y_2 - 2xy_1^2 = 0$$

or $\quad (1-x^2)y_2 - xy_1 = 0$, since $2y_1 \neq 0$.

Now differentiating n times by Leibntiz's theorem, we get

$$(1-x^2)y_{n+2} - (2n+1)xy_{n+1} - n^2y_n = 0.$$

Putting x = 0 in the above relations, we get

$(y)_0 = 0,\ (y_1)_0 = 1,\ (y_2)_0 = 0$

and $(y_{n+2})_0 = n^2\ (y_n)_0.$ ...(1)

Putting n = 1, 2, 3, ... in (1), we get

$(y_3)_0 = 1^2(y_1)_0 = 1^2,\ (y_4)_0 = 2^2(y_2)_0 = 0,\ (y_5)_0 = 3^2\ (y_3)_0 = 3^2.1^2,$

$(y_6)_0 = 4^2(y_4)_0 = 0,\ (y_7)_0 = 5^2(y_5)_0 = 5^2.3^2.1^2$, and so on.

Now by Maclaurin's theorem, we have

$$y = (y)_0 + x(y_1)_0 + \frac{x^2}{2!}(y_2)_0 + \frac{x^3}{3!}(y_3)_0 + \dots$$

$$\therefore \sin^{-1} x = 0 + x.1 + 1 + \frac{x^2}{2!}.0 + \frac{x^3}{3!}.1^2 + \frac{x^4}{4!}.0 + \frac{x^5}{5!}.3^2.1^2$$

$$+ \frac{x^6}{6!}.0 + \frac{x^7}{7!}.5^2.2^2.1^2 + \dots$$

$$= x + \frac{1}{2}.\frac{x^3}{3} + \frac{1}{2}.\frac{3}{4}.\frac{x^5}{5} + \frac{1}{2}.\frac{3}{4}.\frac{5}{6}.\frac{x^7}{7} + \dots$$

Example 15:

Expand $\left\{x + \sqrt{(1 + x^2)}\right\}^m$ *in ascending powers of x and find the general term also.* **(Agra, 1998; Kanpur, 98)**

Solution:

Let $y = \left\{x + \sqrt{(1 + x^2)}\right\}^m$

$(y)_0 = 1,\ (y_1)_0 = m,\ (y_2)_0 = m^2,$

and $(y_{n+2})_0 = (m^2 - n^2)\ (y_n)_0.$ ...(1)

Now putting n = 1, 2, 3, 4, ... in (1), we get

$(y_3)_0 = (m^2 - 1^2)\ (y_1)_0 = (m^2 - 1^2)m,$

$(y_4)_0 = (m^2 - 2^2)\ (y_2)_0 = (m^2 - 2^2)\ m^2,$

$(y_5)_0 = (m^2 - 3^2)\ (y_3)_0 = (m^2 - 3^2)\ (m^2 - 1^2)\ m,$

$(y_6)_0 = (m^2 - 4^2)\ (y_4)_0 = (m^2 - 4^2)\ (m^2 - 2^2)\ m^2$, etc.

In general,

if n is odd, $(y_n)_0 = \{m^2 - (n-2)^2\} \dots (m^2 - 3^2)\ (m^2 - 1^2)\ m$

and if n is even, $(y_n)_0 = \{m^2 - (n-2)^2\} \dots (m^2 - 4^2)\ (m^2 - 2^2)\ m^2$ ...(2)

Now by Maclaurin's theorem, we have

$$y = (y_0) + x(y_1)_0 + \frac{x^2}{2!}(y_2)_0 + \ldots + \frac{x^n}{n!}(y_n)_0 + \ldots$$

$$\therefore \left\{x+\sqrt{(1+x^2)}\right\}^m = 1 + mx + \frac{m^2}{2!}x^2 \frac{m(m^2-1^2)}{3!}x^3$$

$$+\frac{m^2(m^2-2^2)}{4!}x^4 + \frac{m(m^2-1^2)(m^2-3^2)}{5!}x^5 + \ldots$$

The general term $= \frac{x^n}{n!}$ $(y_n)_0$, where $(y_n)_0$ is given by (2).

Example 16:

Expand log {1 – log (1 – x)} in powers of x by Maclaurin's theorem as far as the term x^3.

By substituting x/(1 + x) for x deduce the expansion of g{1 + log (1 + x)} as far as the term in x^3.

Solution:

Let $y = \log\{1 - \log(1 - x)\}$. Then $(y)_0 = 0$.

Now $e^y = 1 - \log(1 - x)$. Differentiating, we get

$$e^x y_1 = (1 - x)^{-1}. \qquad \ldots(1)$$

Putting $x = 0$ in (1), we get $(y_1)_0 = 1$.

Differentiating (1), we get

$$e^y y_1^2 + e^y y_2 = (1-x)^{-2}$$

or $$e^y(y_1^2 + y_2) = (1-x)^{-2} \qquad \ldots(2)$$

Putting $x = 0$ in (2), we get

$$1 + (y_2)_0 = 1 \text{ or } (y_2)_0 = 0. \qquad \ldots(3)$$

Differentiating (2), we get

$$e^y(y_1^3 + 3y_1y_2 + y_3) = 2(1-x)^{-3}$$

Putting $x = 0$ in (3), we get

$$1 + (y_3)_0 = 2 \text{ or } (y_3)_0 = 1.$$

Substituting these values in Maclaurin's theorem, we get

$$\log\{1 - \log(1 - x)\} = 0 + x.1 + \frac{x^2}{2!}.0 + \frac{x^3}{3!}.1 + \ldots$$

$$= x + \left(\frac{x^3}{6}\right) + \ldots \qquad \ldots(A)$$

Now substituting $x/(1 + x)$ for x on both sides of (A), we get

$$\log\left\{1 - \log\left(1 - \frac{x}{1+x}\right)\right\} = \frac{x}{1+x} + \frac{1}{6}\left(\frac{x}{1+x}\right) + \ldots$$

or $\log\{1 + \log(1 + x)\} = x(1 + x)^{-1} + \left(\frac{1}{6}\right)x^3(1 + x)^{-3} + \ldots$

$= x\left\{1 + (-1)x + \frac{(-1)(-2)}{1.2}x^2 + \ldots\right\} + \frac{1}{6}x^3\{1 + (-3)x + \ldots\}$, on expanding by binomial theorem

$$= (x - x^2 + x^3 + ..) + \left(\frac{1}{6}x^3 + \ldots\right)$$

$$= x - x^2 + \frac{7}{6}x^3 + \ldots$$

Example 17:

If $y = \sin\log(x^2 + 2x + 1)$, prove that

$(x + 1)^2 y_{n+2} + (2n + 1)(x + 1)y_{n+1} + (n^2 + 4)y_n = 0.$

Hence or otherwise expand y in ascending powers of x as for as x^6.

(Allahabad, 1999; Agra, 95)

Solution:

Here $y = \sin\log(x^2 + 2x + 1) = \sin\log(x + 1)^2$ ...(1)

$$\therefore y_1 = [\cos\log(x + 1)^2].\frac{1}{(x+1)^2}.2(x+1)$$

$$= [\cos\log(x + 1)^2].\frac{2}{x+1}. \qquad \ldots(2)$$

Squaring both sides of (2), we get

$$(x+1)^2 y_1^2 = 4\cos^2\log(x + 1)^2 = 4[1 - \sin^2\log(x + 1)^2]$$

$$= 4(1 - y^2) \qquad \ldots(3)$$

or $(x + 1)^2 y_1^2 + 4y^2 - 4 = 0.$

Differentiating equation (3), we get

$$(x + 1)^2\, 2y_1y_2 + 2(x + 1)y_1^2 + 8yy_1 = 0$$

or $\quad 2y_1[(x + 1)^2y_2 + (x + 1)y_1 + 4y] = 0 \qquad ...(4)$

or $\quad (x + 1)^2y_2 + (x + 1)y_1 + 4y = 0,$

since $\quad 2y \neq 0.$

Differentiating (4) n times by Leibnitz's theorem, we get

$$(x + 1)^2y_{n+2} + {}^nC_1.y_{n+1}.2(x + 1) + {}^nC_2.y_n.2 + (x + 1)y_{n+1} + {}^nC_1.y_n.1 + 4y_n = 0$$

or $(x + 1)^2y_{n+2} + (2n + 1)(x + 1)y_{n+1} + (n^2 + 4)y_n = 0 \qquad ...(5)$

Putting x = 0 in (5), we get

$$(y)_0 = 0,\ (y_1)_0 = 2,\ (y_2)_0 + (y_1)_0 + 4\,(y)_0 = 0$$

or $\quad (y_2)_0 = -2$

Also putting x = 0 in (5), we get

$$(y_{n+2})_0 + (2n + 1)(y_{n+1})_0 + (n^2 + 4)(y_n)_0 = 0$$

or $\quad (y_{n+2})_0 = -[(2n + 1)(y_{n+1})_0 + (n^2 + 4)(y_n)_0]. \qquad ...(6)$

Now putting n = 1, 2, 3, 4 in (6), we get

$(y_3)_0 = -[3(y_2)_0 + 5(y_1)_0] = -[3.(-2) + 5.2] = -4,$

$(y_4)_0 = -[5(y_3)_0 + 8(y_2)_0] = -[5.(-4) + 8.(-2)] = 36,$

$(y_5)_0 = -[7(y_4)_0 + 13(y_3)_0] = -[7.(36) + 13.(-4)] = -200,$

and $(y_6)_0 = -[9(y_5)_0 + 20(y_4)_0] = -[9.(-200) + 20.(36)] = 1080.$

Now by Maclaurin's theorem, we have

$$y = (y)_0 + \frac{x}{1!}(y_1)_0 + \frac{x^2}{2!}(y_0) + \frac{x^3}{3!}(y_3)_0 + \frac{x^4}{4!}(y_4)_0 + \frac{x^5}{5!}(y_5)_0 + ...$$

$$= 0 + \frac{x}{1}.2 + \frac{x^2}{2}.(-2) + \frac{x^3}{1.2.3}.(-4) + \frac{x^4}{1.2.3.4}.36 + \frac{x^5}{1.2.3.4.5}.(-200) + \frac{x^6}{1.2.3.4.5.6}.(1080) + ...$$

$$= 2x - x^2 - \frac{2}{3}x^3 + \frac{3}{2}x^4 - \frac{5}{3}x^5 + \frac{3}{2}x^6 + ...$$

Example 18:

Use Maclaurin's theorem to show that

$$e^{m\cos^{-1}x} = e^{m\pi/2}\left[1 - mx + \frac{m^2}{2!}x^2 - \frac{m(1^2+m^2)}{3!}x^3 + \frac{m^2(2^2+m^2)}{4!}x^4 - \ldots\right]$$

(Agra, 1993; Delhi, 92)

Solution:

Do yourself.

Example 19:

Expand sin (m sin⁻¹ x) by Maclaurin's theorem as far as h^5. Hence expand sin mθ in powers of sin θ.

(Meerut, 1998, 93; G.N.U. 92; Kanpur, 98; Rohilkhand, 93, 95, 90; Lucknow, 95)

Solution:

Let $y = \sin(m \sin^{-1} x)$, we get

$y(0) = 0,\ y_1(0) = m,\ y_2(0) = 0,$

and $y_{n+2}(0) = (n^2 - m^2)y_n(0).$ ...(1)

Putting $n = 1, 2, 3, \ldots$ in (1), we get

$y_3(0) = (1^2 - m^2)y_1(0) = (1^2 - m^2)m,$

$y_4(0) = (2^2 - m^2)y_2(0) = 0,$

$y_5(0) = (3^2 - m^2)y_3(0) = (3^2 - m^2)(1^2 - m^2)m$, etc.

Substituting these values in Maclaurin's theorem, we get

$$\sin(m \sin^{-1} x) = y(0) + xy_1(0) + \frac{x^2}{2!}y_2(0) + \ldots$$

$$= mx + \frac{m(1^2 - m^2)}{3!}x^3 + \frac{m(1^2 - m^2)(3^2 - m^2)}{5!}x^5 + \ldots$$

Putting $x = \sin\theta$ on both sides, we get

$$\sin m\theta = m\sin\theta + \frac{m(1^2 - m^2)}{3!}\sin^3\theta + \frac{m(1^2 - m^2)(3^2 - m^2)}{5!}\sin^5\theta + \ldots$$

5.3 NORMAL EXPANSIONS OF FUNCTIONS

Maclaurin's Theorem

(Gorakhpur, 97; Meerut, 94; Magadh, 94; Bihar, 92; Kashmir, 93)

Let f(x) be a function of x which possesses continues derivatives of all orders in the interval [0, x]. Assuming that f(x) can be expanded as an infinite power series in x, we have

$$f(x) = f(0) + +\frac{x}{1!}f'(0)+\frac{x^2}{2!}f''(0)+...+\frac{x^n}{n!}f^{(n)}(0)+ ...$$

Proof:

Suppose $f(x) = A_0 + A_1x + A_2x^2 + A_3x^3 + ...$ ...(1)

Let the expansion (1) be differentiable term by term any number of times. Then by successive differentiation, we have

$f'(x) = A_1 + 2A_2 + 3A_3x^2 + 4A_4x^3 + ...,$

$f''(x) = 2.1\ A_2 + 3.2\ A_3x + 4.3\ A_4x^2 + ...,$

$f'''(x) = 3.2.1\ A_3 + 4.3.2\ A_4x + ...,$ and so on.

Putting $x = 0$ in each of these relations, we get

$f(0) = A_0,\ f'(0) = A_1,\ f''(0) = 2\ !\ A_2,\ f'''(0) = 3\ !\ A_3, ...$

Substituting theses values of $A_0, A_1, A_2, ...$ in (1), we get

$$f(x) = f(0) + xf' + \frac{x^2}{2!}f^{(n)}(0) + ...$$

This is **Maclaurin's Theorem:** If we denote f(x) by y, then Maclaurin's theorem can be written in either of the following ways:

$$y = (y)_0 + \frac{x}{1!}(y_1)_0 + \frac{x^2}{2!}(y_2)_0 + \frac{x^3}{3!}(y_3)_0 + ... + \frac{x^n}{n!}(y_n)_0 + ...$$

or $$y = y(0) + \frac{x}{1!}y_1(0) + \frac{x^2}{2!}y_2(0) + ... + \frac{x^n}{n!}y_n(0) + ...$$

Taylor's Theorem **Gorakhpur, 1999; Bihar, 92; Vikram, 98; Meerut, 95; Magadh, 97; Jiwaji, 90)**

Let f(x) be a function of x which possesses continuous derivatives of all orders in the interval [a, a + h]. Assuming that f(a + h) can be expanded as an infinite power series in h, we have

$$f(a + h) = f(a) + hf'(a) + \frac{h^2}{2!}f''(a) + ... + \frac{h^n}{n!}f^{(n)}(a) + ...$$

Proof:

Suppose $f(a + h) = A_0 + A_1h + A_2h^2 + A_3h^3 + ...$ ...(1)

Let the expansion (1) be differentiable term by term any number of times w.r.t. h. Then by successive differentiation w.r.t. h, we have

$f'(a + h) = A_1 + 2A_2h + 3A_3h^2 + ...,$

$f''(a + h) = 2.1A_2 + 3.2\ A_3h + ...,$

$f'''(a + h) = 3.2.1\ A_3 + ...,$ and so on.

Putting h = 0 in each of the above relations, we get

$f(a) = A_0,\ f'(a) = A_1, f''(a) = 2\ !\ A_2,\ f'''(a) = 3\ !\ A_3$, and so on.

$\therefore\ A_0 = f(a),\ A_1,\ f'(a),\ A_2 = \frac{1}{2!}f''(a),\ A_3 = \frac{1}{3!}f'''(a)$, and so on.

Substituting these values of $A_0, A_1, A_2, A_3, ...$ in (1), we get

$$f(a + h) = f(a) + hf'(a) + \frac{h^2}{2!}f'''(a) + ... + \frac{h^n}{n!}f^{(n)}(a) + ...$$

This is **Taylor's Theorem.** Another useful form is obtained on replacing h by (x – a). Thus

$$f(x) = f(a) + (x - a)\ f'(a) + \frac{(x-a)^2}{2!}f''(a) + ... + \frac{(x-a)^n}{n!}f^{(n)}(a) + ...,$$

which is an expansion of f(x) as a power series in (x – a).

Note: If we expand f(x + h), by Taylor's theorem, as a power series in h, then the result is as follows:

$$f(x + h) = f(x) + hf'(x) + \frac{h^2}{2!}f''(x) + ... + \frac{h^n}{n!}f^{(n)}(x) + ...$$

5.4 SOME IMPORTANT EXPANSIONS

***Expansion of e^x:* (Exponential Series)**

Let $f(x) = e^x$. Then $f(0) = e^0 = 1$;

$f^{(n)}(x) = e^x$ so that $f^{(n)}(0) = e^0 = 1$, where n = 1, 2, 3, ...

Substituting these values in Maclaurin's series.

Example 1:

(a) If $y = (\sin^{-1} x)/\sqrt{(1-x)^2}$, where $-1 < x < 1$

and $-\pi/2 < \sin^{-1} x < \pi/2$.

prove that $(1 - x^2)y_{n+1} - (2n + 1)xy_n - n^2y_{n-1} = 0$.

Also if $y = a_0 + a_1x + a_2x^2 + ... + a_nx^n + ...,$

prove that $(n + 1)a_{n+1} = n\ a_{n-1}$ and hence obtain the general term of the expansion. **(Lucknow, 1996)**

(b) Expand $(\sin^{-1} x)/\sqrt{(1-x^2)}$ in powers of x upto three terms.

Solution:

(a) Here $y = (\sin^{-1} x)\sqrt{(1-x^2)}$.

$\therefore \quad y^2(1 - x^2) = (\sin^{-1} x)^2$.

Differentiating w.r.t. x, we ger

$$2yy_1(1 - x^2) - 2xy^2 = 2(\sin^{-1} x)/\sqrt{(1-x^2)} = 2y.$$

Since $2y \neq 0$, therefore $y_1(1 - x^2) - xy = 1$

i.e., $\quad y_1(1 - x^2) - xy - 1 = 0.$ ...(2)

Differentiating (2) n times by Leibnitz's theorem, we get

$$y_{n+1}(1 - x^2) + ny_n(-2x) + \frac{n(n-1)}{2!}y_{n-1}.(-2) - xy_n - ny_{n-1} = 0$$

or $\quad (1 - x^2)y_{n+1} - (2n + 1)xy_n - n^2y_{n-1} = 0.$ ...(3)

Now putting x = 0 in (1), (2) and (3), we get

$$(y)_0 = 0,\ (y_1)_0 = 1 \text{ and } (y_{n+1})_0 = n^2(y_{n-1})_0$$

By Maclaurin's theorem, we have

$$y = (y)_0 + x(y_1)_0 + (x^2/2!)\,(y_2)_0 + ... + (x^n/n!\ (y_n)_0 + ...$$

Also we are given that

$$y = a_0 + a_1x + a_2x^2 + ... + a_nx^n + ...$$

Comparing the coefficients of x^n in the two expansions for y, we get $a_n = (y_n)_0/n!$.

$$\therefore \quad \frac{a_{n+1}}{a_{n-1}} = \frac{(y_{n+1})_0}{(n+1)!} \div \frac{(y_{n-1})_0}{(n-1)!} = \frac{(y_{n+1})_0}{(y_{n-1})_0}.\frac{(n-1)!}{(n+1)!}$$

$$= n^2.\frac{1}{n(n+1)}, \qquad \text{from (4)}$$

$$= \frac{n}{n+1}.$$

$\therefore \quad (n + 1)a_{n+1} = n\, a_{n-1}$ **Proved.**

or $\quad a_{n+1} = \dfrac{n}{n+1}a_{n-1}$ ...(5)

Example 2:

(a) By Maclaurin's theorem or otherwise find the expansion of $y = \sin(e^x - 1)$ upto and including the term in x^4.

(Gorakhpur, 1992; 98)

(b) Also show that $x = y - \frac{1}{2}y^2 + ...$ **(Allahabad, 1993)**

Solution:

(a) Let $y = \sin(e^x - 1)$. Then $(y)_0 = \sin 0 = 0$,

$y_1 = [\cos(e^x - 1)].e^x$ so that $(y_1)_0 = (\cos 0).e^0 = 1$,

$y_2 = [\cos(e^x - 1)].e^x - [\sin(e^x - 1)].e^{2x} = y_1 - ye^{2x}$

so that $(y_2)_0 = (y_1)_0 - (y)_0\, e^0 = 1 - 0 = 1$,

$y_3 = y_2 - y_1\, e^{2x} - 2y\, e^{2x}$ so that $(y_3)_0 = 1 - 1 - 0 = 0$,

$y_4 = y_3 - y_2\, e^{2x} - 4y_1\, e^{2x} - 4y\, e^{2x}$ so that $(y_4)_0 = -5$, etc.

Hence by Maclaurin's theorem, we get

$$\sin(e^x - 1) = (y)_0 + x(y_1)_0 + \frac{x^2}{2!}(y_2)_0 + \frac{x^3}{3!}(y_3)_0 + ...$$

$$= 0 + x.1 + \frac{x^2}{2!}.1 + \frac{x^3}{3!}.0 + \frac{x^4}{4!}.(-5) + ...$$

$$= x + \frac{1}{2}x^2 - \frac{5}{24}x^4 + ...$$

(b) We have

$y = \sin(e^x - 1) \Rightarrow e^x - 1 = \sin^{-1}y \Rightarrow e^x = 1 + \sin^{-1} y.$ ...(1)

Differentiating (1) w.r.t. 'y', we get

$$e^x.x_1 = 1/\sqrt{(1-y^2)}, \text{ where } x_1 = \frac{dx}{dy}. \qquad ...(2)$$

From equation (2), we get $(1 - y^2)x_1^2 = e^{-2x}$.

Differentiating it w.r.t 'y', we get

$$(1-y^2)2x_1x_2 - 2yx_1^2 = e^{-2x}(-2x_1)$$

or $(1 - y^2)x_2 - yx_1 = -e^{-2x}$, since $2x_1 \neq 0$. ...(3)

From (1), we have $x = \log(1 + \sin^{-1} y)$. ...(4)

Now putting y = 0 in (4), (2) and (3), we get

$(x)_0 = \log(1 + 0) = 0, e^0.(x_1)_0 = 1/\sqrt{(1-0)}$ giving $(x_1)_0 = 1$,

$(x_2)_0 = -e^0 = -1.$ [Note that $(x)_{y=0} = 0$]

Hence by Maclaurin's theorem, we get

$x = (x)_0 + y(x_1)_0 + (y^2/2\,!)\,(x_2)_0 + ...$

$= 0 + y.1 + (y_2/2\,!)\,(-1) + ... = y - \frac{1}{2}y^2 + ...$

Now $a_0 = (y)_0 = 0$, $a_1 = (y_1)_0 = 1$. Putting n = 1, 3, 5, ... in (5), we get $a_2 = \frac{1}{2}a_0 = 0$, $a_4 = \frac{3}{4}a_2 = 0$, $a_6 = \frac{5}{6}a_4 = 0$, etc.

Thus, $a_n = 0$ if n is even i.e., $a_{2m} = 0$.

Again putting n = 2, 4, 6, ... in (5), we get

$$a_3 = \frac{2}{3}a_1 = \frac{2}{3}, a_5 = \frac{4}{5}.a_3 = \frac{4}{5}.\frac{2}{3}, a_7 = \frac{6}{7}a_5 = \frac{6}{7}.\frac{4}{5}.\frac{2}{3}, \text{ etc.}$$

In general, if n is odd, we have

$$a_n = \frac{n-1}{n}.\frac{n-3}{n-2}...\frac{4}{5}.\frac{2}{3}.$$

$$\text{Thus } a_{2m+1} = \frac{2m}{2m+1}.\frac{2m-2}{2m-1}...\frac{4}{5}.\frac{2}{3}.$$

(b) As found in part (a), we have $a_0 = 0$, $a_1 = 1$, $a_2 = 0$, $a_3 = \frac{2}{3}$, $a_4 = 0$, $a_5 = \frac{2}{3}.\frac{4}{5}$, etc.

$$\therefore \frac{\sin^{-1} x}{\sqrt{(1-x^2)}} = x + \frac{2}{3}x^3 + \frac{2.4}{3.5}x^5 + ...$$

Example 3:

Show that

$$(\sin^{-1} x)^2 = \frac{2}{2!}x^2 + \frac{2.2^2}{4!}x^4 + \frac{2.2^2.4^2}{6!}x^6 + ...$$

(Meerut, 1993, 94, 96)

Deduce that

$$\theta^2 = 2.\frac{\sin^2\theta}{2!} + 2^2.\frac{2\sin^4\theta}{4!} + 2^2.4^2\frac{2\sin^6\theta}{6!} + ...$$

(Rajasthan, 1997; Meerut, 96 P)

Solution:

Let $y = (\sin^{-1} x)^2$, we get

$y(0) = 0, y_1(0) = 0, y_2(0) = 2,$

and $y_{n+2}(0) = n^2 y_n(0)$. ...(1)

Putting n = 1, 2, 3, 4, ... in (1), we get

$y_3(0) = 1^2 y_1(0) = 0, y_4(0) = 2^2 y_2(0) = 2^2.2,$

$y_5(0) = 3^2 y_5(0) = 0,\ y_6(0) = 4^2 y_4(0) = 4^2.2^2.2$, etc.

Hence by Maclaurin's theorem, we get

$$(\sin^{-1} x)^2 = y(0) + xy_1(0) + \frac{x^2}{2!}y_2(0) + \dots$$

$$= \frac{2}{2!}x^2 + \frac{2.2^2}{4!}x^4 + \frac{2.2^2.4^2}{6!}x^6 + \dots$$

Now putting $\sin^{-1} x = \theta$ or $x = \sin\theta$ on both sides, we get the required expansion for θ^2.

Example 4:

If $y = \sin^{-1} x = a_0 + a_1x + a_2x^2 + \dots$, *prove that*

$(n + 1)(n + 2)a_{n+2} = n^2a_n$. **(Meerut, 1990 P)**

Solution:

Let $y = \sin^{-1} x$. ... (1)

Then $y_1 = \dfrac{1}{\sqrt{(1-x^2)}}$. ...(2)

$\therefore$ $(1-x^2)y_1^2 - 1 = 0$.

Differentiating again, we get

$$(1-x^2)2y_1y_2 - 2xy_1^2 = 0 \text{ or } 2y_1[(1 - x^2)y_2 - xy_1] = 0$$

or $(1 - x^2)y_2 - xy_1 = 0$, ...(3)

since $2y_1 \neq 0$.

Now differentiating (3) n times by Leibnitz's theorem, we get

$$(1 - x^2)y_{n+2} + n.y_{n+1}.(-2x) + \frac{n(n-1)}{1.2}y_n.(-2) - y_{n+1}.x - ny_n.1 = 0$$

or $(1 - x^2)y_{n+2} - (2n + 1) - n^2y_n = 0$. ...(4)

Putting $x = 0$ in (4), we get

$$(y_{n+2})_0 = n^2(y_n)_0 \quad ...(5)$$

By Maclaurin's theorem, we have

$$y = (y)_0 + \frac{x}{1!}(y_1)_0 + \frac{x^2}{2!}(y_2)_0 + \frac{x^3}{3!}(y_3)_0 + \dots + \frac{x^n}{n!}(y_n)_0 + \dots$$

Also we are given that

$$y = \sin^{-1} x = a_0 + a_1x + a_2x^2 + \dots + a_nx^n + \dots$$

Equating the coefficients of x^n in the two expansions for y, we get

$$a_n = \frac{(y_n)_0}{n!}.$$

$$\therefore \frac{a_{n+2}}{a_n} = \frac{(y_{n+2})_0}{(n+2)!} \cdot \frac{n!}{(y_n)_0} = \frac{(y_{n+2})_0}{(y_n)_0} \cdot \frac{1}{(n+2)(n+1)}$$

$$= \frac{n^2}{(n+2)(n+1)}, \text{ substituting for } \frac{(y_{n+2})_0}{(y_n)_0} \text{ from (5).}$$

Hence $(n + 1)(n + 2)a_{n+2} = n^2 a_n$.

Example 5:

Expand $\sin^{-1}(x + h)$ in powers of x as far as the term in x^3.

(Ranchi, 1991; Kashmir, 1991)

Solution:

First we observe that we are to expand $\sin^{-1}(x + h)$ in ascending powers of x. So let $f(h) = \sin^{-1} h$. Then

$$f(h + x) = \sin^{-1}(h + x).$$

Thus, we are to expand f(h + x) in powers of x. So by Taylor's theorem, we have

$$f(h + x) = f(h) + xf'(h) + \frac{x^2}{2!}f''(h) + \frac{x^3}{3!}f'''(h) + \ldots \qquad \ldots(1)$$

Now $f(h) = \sin^{-1}h$. Therefore

$$f'(h) = \frac{1}{\sqrt{(1-h^2)}} = (1-h^2)^{-1/2}, \ f''(h) = h(1 - h^2)^{-3/2},$$

$$f''(h) = h(1 - h^2)^{-3/2},$$

$$f'''(h) = (1 - h^2)^{-3/2} + h(-3/2)(1 - h^2)^{-5/2}(-2h)$$

$$= (1 - h^2)^{-3/2} + 3h^2(1 - h^2)^{-5/2} = (1 - h^2)^{-5/2}[(1 - h^2) + 3h^2]$$

$$= (1 - h^2)^{-5/2}(1 + 2h^2), \text{ etc.}$$

Substituting these values in (1), we have

$$\sin^{-1}(h + x) = \sin^{-1} h + (1 - h^2)^{-1/2}x + (x^2/2!)\, h(1 - h^2)^{-3/2}$$
$$+ (x^3/3!)(1 - h^2)^{-5/2}(1 + 2h^2) + \ldots$$

Example 6:

Expand $2x^3 + 7x^2 + x - 1$ 9n powers of $(x - 2)$.

(Meerut, 1991 S; Jhansi, 99)

Solution:

Let $f(x) = 2x^3 + 7x^2 + x - 1$. We can write

$f(x) = f[2 + (x - 2)]$.

Now expanding $f[2 + (x - 2)]$ by Taylor's Theorem in powers of $x - 2$, we get

$$f(x) = f[2 + (x - 2)] = f(2) + (x - 2)f'(2) + (1/2\,!)\,(x - 2)^2 f''(2) + \ldots \quad \ldots(1)$$

Now $f(x) = 2x^3 + 7x^2 + x - 1$

so that $f(2) = 2.2^3 + 7.2^2 + 2 - 1 = 45$, $f'(x) = 6x^2 + 14x + 1$

so that $f'(2) = 53$, $f''(x) = 12x + 14$ so that $f''(2) = 38$,

$f'''(x) = 12$, so that $f'''(2) = 12$,

$f^{iv}(x) = 0$ so that $f^{iv}(2) = 0$. Obviously $f^n(2) = 0$ when $n \geq 4$. Now substituting these values in (1), we get

$$f(x) = 45 + (x - 2).53 + \frac{(x-2)^2}{2!}.38 + \frac{(x-2)^3}{3!}.12$$

$$= 45 + 53(x - 2) + 19(x - 2)^2 + 2(x - 2)^3.$$

Example 7:

Show that $\log(x + h) = \log h + \frac{x}{h} - \frac{x^2}{2h^2} + \frac{x^3}{3h^3} - \ldots$

(Rohilkhand, 1991; Gorakhpur, 90; Vikram, 95; Indore, 92)

Solution:

First we observe that we are to expand $\log(x + h)$ in ascending powers of x. So let $f(h) = \log h$. Then

$$f(h + x) = \log(h + x).$$

Here $\quad f(h) = \log h$, $f'(h) = 1/h$, $f''(h) = -\frac{1}{h^2}$,

$f'''(h) = \frac{2}{h^3}$, ... etc.

Substituting these values in Taylor's expansion, we get

$$\log(h + x) = \log h + \frac{x}{h} - \frac{x^2}{2h^2} + \frac{x^3}{3h^3} - \ldots$$

Example 8:

Expand $\tan^{-1} x$ *in powers of* $\left(x - \frac{1}{4}\pi\right)$. **(Meerut, 1990; 95, 96P)**

Solution:

Let $f(x) = \tan^{-1} x$. Then, we have

$$\tan^{-1} x = f(x) = f\left[\frac{1}{4}\pi + \left(x - \frac{1}{4}\pi\right)\right],$$

$$\left[\because \text{ we have to expand } f(x) \text{ in powers of } \left(x - \frac{1}{4}\pi\right)\right]$$

$$= f\left(\frac{\pi}{4}\right) + \left(x - \frac{1}{4}\pi\right) f'\left(\frac{\pi}{4}\right) + \frac{1}{2!}\left(x - \frac{1}{4}\pi\right)^2 f''\left(\frac{\pi}{4}\right) + \ldots,$$

on expanding $f\left[\frac{1}{4}\pi + \left(x - \frac{1}{4}\pi\right)\right]$ by Taylor's theorem in powers of $\left(x - \frac{1}{4}\pi\right)$.

Now $f(x) = \tan^{-1}x$. Therefore

$$f\left(\frac{\pi}{4}\right) = \tan^{-1}\left(\frac{\pi}{4}\right),\ f'(x) = \frac{1}{1 + x^2}$$

so that $f'\left(\frac{\pi}{4}\right) = 1/\left(1 + \frac{\pi^2}{16}\right)$, $f''(x) = -2x/(1 + x^2)^2$ so that

$$f''\left(\frac{\pi}{4}\right) = -\pi/\{2(1 + \pi^2/16)^2\} \text{ and so on.}$$

Substituting these values in the above expansion, we get

$$\tan^{-1} x = \tan^{-1}\left(\frac{\pi}{4}\right) + \left(x - \frac{1}{4}\pi\right)/\left(1 + \frac{\pi^2}{16}\right)$$

$$-\pi\left(x - \frac{1}{4}\pi\right)^2/\left\{4\left(1 + \frac{\pi^2}{16}\right)^2\right\} + \ldots$$

Example 9:

Expand log sin (x + h) in powers of h by Taylor's theorem.

(Agra, 1992; 98; Kanpur, 99; Meerut, 98)

Solution:

First we observe that we are to expand log sin (x + h) in powers of h. So let $f(x) = \log \sin x$. Then

$$f(x + h) = \log \sin (x + h).$$

Expanding $f(x + h)$ by Taylor's theorem in powers of h, we have

$$f(x+h) = f(x) = hf'(x) + \frac{h^2}{2!}f''(x)+\frac{h^3}{3!}f'''(x)+ \ldots \qquad \ldots(1)$$

Now $f(x) = \log \sin x$. Therefore $f'(x) = \frac{1}{\sin x}.\cos x = \cot x$,

$f''(x) = -\operatorname{cosec}^2 x$, $f'''(x) = 2\operatorname{cosec}^2 x \cot x$, etc.

Substituting these values in (1), we get

$$\log \sin (x+h) = \log \sin x + h \cot x - (h^2/2!) \operatorname{cosec}^2 x + (2h^3/3!) \operatorname{cosec}^2 x \cot x + \ldots$$

Example 10:

Expand sin x in powers of $\left(x - \frac{1}{2}\pi\right)$.

(Rohilkhand, 1992; Agra, 94; Kanpur, 90, 99; Vikram, 97; Meerut, 98 P; Gorakhpur, 96)

Solution:

Let $f(x) = \sin x$. We want to expand f(x) in powers of $x - \frac{1}{2}\pi$. We can write $f(x) = f\left[\frac{1}{2}\pi + \left(x - \frac{1}{2}\pi\right)\right]$. Now expanding $f\left[\frac{1}{2}\pi + \left(x - \frac{1}{2}\pi\right)\right]$ by Taylor's theorem in powers of $\left(x - \frac{1}{2}\pi\right)$, we get

$$f(x) = f\left[\frac{1}{2}\pi + \left(x - \frac{1}{2}\pi\right)\right] = f\left(\frac{\pi}{2}\right) + \left(x - \frac{1}{2}\pi\right)f'\left(\frac{\pi}{2}\right)$$

$$+\frac{1}{2!}\left(x - \frac{1}{2}\pi\right)^2 f''\left(\frac{\pi}{2}\right) + \frac{1}{3!}\left(x - \frac{1}{2}\pi\right)^3 f'''\left(\frac{\pi}{2}\right) + \ldots \qquad \ldots(1)$$

Now $f(x) = \sin x$. Therefore $f\left(\frac{\pi}{2}\right) = \sin\frac{\pi}{2} = 1$,

$f'(x) = \cos x$ giving $f'\left(\frac{\pi}{2}\right) = \cos\frac{\pi}{2} = 0$,

$f''(x) = -\sin x$ so that $f''\left(\frac{\pi}{2}\right) = -\sin\frac{\pi}{2} = -1$,

$f'''(x) = -\cos x$ so that $f'''\frac{\pi}{2} = -\cos\frac{\pi}{2} = 0$, etc.

Substituting these values in (1), we get

$$\sin x = 1 + \left(x - \frac{1}{2}\pi\right).0 + \frac{1}{2!}\left(x - \frac{1}{2}\pi\right)^2.(-1) + \frac{1}{3!}\left(x - \frac{1}{2}\pi\right)^3.0 + \ ...$$

$$= 1 - \left(\frac{1}{2!}\right)\left(x - \frac{1}{2}\pi\right)^2 + ...$$

Example 11:

Prove that

$$sin(x + h) = sin\ x + h\ cos\ x - \frac{h^2}{2!} sin - ...$$ **(Gorakhpur, 1997)**

Solution:

First we observe that we are to expand sin(x + h) in powers of h. So let f(x) = sin x. Then

f(x + h) = sin (x + h).

Expanding f(x + h) by Taylor's theorem in powers of h, we have

$$\sin (x + h) = f(x + h) = f(x) + hf'(x) + \frac{h^2}{2!} f''(x) + \frac{h^3}{3!} f'''(x) + ...$$

$$= \sin x + h \cos x + \frac{h^2}{2!}(- \sin x) + \frac{h^3}{3!}(- \cos x) + ...$$

$$= \sin x + h \cos - \frac{h^2}{2!} \sin x - \frac{h^3}{3!} \cos x + ...$$

Example 12:

Expand e^x in powers of (x – 1).

Solution:

Let f(x) = e^x. Then, we have

e^x = f(x) = f[1 + (x – 1)]

[∵ we are to expand f(x) in powers of (x – 1)]

= f(1) + (x – 1) f '(1) + (1/2!) $(x - 1)^2$ f "(1) + (1/3 !) $(x - 1)^3$ f "' (1) + ...,

on expanding f[1 + (x – 1)] by Taylor's theorem in powers of (x – 1).

Now f(x) = e^x. Therefore f(1) = e^1 = e, f '(x) = e^x so that f '(1) = e,

f "(x) = e^x so that f " (1) = e, and so on.

Substituting these values in the above expansion, we get

$e^x = e + (x - 1)\,e + (1/2\,!)\,(x - 1)^2\,e + (1/3\,!)\,(x - 1)^3\,e + \ldots$

$= e[1 + (x - 1) + (1/2\,!)\,(x - 1)^2 + (1/3\,!)\,(x - 1)^3 + \ldots].$

Example 13:

Prove that

(i) $$f\left(\frac{x^2}{1+x}\right) = f(x) - \frac{x}{1+x}f'(x) + \frac{x^2}{(1+x)^2}\,\frac{1}{2!}f''(x) - \ldots$$

(Rohilkhand, 1997)

(ii) $$f(x) = f(0) + xf'(x) - \frac{x^2}{2!}f''(x) + \frac{x^3}{3!}f'''(x) - \ldots$$

Solution:

(i) First we observe that we are to expand $f\left(\frac{x^2}{1+x}\right)$ in powers of $\left(-\frac{x}{1+x}\right)$.

We can write $f\left(\frac{x^2}{1+x}\right) = f\left[x + \left(-\frac{x}{1+x}\right)\right]$.

Now expanding $f\left[x + \left(-\frac{x}{1+x}\right)\right]$ by Taylor's theorem in powers of $-\frac{x}{1+x}$, we get

$$f\left(\frac{x^2}{1+x}\right) = f(x) + \left(-\frac{x}{1+x}\right)f'(x) + \frac{1}{2!}\left(-\frac{1}{1+x}\right)^2 f''(x) + \ldots$$

$$= f(x) - \frac{x}{1+x}f'(x) + \frac{x^2}{(1+x)^2}\,\frac{1}{2!}f''(x) - \ldots$$

(ii) We can write $f(0) = f[x + (-x)]$. Now expanding $f[x + (-x)]$ by Taylor's theorem in powers of $-x$, we get

$$f(0) = f[x + (-x)] = f(x) + (-x)\,f'(x) + \frac{(-x)^2}{2!}f''(x) + \frac{(-x)^3}{3!}f'''(x) + \ldots$$

or $$f(0) = f(x) - xf'(x) + \frac{x^2}{2!}f''(x) - \frac{x^3}{3!}f'''(x) + \ldots$$

$\therefore$ $f(x) = f(0) + xf'(x) - \frac{x^2}{2!}f''(x) + \frac{x^3}{3!}f'''(x) - \ldots$, by transposition.

Example 14:

Use Taylor's theorem to prove that

$$\tan^{-1}(x+h) = \tan^{-1} x + h \sin\theta \frac{\sin\theta}{1} - (h\sin\theta)^2 \frac{\sin 2\theta}{2}$$

$$+ (h\sin\theta)^3 \frac{\sin 3\theta}{3} - \ldots + (-1)^{n-1}(h\sin\theta)^n \frac{\sin n\theta}{n} + \ldots,$$

where $\theta = \cot^{-1} x$.

(Allahabad, 1990; Rohilkhand, 96, 98; Jhansi, 98; Lucknow, 94; Kanpur, 91; Meerut, 94P)

Solution:

First we observe that we are to expand $\tan^{-1}(x + h)$ in ascending powers of h. So let $f(x) = \tan^{-1} x$. Then $f(x + h) = \tan^{-1}(x + h)$. Expanding $f(x + h)$ in powers of h by Taylor's theorem we have

$$f(x+h) = f(x) + \frac{h}{1!} f'(x) + \frac{h^2}{2!} f''(x) + \ldots + \frac{h^n}{n!} f^n(x) + \ldots \qquad \ldots(1)$$

Now $f(x) = \tan^{-1} x$. Therefore $f^n(x) = D^n \tan^{-1} x$

$= (-1)^{n-1}(n-1)!\, \sin^n\theta \sin n\theta$, where $\theta = \cot^{-1} x$.

Putting n = 1, 2, 3, ... in it, we get

$f'(x) = \sin\theta \sin\theta$, $f''(x) = -1!\, \sin^2\theta \sin 2\theta$,

$f'''(x) = 2!\, \sin^3\theta \sin 3\theta$, etc.

Substituting these values in (1), we have

$$\tan^{-1}(x+h) = \tan^{-1} x + h\sin\theta\sin\theta - \frac{h^2}{2!}\sin^2\theta\sin 2\theta$$

$$+ \frac{h^3}{3!} 2!\sin^3\theta \sin 3\theta - \ldots + \frac{h^n}{n!}(-1)^{n-1}(n-1)!\, \sin^n\theta \sin n\theta + \ldots$$

$$= \tan^{-1} x + h\sin\theta \frac{\sin\theta}{1} - (h\sin\theta)^2 \frac{\sin 2\theta}{2}$$

$$+ (h\sin\theta)^3 \frac{\sin 3\theta}{3} - \ldots + (-1)^{n-1}(h\sin\theta)^n \frac{\sin n\theta}{n} + \ldots$$

Example 15:

Prove that f(mx) is equal to

$f(x) + (m-1)\, xf'(x) + (1/2!)(m-1)^2 x^2 f''(x) + \ldots$

(Kanpur, 1999; Agra, 96)

Solution:

First we observe that we are to expand f(mx) in powers of (m – 1)x. We can write f(mx) = f{x + (m – 1)x}.

Expanding f{x + m – 1)x} in powers of (m – 1)x by Taylor's theorem, we get

$f\{x + (m - 1)x\} = f(x) + (m-1)\, xf'(x) + (1/2\,!)\,(m-1)^2x^2f''(x) + \ldots$

$\therefore\ f(mx) = f(x) + (m - 1)\, xf'(x) + (1/2!)\,(m - 1)^2x^2f''(x) + \ldots$

EXERCISES

1. Apply Maclaurin's Theorem to prove that

   $\log \sec x = \frac{1}{2}x^2 + \frac{11}{12}x^4 + \frac{1}{4s}x^6 + \ldots$

2. Show that the first time terms in the power series for log (1 + sin x)

   $= x - \frac{1}{2}x^2 + \frac{1}{6}x^3 - \frac{1}{12}x^4 + \frac{1}{14}x^5 + \ldots$

3. Use Taylor's Theorem to prove that

   $\tan^{-1}(x + h) - \tan^{-1}x + h \sin z.\frac{\sin z}{1} - (h \sin h)^2\, \frac{\sin z}{2} + (h \sin z)^3$

   $\frac{\sin 3z}{3} + \ldots$

   where $z = \cos^{-1} x$.

4. Expand log sin x in power of x – 2.
5. Expand log sin (x + h) in power of h by Taylor's theorem.
6. Apply Maclaurin's theorem to obtain the expansion of the function e^{ax} cos bx in an infinite series of power of x giving the general term.

SOME MORE SOLVED EXAMPLES

Example 1:

If $y = \sin^{-1} x$, *find* $(y_n)_0$.

(Allahabad, 1992; Meerut, 91; Agra, 90; Vikram, 98)

Solution:

We have $y = \sin^{-1} x$. ...(1)

$$\therefore\ y_1 = \frac{1}{\sqrt{(1-x^2)}} \qquad ...(2)$$

or $\left(1-x^2\right)y_1^2-1=0$ ...(3)

Differentiating (3) w.r.t. x, we get

$\left(1-x^2\right)2y_1y_2-2xy_1^2=0$ or $2y_1[(1-x^2)y_2-xy_1]=0$.

Cancelling $2y_1$, since $2y_1$ is not identically equal to zero, we get $(1-x^2)y_2-xy_1=0$. ...(4)

Differentiating (4) n times by Leibntiz's theorem, we get

$y_{n+2}(1-x^2)+{}^nC_1y_{n+1}.(-2x)+{}^nC_2y_n.(-2)-y_{n+1}.x-{}^nC_1y_n.1=0$ ... (5)

Putting x = 0 in (1), (2) and (4), we get

$(y)_0=0$, $(y_1)_0=1$ and $(y_2)_0-0\,(y_1)_0=0$ i.e., $(y_2)_0=0$.

Also putting x = 0 in (5), we get

$(y_{n+2})_0=n^2(y_n)_0$. ...(6)

(6) is reduction formula which expresses $(y_{n+2})_0$ in terms of $(y_n)_0$.

Putting n – 2 in place of n in (6), we get

$(y_n)_0=(n-2)^2\,(y_{n-2})_0$
$=(n-2)^2\,(n-4)^2\,(y_{n-4})_0$,
[$\because$ from (6), $(y_{n-2})_0=(n-4)^2\,(y_{n-4})_0$]

Now there arise two cases.

Case I: When n is odd. Putting n = 1, 3, 5, ... in (6), we have

$(y_3)_0=1^2(y_1)_0=1^2.1$, [$\because$ $(y_1)_0=1$]

$(y_5)_0=3^2(y_3)_0=3^2.1^2.1$, $(y_7)_0=5^2(y_5)_0=5^2.3^2.1^2.1$, and so on.

Thus if n is odd, we have

$(y_n)_0=(n-2)^2\,(n-4)^2\,...\,5^2.3^2.1^2.1$.

Case II: When n is even: Putting n = 2, 4, 6, ... in (6), we have

$(y_4)_0=2^2(y_2)_0=0$, [$\because$ $(y_2)_0=0$]

$(y_6)_0=4^2(y_4)_0=4^2.0=0$, $(y_8)_0=6^2(y_6)_0=0$, and so on.

Thus if n is even, we have $(y_n)_0=0$.

Example 2:

Prove that

$$\frac{d^n}{dx^n}\left(\frac{\log x}{x}\right)=\frac{(-1)^n.(n)!}{x^{n+1}}\left[\log x-1-\frac{1}{2}-\frac{1}{3}-\,...\,-\frac{1}{n}\right].$$

(Lucknow, 1990; Delhi, 91; Kanpur, 99)

Solution:

We have $y = \left(\frac{1}{x}\right).\log x = (x^{-1}).\log x.$ ...(1)

Differentiating (1)n times by Leibnitz's theorem taking x^{-1} as first function, we have

$$y_n = \frac{(-1)^n n!}{x^{n+1}}\log x + {}^nC_1\frac{(-1)^{n-1}(n-1)!}{x^n}.\frac{1}{x}$$

$$+{}^nC_2\frac{(-1)^{n-2}(n-2)!}{x^{n-1}}.(-1).\frac{1}{x^2} + {}^nC_3\frac{(-1)^{n-1}(n-1)!}{x^{n-2}}.(-1)(-2).\frac{1}{x^3}$$

$$+ \ldots +.\frac{(-1)^{n-1}(n-1)!}{x^n}$$

$$= \frac{(-1)^n n!}{x^{n+1}}\log x + n.\frac{(-1)^{n-1}(n-1)!}{x^{n+1}}$$

$$+\frac{n(n-1)}{1.2}.\frac{(-1)^{n-1}(n-2)!}{x^{n-1}} + \frac{n(n-1)(n-2)}{1.2.3}$$

$$\frac{(-1)^{n-1}(n-3)!}{x^{n+1}}.1.2 + \ldots \frac{(-1)^{n-1}(n-1)!}{x^{n+1}}$$

$$= \frac{(-1)^n n!}{x^{n+1}}\left[\log x - 1 - \frac{1}{2} - \frac{1}{3} - \ldots - \frac{1}{n}\right].$$

Note: Solve the question also by differentiating (1) n times by Leibnitz theorem taking log x as first function.

Example 3:

If $y = (\sin^{-1} x)^2$, *find* $(y_n)_0$. **(Delhi, 1982; Lucknow, 92)**

Solution:

We have $y = (\sin^{-1} x)^2$. ...(1)

Differentiating both sides of (1) w.r.t. x, we get

$$y_1 = (2\sin^{-1} x).\frac{1}{\sqrt{(1-x^2)}}$$ ...(2)

Squaring both sides of (2) and multiplying by $(1 - x^2)$,

we have $(1 - x^2)y_1^2 = 4(\sin^{-1} x)^2 = 4y.$ $[\because y = (\sin^{-1} x)^2]$

or $(1-x^2)y_1^2 - 4y = 0.$ ...(3)

Differentiating both sides of (3) w.r.t. x, we get

$$(1 - x^2)\, 2y_1y_2 - 2xy_1^2 - 4y_1 = 0$$

or $$2y_1[(1 - x^2)y_2 - xy_1 - 2] = 0.$$

Cancelling $2y_1$, since $2y_1 \neq 0$, we get

$$(1 - x^2)y_2 - xy_1 - 2 = 0. \quad ...(4)$$

Differentiating (4) n times by Leibnitz's Theorem, we get

$$[(1 - x^2)y_{n+2} + {}^nC_1y_{n+1}(-2x) + {}^nC_2y_n(-2)]$$
$$- [xy_{n+1} + {}^nC_1y_1] - 0 = 0$$

or $$y_{n+2}(1 - x^2)y_{n+2} - (2n + 1)xy_{n+1} - n^2y_n = 0. \quad ...(5)$$

Putting x = 0 in (1), we get $(y)_0 = (\sin^{-1} 0)^2 = 0^2 = 0$.

Putting x = 0 in (2), we get $(y_1)_0 = (2 \sin^{-1} 0).1 = 0$.

Putting x = 0 in (4), we get $(y_2)_0 - 2 = 0$ i.e., $(y_2)_0 = 2$.

[Note that y_2 is a function of x. So on putting x = 0 in (4), y_2 becomes $(y_2)_0$].

Also putting x = 0 in (5), we get

$$(y_{n+2})_0 = n^2 (y_n)_0.$$

Now there arise two cases.

Case I: When n is odd: Putting n = 1, 3, 5, ... in (6), we get

$$(y_3)_0 = 1^2(y_1)_0 = 1^2.0 = 0, (y_5)_0 = 3^2(y_3)_0 = 3^2.0 = 0,$$

$(y_7)_0 = 5^2(y_5)_0 = 5^2.0 = 0$, and so on. Thus, when n is odd, $(y_n)_0 = 0$.

Case II: When n is even: Putting n = 2, 3, 4, 6, ... in (6), we get

$$(y_4)_0 = 2^2(y_2)_0 = 2^2.2, (y_6)_0 = 4^2(y_4)_0$$
$$= 4^2.2^2.2, (y_8)_0 = 6^2(y_6)_0$$
$$= 6^2.4^2.2^2.2, \text{ and so on. Thus, when n is even, we have}$$

$$(y_n)_0 = (n - 2)^2 (n - 4)^2 \ldots 6^2.4^2.2^2.2.$$

Example 4:

Find the nth derivative of x^3 log x set by Leibnitz's theorem

We have u = log x and v = x^3

$D^nu = (-1)^{n-1} (n-1)!/x^n$,

$D^{n-1} u = (-1)^{n-2} (n-2)!/x^{n-1}$ and $Dv = 3x^2$,

$D^{n-2} u = (-1)^{n-3} (n-3)!/x^{n-2}$ and $D^2v = 6x$,

$D^{n-3} u = (-1)^{n-4} (n-4)!/x^{n-3}$ and $D^3v = 6$.

Therefore, by Leibnitz's Theorem, we have

$$D^n(x^3 \log x) = \frac{(-1)^{n-1}(n-1)!}{x^n}x^3 + {}^nC_1\frac{(-1)^{n-2}(n-2)!}{x^{n-1}}3x^2$$

$+{}^nC_2\dfrac{(-1)^{n-3}(n-3)!}{x^{n-2}}6x + {}^nC_3\dfrac{(-1)^{n-4}(n-4)!}{x^{n-3}}6$, since all other terms become zero

$$= \frac{(-1)^{n-1}(n-4)!}{x^{n-3}}[(n-1)(n-2)(n-3) - 3n(n-2)(n-3)$$
$$+ 3n(n-1)(n-3) - n(n-1)(n-2)]$$
$$= (-1)^{n-1}(n-4)!\ x^{3-n}[(n-1)(n-2)(n-3-n)$$
$$+ 3n(n-3)(n-1-n+2)]$$
$$= 6(-1)^n(n-4)!x^{3-n}.$$

Example 5:

If $y = a \cos(\log x) + b \sin(\log x)$.

$$x^2y_2 + xy_1 + y = 0,$$

and $$x^2y_{n+2} + (2n+1)xy_{n+1} + (n^2+1)y_n = 0,$$

(Rohilkhand, 1997; Delhi, 92; Meerut, 94P, 91P, 90S)

Solution:

We have $y = a\cos(\log x) + b\sin(\log x)$, ...(1)

$\therefore$ $y_1 = (a/x)\cos(\log x) - (b/x)\sin(\log x)$

or $xy_1 = -a\sin(\log x) + b\cos(\log x)$.

Defferentiating (2) with respect to x, we have

or $x^2y_2 + xy_1 = -y$ [from (1)]

or $x^2y_2 + xy_1 + y = 9$. ...(3)

Differentiating equation (3) n times by Leibnitz's theorem, we have

$$D^n(x^2y_2) + D^n(xy_1) + D^n(y) = 0$$

or $$(D^ny_2).x^2 + {}^nC_1(D^{n-1}y_2).(Dx^2) + {}^nC_2(D^{n-2}y_2).(D^2x^2)$$
$$+ (D^ny_1).x + {}^nC_1(D^{n-1}y_1).(Dx) + D^ny = 0$$

or $$\left[y_{n+2}x^2 + n.y_{n+1}.2x + \frac{n(n-1)}{2!}y_n.2\right]$$
$$+ [y_{n+1}.x + ny_n.1] + y_n = 0$$

or $$x^2y_{n+2} + (2n+1)xy_{n+1} + (n^2+1)y_n = 0.$$

Example 6:

Find the nth differential coefficient of $e^x \log x$. **(Garhwal, 1993)**

Solution:

By Leibnitz's theorem, we have

$$D^n(e^x \log x) = (D^n e^x).\log x + {}^nC_1 (D^{n-1} e^x).(D \log x)$$
$$+ {}^nC_2(D^{n-2} e^x)(D^2 \log x) + ... + e^x D^n \log x$$

$$= e^x.\log x + {}^nC_1 e^x.\left(\frac{1}{x}\right) + {}^nC_2 e^x\left(-\frac{1}{x^2}\right) + {}^nC_3 e^x.\frac{2!}{x^3} + ...$$
$$+ ... + e^x(-1)^{n-1}.(n-1)!.x^n$$

$$= e^x[\log x + {}^nC_1 x^{-1} - {}^nC_2 x^{-2} + {}^nC_3 2!\, x^{-3} - ...$$
$$... + (-1)^{n-1}.(n-1)!\, x^{-n}].$$

Example 7:

If $y = x^2e^x$, prove that

$$\frac{d^n y}{dx^n} = \frac{1}{2}n(n-1)\frac{d^2 y}{dx^2} - n(n-2)\frac{dy}{dx} + \frac{1}{2}(n-1)(n-1)y.$$

(Agra, 1993; Gorakhpur, 91)

Solution:

We have $y = x^2e^x$. ...(1)

Differentiating (1) n times by Leibnitz's theorem, we get

$$y_n = e^x.x^2 + {}^nC_1 e^x.2x + {}^nC_2 e^x.2$$
$$= x^2e^x + 2nx\, e^x + n(n-1)\, e^x. \qquad ...(2)$$

Also differentiating (1) only once w.r.t. x, we have

$$y_1 = e^x.x^2 + 2xe^x \text{ or } y_1 = y + 2xe^x. \qquad ...(3)$$

Now differentiating (3) w.r.t. x, we have

$$y = y_1 + 2xe^x + 2e^x$$
$$= y_1 + (y_1 - y) + 2e^x \quad [\because \text{ from (3), } y_1 - y = 2xe^x]$$
$$y_2 = 2y_1 - y + 2e^x. \qquad ...(4)$$

Hence substituting for x^2e^x, $2xe^x$ and e^x respectively from (1), (3) and (4) in (2), we get

$$y_n = y + n(y_1 - y) + \frac{1}{2}n(n-1)(y_2 - 2y_1 + y)$$
$$= \frac{1}{2}n(n-1)y_2 - n(n-2)y_1 + \frac{1}{2}(n-1)(n-2)y$$

i.e., $$\frac{d^n y}{dx^n} = \frac{1}{2}n(n-1)\frac{d^2 y}{dx^2} - n(n-2)\frac{dy}{dx} + \frac{1}{2}(n-1)(n-2)y.$$

Example 8:

If $y = x^n \log x$, *show that* $y_{n+1} = n!/x$.

Solution:

We have $y = x^n \log x$

$$\therefore \qquad y_1 = x^n\left(\frac{1}{x}\right) + nx^{n-1} \log x \text{ or } xy_1 = x^n + nx^n \log x$$

$$\text{or} \qquad xy_1 = x^n + ny.$$

Now differentiating both sides n times and using Leibnitz's theorem, we get

$$D^n(y_1x) = D^nx^n + nD^ny$$

$$\text{or} \qquad (D^ny_1).x + {}^nC_1(D^{n-1}y_1).(Dx) + n! + ny_n$$

$$\text{or} \qquad y_{n+1}.x + ny_n.1 = n! + ny_n \text{ or } y_{n+1}.x = n!.$$

Therefore $y_{n+1} = n!/x$.

Example 9:

If $y = e^{a \sin^{-1}x}$, *show that*

$$(1 - x^2)y_{n+2} - (2n + 1)xy_{n+1} - (n^2 + a^2)y_n = 0.$$

(Rohilkhand, 1992; Meerut, 90, 94, 98; Garhwal, 97; Bundelkhand, 94; Indore, 93; Gorakhpur, 93; Allahabad, 90; Bihar, 98)

Solution:

We have $y = e^{a \sin^{-1}x}$.

Therefore $y_1 = e^{a \sin^{-1}x}.a/\sqrt{(1-x^2)}$

$$\text{or} \qquad y_1 \cdot \sqrt{(1-x^2)} = ae^{a \sin^{-1} x}\ ay, \qquad \text{[replacing } e^{a \sin^{-1} x} \text{ by y]}$$

$$\text{or} \qquad y_1^2(1-x^2) = a^2y^2. \qquad ...(1)$$

Differentiating (1) w.r.t. 'x', we have

$$2y_1y_2(1 - x^2) + (-2x) = 2a^2yy_1$$

$$\text{or} \qquad 2y_1[y_2(1 - x^2) - y_1x - a^2y] = 0.$$

Cancelling $2y_1$, since $2y_1 \neq 0$, we get

$$y_2(1 - x^2) - y_1x - a^2y = 0. \qquad ...(2)$$

Differentiating (2) n times by Leibnitz's theorem, we have

$$D^n[y_2(1 - x^2)] - D^n(y_1x) - a^2D^ny = 0$$

$$\text{or } \left[y_{n+2}\cdot\left(1-x^2\right)+ny_{n+1}(-2x)+\frac{n(n-1)}{2!}y_n(-2)\right]$$

$$- [y_{n+1}x + ny_n.1] - a^2y_n = 0$$

or $(1 - x^2)y_{n+2} - (2n + 1)xy_{n+1} - (n^2 + a^2)y_n = 0.$

Example 10:

If $y = \frac{1}{2}(\tan^{-1} x)^2$, *show that*

$(y_{n+2})_0 + 2n^2(y_n)_0 + n(n - 1)^2 (n - 2) (y_{n-2})_0 = 0.$

Solution:

Here $\quad y = \frac{1}{2}(\tan^{-1} x)^2; \qquad \therefore\ y_1 = \frac{\tan^{-1} x}{1+x^2}.$

$\therefore \quad (1 + x^2)y_1^2 = 2y.$

Differentiating again, we have

$$(1 + x^2)^2\, 2y_1y_2 = 2(1 + x^2).2xy_1^2 = 2y_1$$

or $(1 + 2x^2 + x^4)y_2 + 2(x + x^3)y_1 - 1 = 0$, since $2y_1 \neq 0$.

Differentiating n times by Leibnitz's theorem, we have

$$y_{n+2}(1 + 2x^2 + x^4) + ny_{n+1}(4x + 4x^3) + \frac{n(n-1)}{2}y_n\left(4+12x^2\right)$$

$$+\frac{n(n-1)(n-2)}{6}y_{n-1}(24x)+\frac{n(n-1)(n-2)(n-3)}{24}y_{n-2}\cdot(24)$$

$$+ 2y_{n+1}.(x + x^3) + 2ny_n.(1 + 3x^2) + \frac{2n(n-1)}{2}y_{n-1}\cdot(6x)$$

$$+\frac{2n(n-1)(n-2)}{6}y_{n-2}\cdot(6)=0$$

or $\quad (1 + x^2)^2y_{n+2} + (4n + 2)(x + x)^3y_{n+1} + 2n^2(3x^2 + 1)y_n$

$\qquad + 2n(n - 1)(2n - 1)xy_{n-1} + n(n - 1)^2(n - 2)y_{n-2} = 0. \quad ...(1)$

Putting $x = 0$ in (1), we get

$$(y_{n+2})_0 + 2n^2(y_n)_0 + n(n - 1)^2 (n - 2)(y_{n-2})_0 = 0.$$

Example 11:

If $y = (x^2 - 1)^n$, *prove that*

$(x^2 - 1)y_{n+2} + 2xy_{n+1} - n(n+1)y_n = 0.$

(Allahabad, 1991; Gorakhpur, 96; Rohilkhand, 92; Agra, 96; Jhansi, 99)

Hence if $P_n = \dfrac{d^n}{dx^n}(x^2 - 1)^n$, *show that*

$$\frac{d}{dx}\left\{\left(1-x^2\right)\frac{dP_n}{dx}\right\} + n(n+1)P_n = 0.$$ **(Kanpur, 1999)**

Solution:

We have $y = (x^2 - 1)^n$. Therefore, $y_1 = n(x^2 - 1)^{n-1}.2x$

or $(x^2 - 1)y_1 = n(x^2 - 1)^n.2x = 2nxy$, [replacing $(x^2 - 1)^n$ by y]

or $(x^2 - 1)y_1 - 2nxy = 0.$ ...(1)

Differentiating (1) (n + 1) times by Leibnitz's theorem, we have

$$D^{n+1}[y_1(x^2 - 1)] - 2n\, D^{n+1}(yx) = 0$$

or $y_{n+2}(x^2 - 1) + (n+1)y_{n+1}.2x + \dfrac{(n+1)n}{2!}.y_n.2$

$- 2ny_{n+1}.x - 2n(n+1)y_n.1 = 0$

or $(x^2 - 1)y_{n+2} + 2xy_{n+1} - n(n+1)y_n = 0,$ ...(2)

giving the first result. From (2), we get

or $(x^2 - 1)D^2y_n + 2xDy_n - n(n+1)y_n = 0,$ ...(3)

Putting $y_n = \dfrac{d^n}{dx^n}(x^2 - 1)^n = P_n$, (3) becomes

$$(x^2 - 1)\, D^2P_n + 2x\, D(P_n) - n(n+1)P_n = 0$$

or $-(1 - x^2)\, D^2(P_n) + 2x\, D(P_n) - n(n+1)\, P_n = 0$

or $-\dfrac{d}{dx}\{(1 - x^2)\, D(P_n)\} - n(n+1)\, P_n = 0$

or $\dfrac{d}{dx}\left\{\left(1-x^2\right)\dfrac{d}{dx}P_n\right\} + n(n+1)\, P_n = 0.$

Example 12:

If $\cos^{-1}\left(\dfrac{y}{b}\right) = \log\left(\dfrac{x}{n}\right)^n$, *prove that*

$x^2y_{n+2} + (2n+1)y_{n+1} + 2n^2y_n = 0.$

(Kanpur, 1996; Gorakhpur, 94; Jhansi, 98; Indore, 90; Meerut, 96)

Solution:

Here $\cos^{-1}(y/b) = \log(x/n)^n = n \log(x/n) = n(\log x - \log n)$. Differentiating both sides with respect to x, we have

$$-\frac{1}{\sqrt{\{1-(y^2/b^2)\}}}\frac{y_1}{b} = \frac{n}{x} \text{ or } -\frac{y_1}{\sqrt{(b^2-y^2)}} = \frac{n}{x}$$

or $\quad y_1^2 x^2 = n^2(b^2 - y^2)$.

Differentiating again, we get

$$2y_1y_2x^2 + 2xy_1^2 = -2n^2yy_1$$

or $\quad y_2x^2 + y_1x + n^2y = 0$, since $2y_1 \neq 0$. ...(1)

Differentiating (1) n times by Leibnitz's theorem, we get

$$y_{n+2}.x^2 + {}^nC_1y_{n+1}.(2x) + {}^nC_1y_n.(2) + y_{n+1}.x + {}^nC_1y_n + n^2y_n = 0$$

or $x^2y_{n+2} + (2n+1)xy_{n+1} + 2n^2y_n = 0$.

Example 13:

If $y^{1/m} + y^{-1/m} = 2x$, *prove that*

$(x^2 - 1)y_{n+2} + (2n+1)xy_{n+1} + (n^2 - m^2)y_n = 0$.

(Rohilkhand, 1991; Meerut, 83S; Lucknow, 81; Kanpur, 95)

Solution:

We have $y^{1/m} + y^{-1/m} = 2x$.

Multiplying both sides by $y^{1/m}$, we get

$y^{2/m} + 1 = 2xy^{1/m}$ or $y^{2/m} - 2xy^{1/m} + 1 = 0$.

$$\therefore \quad y^{1/m} = \frac{2x \pm \sqrt{(4x^2-4)}}{2} = x \pm \sqrt{(x^2-1)} \qquad ...(1)$$

or
$$y = \left[x \pm \sqrt{(x^2-1)}\right]^{m-1}\left\{1 \pm \frac{x}{\sqrt{(x^2-1)}}\right\}$$

$$= \pm\frac{my}{\sqrt{(x^2-1)}}\left[x \pm \sqrt{(x^2-1)}\right]^m$$

$$= \pm\frac{my}{\sqrt{(x^2-1)}}, \text{ from (1).}$$

Squaring both sides, we get

$y_1^2(x^2-1) = m^2y^2$. Differentiating again, we get

$2y_1y_2(x^2 - 1) + 2xy_1^2 = 2m^2yy_1$

or $2y_1[y_2(x^2 - 1) + xy_1 - m^2y] = 0$

or $y_2(x^2 - 1) + xy_1 - m^2y = 0$, since $2y_1 \neq 0$. ...(2)

Differentiating (2) n times by Leibnitz's theorem, we get

$D^n\{y_2(x^2 - 1)\} + D^n(y_1x) - m^2D^ny = 0$

or $y_{n+2}.(x^2 - 1) + {}^nC_1y_{n+1}\,2x + {}^nC_2y_n.2 + y_{n+1}.x + {}^nC_1y_n.1 - m^2y_n = 0$

or $(x^2 - 1)y_{n+2} + (2n + 1)xy_{n+1} + (n^2 - m^2)y_n = 0$.

Example 14:

If $I_n = \dfrac{d^n}{dx^n}(x^n \log x)$, *prove that*

$I_n = n\,I_{n-1} + (n-1)!;$

(Meerut, 1993, 97, 95, 97; Agra, 97)

hence show that

$$I_n = n!\left(\log x + 1 + \frac{1}{2} + \frac{1}{3} + \ldots + \frac{1}{n}\right).$$

Solution:

We have,

$$I_n = \frac{d^n}{dx^n}\left[x^n \log x\right] = \frac{d^{n-1}}{dx^{n-1}}\left[\frac{d}{dx}\left(x^n \log x\right)\right]$$

$$= \frac{d^{n-1}}{dx^{n-1}}\left[nx^{n-1}\log x + x^n.\frac{1}{x}\right]$$

$$= n\frac{d^{n-1}}{dx^{n-1}}\left(x^{n-1}\log x\right) + \frac{d^{n-1}}{dx^{n-1}}\left(x^{n-1}\right)$$

$= n\,I_{n-1} + (n-1)!$. **Proved.** ...(1)

We have just proved that $I_n = n\,I_{n-1} + (n-1)!$.

Dividing both sides by n!, we have

$$\frac{I_n}{n!} = \frac{I_{n-1}}{(n-1)!} + \frac{1}{n}. \qquad ...(2)$$

Changing n to n – 1 in the above relation (2), we have

$$\frac{I_{n-1}}{(n-1)!} = \frac{I_{n-2}}{(n-2)!} + \frac{1}{n-1}.$$

Putting this value of $\frac{I_{n-1}}{(n-1)!}$ in (2), we have

$$\frac{I_n}{n!} = \frac{I_{n-2}}{(n-2)!} + \frac{1}{n-1} + \frac{1}{n}.$$

Thus, making repeated use of the reduction formula (2), we have ultimately have

$$\frac{I_n}{n!} = \frac{I_1}{1!} + \frac{1}{2} + \frac{1}{3} + \ldots + \frac{1}{n}.$$

But $I_1 = \frac{d}{dx}(x \log x) = x.\frac{1}{x} + \log x = \log x + 1.$

$$\therefore \frac{I_n}{n!} = \log x + 1 + \frac{1}{2} + \frac{1}{3} + \ldots + \frac{1}{n}$$

or $I_n = n!\left(\log x + 1 + \frac{1}{2} + \frac{1}{3} + \ldots + \frac{1}{n}\right).$

Example 15:

Prove that

$$D^n\left(\frac{\sin x}{x}\right) = \left\{P \sin\left(x + \frac{1}{2}n\pi\right) + Q \cos\left(x + \frac{1}{2}np\right)\right\}/x^{n+1},$$

where $P = x^n - n(n-1)x^{n-2} + n(n-1)(n-2)(n-3)x^{n-4} - \ldots$
and $Q = nx^{n-1} - n(n-1)(n-2)x^{n-3} + \ldots$

Solution:

Differentiating n times by Leibnitz's theorem taking sin x as first function and x^{-1} as second function, we get

$$D^n[(\sin x).x^{-1} = \left[\sin\left(x + \frac{1}{2}n\pi\right)\right].(x^{-1})$$

$$+{}^nC_1\left[\sin\left\{x + \frac{1}{2}(n-1)\pi\right\}\right][(-1)x^{-2}] + {}^nC_2\left[\sin\left\{x + \frac{1}{2}(n-2)\pi\right\}\right].$$

$$[(-1)(-2)x^{-3}] + {}^nC_3\left[\sin\left\{x + \frac{1}{2}(n-3)\pi\right\}\right].[(-1)(-2)(-3)x^{-4}]$$

$$+{}^{n}C_{4}\left[\sin\left\{x+\frac{1}{2}(n-4)\pi\right\}\right].\times[(-1)+(-2)(-3)(-4)x^{-5}]+\ldots$$

$$=\frac{1}{x^{n}+1}\left[x^{n}\sin\left(x+\frac{1}{2}n\pi\right)+nx^{n-1}\cos\left(x+\frac{1}{2}n\pi\right)-n(n-1)x^{n-2}\right.$$

$$\times\sin\left(x+\frac{1}{2}n\pi\right)-n(n-1)(n-2)x^{n-3}\cos\left(x+\frac{1}{2}n\pi\right)$$

$$\left.+\,n(n-1)(n-2)(n-3)x^{n-4}\sin\left(x+\frac{1}{2}n\pi\right)+\ldots\right]$$

$$=\frac{1}{x^{n+1}}[\{x^{n}-n(n-1)x^{n-2}+n(n-1)(n-2)(n-3)x^{n-4}-\ldots\}$$

$$\times\sin\left(x+\frac{1}{2}n\pi\right)+\{nx^{n-1}-n(n-1)(n-2)x^{n-3}+\ldots\}\times\cos\left(x+\frac{1}{2n\pi}\right)\Big]$$

$$=\frac{1}{x^{n+1}}\left[P\sin\left(x+\frac{1}{2}n\pi\right)+Q\cos\left(x+\frac{1}{2}n\pi\right)\right].$$

Example 16:

By forming in two different ways the nth derivative of x^{2n}, prove that

$$1+\frac{n^2}{1^2}+\frac{n^2(n-1)^2}{1^2.2^2}+\frac{n^2(n-1)^2(n-2)^2}{1^2.2^2.3^2}+\ldots=\frac{(2n)!}{(n!)^2}.$$ **(I.C.S., 1995)**

Solution:

$$\because y=x^{2n},\ \therefore y_n=\frac{(2n)!}{(2n-n)!}.x^{2n-n}=\frac{(2n)!}{n!}.x^{n}. \qquad ...(1)$$

Again, $y = x^n.x^n$. ...(2)

Differentiating equation (2) n times by Leibnitz's theorem, we get

$$y_n=n!\,x^n+{}^{n}C_1\frac{n!}{1!}x.nx^{n-1}+{}^{n}C_2\frac{n!}{2!}x^2.n(n-1)x^{n-2}$$

$$+{}^{n}C_3\frac{n!}{3!}x^3.n(n-1)(n-2)x^{n-3}+\ldots+{}^{n}C_n x^n.n!$$

$$=n!\,x^n\left[1+\frac{n^2}{1^2}+\frac{n^2(n-1)^2}{1^2.2^2}+\frac{n^2(n-1)^2(n-2)^2}{1^2.2^2.3^2}+\ldots\right] \qquad ...(3)$$

Equating the two values of y_n obtained in (1) and (3), we get

$$1+\frac{n^2}{1^2}+\frac{n^2(n-1)^2}{1^2.2^2}+\frac{n^2(n-1)^2(n-2)^2}{1^2.2^2.3^2}+...=\frac{(2n)!}{(n!)^2}.$$

To find the nth derivative for x = 0.

Example 17:

If $y = e^{tan^{-1}x}$, *prove that*

$(1 + x^2)y_{n+2} + [2(n + 1)x - 1]_{n+1} + n(n + 1)y_n = 0$

(Agra, 1995; Meerut, 96, 96BP; Ranchi, 92)

Solution:

We have $y = e^{\tan^{-1}x}$.

Therefore $y_1 = e^{\tan^{-1}x}.\frac{1}{(1+x^2)}=\frac{1}{(1+x^2)}$

or $\quad y_1(1 + x^2) - y = 0.$ ...(1)

Differentiating (1) (n + 1) times by Leibnitz's theorem, we get

$$y_{n+2}(1 + x^2) + {}^{n+1}C_1y_{n+1}.2x + {}^{n+1}C_2y_n.2 - y_{n+1} = 0$$

or $\quad (1 + x^2)y_{n+2} + [2(n + 1)x - 1]y_{n+1} + (n + 1)ny_n = 0.$

Example 18:

If $y=\left[x+\sqrt{(1+x^2)}\right]^m$, *find the value of the* n^{th} *differential coefficient of y for x = 0.*

(Rohilkhand, 1990; Agra, 95; Gorakhpur, 96; Jiwaji, 98; Indore, 93; Kanpur, 90; Lucknow, 96; Meerut, 95, 92)

Solution:

Here $y=\left[x+\sqrt{(1+x^2)}\right]^m$. ...(1)

$$\therefore\ y_1 = m\left[x+\sqrt{(1+x^2)}\right]^{m-1}.\left[1+\frac{1}{2}\frac{2x}{\sqrt{(1+x^2)}}\right]$$

$$=\frac{m}{\sqrt{(1+x^2)}}\left[x+\sqrt{(1+x^2)}\right]^m=\frac{my}{\sqrt{(1+x^2)}} \quad ...(2)$$

or $\; y_1^2(1 + x^2) - m^2y^2 = 0.$

Differentiating again, we get

$$2y_1y_2(1 + x^2) + \quad = 0$$

or $\quad y_2(1 + x^2) + xy_1 - m^2y = 0,$ ...(3)

cancelling $2y_1$, since $2y_1$ ¹ 0.

Again differentiating (3) n times, we get

$y_{n+2}(1 + x^2) + {}^nC_1.2xy_{n+1} + {}^nC_2.2y_n + xy_{n+1} + {}^nC_1.y_n - m^2y_n = 0$

or $y_{m+2}(1 + x^2) + (2n + 1)xy_{n+\_1} + (n^2 - m^2)y_n = 0.$ ...(4)

Putting x = 0 in (1), (2), (3) and (4), we have

$$(y)_0 = 1, (y_1)_0 = m\ (y_2)_0 = m^2,$$

and $\quad (y_{n+2}) + (n^2 - m^2).(y_n)_0 = 0$

i.e., $\quad (y_{n+2})_0 = (n^2 - m^2).(y_n)_0.$ ..(5)

Case I: When n is odd.

Putting n = 1, 3, 5, 7, ... in (5), we get

$$(y_3)_0 = (m^2 - 1^2)\ (y_1)_0 = (m^2 - 1^2).m,$$

$$(y_5)_0 = (m^2 - 3^2)\ (y_3)_0 = (m^2 - 3^2)\ (m^2 - 1^2).m, \text{ and so on.}$$

Hence when n is odd, we have

$$(y_n)_0 = \{m^2 - (n - 2)^2\}\ \{m^2 - (n - 4)^2\}\ ...\ (m^2 - 3^2)\ (m^2 - 1^2)m.$$

Case II: When n is even.

Putting n = 2, 4, 6, ... in (5), we get

$$(y_4)_0 = (m^2 - 2^2)\ (y_2)_0 = (m^2 - 2^2).m^2,$$

$$(y_6)_0 = (m^2 - 4^2)\ (y_4)_0 = (m^2 - 4^2)\ (m^2 - 2^2).m^2, \text{ and so on.}$$

Hence when n is even, we have

$$(y_0)_0 = \{m^2 - (n - 2)^2\}\ \{m^2 - (n - 4)^2\}\ ...\ (m^2 - 2^2).m^2.$$

Example 19:

If $y = \sin(m \sin^{-1} x)$, find $(y_n)_0$.

(Kashmir, 1993; Avadh, 97; Rohilkhand, 93; 91; Meerut, 98P; 89P, 91, 93; Gorakhpur, 92; Lucknow, 90; Delhi, 91)

Solution:

We have $y = \sin(m \sin^{-1} x)$. ...(1)

Differentiating we get $y_1 = [\cos(m \sin^{-1} x)].\dfrac{m}{\sqrt{(1-x^2)}}$ ...(2)

Squaring both sides (2) and multiplying by $(1 - x^2)$, we get

$$\left(1-x^2\right)y_1^2 = m^2\cos^2(m \sin^{-1} x)$$

or $$\left(1-x^2\right)y_1^2 = m^2[1 - \sin^2(m \sin^{-1} x)]$$

or $$\left(1-x^2\right)y_1^2 = m^2\left(1-y^2\right), \qquad [\because y = \sin(m \sin^{-1} x)]$$

or $$(1 - x^2)y_1^2 + m^2y^2 - m^2 = 0. \qquad ...(3)$$

Differentiating (3), we get $\left(1-x^2\right)2y_1y_1 - 2xy_1^2 + 2m^2yy_1 = 0$.

Cancelling $2y_1$, since $2y_1 = 0$, we get

$$(1 - x^2)y_2 - xy_1 + m^2y = 0. \qquad ...(4)$$

Differentiating (4) n times by Leibnitz's theorem, we get

$(1 - x^2)y_{n+2} + {}^nC_1y_{n+} (-2x) + {}^nC_2y_n(-2) - xy_{n+1} - {}^nC_1y_n + m^2y_n = 0$

or $(1 - x^2)y_{n+2} - (2n + 1)xy_{n+1} - (n^2 - m^2)y_n = 0$. ...(5)

Putting x = 0 in (1), (2) and (4), we get

$$(y)_0 + \sin(m \sin^{-1} 0)_0 = \sin 0 = 0, (y_1)_0 = (\cos 0).m = m,$$

$$(y_2)_0 + m^2(y)_0 = 0 \text{ i.e., } (y_2)_0 = -m^2 (y)_0 = -m^2.0 = 0.$$

Also putting x = 0 in (5), we get

$$(y_{n+2})_0 = (n^2 - m^2)(y_n)_0. \qquad ...(6)$$

Putting n – 2 in place of n in the reduction formula (6), we get

$$(y_n)_0 = \{(n-2)^2 - m^2\}(y_{n-2})_0$$

$$= \{(n-2)^2 - m^2\}\{(n-4)^2 - m^2\}(y_{n-4})_0,$$ since from (6) we have $(y_{n-2})_0 = \{(n-4)^2 - m^2\}(y_{n-4})_0$.

Now there arise two cases:

Case I: When n is odd: Putting n = 1, 3, 5, ... in (6), we have

$(y_3)_0 = (1^2 - m^2)(y_1)_0 = (1^2 - m^2)m,$

$(y_5)_0 = (3^2 - m^2)(y_3)_0 = (3^2 - m^2)(1^2 - m^2)m,$

$(y_7)_0 = (5^2 - m^2)(y_5)_0 = (5^2 - m^2)(3^2 - m^2)(1^2 - m^2)m$, and so on.

Thus if n odd, we have

$$(y_n)_0 = \{(n-2)^2 - m^2\}\{(n-4)^2\} - m^2\} \dots (3^2 - m^2)(1^2 - m^2)m.$$

Case II: When n is even. Putting n 2, 4, 6, ... in (6), we have

$(y_4)_0 = (2^2 - m^2)(y_2)_0 = 0, (y_6)_0 = (4^2 - m^2)(y_4)_0 = 0$, and so on.

Thus, if n is even, we have $(y_n)_0 = 0$.

Example 20:

If $x + y = 1$, prove that

$$\frac{d^n}{dx^n}(x^n y^n) = n!\{y^n - ({}^nC_1)^2 y^{n-1}.x + ({}^nC_2)^2.y^{n-2}.x^2 + \dots + (-1)^n x^n\}.$$

(Kanpur, 1997; Agra, 98; Gorakhpur, 94)

Solution:

Given $x + y = 1$; $\therefore$ $y^n = (1 - x)^n$.

Thus $\frac{d^n}{dx^n}(x^n y^n) = \frac{d^n}{dx^n}\{x^n(1-x)^n\} = D^n[x^n.(1-x)^n]$

$= (D^n x^n)(1 - x)^n + {}^nC_1 (D^{n-1}x^n).D(1 - x)^n$

$+ {}^nC_2(D^{n-2}x^n).D^2(1 - x)^n + \dots + x^n D^n (1 - x)^n$

$= n!(1 - x)^n + {}^nC_1 \frac{n!}{1!} x \{- n (1 - x)^{n-1}\}$

$+ {}^nC_2 \frac{n!}{2!}.x^2\{(-1)^2.n(n-1)(1-x)^{n-2}\} + \dots + x^n(-1)^n.n!$

$\left[\text{Note that } D^r x^n = \frac{n!}{(n-r)!}x^{n-r}\right]$

$= n!\Big[(1-x)^n - n.{}^nC_1 x(1-x)^{n-1}$

$+ \frac{n(n-1)}{2!}.{}^nC_2 x^2(1-x)^{n-2} + \dots + (-1)^n x^n\Big]$

$= n![(1 - x)^n - ({}^nC_1)^2 x (1 - x)^{n-1} + ({}^nC_2)^2 x^2(1 - x)^{n-2} + \dots + (-1)^n x^n]$,

since $n = {}^nC_1, \frac{n(n-1)}{2!} = {}^nC_2$, etc.

$= n![y^n - ({}^nC_1)^2 xy^{n-1} + ({}^nC_2)^2 x^2y^{n-2} - \dots + (-1)^n x^n]$.

Example 21:

If $y = \tan^{-1} x$, find the value of $(y_7)_0$ and $(y_8)_0$.

(Rohilkhana, 1985)

Solution:

Putting $n = 1, 2, 3, 4, 5, 6$, in (6) we get

$(y_3)_0 = -(2 . 1)(y_1)_0 = -2!$, $(y_4)_0 = -(3 . 2)(y_2)_0 = 0$,

$(y_5)_0 = -(4 . 3)(y_3)_0 = -(4 . 3) . (-2!) = 4!$,

$(y_6)_0 = -(5\ .\ 4)\ (y_4)_0 = 0,$

$(y_7)_0 = -(6\ .\ 5)\ (y_5)_0 = -(6\ .\ 5)\ (4!) = -6!,$

$(y_8)_0 = -(7\ .\ 6)\ (y_6)_0 = 0.$

$\therefore\ (y_7)_0 = -6! = -720$ and $(y_8)_0 = 0.$

Example 22:

Find $y_n(0)$, *when* $y = log\left[x+\sqrt{(1+x)}\right].$ **(Meerut, 1999S)**

Solution:

Here $y = \log\left[x+\sqrt{\left(1+x^2\right)}\right].$ ...(1)

$$\therefore\ y_1 = \frac{1}{x+\sqrt{\left(1+x^2\right)}}.\left[1+\frac{1}{2}.\frac{1}{\sqrt{\left(1+x^2\right)}}.2x\right] = \frac{1}{\sqrt{\left(1+x^2\right)}} \quad ...(2)$$

Squaring both sides of (2), we get

$$(1 + x^2)y_1^2 - 1 = 0.$$

Differentiating again, we get

$$2y_1y_1(1 + x^2) + 2xy_1^2 = 0$$

or $\quad 2y_1[y_2(1 + x^2) + xy_1] = 0$

or $\quad y_2(1 + x^2) + xy_1 = 0,$ ...(3)

since $\quad y_1 \neq 0.$

Differentiating (3) n times by Leibnitz's theorem, we get

$$y_{n+2}\ (1 + x^2) + {}^nC_1.y_{n+1}\ 2x + {}^nC_2.y_n.2 + y_{n+1}.x + {}^nC_1.y_n.1 = 0$$

or $(1 + x^2)y_{n+2} + (2n + 1)xy_{n+1} + n^2y_n = 0.$ ...(4)

Putting x = 0 in (1), (2), (3) and (4), we have

$(y)_0 = 0,\ (y_1)_0 = 1,\ (y_2)_0 = 0,$

and $(y_{n+2})_0 + n^2(y_n)_0 = 0$

i.e., $(y_{n+2})_0 = -n^2(y_n)_0.$ ...(5)

Putting n – 2 in place of n on both sides of (5), we get

$(y_n)_0 = -(n-2)^2\ (y_{n-2})_0$

$\quad = [-(n-2)^2]\ [-(n-4)^2]\ (y_{n-4})_0$, since from (5),

we have $(y_{n-2})_0 = -(n-4)^2\ (y_{n-4})_0.$

Now there arise two cases:

Case I: When n is odd, we have

$(y_n)_0 = [-(n-2)^2]\ [-(n-4)^2]\ [-(n-6)^2]\ [-3^2]\ [-1^2]\ (y_1)_0$

$= (-1)^{(n-1)/2}\ (n-2)^2\ (n-4)^2 \ldots 3^2.1^2$, since $(y_1)_0 = 1$.

Case II: When n is even, we have

$(y_n)_0 = [-(n-2)^2]\ [-(n-4)^2] \ldots (y_2)_0$

$= 0$, since $(y_2)_0 = 0$.

Example 23:

If $y = e^{a \sin^{-1} x}$, *find* y_n *at* $x = 0$.

(Agra, 1992; Jiwaji, 98; Meerut, 95, 98, 90; Rohilkhand, 98)

Solution:

We have

$$y = e^{a \sin^{-1} x} \qquad \ldots(1)$$

$$y_1 = e^{a \sin^{-1} x}.a/\sqrt{(1-x^2)}, \qquad \ldots(2)$$

$$(1-x^2)y_2 - xy_1 - a^2r = 0, \qquad \ldots(3)$$

and $$(1-x^2)y_{n+2} - (2n+1)xy_{n+1} - (n^2+a^2)y_n = 0. \qquad \ldots(4)$$

Putting $x = 0$ in (1), (2), (3), and (4), we have

$(y)_0 = 1,\ (y_1)_0 = a,\ (y_2)_0 = a^2,$

and $$(y_{n+2}) = (n^2+a^2)\ (y_n)_0 \qquad \ldots(5)$$

Now there arise two cases.

Case I: When n is odd.

Putting $n = 1, 3, 5, \ldots$ in (5), we have

$(y_3)_0 = (1^2+a^2)\ (y_1)_0 = (1^2+a^2).a,$

$(y_5)_0 = (3^2+a^2)\ (y_3)_0 = (3^2+a^2)\ (1^2+a^2).a,$

$(y_7)_0 = (5^2+a^2)\ (y_5)_0 = (5^2+a^2)\ (3^2+a^2)\ (1^2+a^2).a$, and so on.

Thus, if n is odd, we have

$(y_n)_0 = [(n-2)^2+a^2] \ldots (3^2+a^2)\ (1^2+a^2)a.$

Case II: When n is even.

Putting $n = 2, 4, 6, \ldots$ in (5), we have

$(y_4)_0 = (2^3+a^2)\ (y_2)_0 = (2^2+a^2).a^2,$

$(y_6)_0 = (4^2+a^2)\ (y_4)_0 = (4^2+a^2)\ (2^2+a^2).a^2,$

$(y_8)_0 = (6^2+a^2)\ (y_6)_0 = (6^2+a^2)\ (4^2+a^2)\ (2^2+a^2)a^2$, ans. so on.

Thus, if n is even, we have

$$(y_n)_0 = [(n-2)^2 + a^2] \ldots (4^2 + a^2)(2^2 + a^2)a^2.$$

Example 24:

$y = e^{m \cos^{-1} x}$, *find* $(y_n)_0$.

(Meerut, 1990, 94, 98S, 94; Delhi, 94; Gorakhpur, 93)

Solution:

Proceed as in Ex. 23. The answer is:

When n is odd,

$(y_n)_0 = [(n-2)^2 + m^2] \ldots [3^2 + m^2][1^2 + m^2].(-me^{m\pi/2})$; and when n is even.

$$(y_n)_0 = [(n-2)^2 + m^2] \ldots (2^2 + m^2)(m^2 e^{m\pi/2}).$$

Example 25:

Prove that the nth differential coefficient of $x^n (1-x)^n$ *is equal to*

$$n!\,(1-x)^n \left\{1 - \frac{n^2}{1^2}\frac{x}{1-x} + \frac{n^2(n-1)^2}{1^2.2^2}\frac{x^2}{(1-x)^2} - \ldots\right\}.$$

(Rohilkhand, 1999)

Solution:

Do yourself.

Example 26:

If $Y = sX$ *and* $Z = tX$, *all the variables being functions of x, prove that*

$$D \equiv \begin{vmatrix} X & Y & Z \\ X_1 & Y_1 & Z_1 \\ X_2 & Y_2 & Z_2 \end{vmatrix} = X^3 \begin{vmatrix} s_1 & t_1 \\ s_2 & t_2 \end{vmatrix},$$

where the suffices denote the order of differentiation with respect to x.

Solution:

We have $Y_1 = sX_1 + s_1X$

and $\quad Y_2 = sX_2 + 2s_1X_1 + s_2X.$

Similarly $Z_1 = tX_1 + t_1X$

and $\quad Z_2 = tX_2 + 2t_1X_1 + t_2X.$

Hence $\begin{vmatrix} X & Y & Z \\ X_1 & Y_1 & Z_1 \\ X_2 & Y_2 & Z_2 \end{vmatrix}$

$$= \begin{vmatrix} X & sX & tX \\ X_1 & sX_1 + s_1X & tX_1 + t_1X \\ X_2 & sX_2 + 2s_1X_1 & tX_2 + 2t_1X_1 + t_2X \end{vmatrix}$$

$$= \begin{vmatrix} X & 0 & 0 \\ X_1 & s_1X & t_1X \\ X_2 & 2s_1X_1 + s_2X & 2t_1X_1 + t_2X \end{vmatrix},$$

operating $C_2 - sC_1$ and $C_3 - tC_1$

$$= X^2 \begin{vmatrix} s_1X & t_1X \\ 2s_1X_1 + s_2X & 2t_1X_1 + t_2X \end{vmatrix}$$

$$= X^2 \begin{vmatrix} s_1 & t_1 \\ 2s_1X_1 + s_2X & 2t_1X_1 + t_2X \end{vmatrix}$$

$$= X^2 \begin{vmatrix} s_1 & t_1 \\ s_2X & t_2X \end{vmatrix}, \text{ operating } R_2 - 2X_1R_1$$

$$= X^3 \begin{vmatrix} s_1 & t_1 \\ s_2 & t_2 \end{vmatrix}.$$

Example 27:

If $y = \cos(m \sin^{-1} x)$ find $(y_n)_0$

Solution:

We have $y = \cos(m \sin^{-1} x)$. ...(1)

Differentiating, we get

$$y_1 = [-\sin(m \sin^{-1} x)].\left[m/\sqrt{(1-x^2)}\right]. \quad ...(2)$$

Squaring both sides of (2) and multiplying by $(1 - x^2)$, we get

$$(1 - x^2)y_1^2 = m^2\sin^2(m \sin^{-1} x) = m^2[1 - \cos^2(m \sin^{-1} x)]$$

$$= m^2(1 - y^2).$$

$\therefore \qquad (1 - x^2)y_1^2 + m^2y^2 - m^2 = 0.$ ...(3)

Differentiating (3), we get

$$(1 - x^2)2y_1y_2 - 2xy_1^2 + 2m^2yy_1 = 0$$

or $\qquad 2y_1[(1 - x^2)y_2 - xy_1 + m^2y] = 0.$

Cancelling $2y_1$, since $2y_1$ is not identically equal to zero, we get

$$(1 - x^2)y_2 - xy_1 + m^2y = 0. \qquad ...(4)$$

Differentiating (4) n times by Leibnitz's theorem, we get

$y_{n+2}(1-x^2) + {}^nC_1y_{n+1}(-2x) + {}^nC_2y_n(-2) - xy_{n+1} - {}^nC_1y_n + m^2y_n = 0$

or $(1 - x^2)y_{n+2} - (2n + 1)xy_{n+1} - (n^2 - m^2)y_n = 0.$ ...(5)

Putting x = 0 in (1), (2) and (4), we get

$(y)_0 = \cos 0 = 1$, $(y_1)_0 = 0$, $(y_2)_0 + m^2(y)_0 = 0$ i.e., $(y_2)_0 = -m^2(y)_0$
$= -m^2.$

Also putting x = 0 in (5), we get

$$(y_{n+2})_0 = (n^2 - m^2)(y_n)_0. \qquad ...(6)$$

Putting n – 2 in place of n in the reduction formula (6), we get

$(y_n)_0 = \{(n - 2)^2 - m^2\}(y_{n-2})_0$

$\qquad = \{(n - 2)^2 - m^2\}\{(n - 4)^2 - m^2\}(y_{n-4})_0$, since from (6),

we have $(y_{n-2})_0 = \{(n - 4)^2 - m^2\}(y_{n-4})_0.$

Now there arise two cases.

Case I: What n is odd, we have

$(y_n)_0 = \{(n - 2)^2 - m^2\}\{(n - 4)^2 - m^2\} \ldots (3^2 - m^2)(1^2 - m^2)(y_1)_0$

$\qquad = 0$, since $(y_1)_0 = 0.$

Case II: When n is even, we have

$(y_n)_0 = \{(n - 2)^2 - m^2\}\{(n - 4)^2 - m^2\} \ldots (4^2 - m^2)(2^2 - m^2)(y_2)_0$

$\qquad = -\{(n - 2)^2 - m^2\}\{(n - 4)^2 - m^2\} \ldots (4^2 - m^2)(2^2 - m^2)m^2,$

since $(y_2)_0 = -m^2.$

Example 28:

If $= \left[\log\left\{x + \sqrt{(1 + x^2)}\right\}\right]^2$, *prove that*

$(y_{n+2})_0 = -n^2(y_n)_0$, *hence find* $(y_n)_0$.

(Kanpur, 1993; Rohilkhand, 92; Meerut, 93, 95, 97)

Solution:

Here $y = \left[\log\left\{x+\sqrt{(1+x^2)}\right\}\right]^2$.

$$\therefore\ y_1 \left[2\log\left\{x+\sqrt{(1+x^2)}\right\}\right]^2 . \frac{1}{x+\sqrt{(1+x^2)}} . \left[1+\frac{1}{2}.\frac{1}{\sqrt{(1\_x^2)}}.2x\right]$$

$$= \frac{2}{\sqrt{(1+x^2)}} \log\left\{x+\sqrt{(1+x^2)}\right\}. \quad ...(2)$$

Squaring both sides of (2), we get

$$(1 + x^2)y_1^2 = 4\left[\log\left\{x+\sqrt{(1+x^2)}\right\}\right]^2 = 4y, \text{ from (1)}$$

or $\quad (1+x^2)y_1^2 - 4y = 0.$

Differentiating again, we get

$$2y_1y_2(1 + x^2) + 2xy_1^2 - 4y_1 = 0$$

or $\quad 2y_1[y_2(1 + x^2) + xy_1 - 2] = 0$

or $\quad y_2(1 + x^2) + xy_1 - 2 = 0. \quad ...(3)$

since $\quad y_1 \neq 0.$

Differentiating (3) n times by Leibnitz's theorem, we get

$$y_{n+2}(1 + x^2) + {}^nC_1.y_{n+1}.2x + {}^nC_2y_n.2 + xy_{n+1} + {}^nC_1y_n.1 = 0$$

or $\quad (1 + x^2)y_{n+2} + (2n + 1)xy_{n+1} + n^2y_n = 0 \quad ...(4)$

Putting x = 0 in (1), (2), (3) and (4), we get

$$(y)_0 = 0,\ (y_1)_0 = 0,\ (y_2)_0 = 2,$$

and $\quad (y_{n+2})_0 + n^2(y_n)_0 = 0$

i.e., $\quad (y_{n+2})_0 = -n^2(y_n)_0. \quad ...(5)$

Putting n – 2 in place of n on both sides of (5), we get

$$(y_n)_0 = -(n-2)^2 (y_{n-2})_0$$

$$= [-(n-2)^2]\,[-(n-4)^2]\,(y_{n-4})_0,$$

since from (5), we have $(y_{n-2})_0 = -(n-4)^2 (y_{n-4})_0$.

Now there arise two cases:

Case I: When n is odd, we have

$$(y_n)_0 = [-1(n-2)^2]\,[-(n-4)^2] \ldots (y_1)_0 = 0, \text{ since } (y_1)_0 = 0.$$

Case II: When n is even, we have

$$(y_n)_0 = [-(n-2)^2][-(n-4)^2][-(n-6)^2] \ldots [-4^2][-2^2](y_2)_0$$
$$= -(-1)^{(n-2)/2}(n-2)^2(n-4)^2(n-6)^2 \ldots 4^2.2^2.2.$$

Example 29:

If $y = \tan^{-1} x$, find $(y_n)_0$. **(Delhi, 1991; Meerut, 90P, 94P)**

Solution:

We have $y = \tan^{-1} x$. ...(1)

$\therefore \quad y_1 = 1/(1+x^2)$, ...(2)

or $\quad (1+x^2)y_1 - 1 = 0$. ...(3)

Differentiating (3), we get

$$(1+x^2)y_2 + 2xy_1 = 0. \quad \ldots(4)$$

Differentiating equation (4) n times by Leibnitz's theorem, we get

$$y_{n+2}(1+x^2) + ny_{n+1}(2x) + \frac{n(n-1)}{2!}y_n.2 + 2xy_{n+1} + 2ny_n = 0$$

or $\quad (1+x^2)y_{n+2} + 2(n+1)xy_{n+1} + n(n+1)y_n = 0$. ...(5)

Putting x = 0 in (1), (2) and (4), we get

$$(y)_0 = 0, (y_1)_0 = 1, (y_2)_0 = 0.$$

Also putting x = 0 in (5), we get

$$(y_{n+2})_0 = -\{(n+1)n\}(y_n)_0. \quad \ldots(6)$$

Putting n – 2 in place of n in the reduction formula (6), we get

$$(y_n)_0 = -1\{(n-1)(n-2)\}(y_{n-2})_0$$
$$= [-\{(n-1)(n-2)\}][-\{(n-3)(n-4)\}](y_{n-4})_0,$$

since from (6), we have $(y_{n-2})_0 = -\{(n-3)(n-4)\}(y_{n-4})_0$.

Now there arise two cases.

Cases I: When n is even, we have

$$(y_n)_0 = [-\{(n-1)(n-2)\}][-\{(n-3)(n-4)\}] \ldots [-(3)(2)](y_2)_0$$
$$= 0, \text{ since } (y_2)_0 = 0.$$

Case II: When n is odd, we have

$$(y_n)_0 = [-1\{(n-1)(n-2)\}][-\{(n-3)(n-4)\}]$$
$$\ldots [-(4)(3)][-(2)(1)](y_1)_0$$
$$= (-1)^{(n-1)/2}(n-1)!, \text{ since } (y_1)_0 = 1.$$

1. Expansion of $(1 + x)^n$ *(Binomial Series)*

Let $f(x) = (1 + x)^n$. Then $f(0) = 1$;

$f^m(x) = n(n-1)(n-2) \ldots (n-m+1)(1+x)^{n-m}$

so that $f^m(0) = n(n-1) \ldots (n-m+1)$, where $m = 1, 2, 3, \ldots$

$\therefore$ $f'(0) = n$, $f''(0) = n(n-1)$, $f'''(0) = n(n-1)(n-2)$ and so on.

Substituting the values of $f(0)$, $f'(0)$, $f''(0)$ etc. in Maclaurin's series for $f(x)$, we get

$$(1+x)^n = 1 + nx + \frac{n(n-1)}{2!}x^2 + \ldots + \frac{n(n-1)\ldots(n-m+1)}{m!}x^m + \ldots$$

2. Expansion of $\log(1 + x)$ **(Meerut, 1996, 99; Lucknow, 93)**

Let $f(x) = \log(1 + x)$. Then $f(0) = \log 1 = 0$;

$$f^n(x) = \frac{(-1)^{n-1}(n-1)!}{(x+1)^n}$$

so that $f^n(0) = (-1)^{n-1}(n-1)!$, where $n = 1, 2, 3, \ldots$

$\therefore$ $f'(0) = (-1)^{1-1}(1-1)! = 1.$

$f''(0) = (-1)^{2-1}(1-1)! = -1!,$

$f'''(0) = (-1)^{3-1}(3-1)! = 2!,$

$f^{iv}(0) = (-1)^{4-1}(4-1)! = 3!$, and so on,

Substituting the values of $f(0)$, $f'(0)$, $f''(0)$, etc. in Maclaurin's series

$$f(x) = f(0) + xf'(0) + \frac{x^2}{2!}f''(0) + \ldots + \frac{x^n}{n!}f^n(0) + \ldots,$$

we get $\log(1+x) = 0 + x \cdot 1 - \frac{x^2}{2!}.1! + \frac{x^3}{3!}.2! - \frac{x^4}{4!}.3! + \ldots$

$$+ \frac{x^n}{n!}(-1)^{n-1}(n-1)! + \ldots$$

or $\log(1+x) = x - \frac{x^2}{2} + \frac{x^3}{3} - \frac{x^4}{4} + \ldots + (-1)^{n-1}\frac{x^n}{n} + \ldots$

Note: the function $\log x$ does not passes Maclaurin's series expansion because it is not defined at $x = 0$.

3. Expansion of $\sin x$ *(Sine Series)*

Let $f(x) = \sin x$. Then $f(0) = 0$,

$f'(x) = \cos x$ so that $f'(0) = 1$,

$f''(x) = -\sin x$ so that $f''(0) = 0$,

$f'''(x) = -\cos x$ so that $f'''(0) = -1$, and so on.

In general, $f^n(x) = \sin$ so that $f^n(x) = \sin\left(\frac{1}{2}n\pi\right)$.

When $n = 2m$, $f^n(0) = \sin m\pi = 0$ and when $n = 2m + 1$,

$$f^n(0) = \sin\left\{\frac{1}{2}(2m+1)\pi\right\} = \sin\left(m\pi + \frac{1}{2}\pi\right)$$

$$= (-1)^m \sin\left(\frac{1}{2}\pi\right) = (-1)^m.$$

Substituting these values in Maclaurin's series, we get

$$\sin x = 0 + x.1 + 0 + \frac{x^3}{3!}(-1) + 0 + \ldots + 0 + (-1)^m \frac{x^{2m+1}}{(2m+1)!} + \ldots$$

$$\text{or } \sin x = x - \frac{x^3}{3!} + \frac{x^4}{4!} - \frac{x^6}{6!} + \ldots + (-1)^m \frac{x^{2m}}{(2m)!} + \ldots$$

6

PARTIAL DIFFERENTIATION

6.1 FUNCTIONS OF TWO OR MORE INDEPENDENT VARIABLES

We after come across which depend two or more variables. For example, volume of a right circular cylinder of radius r and height h is given by $v = \pi r^2 h$ for a given value of r and h V has a definite values.

A function of x, y and z may be written as f(x, y, z) or $\phi(x, y, z)$, etc. Similarly, a function of x and y is generally denoted by the symbols f(x, y) or $\phi(x, y)$ etc.

If a derivative of a function of several independent variables be found with respect to any of them, keeping the others as constants, it is said to be a **Partial Derivative.** The operation of finding the partial derivatives of a function of more than one independent variables is called **Partial Differentiation.** The symbols $\partial/\partial x$, $\partial/\partial y$, etc., areused to denote such differentiations and the represents $\partial u/\partial x$, $\partial u/\partial y$, etc., are respectively called the **Partial Differential Coefficients** of u w.r.t. x, y, etc. Thus, if u = f(x, y, z), then the partial differential coefficient of u w.r.t. x *i.e.* $\partial u/\partial x$ is obtained by differentiating u w.r.t. x keeping y and z as constants. Sometimes $\partial f/\partial x$ is also denoted by f_x.

Second Order Partial Differential Coefficiens: If u = f(x, y), then $\partial u/\partial x$ or f_x and $\partial u/\partial y$ or f_y are themselves functions of x and y and can be again differentiated partially.

We call $\frac{\partial}{\partial x}\left(\frac{\partial u}{\partial y}\right)$, $\frac{\partial}{\partial y}\left(\frac{\partial u}{\partial x}\right)$, $\frac{\partial}{\partial x}\left(\frac{\partial u}{\partial x}\right)$ and $\frac{\partial}{\partial y}\left(\frac{\partial u}{\partial y}\right)$

as the second order partial derivatives are u and these are respectively as denoted by $\frac{\partial^2 u}{\partial x\, \partial y}, \frac{\partial^2 u}{\partial y\, \partial x}, \frac{\partial^2 u}{\partial x^2}$ and $\frac{\partial^2 u}{\partial y^2}$

Note: If u = f(x, y) and its partial derivatives are continuous (as is true in all ordinary cases), the order of differentiation is immaterial *i.e.*, $\frac{\partial^2 u}{\partial x\, \partial y} = \frac{\partial^2 u}{\partial y\, \partial x}$. The partial differentiation co-efficient of fx and fy are, fxu, fxy, fyx, fyy or

$\frac{\partial^2 f}{\partial x^2}, \frac{\partial^2 f}{\partial y\, \partial x}, \frac{\partial^2 f}{\partial x\, \partial y}, \frac{\partial^2 f}{\partial y^2}$ respectively. It should be specially noted that

$$\frac{\partial^2 f}{\partial y\, \partial x} \text{ means } \frac{\partial}{\partial y}\left(\frac{\partial f}{\partial x}\right) \text{ and } \frac{\partial^2 f}{\partial x\, \partial y} \text{ means } \frac{\partial}{\partial x}\left(\frac{\partial f}{\partial y}\right).$$

The student will be able to convince himself that in all ordinary cases.

$$\frac{\partial^2 f}{\partial y\, \partial x} = \frac{\partial^2 f}{\partial x\, \partial y}.$$

Example 1:

If $z = \sin\left(\frac{y}{x}\right)$, *then prove that* $\frac{\partial^2 z}{\partial x\, \partial y} = \frac{\partial^2 z}{\partial y\, \partial x}$.

(D.U. B.Sc. (G), 1995)

Solution:

We have

$$\frac{\partial z}{\partial x} = \cos\left(\frac{y}{x}\right)\left(\frac{-y}{x^2}\right), \frac{\partial z}{\partial y} = \cos\left(\frac{y}{x}\right)\left(\frac{1}{x}\right)$$

$$\frac{\partial^2 z}{\partial x\, \partial y} = \frac{\partial}{\partial x}\left[\frac{\partial z}{\partial y}\right] = \frac{\partial}{\partial x}\left[\frac{1}{z}\cos\left(\frac{y}{x}\right)\right]$$

$$= -\frac{1}{x^2}\cos\left(\frac{y}{x}\right) - \frac{1}{x}\sin\left(\frac{y}{x}\right)\left(\frac{-y}{x^2}\right)$$

$$= -\frac{1}{x^2}\cos\left(\frac{y}{x}\right) + \frac{y}{x^2}\sin\left(\frac{y}{x}\right) \qquad \text{...(1)}$$

Now $\frac{\partial^2 z}{\partial y\, \partial x} = \frac{\partial}{\partial y}\left(\frac{\partial z}{\partial x}\right) = \frac{\partial}{\partial y}\left(-\frac{y}{x^2}\cos\frac{y}{x}\right)$

$$= -\frac{1}{x^2}\left[1.\cos\left(\frac{y}{x}\right) - y\,\sin\left(\frac{y}{x}\right).\frac{1}{x}\right]$$

$$= -\frac{1}{x^2}\cos\left(\frac{y}{x}\right) + \frac{y}{x^2}\sin\left(\frac{y}{x}\right) \qquad ...(2)$$

from eqaution (1) and (2) we get

$$\frac{\partial^2 z}{\partial x\,\partial y} = \frac{\partial^2 z}{\partial y\,\partial x}.$$

Example 2:

If $z = f(y/x)$, show that $x\,(\partial z/\partial x) = +\,y(\partial z/\partial y) = 0$.

(Garhwal, 1993; Budelkhand, 92; U.P.P.C.S., 94)

Solution:

$$\frac{\partial z}{\partial x} = \left[f'\left(\frac{y}{x}\right)\right]\left(-\frac{y}{x^2}\right), \text{ (diff. particularly w.r.t. x).}$$

$$\therefore \qquad x\left(\frac{\partial z}{\partial x}\right) = -\left(\frac{y}{x}\right)f'\left(\frac{y}{x}\right). \qquad ...(1)$$

Again $\quad \dfrac{\partial z}{\partial y} = \left[f'\left(\dfrac{y}{x}\right)\right].\left(\dfrac{1}{x}\right)$, (diff. partially w.r.t. y)

$$\therefore \qquad y\left(\frac{\partial z}{\partial y}\right) = \left(\frac{y}{x}\right)f'\left(\frac{y}{x}\right). \qquad ...(2)$$

Adding equations (1) and (2), we get $x\left(\dfrac{\partial z}{\partial x}\right) + y\left(\dfrac{\partial z}{\partial y}\right) = 0$.

Example 3:

If $u = \sin^{-1}\dfrac{x}{y} + \tan^{-1}\dfrac{y}{x}$, show that $\dfrac{\partial u}{\partial x} + y\,\dfrac{\partial u}{\partial y} = 0$.

(Meerut, 1992S, 94P, 95, 96P; Agra, 92; Allahabad, 94; Gorakhpur, 95; Lucknow, 92; Kashmir, 94; Rohilkhand, 91)

Solution:

Here we have

$$\frac{\partial u}{\partial x} = \frac{1}{\sqrt{\left\{1-\left(\frac{x}{y}\right)^2\right\}}}.\frac{1}{y} + \frac{1}{1+\left(\frac{y}{x}\right)^2}.\left(-\frac{y}{x^2}\right), \text{ (treating as constant)}$$

$$= \frac{1}{\sqrt{(y^2 - x^2)}} - \frac{1}{(x^2 + y^2)}.$$

$$\therefore \qquad x\frac{\partial u}{\partial x} = \frac{x}{\sqrt{(y^2 - x^2)}} - \frac{xy}{x^2 + y^2}. \qquad ...(1)$$

Again

$$\frac{\partial u}{\partial y} = \frac{1}{\sqrt{\left\{1 - \left(\frac{x}{y}\right)^2\right\}}}\left(1 - \frac{x}{y^2}\right) + \frac{1}{1 + \left(\frac{y}{x}\right)^2} \cdot \frac{1}{x}, \text{ (treating x as constant)}$$

$$= -\frac{1}{y\sqrt{(y^2 - x^2)}} + \frac{x}{x^2 + y^2}.$$

$$\therefore \qquad y\frac{\partial u}{\partial y} = -\frac{r}{\sqrt{(y^2 - x^2)}} + \frac{xy}{x^2 + y^2}. \qquad ...(2)$$

Adding (1) and (2), we have $x\left(\frac{\partial u}{\partial x}\right) + y\left(\frac{\partial u}{\partial y}\right) = 0$.

Example 4:

Verify that $\frac{\partial^2 z}{\partial x\, \partial y} = \frac{\partial^2 z}{\partial y\, \partial x}$ *for the following functions:*

(i) $z = x \log xy$ **(Kanpur, 1998)**

(ii) $z = \sin^{-1}\left(\frac{x}{y}\right)$

(iii) $z = a \tan^{-1}\left(\frac{x}{y}\right)$ **(D.U. B.Sc. (G), 1997)**

(iv) $z = x^y + y^x$

(v) $z = a^{ax} \sin by$

(vi) $z = \log (y \sin x + x \sin y)$.

Solution:

(i) We have z = x log y.

Differentiating z partially w.r.t. a taking y as constant, we have

$$\frac{\partial z}{\partial x} = \log y.$$

Now $$\frac{\partial^2 z}{\partial y\,\partial x}=\frac{\partial}{\partial y}\left(\frac{\partial z}{\partial x}\right)=\frac{\partial}{\partial y}(\log y)=\frac{1}{y}.$$

Again differentiating z partially w.r.t. y taking x as constant, we have

$$\frac{\partial z}{\partial y}=\frac{x}{y}.$$

Now $\dfrac{\partial^2 z}{\partial x\,\partial y}=\dfrac{\partial}{\partial x}\left(\dfrac{\partial z}{\partial y}\right)=\dfrac{\partial}{\partial x}\left(\dfrac{x}{y}\right)=\dfrac{1}{y}$, treating y as constant.

Hence $\dfrac{\partial^2 z}{\partial x\,\partial y}=\dfrac{\partial^2 z}{\partial y\,\partial x}$.

(ii) We have $z=\sin^{-1}\left(\dfrac{x}{y}\right)$.

$$\therefore\quad \frac{\partial z}{\partial x}=\frac{1}{\sqrt{\left\{1-\left(\frac{x}{y}\right)^2\right\}}}.\frac{1}{y}=\frac{1}{\sqrt{(y^2-x^2)}}=(y^2-x^2)^{-1/2},$$

Now $$\frac{\partial^2 z}{\partial y\,\partial x}=\frac{\partial}{\partial y}\left(\frac{\partial z}{\partial x}\right)=\frac{\partial}{\partial y}\left\{(y^2-x^2)^{-1/2}\right\}$$

$$=-\frac{1}{2}(y^2-x^2)^{-3/2}.2y=\frac{-y}{(y^2-x^2)^{3/2}}.$$

Again $$\frac{\partial z}{\partial y}=\frac{1}{\sqrt{\left\{1-\left(\frac{x}{y}\right)^2\right\}}}.\left(-\frac{x}{y^2}\right)=-\frac{x}{y}(y^2-x^2)^{-1/2},$$

and $$\frac{\partial^2 z}{\partial x\,\partial y}=\frac{\partial}{\partial x}\left(\frac{\partial z}{\partial y}\right)$$

$$=-\frac{1}{y}(y^2-x^2)^{-1/2}-\left(\frac{y}{x}\right).\left\{-\frac{1}{2}(y^2-x^2)^{-3/2}\right\}.(-2x)$$

$$=-\frac{1}{y(y^2-x^2)^{1/2}}-\frac{x^2}{y(y^2-x^2)^{2/2}}=-\frac{(y^2-x^2)+x^2}{y(y^2-x^2)^{3/2}}$$

$$=\frac{-y}{(y^2-x^2)^{3/2}}.$$

Hence $\frac{\partial^2 z}{\partial x\, \partial y} = \frac{\partial^2 z}{\partial y\, \partial x}$.

(iii) We have $z = a \tan^{-1}\left(\frac{x}{y}\right)$.

$\therefore\ \frac{\partial z}{\partial x} = a\frac{1}{1+\left(\frac{x^2}{y^2}\right)}.\frac{1}{y} = \frac{ay}{x^2+y^2},$

and $\frac{\partial^2 z}{\partial y\, \partial x} = a.\frac{1.\left(x^2+y^2\right)-y.2y}{\left(x^2+y^2\right)^2} = \frac{a\left(x^2+y^2\right)}{\left(x^2+y^2\right)^2}.$

Again $\frac{\partial z}{\partial y} = a.\frac{1}{1+\left(\frac{x^2}{y^2}\right)}.\left(-\frac{x}{y^2}\right) = \frac{-ax}{x^2+y^2},$

and $\frac{\partial^2 z}{\partial x\, \partial y} = -a.\frac{1.\left(x^2+y^2\right)-x.2x}{\left(x^2+y^2\right)^2} = -a.\frac{y^2-x^2}{\left(x^2+y^2\right)^2} = \frac{a\left(x^2-y^2\right)}{\left(x^2+y^2\right)^2}$

Thus, $\frac{\partial^2 z}{\partial x\, \partial y} = \frac{\partial^2 z}{\partial y\, \partial x}$.

(iv) We have $z = x^y + y^x$.

$\therefore = \frac{\partial z}{\partial x} = yx^{y-1} + y^x.\log y$

and $\frac{\partial^2 z}{\partial y\, \partial x} = \frac{\partial}{\partial y}\ [yx^{y-1} + y^x \log y]$

$= 1.xy^{-1} + yx^{y-1}.\log x + xy^{x-1} \log y + y^x.\left(\frac{1}{y}\right)$

$= xy^{-1}\ (1 + y \log x) + y^{x-1}\ (1 + x \log y).$

Again $\frac{\partial z}{\partial y}\ x^y \log x + xy^{x-1}$ and $\frac{\partial^2 z}{\partial x\, \partial y} = \frac{\partial}{\partial x}\ [x^y \log x + xy^{x-1}]$

$= yx^{y-1}.\log x + x^y.\left(\frac{1}{x}\right) + 1.y^{x-1} + xy^{x-1} \log y$

$= x^{y-1}\ (1 + y \log x) + y^{z-1}\ (1 + x \log y).$

Thus $\frac{\partial^2 z}{\partial x\, \partial y} = \frac{\partial^2 z}{\partial y\, \partial x}$.

(v) We have $z = e^{ax} \sin by$.

$$\therefore \frac{\partial z}{\partial x} = ae^{ax} \sin by,$$

$$\frac{\partial^2 z}{\partial y\, \partial x} = \frac{\partial}{\partial y}(ae^{ax} \sin by) = ae^{ax}.b \cos by = ab\, e^{ax} \cos by.$$

Again $\frac{\partial z}{\partial y} = e^{ax}.b \cos by;$

$$\frac{\partial^2 z}{\partial x\, \partial y} = \frac{\partial}{\partial x}(be^{ax} \cos by) = ab\, e^{ax} \cos by.$$

Thus $\frac{\partial^2 z}{\partial y\, \partial x}\, \partial x = \frac{\partial^2 z}{\partial x\, \partial y}$.

(vi) We have $z = \log (y \sin x + x \sin y)$.

$$\therefore \qquad \frac{\partial z}{\partial x} = \frac{y \cos x + \sin y}{y \sin x + x \sin y},$$

and $\frac{\partial^2 z}{\partial y\, \partial x}$

$$= \frac{(y \sin x + x \sin y)(\cos x + \cos y) - (y \cos x + \sin y)(\sin x + x \cos y)}{(y \sin x + x \sin y)^2}$$

$$= \frac{x \sin y \cos + y \sin x \cos y - \sin x \sin y - xy \cos x \cos y}{(y \sin x + x \sin y)^2}$$

Again $\frac{\partial z}{\partial y} = \frac{\sin x + x \cos y}{y \sin x + x \sin y}$

and $\frac{\partial^2 z}{\partial x\, \partial y}$

$$= \frac{(y \sin x + x \sin y)(\cos x + \cos y) - (\sin x + x \cos y)(y \cos x + \sin y)}{(y \sin x + x \sin y)^2}$$

$$= \frac{x \sin y \cos x + y \sin x \cos y - \sin x \sin y - xy \cos x \cos y}{(y \sin x + x \sin y)^2}.$$

$$\therefore \frac{\partial^2 z}{\partial x\, \partial y} = \frac{\partial^2 z}{\partial y\, \partial x}.$$

Example 5:

If $z = f(x + ay) + \phi(x - ay)$, prove that

$$\frac{\partial^2 z}{\partial y^2} = a^2\left(\frac{\partial^2 z}{\partial x^2}\right).$$

(Rohilkhand, 1991; Bihar, 97; Gorakhpur, 90; Meerut, 90, 95)

Solution:

We have z = f(x + ay) + f(x − ay).

$\therefore \quad \frac{\partial z}{\partial x} = f'(x + ay) + \phi'(x - ay)$, (diff. partially w.r.t. 'x')

and $\frac{\partial^2 z}{\partial x^2} = f''(x - y) + \phi''(x - y)$ ...(1)

Again $\frac{\partial z}{\partial y} = af'(x + ay) - a\phi'(x - ay)$.

$\therefore \quad \frac{\partial^2 z}{\partial y^2} = a^2 f''(x + ay) + a^2\phi''(x - ay)$. ...(2)

From (1) and (2), we get $\frac{\partial^2 z}{\partial y^2} = a^2\left(\frac{\partial^2 z}{\partial x^2}\right)$.

Example 6:

If $z = x f\left(\frac{y}{x}\right)$ *prove that*

$x\frac{\partial z}{\partial x} + y\frac{\partial z}{\partial y} = 2z.$ **(D.U. B.A. (P), 1999, 1995)**

Solution:

We have $\frac{\partial z}{\partial x} = y\left\{f\left(\frac{y}{x}\right) + xf'\left(\frac{y}{x}\right)\left(-\frac{y}{x^2}\right)\right\}$

$\frac{\partial z}{\partial y} = x\left\{f\left(\frac{y}{x}\right) + yf'\left(\frac{y}{x}\right)\frac{1}{x}\right\}.$

Hence $x\frac{\partial z}{\partial x} + y\frac{\partial z}{\partial y} = 2xy\ f\left(\frac{y}{x}\right) = 22$.

Example 7:

If $u = \sin^{-1}\left(\frac{x^2 + y^2}{x + y}\right)$, *show that*

$$x\left(\frac{\partial u}{\partial x}\right)+y\left(\frac{\partial u}{\partial y}\right)=\tan u.$$ **(Kanpur, 1990; Allahabad, 94)**

Solution:

We have sin u = $(x^2 + y^2)/(x + y)$.

$\therefore$ $\log \sin u = \log (x^2 + y^2) - \log (x + y)$. ...(1)

Differentiating (1) partially w.r.t. x, we get

$$\frac{1}{\sin u}\cos u.\frac{\partial u}{\partial x}=\frac{2x}{x^2+y^2}-\frac{1}{x+y},$$

$$\therefore \quad (\cot u)\, x\frac{\partial u}{\partial x}=\frac{2x^2}{x^2+y^2}-\frac{x}{x+y}. \quad ...(2)$$

Again differentiating (1) partially w.r.t. y, we get

$$\frac{\cos u}{\sin u}.\frac{\partial u}{\partial y}=\frac{2y}{x^2+y^2}-\frac{1}{x+y}.$$

$$\therefore \quad (\cot u).y\frac{\partial u}{\partial y}=\frac{2y^2}{x^2+y^2}-\frac{y}{x+y}. \quad ...(3)$$

Adding (2) and (3), we get

$$(\cot u)\left(x\frac{\partial u}{\partial x}+y\frac{\partial u}{\partial y}\right)=\frac{2x^2+2y^2}{x^2+y^2}-\frac{x+y}{x+y}=2-1=1.$$

$$\therefore \quad x\left(\frac{\partial u}{\partial x}\right)+y\left(\frac{\partial u}{\partial y}\right) = 1/\cot u = \tan u.$$

Example 8:

If $z = xy\, f(y/x)$

prove that $x\dfrac{\partial z}{\partial x}+y\dfrac{\partial z}{\partial y}=2z$ **(D.U., B.A. (P), 1999, 96, 95)**

Solution:

We have, $\dfrac{\partial z}{\partial x} = y\,[f(y/x) + xf'(y/x)\,(-y/x^2)]$

$\dfrac{\partial z}{\partial y} = x\,[f(y/x) + yf'(y/x)\,1/x]$

Hence, $x\dfrac{\partial z}{\partial x}+y\dfrac{\partial z}{\partial y} = xy\, f(y/x) = 22$

Example 9:

If $z = tan^{-1}\dfrac{x^2 + y^2}{x - y}$. *Prove that*

$x\dfrac{\partial z}{\partial u} + y\dfrac{\partial z}{\partial y} = sin\ 2z.$ **(D.U., B.Sc. (H) 1993, 1998)**

Solution:

We have

$$z = \tan^{-1}\frac{x^2 + y^2}{x - y}$$

$$\Rightarrow \qquad \tan z = \frac{x^2 + y^2}{x - y}$$

Let $\quad u = \tan z$ ...(1)

$$\therefore \qquad u = \frac{x^2 + y^2}{x - y} = \frac{x^2\left\{1 + \left(\dfrac{y}{x}\right)^2\right\}}{x\left\{1 - \left(\dfrac{y}{x}\right)\right\}} = x^2 f\left(\frac{y}{x}\right).$$

Thus, u is a homogeneous function of degree z.

Hence, the Euler's theorem;

$$x\frac{\partial u}{\partial x} + y\frac{\partial u}{\partial y} = 2u \qquad \text{...(2)}$$

From (1) $\dfrac{\partial u}{\partial u} = \sec^2 z\dfrac{\partial z}{\partial u}, \dfrac{\partial u}{\partial y} = \sec^2 z\dfrac{\partial z}{\partial y}$ ...(3)

Putting (3) and (1) in (2) we get

$$\sec^2 z\left(x\frac{\partial z}{\partial x} + y\frac{\partial z}{\partial y}\right) = 2\tan z$$

or $$x\frac{\partial z}{\partial x} + y\frac{\partial z}{\partial y} = 2\frac{\sin z}{\cos z}.\cos^2 z$$

$$2\sin z \cos z = \sin 2z.$$

Example 10:

If $u = x^2\ tan^{-1}\dfrac{y}{x} - y^2 tan^{-1}\dfrac{x}{y}$, *find* $\dfrac{\partial^2 u}{\partial x\, \partial y}$.

(Andhra, 1996; Meerut, 99P; Kashmir, 98; Utkal, 99)

Solution:

We have

$$\frac{\partial u}{\partial y} = x^2 . \frac{1}{1+\left(\frac{y}{x}\right)^2} . \frac{1}{x} - 2y \tan^{-1}\frac{x}{y} - y^2 \frac{1}{1+\left(\frac{x}{y}\right)^2} . \left(-\frac{x}{y^2}\right)$$

$$= \frac{x^3}{x^2+y^2} - 2y \tan^{-1}\frac{x}{y} + \frac{xy^2}{x^2+y^2}$$

$$= \frac{x(x^2+y^2)}{x^2+y^2} - 2y \tan^{-1}\frac{x}{y} = x - 2y \tan^{-1}\frac{x}{y}.$$

Now $$\frac{\partial^2 u}{\partial x\, \partial y} = \frac{\partial}{\partial x}\left(\frac{\partial u}{\partial y}\right) = 1 - 2y \frac{1}{1+\left(\frac{x}{y}\right)^2} . \frac{1}{y} = 1 - \frac{2y^2}{x^2+y^2}$$

$$= \frac{x^2+y^2-2y^2}{x^2+y^2} = \frac{x^2-y^2}{x^2+y^2}.$$

Hence $$\frac{\partial^2 u}{\partial x\, \partial y} = \frac{x^2-y^2}{x^2+y^2}.$$

Example 11:

Find the value of

$$\frac{1}{a^2}\frac{\partial^2 z}{\partial x^2} + \frac{1}{b^2}\frac{\partial^2 z}{\partial y^2} \text{ when } a^2x^2 + b^2y^2 - c^2z^2 = 0.$$ **(Meerut, 1991)**

Solution:

We have $$z^2 = \frac{a^2}{c^2}x^2 + \frac{b^2}{c^2}y^2. \qquad ...(1)$$

Differentiating (1) partially with respect to x taking y as constant, we have

$$2z\frac{\partial z}{\partial x} = 2\frac{a^2}{c^2}x \text{ or } \frac{\partial z}{\partial x} = \frac{a^2}{c^2}.\frac{x}{z}.$$

$$\therefore \quad \frac{\partial^2 z}{\partial x^2} = \frac{\partial}{\partial x}\left(\frac{a^2}{c^2}x.\frac{1}{z}\right) = \frac{a^2}{c^2}.\frac{1}{z} + \frac{c^2}{c^2}x.\left(-\frac{1}{z^2}\right)\frac{\partial z}{\partial x}$$

$$= \frac{a^2}{c^2 z} - \frac{a^2 x}{c^2 z^2} \cdot \left(\frac{a^2 x}{c^2 z} \right) = \frac{a^2}{c^2 z} - \frac{a^4 x^2}{c^4 z^3}.$$

$$\therefore \quad \frac{1}{a^2} \frac{\partial^2 z}{\partial x^2} = \frac{1}{c^2 z} - \frac{a^2 x^2}{c^4 z^3} \qquad ...(2)$$

Again differentaiting (1) partially with respect to y taking x as constant, we have

$$2z \frac{\partial z}{\partial y} = 2 \frac{b^2}{c^2} y \text{ or } \frac{\partial z}{\partial y} = \frac{b^2}{c^2} \cdot \frac{y}{z}.$$

$$\therefore \quad \frac{\partial^2 z}{\partial y^2} = \frac{b^2}{c^2} \cdot \frac{1}{z} + \frac{b^2}{c^2} y \cdot \left(-\frac{1}{z^2} \right) \frac{\partial z}{\partial y}$$

$$= \frac{b^2}{c^2 z} - \frac{b^2 y}{c^2 z^2} \cdot \left(\frac{b^2 y}{c^2 z} \right) = \frac{b^2}{c^2 z} - \frac{b^4 y^2}{c^4 z^3}.$$

$$\therefore \quad \frac{1}{b^2} \frac{\partial^2 z}{\partial y^2} = \frac{1}{c^2 z} - \frac{b^2 y^2}{c^4 z^3}. \qquad ...(3)$$

Adding (2) and (3), we get

$$\frac{1}{a^2} \frac{\partial^2 z}{\partial x^2} + \frac{1}{b^2} \frac{\partial^2 z}{\partial y^2} = \frac{2}{c^2 z} - \frac{a^2 x^2 + b^2 y^2}{c^4 z^3}$$

$$= \frac{2}{c^2 z} - \frac{c^2 z^2}{c^4 z^3} - \frac{1}{c^2 z} - \frac{1}{c^2 z} = \frac{1}{c^2 z}.$$

Example 12:

If $z = xf(x + y) + y\phi(x + y)$, prove that

$$\left(\frac{\partial^2 z}{\partial x^2} \right) + \left(\frac{\partial^2 z}{\partial y^2} \right) = 2 \left(\frac{\partial^2 z}{\partial x \, \partial y} \right).$$

Solution:

We have $z = xf(x + y) + y\phi(x + y)$.

$$\therefore \quad \frac{\partial z}{\partial x} = f(x + y) + xf'(x + y) + y\phi(x + y),$$

$$\text{and } \frac{\partial^2 z}{\partial x^2} = f'(x + y) + f'(x + y) + xf''(x + y) + y\phi''(x + y)$$

$$= 2f'(x + y) + xf''(x + y) + y\phi''(x + y).$$

$$\text{Also } \frac{\partial y}{\partial z} = xf'(x + y) + \phi(x + y) + y\phi'(x + y)$$

$$\text{and } \frac{\partial^2 z}{\partial y^2} = xf''(x + y) + \phi'(x + y) + \phi'(x + y) + y\phi''(x + y)$$

$$= 2f'(x + y) + xf''(x + y) + yf''(x + y).$$

$$\text{Again } \frac{\partial^2 z}{\partial x\, \partial y} = \frac{\partial}{\partial x}\left(\frac{\partial z}{\partial y}\right)$$

$$= \frac{\partial}{\partial x}\{xf'(x + y) + \phi(x + y) + y\phi'(x + y)\}$$

$$= f'(x + y) + xf''(x + y) + \phi'(x + y) + y\phi''(x + y).$$

$$\text{Now } \left(\frac{\partial^2 z}{\partial x^2}\right) + \left(\frac{\partial^2 x}{\partial y^2}\right)$$

$$= 2[f'(x + y) + f''(x + y) + xf'(x + y) + y\phi''(x + y)]$$

$$= 2\left(\frac{\partial^2 z}{\partial x\, \partial y}\right).$$

Example 13:

If $u = \tan^{-1} \frac{xy}{\sqrt{(1 + x^2 + y^2)}}$, *show that*

$$\frac{\partial^2 u}{\partial x\, \partial y} = \frac{1}{(1 + x^2 + y^2)^{3/2}}.$$ **(Agra, 1997)**

Solution:

We have

$$\frac{\partial u}{\partial x} = \frac{1}{1 + \frac{x^2 y^2}{1 + x^2 + y^2}}$$

$$y.\frac{1.\sqrt{(1 + x^2 + y^2) - x.\frac{1}{2}(1 + x^2 + y^2)^{-1/2}(-2x)}}{1 + x^2 + y^2}$$

$$= \frac{1 + x^2 + y^2}{1 + x^2 + y^2 + x^2 y^2} \cdot \frac{(1 + x^2 + y^2) - x^2}{(1 + x^2 + y^2)(1 + x^2 + y^2)^{1/2}}$$

$$= \frac{y\left(1+y^2\right)}{\left(1+x^2\right)\left(1+y^2\right)\left(1+x^2+y^2\right)^{1/2}} = \frac{1}{1+x^2}\cdot\frac{y}{\left(1+x^2+y^2\right)^{1/2}}.$$

$$\therefore \qquad \frac{\partial^2 u}{\partial x\,\partial y} = \frac{\partial}{\partial y}\left(\frac{\partial u}{\partial x}\right)$$

$$= \frac{1}{1+x^2}\cdot\frac{1.\left(1+x^2+y^2\right)^{1/2} - y.\frac{1}{2}\left(1+x^2+y^2\right)^{-1/2}.2y}{1+x^2+y^2}$$

$$= \frac{1}{1+x^2}\cdot\frac{1+x^2+y^2-y^2}{\left(1+x^2+y^2\right)\left(1+x^2+y^2\right)^{1/2}}$$

$$= \frac{1}{1+x^2}\cdot\frac{1+x^2}{\left(1+x^2+y^2\right)^{3/2}} = \frac{1}{\left(1+x^2+y^2\right)^{3/2}}.$$

Example 14:

If $\dfrac{x^2}{a^2+u} + \dfrac{y^2}{b^2+u} + \dfrac{z^2}{c^2+u} = 1$, *prove that*

$$\left(\frac{\partial u}{\partial x}\right)^2 + \left(\frac{\partial u}{\partial y}\right)^2 + \left(\frac{\partial u}{\partial z}\right)^2 = 2\left(x\frac{\partial u}{\partial x} + y\frac{\partial u}{\partial y} + z\frac{\partial u}{\partial z}\right).$$

(Meerut, 1993; Kanpur, 92; Gorakhpur, 97)

Solution:

From the given equation we observe that u is a function of three independent variables x, y and z. Differentiating the given equation partially w.r.t. 'x', we get

$$\frac{2x}{a^2+u} - \left\{\frac{x^2}{\left(a^2+u\right)^2} + \frac{y^2}{\left(b^2+u\right)^2} + \frac{z^2}{\left(c^2+u\right)^2}\right\}\frac{\partial u}{\partial x} = 0.$$

$$\therefore \quad \frac{\partial u}{\partial x} = \frac{2x/\left(a^2+u\right)}{\Sigma\left\{x^2/\left(a^2+u\right)^2\right\}}.$$

Similarly, by symmetry, we can write the values of $\dfrac{\partial u}{\partial y}$ and $\dfrac{\partial u}{\partial z}$.

$$\text{Now } \left(\frac{\partial u}{\partial x}\right)^2 = \frac{4x^2/\left(a^2+u\right)^2}{\left[\Sigma\left\{x^2/(a^2+u)^2\right\}\right]}.$$

$$\therefore \quad \Sigma\left(\frac{\partial u}{\partial x}\right)^2 = \frac{4\Sigma\left\{x^2/\left(a^2+u\right)^2\right\}}{\left[\left\{x^2/\left(a^2+u\right)^2\right\}\right]}$$

$$= \frac{4}{\Sigma\left\{x^2/\left(a^2+u\right)^2\right\}} \qquad ...(1)$$

$$\text{Again } 2x\frac{\partial u}{\partial x} = \frac{4x^2/\left(a^2+u\right)}{\Sigma\left\{x^2/\left(a^2+u\right)^2\right\}}.$$

$$\therefore \quad 2\Sigma x\left(\frac{\partial u}{\partial x}\right) = \frac{4\Sigma\left\{x^2/\left(a^2+u\right)\right\}}{\Sigma\left\{x^2/\left(a^2+u\right)^2\right\}} = \frac{4}{\Sigma\left\{x^2/\left(a^2+u\right)^2\right\}}. \qquad ...(2)$$

[$\because$ Σ $\{x^2/(a^2 + u)\} = 1$, from the given relation]

Now from (1) and (2), we have

$$\Sigma\left(\frac{\partial u}{\partial x}\right)^2 = 2\Sigma\left\{x\left(\frac{\partial u}{\partial x}\right)\right\}$$

Example 15:

If $z = (x^2 + y^2)/(x + y)$, show that

$$\left(\frac{\partial z}{\partial x} - \frac{\partial z}{\partial y}\right)^2 = 4\left(1 - \frac{\partial z}{\partial x} - \frac{\partial z}{\partial y}\right).$$

(Kanpur, 1998; Meerut, 95; Agra, 90; Allahabad, 98)

Solution:

We have $z = (x^2 + y^2)/(x + y)$.

$$\therefore \quad \frac{\partial z}{\partial x} = \frac{(x+y).2x - \left(x^2+y^2\right).1}{(x+y)^2} = \frac{x^2 - y^2 + 2xy}{(x+y)^2}$$

and $\frac{\partial z}{\partial y}=\frac{(x+y)2y-(x^2+y^2).1}{(x+y)^2}=\frac{y^2-x^2+2xy}{(x+y)^2}.$

$$\therefore \left(\frac{\partial z}{\partial x}-\frac{\partial z}{\partial y}\right)^2=\left[\frac{(x^2-y^2+2xy)-(y^2-x^2+2xy)}{(x+y)^2}\right]^2$$

$$=\left[\frac{2(x^2-y^2)^2}{(x^2+y^2)}\right]=4\left[\frac{x-y}{x+y}\right]^2.$$

Also $1-\frac{\partial z}{\partial x}-\frac{\partial z}{\partial y}=1-\frac{x^2-y^2+2xy}{(x+y^2)}-\frac{y^2-x^2+2xy}{(x+y)^2}$

$$=\frac{(x^2+y^2+2xy)-x^2+y^2-2xy-y^2+x^2-2xy}{(x+y)^2}=\frac{x^2-2xy+y^2}{(x+y)^2}$$

$$=\left(\frac{x-y}{x+y}\right)^2.$$

Hence $\left(\frac{\partial z}{\partial x}-\frac{\partial z}{\partial y}\right)^2=4\left(1-\frac{\partial z}{\partial x}-\frac{\partial z}{\partial y}\right).$

Example 16:

If $u = \log r$, where $r^2 = (x-a)^2 + (y-b^2) + (z-c)^2$, show that

$$\frac{\partial^2 u}{\partial x^2}+\frac{\partial^2 u}{\partial y^2}+\frac{\partial^2 u}{\partial z^2}=\frac{1}{r^2}.$$

Solution:

We have $r^2 = (x-a)^2 + (y-b)^2 + (z-c)^2$. ...(1)

Differentiating (1) partially wrt x we have

$$2r\frac{\partial r}{\partial x}=2(x-a) \text{ or } \frac{\partial r}{\partial x}=\frac{1}{r}\left(\frac{x-a}{r}\right) \quad ...(2)$$

Now $u = \log r$. $\therefore \frac{\partial u}{\partial x}=\frac{1}{r}$

$$\therefore \frac{\partial r}{\partial x}=\frac{1}{r}\left(\frac{x-a}{r}\right). \quad ...[\text{from (2)}]$$

Thus, $\dfrac{\partial u}{\partial x} = \dfrac{(x-a)}{r^2}$

$$\therefore \frac{\partial^2 u}{\partial x^2} = \frac{\partial}{\partial x}\left(\frac{\partial u}{\partial x}\right) = \frac{\partial}{\partial x}\left(\frac{x-a}{r^2}\right) = \frac{r^2(1)-(x-a).2r\left(\frac{\partial r}{\partial x}\right)}{r^4}$$

$$= \frac{r^2-2(x-a)^2}{r^4}, \qquad \left[\because \text{ from (2), } \frac{\partial r}{\partial x} = (x-a)/r\right]$$

Similarly, by symmetry

$$\frac{\partial^2 u}{\partial y^2} = \frac{r^2-2(y-b)^2}{r^4} \text{ and } \frac{\partial^2 u}{\partial z^2} = \frac{r^2-2(z-c)^2}{r^4}.$$

Hence $\dfrac{\partial^2 u}{\partial x^2} + \dfrac{\partial^2 u}{\partial y^2} + \dfrac{\partial^2 u}{\partial z^2} = \dfrac{3r^2-2\{(x-a)^2+(y-b)^2+(z-c)^2\}}{r^4}$

$$= \frac{3r^2-2r^2}{r^4}, \text{ \{using (1)\} } = \frac{r^2}{r^4} = \frac{1}{r^2}.$$

Example 17:

If $1/u = \sqrt{(x^2+y^2+z^2)}$, *show that*

$$x\frac{\partial u}{\partial x} + y\frac{\partial u}{\partial y} + z\frac{\partial u}{\partial z} = -u,$$

and
$$\frac{\partial^2 u}{\partial x^2} + \frac{\partial^2 u}{\partial y^2} + \frac{\partial^2 u}{\partial z^2} = 0.$$

(Allahabad, 1991; Meerut, 95, 97, 98; Agra, 99; Garhwal, 97; Kumaun, 93; Gorakhpur, 93; Vikram, 95; Jhansi, 99)

Solution:

We have $u = (x^2 + y^2 + z^2)^{-1/2}$.

$$\therefore \frac{\partial u}{\partial x} = -\frac{1}{2}(x^2 + y^2 + z^2)^{-3/2}.(2x) = -x(x^2 + y^2 + z^2)^{-3/2},$$

and
$$\frac{\partial^2 u}{\partial x^2} = x.\frac{3}{2}(x^2 + y^2 + z^2)^{-5/2}(2x) - 1(x^2 + y^2 + z^2)^{-3/2}$$
$$= (x^2 + y^2 + z^2)^{-5/2}[3x^2 - (x^2 + y^2 + z^2)]$$
$$= u^5(2x^2 - y^2 - z^2).$$

Similarly, by symmetry $\dfrac{\partial u}{\partial y} = -y(x^2+y^2+z^2)^{-3/2}$,

$$\frac{\partial^2 u}{\partial y^2} = u^5(2y^2 - x^2 - z^2), \left(\frac{\partial u}{\partial z}\right) = -z(x^2+y^2+z^2)^{-3/2},$$

Now $x\dfrac{\partial u}{\partial x} + y\dfrac{\partial u}{\partial y} + z\dfrac{\partial u}{\partial z}$

$$= -x^2(x^2+y^2+z^2)^{-3/2} - y^2(x^2+y^2+z^2)^{-3/2} - z^2(x^2+y^2+z^2)^{-3/2}$$

$$= -(x^2+y^2+z^2)^{-3/2}(x^2+y^2+z^2) = -(x^2+y^2+z^2)^{-1/2} = -u.$$

Also $\dfrac{\partial^2 u}{\partial x^2} + \dfrac{\partial^2 u}{\partial y^2} + \dfrac{\partial^2 u}{\partial z^2}$

$$= u^5(2x^2 - y^2 - z^2 + 2y^2 - x^2 - z^2 - x^2 - y^2) = u^5.0 = 0.$$

Example 18:

If $u = e^{xyz}$, show that

$$\frac{\partial^3 u}{\partial x\, \partial y\, \partial z} = (1 + 3xyz + x^2y^2z^2)e^{xyz}.$$

(Lucknow, 1993; Meerut, 94; Gorakhpur, 91)

Solution:s

Here $u = e^{xyz}$. $\quad \therefore \dfrac{\partial u}{\partial z} = xy\, e^{xyz}$.

$$\text{Now } \frac{\partial^2 u}{\partial y\, \partial z} = \frac{\partial}{\partial y}\left(\frac{\partial u}{\partial z}\right) = \frac{\partial}{\partial y}\left(xy\, e^{xyz}\right) = x\frac{\partial}{\partial y}\left(y\, e^{xyz}\right)$$

$$= x[y.xz\, e^{xyz} + e^{xyz}] = e^{xyz}(x^2yz + z).$$

$$\text{Again } \frac{\partial^3 u}{\partial x\, \partial y\, \partial z} = \frac{\partial}{\partial x}\left(\frac{\partial^2 u}{\partial y\, \partial z}\right) = \frac{\partial}{\partial x}\left[e^{xyz}\left(x^2yz + x\right)\right]$$

$$= e^{xyz}(2xyz + 1) + yz\, e^{xyz}(x^2yz + x)$$

$$= e^{xyz}[2xyz + 1 + x^2y^2z^2 + xyz]$$

$$= e^{xyz}[1 + 3xyz + x^2y^2z^2].$$

Example 19:

If $u = \log(x^3 + y^3 + z^3 - 3xyz)$, show that

$$\frac{\partial u}{\partial x} + \frac{\partial u}{\partial y} + \frac{\partial u}{\partial z} = \frac{3}{x+y+z}$$

and $$\left(\frac{\partial}{\partial x}+\frac{\partial}{\partial y}+\frac{\partial}{\partial z}\right)^2 u=\frac{-9}{(x+y+z)^2}.$$

(Meerut, 1996, 94P, 93S, 91; Bundelkhand, 98; Gorakhpur, 95; Rohilkhand, 93, 98, 99; Lucknow, 90; Allahabad, 90; Kanpur, 96)

Solution:

We have $u = \log(x^3 + y^3 + z^3 - 3xyz)$.

$$\therefore \quad \frac{\partial u}{\partial x}=\frac{3x^2-3yz}{x^3+y^3+z^3-3xyz},\ \frac{\partial u}{\partial y}=\frac{3y^2-3zx}{x^3+y^3+z^3-3xyz}$$

and $$\frac{\partial u}{\partial z}=\frac{3z^2-3xy}{x^3+y^3+z^3-3xyz}$$

$$\therefore \quad \frac{\partial u}{\partial x}+\frac{\partial u}{\partial y}+\frac{\partial u}{\partial z}=\frac{3(x^2+y^2+z^2-yz-zx-xy)}{x^3+y^3+z^3-3xyz}$$

$$=\frac{3(x^2+y^2+z^2-yz-zx-xy)}{(x+y+z)(x^2+y^2+z^2-yz-zx-xy)}=\frac{3}{x+y+z}. \quad ...(1)$$

Now $$\left(\frac{\partial}{\partial x}+\frac{\partial}{\partial y}+\frac{\partial}{\partial z}\right)^2 u=\left(\frac{\partial}{\partial x}+\frac{\partial}{\partial y}+\frac{\partial}{\partial z}\right)\left(\frac{\partial u}{\partial x}+\frac{\partial u}{\partial y}+\frac{\partial u}{\partial z}\right)$$

$$=\left(\frac{\partial}{\partial x}+\frac{\partial}{\partial y}+\frac{\partial}{\partial z}\right)\left(\frac{3}{x+y+z}\right), \quad \text{from (1)}$$

$$=3\left[\frac{\partial}{\partial x}\left(\frac{1}{x+y+z}\right)+\frac{\partial}{\partial y}\left(\frac{1}{x+y+z}\right)+\frac{\partial}{\partial z}\left(\frac{1}{x+y+z}\right)\right]$$

$$=3\left[\frac{-1}{(x+y+z)^2}+\frac{-1}{(x+y+z)^2}+\frac{-1}{(x+y+z)^2}\right]=\frac{-9}{(x+y+z)^2}.$$

Example 20:

If u $\log(x^2 + y^2 + z^2)$, show that

$$x\frac{\partial^2 u}{\partial y\,\partial z}=y\frac{\partial^2 u}{\partial z\,\partial x}=z\frac{\partial^2 u}{\partial x\,\partial y}.$$

Solution:

We have $u = \log(x^2 + y^2 + z^2)$.

$$\therefore \quad \frac{\partial u}{\partial z} = \frac{2z}{x^2 + y^2 + z^2}, \text{ [treating x and y as constants]}$$

and $$\frac{\partial^2 u}{\partial y\, \partial z} = \frac{\partial}{\partial y}\left(\frac{\partial u}{\partial z}\right) = 2z.\left[-\left(x^2 + y^2 + z^2\right)^{-2}.2y\right]$$

$= -\,4yz/(x^2 + y^2 + z^2)^2.$

Now $x\dfrac{\partial^2 u}{\partial y\, \partial z} = -\,4xyz/(x^2 + y^2 + z^2)^2.$

By symmetry, $y\dfrac{\partial^2 u}{\partial z\, \partial x} = z\dfrac{\partial^2 u}{\partial x\, \partial y} = -\dfrac{4xyz}{\left(x^2 + y^2 + z^2\right)^2}.$

Hence $x\dfrac{\partial^2 u}{\partial y\, \partial z} = y\dfrac{\partial^2 u}{\partial z\, \partial x} = z\dfrac{\partial^2 u}{\partial x\, \partial y}.$

Example 21:

If $x^x y^y z^z = c$ show that $x = y - z$,

$$\frac{\partial^2 z}{\partial x\, \partial y} = -\,\{x \log (ex)\}^{-1}.$$

(Meerut, 1992; Meerut, 1992, 94, 98 P, Allahabad, 90, 96)

Solution:

We have $x^x.y^y.z^z = c$. From this equation we observe that we can regard z as a function of two independent variables x and y. Taking logarithms of both sides of the given equation, we get

$$x \log x + y \log y + z \log z = \log c. \qquad ...(1)$$

Now differentiating (1) partially w.r.t. x taking y as constant, we have

$$x.\frac{1}{x} + 1.\log x + \left[z.\frac{1}{z} + 1.\log z\right]\frac{\partial z}{\partial x} = 0.$$

[Note that z is not a constant but is a function of x and y]

$$\therefore \quad \frac{\partial z}{\partial x} = -\frac{(1 + \log x)}{(1 + \log z)}. \qquad ...(2)$$

Similarly differentiating (1) partially w.r.t. y, we have

$$\frac{\partial z}{\partial y} = -\frac{(1 + \log y)}{(1 + \log z)}. \qquad ...(3)$$

Now $$\frac{\partial^2 z}{\partial x\, \partial y} = \frac{\partial}{\partial x}\left(\frac{\partial z}{\partial y}\right) = \frac{\partial}{\partial x}\left[-\left(\frac{1 + \log y}{1 + \log z}\right)\right], \qquad \text{[from (3)]}$$

$$= -(1 + \log y).\frac{\partial}{\partial x}\ [(1 + \log z)^{-1}]$$

$$= -(1 + \log y).\left[-(1+\log z)^{-2}.\frac{1}{z}.\frac{\partial z}{\partial x}\right]$$

$$= \frac{(1+\log y)}{z(1+\log z)^2}.\left[-\left(\frac{1+\log x}{1+\log z}\right)\right], \qquad \text{[from (2)].}$$

Hence, when x = y = z, we have

$$\frac{\partial^2 z}{\partial x\, \partial y} = -\frac{(1+\log x)^2}{x(1+\log x)^3}, \quad \left[\text{putting } y = z = x \text{ in the value of } \left(\frac{\partial^2 z}{\partial x\, \partial y}\right)\right]$$

$$= -\frac{1}{x(1+\log x)} = -\frac{1}{x(\log e + \log x)}, \qquad [\because \log e = 1]$$

$$= -\frac{1}{x \log (ex)} = -\{x \log (ex)\}^{-1}.$$

Example 22:

If $z = \tan (y + ax) + (y - ax)^{3/2}$, *find the value of*

$$\left(\frac{\partial^2 z}{\partial x^2}\right) - a^2\left(\frac{\partial^2 z}{\partial y^2}\right).$$

Solution:

Here $z = \tan (y + ax) + (y - ax)^{3/2}$.

$$\therefore \qquad \left(\frac{\partial z}{\partial x}\right) = \{\sec^2 (y + ax)\}.a + \frac{3}{2}(y - ax)^{1/2}.(-a),$$

$$\text{and} \qquad \left(\frac{\partial^2 z}{\partial x^2}\right) = 2a^2 \tan(y + ax) \sec^2 (y + ax) + \frac{3}{4}a^2 (y - ax)^{-1/2}.$$

$$\text{Again} \qquad \left(\frac{\partial z}{\partial y}\right) = \sec^2 (y + ax) + \frac{3}{2}(y - ax)^{1/2}$$

$$\text{and} \qquad \left(\frac{\partial^2 z}{\partial y^2}\right) = 2\sec^2 (y + ax) \tan (y + ax) + \frac{3}{4}(y - ax)^{-1/2}.$$

$$\text{Thus,} \qquad \left(\frac{\partial^2 z}{\partial x^2}\right) - a^2\left(\frac{\partial^2 z}{\partial y^2}\right) = 0.$$

Example 23:

If $u = x\phi\,(y/x) = \psi(y/x)$, *prove that*

$$x^2\frac{\partial^2 u}{\partial x^2}+2xy\frac{\partial^2 u}{\partial x\,\partial y}+y^2\frac{\partial^2 u}{\partial y^2}=0.$$

(Agra, 1998; Allahabad, 94)

Solution:

We have $u = x\phi\,(y/x) + \psi\,(y/x)$. ...(1)

Differentiating (1) partially w.r.t. x and y, we get

$$\left(\frac{\partial u}{\partial x}\right)=x\left\{\phi'\left(\frac{y}{x}\right)\right\}.\left(-\frac{y}{x^2}\right)+\phi\left(\frac{y}{x}\right)+\left\{\psi'\left(\frac{y}{x}\right)\right\}.\left(-\frac{y}{x^2}\right),$$

and $$\left(\frac{\partial u}{\partial x}\right)=x\left\{\phi'\left(\frac{y}{x}\right)\right\}.\left(\frac{1}{x}\right)+\phi\left(\frac{y}{x}\right)+\left\{\psi'\left(\frac{y}{x}\right)\right\}.\left(\frac{1}{x}\right).$$

$$\therefore\ x\left(\frac{\partial u}{\partial x}\right)+y\left(\frac{\partial u}{\partial y}\right)=x\phi\left(\frac{y}{x}\right) \qquad ...(2)$$

Now differentiating (2) partially w.r.t. x and y respectively, we get

$$x\frac{\partial^2 u}{\partial x^2}+\frac{\partial u}{\partial y}+y\frac{\partial^2 u}{\partial x\,\partial x}=x\left\{\phi'\left(\frac{y}{x}\right)\right\}\left(-\frac{y}{x^2}\right)+\phi\left(\frac{y}{x}\right),$$

and $$x\frac{\partial^2 u}{\partial x\,\partial y}+y\frac{\partial^2 u}{\partial y^2}+\frac{\partial u}{\partial y}=x\left\{\phi'\left(\frac{y}{x}\right)\right\}.\frac{1}{x}.$$

Multiplying these equations by x and y respectively and adding, we get

$$x^2\frac{\partial^2 u}{\partial x^2}+2xy\frac{\partial^2 u}{\partial x\,\partial y}+y^2\frac{\partial^2 u}{\partial y^2}+x\frac{\partial u}{\partial x}+y\frac{\partial u}{\partial y}=x\phi\left(\frac{y}{x}\right)$$

or $$x^2\frac{\partial^2 u}{\partial x^2}+2xy\frac{\partial^2 u}{\partial x\,\partial y}+y^2\frac{\partial^2 u}{\partial y^2}=0,$$ [from (2)]

Example 24:

If $u = f(r)$ *where* $r^2 = x^2 + y^2$, *show that*

$$\frac{\partial^2 u}{\partial x^2}+\frac{\partial^2 u}{\partial y^2}=f''(r)+\frac{1}{r}f'(r).$$

(Delhi, 1992; Lucknow, 92; Rohilkhand, 92; Kanpur, 95, 99; Meerut, 97, 94, 92 S, 97; Allahabad, 92, 97)

Solution:

Differentiating $r^2 = x^2 + y^2$ partially w.r.t. x and y, we get

$$2r\frac{\partial r}{\partial x} = 2x \text{ or } \frac{\partial r}{\partial x} = \frac{x}{r};\ 2r\frac{\partial r}{\partial y} = 2y \text{ or } \frac{\partial r}{\partial y} = \frac{y}{r}. \quad ...(1)$$

Now u = f(r). Therefore $\frac{\partial u}{\partial x} = \{f'(r)\} = \frac{x}{r} f'(r)$ [from (1)]

and $\frac{\partial^2 u}{\partial x^2} = \frac{\partial}{\partial x}\left(\frac{\partial u}{\partial x}\right) = \frac{\partial}{\partial x}\left[x.\frac{1}{r}.f'(r)\right]$

$$= 1.\frac{1}{r}.f'(r) + \{xf'(r)\}\left(-\frac{1}{r^2}\frac{\partial r}{\partial x}\right) + \frac{x}{r}\{f''(r)\}\frac{\partial r}{\partial x}$$

$$= \frac{1}{r}f'(r) - \frac{x}{r^2}.\frac{x}{r}f'(r) + \frac{x^2}{r^2}f''(r), \quad \left[\because \text{ from } (1), \frac{\partial x}{\partial r} = \frac{x}{r}\right]$$

$$= \left(\frac{1}{r}\right)f'(r) - \left(\frac{x^2}{r^3}\right)f'(r) + \left(\frac{x^2}{r^2}\right)f''(r). \quad ...(2)$$

Similarly, by symmetry, we have

$$\frac{\partial^2 u}{\partial y^2} = \frac{1}{r}f'(r) - \frac{y^2}{r^3}f'(r) + \frac{y^2}{r^2}f''(r). \quad ...(3)$$

Adding (2) and (3), we get

$$\frac{\partial^2 u}{\partial x^2} + \frac{\partial^2 u}{\partial y^2} = \frac{2}{r}f'(r) - \frac{x^2+y^2}{r^3}f'(r) + \frac{x^2+y^2}{r^2}f''(r)$$

$$= \left(\frac{2}{r}\right)f'(r) - \left(\frac{r^2}{r^3}\right)f'(r) + \left(\frac{r^2}{r^2}\right)f''(r), \quad [\because r^2 = x^2 + y^2]$$

$$= \left(\frac{2}{r}\right)f'(r) - \left(\frac{1}{r}\right)f'(r) + f''(r) = f''(r) + \left(\frac{1}{r}\right)f'(r).$$

Example 25:

If $x = r\cos\theta$, $y = r\sin\theta$, *prove that*

(a) $$\frac{\partial^2 r}{\partial x^2} + \frac{\partial^2 r}{\partial y^2} = \frac{1}{r}\left[\left(\frac{\partial r}{\partial x}\right)^2 + \left(\frac{\partial r}{\partial y}\right)^2\right],$$

(Kanpur, 91; Vikram, 93; Gorakhpur, 92)

(b) $$\frac{\partial^2 r}{\partial x^2}.\frac{\partial^2 r}{\partial y^2} = \left(\frac{\partial^2 r}{\partial x\,\partial y}\right)^2,$$

(c) $\left(\frac{\partial r}{\partial x}\right)^2 + \left(\frac{\partial r}{\partial y}\right)^2 = 1.$ **(Meerut, 1993)**

Solution:

(a) We have $x = r\cos\theta,\ y = r\sin\theta.$

Therefore $r^2 = x^2 + y^2$ ...(1)

Now $2r\left(\frac{\partial r}{\partial x}\right) = 2x;$ [diff. (1) partially w.r.t 'x']

$$\therefore\ \frac{\partial r}{\partial x} = \frac{x}{r}.$$

$$\frac{\partial^2 r}{\partial x^2} = \frac{r.1 - x.\frac{\partial r}{\partial x}}{r^2} = \frac{r - x.\frac{x}{r}}{r^2}, \quad \text{...[using (2)]}$$

$$= \frac{r^2 - x^2}{r^3} = \frac{(x^2 + y^2) - x^2}{r^3} = \frac{y^2}{r^3}. \quad \text{...(3)}$$

Again differentiating (1) partially w.r.t. 'y', we get

$$2r\frac{\partial r}{\partial y} = 2y; \qquad \therefore\ \frac{\partial r}{\partial y} = \frac{y}{r} \quad \text{...(4)}$$

Differentiating (4) partially w.r.t. y, we get

$$\frac{\partial^2 r}{\partial y} = \frac{r.1 - y.\frac{\partial r}{\partial y}}{r^2} = \frac{r - y.\frac{y}{r}}{r^2} \quad \left[\because \text{ from (4), } \frac{\partial r}{\partial r} = \frac{y}{r}\right]$$

$$= \frac{r^2 - y^2}{r^3} = \frac{(x^2 + y^2) - y^2}{r^3} = \frac{x^2}{r^3}. \quad \text{...(5)}$$

Adding (3) and (5), we get

$$\frac{\partial^2 r}{\partial x^2} + \frac{\partial^2 r}{\partial y^2} = \frac{y^2}{r^3} + \frac{x^2}{r^3} = \frac{x^2 + y^2}{r^3} = \frac{1}{r}.$$

Also $\frac{1}{r}\left[\left(\frac{\partial r}{\partial x}\right)^2 + \left(\frac{\partial r}{\partial y}\right)^2\right] = \frac{1}{r}\left[\frac{x^3}{r^2} + \frac{y^2}{r^2}\right] = \frac{x^2 + y^2}{r^3} = \frac{r^2}{r^3} = \frac{1}{r}.$

$$\therefore\ \frac{\partial^2 r}{\partial x^2} + \frac{\partial^2 r}{\partial y^2} = \frac{1}{r}\left[\left(\frac{\partial r}{\partial x}\right)^2 + \left(\frac{\partial r}{\partial y}\right)^2\right].$$

(b) Differentiating (4) partially w.r.t. 'x', we get

$$\frac{\partial^2 r}{\partial x\,\partial y} = -\frac{y}{r^2}.\frac{\partial r}{\partial x} = -\frac{xy}{r^3}, \qquad \left[\because \text{ from } (2), \frac{\partial r}{\partial x} = \frac{x}{r}\right]$$

$$\text{Now } \frac{\partial^2 r}{\partial x^2}.\frac{\partial^2 r}{\partial y^2} = \frac{y^2}{r^3}.\frac{x^2}{r^3}, \qquad [\text{from (3) and (5)}]$$

$$= \frac{x^2 y^2}{r^6} = \left(\frac{-xy}{r^3}\right)^2 = \left(\frac{\partial^2 r}{\partial x\,\partial y}\right)^2.$$

(c) From (2) and (4), on squaring and adding, we get

$$\left(\frac{\partial r}{\partial x}\right)^2 + \left(\frac{\partial r}{\partial y}\right)^2 = \frac{x^2}{r^2} + \frac{y^2}{r^2} = \frac{x^2+y^2}{r^2} = \frac{r^2}{r^2} = 1.$$

Example 26:

If $u = (1 - 2xy + y^2)^{-1/2}$, prove that

$$\frac{\partial}{\partial x}\left\{\left(1-x^2\right)\frac{\partial u}{\partial x}\right\} + \frac{\partial}{\partial y}\left\{y^2\frac{\partial u}{\partial y}\right\} = 0. \qquad \textbf{(Meerut, 1992)}$$

Solution:

Here $u = (1 - 2xy + y^2)^{-1/2}$.

$$\therefore \qquad \frac{\partial u}{\partial x} = -\frac{1}{2}\left(1-2xy+y^2\right)^{-3/2}(-2y) = yu^3$$

$$\text{and} \qquad \frac{\partial u}{\partial y} = -1/2\,(1 - 2xy + y^2)^{-3/2}\,(-2y + 2y) = (x - y)\,u^3.$$

$$\text{Now} \qquad \frac{\partial}{\partial x}\left\{\left(1-x^2\right)\frac{\partial u}{\partial x}\right\} = \frac{\partial}{\partial x}\left\{\left(1-x^2\right).yu^3\right\}$$

$$= y(-2x)u^3 + y(1 - x^2).3u^2\frac{\partial u}{\partial x} = -2xy\,u^3 + 3y\,(1 - x^2)u^2.yu^3$$

$$= -2xy\,u^3 + 3y^2u^5\,(1 - x)^2. \qquad ...(1)$$

$$\text{Also } \frac{\partial}{\partial y}\left\{y^2\frac{\partial u}{\partial y}\right\} = \frac{\partial}{\partial y}\left\{y^2(x-y)u^3\right\} = \frac{\partial}{\partial y}\left\{\left(y^2x-y^3\right)u^3\right\}$$

$$= (2xy - 3y^2)\,u^3 + (y^2x - y^3).3u^2\frac{\partial u}{\partial y}$$

$$= (2xy - 3y^2)\,u^3 + y^2(x - y).3u^2.(x - y)u^3$$

$$= (2xy - 3y^2)u^3 + y^2(x - y)^2.3u^5$$

$$= 2xy\,u^3 + 3y^2u^5\,[(x - y)^2 - u^{-2}]$$

6.2 HOMOGENEOUS FUNCTIONS

An expression in which every term is of the same degree is called a homogeneous function. Thus

$$a_0x^n + a_1x^{n-1}y + a_2x^{n-2}y^2 + \ldots + a_{n-1}xy^{n-1} + a_ny^n$$

is a homogeneous function of x and y of degree n. This can also be written as

$$x^n\left\{a_0 + a_1\left(\frac{y}{x}\right) + a_2\left(\frac{y}{x}\right)^2 + \ldots + a_{n-1}\left(\frac{y}{x}\right)^{n-1} + a_n\left(\frac{y}{x}\right)^n\right\}$$

or $x^n f\left(\frac{y}{x}\right)$, where $f\left(\frac{y}{x}\right)$ is some function of $\left(\frac{y}{x}\right)$.

Note 1: To test whether a given function f(x, y) is homogeneous or not we put tx for x and ty for y in it.

If we get $f(tx, ty) = t^n f(x, y)$,

the function f(x, y) is homogeneous of degree n; otherwise f(x, y) is not a homogeneous function.

Note 2: If u is a homogeneous function of x and y of degee n then $\frac{\partial u}{\partial x}$ and $\frac{\partial u}{\partial y}$ are also homogeneous functions of x and y each being of degree n – 1.

Let $u = x^n f\left(\frac{y}{x}\right)$.

[$\because$ u is a homogeneous function of x and y of degree n]

Then $\frac{\partial u}{\partial x} = nx^{n-1} f\left(\frac{y}{x}\right) + x^n\left\{f'\left(\frac{y}{x}\right)\right\}.\left(-\frac{y}{x^2}\right)$

$$= x^{n-1}\left[n\, f\left(\frac{y}{x}\right) - \left(\frac{y}{x}\right) f'\left(\frac{y}{x}\right)\right]$$

$= x^{n-1}.\left[\text{some function of } \frac{y}{x}\right]$

= a homogeneous function of x and y of degree (n – 1).

Similarly, $\frac{\partial u}{\partial y} = x^n\left\{f'\left(\frac{y}{x}\right)\right\}.\frac{1}{x} = x^{n-1}f'\left(\frac{y}{x}\right)$

$= x^{n-1}.\left(\text{some function of } \frac{y}{x}\right)$

= a homogeneous function of x and y of degree (n – 1).

6.3 EULER'S THEOREM ON HOMOGENEOUS FUNCTIONS

If u is a homogeneous function of x and y of degree n, then

$$x\frac{\partial u}{\partial x}+y\frac{\partial u}{\partial y}=nu.$$

(Meerut, 1992, 94, 96, 97S, 98, 98S, 99, 92, BP; Agra, 91; Gorakhpur, 96, 98; Kanpur, 98; Lucknow, 99; Allahabad, 92, 98)

Proof:

Since u is a homogeneous function of x and y of degree n, therefore u may be put in the form

$$u=x^n f\left(\frac{y}{x}\right). \qquad ...(1)$$

Differentiating (1) partially w.r.t. 'x', we hae

$$\frac{\partial u}{\partial x}=\frac{\partial}{\partial x}\left[x^n f\left(\frac{y}{x}\right)\right]$$

$$=\left[f\left(\frac{y}{x}\right)\right]nx^{n-1}+x^n\left[f'\left(\frac{y}{x}\right)\right]\left(-\frac{y}{x^2}\right).$$

$$\therefore\ x\frac{\partial u}{\partial x}=nx^n f\left(\frac{y}{x}\right)-x^{n-1}y.f'\left(\frac{y}{x}\right). \qquad ...(2)$$

Again differentiating (1) partially w.r.t. 'y', we have

$$\frac{\partial u}{\partial y}=\frac{\partial}{\partial y}\left[x^n f\left(\frac{y}{x}\right)\right]=x^n\left[f'\left(\frac{y}{x}\right)\right].\frac{1}{x}=x^{n-1}f'\left(\frac{y}{x}\right).$$

$$\therefore\ y\frac{\partial u}{\partial y}=y.x^{n-1}f'\left(\frac{y}{x}\right). \qquad ...(3)$$

Adding (2) and (3), we have

$= 2xy\ u^3 + 3y^2u^5\ [(x - y)^2 - (1 - 2xy + y^2)],$

$[\because u^{-2} = 1 - 2xy + y^2]$

$= 2xy\ u^3 + 3y^2u^5\ [x^2 - 1] = 2xy\ u^3 - 3y^2u^5\ (1 - x^2).$ ...(2)

Adding (1) and (2), we have

$$\frac{\partial}{\partial x}\left\{(1-x^2)\frac{\partial u}{\partial x}\right\}+\frac{\partial}{\partial y}\left\{y^2\frac{\partial u}{\partial y}\right\}=0.$$

Example 1:

If $\theta = t^n e^{-r^2/4t}$, *what value of n will make*

$$\frac{1}{r^2}\frac{\partial}{\partial r}\left(r^2\frac{\partial \theta}{\partial r}\right)=\frac{\partial \theta}{\partial t}?$$ **(Meerut, 1999)**

Solution:

We have $\frac{\partial \theta}{\partial r} = t^n . e^{-r^2/4}.\left(-\frac{2r}{4t}\right) = -\frac{r}{2}t^{n-1}e^{-r^2/4t}$.

$$\therefore \quad r^2\frac{\partial \theta}{\partial r} = -\frac{1}{2}r^3 t^{n-1} e^{-r^2/4t}.$$

$$\therefore \quad \frac{\partial}{\partial r}\left(r^2\frac{\partial \theta}{\partial r}\right) = -\frac{3r^2}{2}t^{n-1}e^{-r^2/4t} - \frac{1}{2}r^3 t^{n-1} e^{-r^2/4t}.\left(-\frac{2r}{4t}\right)$$

$$= -\frac{3}{2}r^2 t^{n-1} e^{-r^2/4t} + \frac{1}{4}r^4 t^{n-2} e^{-r^2/4t}.$$

$$\therefore \quad \frac{1}{r^2}\frac{\partial}{\partial r}\left(r^2\frac{\partial \theta}{\partial r}\right) = -\frac{3}{2}t^{n-1}e^{-r^2/4t} + \frac{1}{2}r^2 t^{n-2} e^{-r^2/4t}$$

Also $\frac{\partial \theta}{\partial r}\ nt^{n-1}e^{-r^2/4t} + t^n e^{-r^2/4t} . \frac{r^2}{4t^2}$

$$= n\ t^{n-1}\ e^{-r2} + 1/4\ r^2\ t^{n-2}\ e^{-r2/4t}.$$

Now $\frac{1}{r^2}\frac{\partial}{\partial r}\left(r^2\frac{\partial \theta}{\partial r}\right) = \frac{\partial \theta}{\partial t}$

$$\Rightarrow -\frac{3}{2}t^{n-1}e^{-r^2/4t} + \frac{1}{4}r^2 t^{n-2} e^{-r^2/4t} = nt^{n-1}e^{-r^2/4t} + \frac{1}{4}r^2 t^{n-2} e^{-r^2/4t}$$

$\Rightarrow -\frac{3}{2}t^{n-1}e^{-r^2/4t} = nt^{n-1}e^{-r^2/4t}$, for all posible values of r and t

$$\Rightarrow n = -\frac{3}{2}.$$

6.4 HOMOGENEOUS FUNCTIONS

An expression in which every term is of the same degree is called a homogeneous function. Thus

$$a_0x^n + a_1x^{n-1}y + a_2x^{n-2}y^2 + ... + a_{n-1}\ xy^{n-1} + a_ny^n$$

is a homogeneous function of x and y of degree n. This can also be written as

$$x^n\left\{a_0+a_1\left(\frac{y}{x}\right)+a_2\left(\frac{y}{x}\right)^2+\ldots+a_{n-1}\left(\frac{y}{x}\right)^{n-1}+a_n\left(\frac{y}{x}\right)^n\right\}$$

or $\quad x^n f\left(\frac{y}{x}\right)$, where $f\left(\frac{y}{x}\right)$ is some function of $\left(\frac{y}{x}\right)$.

Note 1: To test whether a given function f(x, y) is homogeneous or not we put tx for x and ty for y in it.

If we get $f(tx, ty) = t^n f(x, y)$,

the function f(x, y) is homogeneous of degree n; otherwise f(x, y) is not a homogeneous function.

Note 2: If u is a homogeneous function of x and y of degree n then $\frac{\partial u}{\partial x}$ and $\frac{\partial u}{\partial y}$ are also homogeneous functions of x and y each being of degree n – 1.

Let $u = x^n f\left(\frac{y}{x}\right)$.

[$\because$ u is a homogeneous function of x and y of degree n]

Then $\frac{\partial u}{\partial x} = nx^{n-1}f\left(\frac{y}{x}\right)+x^n\left\{f'\left(\frac{y}{x}\right)\right\}.\left(-\frac{y}{x^2}\right)$

$= x^{n-1}\left[n f\left(\frac{y}{x}\right)-\left(\frac{y}{x}\right)f'\left(\frac{y}{x}\right)\right]$

x^{n-1} (some function of y/x)

= a homogeneous function of x and y of degree (n – 1).

Similarly, $\frac{\partial u}{\partial y} = x^n\left\{f'\left(\frac{y}{x}\right)\right\}.\frac{1}{x} = x^{n-1} f'\left(\frac{y}{x}\right)$

$= x^{n-1}.\left(\text{some function of } \frac{y}{x}\right)$

= a homogeneous function of x and y of degree (n – 1).

6.5 EULER'S THEOREM ON HOMOGENEOUS FUNCTIONS

If u is a homogeneous function of x and y of degree n, then

$$x\frac{\partial u}{\partial x}+y\frac{\partial u}{\partial y}=nu.$$

(Meerut, 1992, 94, 96, 97S, 98, 98S, 99 BP; Agra, 91; Gorakhpur, 96, 98; Kanpur, 98; Lucknow, 99; Allahabad, 92, 98)

Proof:

Since u is a homogeneous function of x and y of degree n, therefore u may be put in the form

$$u = x^n f\left(\frac{y}{x}\right). \qquad ...(1)$$

Differentiating (1) partially w.r.t. 'x', we have

$$\frac{\partial u}{\partial x} = \frac{\partial}{\partial x}\left[x^n f\left(\frac{y}{x}\right)\right]$$

$$= \left[f\left(\frac{y}{x}\right)\right] nx^{n-1} + x^n\left[f'\left(\frac{y}{x}\right)\right]\left(-\frac{y}{x^2}\right).$$

$$\therefore \; x\frac{\partial u}{\partial x} = nx^n f\left(\frac{y}{x}\right) - x^{n-1} y.f'\left(\frac{y}{x}\right). \qquad ...(2)$$

Again differentiating (2) partially w.r.t 'y', we have

$$\frac{\partial u}{\partial y} = \frac{\partial}{\partial y}\left[x^n f\left(\frac{y}{x}\right)\right] = x^n\left[f'\left(\frac{y}{x}\right)\right].\frac{1}{x} = x^{n-1} f'\left(\frac{y}{x}\right).$$

$$\therefore \; y\frac{\partial u}{\partial y} = y.x^{n-1} f'\left(\frac{y}{x}\right). \qquad ...(3)$$

Adding (2) and (3), we have

$$x\frac{\partial u}{\partial x} + y\frac{\partial u}{\partial y} = nx^n f\left(\frac{y}{x}\right) = nu. \qquad \text{[from (1)]}$$

Hence $x\frac{\partial u}{\partial x} + y\frac{\partial u}{\partial y} = nu$.

Note: Euler's theorem can be extended to a homogeneous function of any number of variables. *Thus if $f(x_1, x_2, ..., x_n)$ be a homogeneous function of $x_1, x_2, ..., x_n$ of degree n, then*

$$x_1\frac{\partial f}{\partial x_1} + x_2\frac{\partial f}{\partial x_2} + ... + x_n\frac{\partial f}{\partial x_n} = nf.$$

Example 1:

If $u \tan^{-1} \frac{x^3 + y^3}{x - y}$, find the value of

$$x^2\frac{\partial^2 u}{\partial x^2} + 2xy\frac{\partial^2 u}{\partial x\,\partial y} + y^2\frac{\partial^2 u}{\partial y^2}. \qquad \textbf{(U.P.P.C.S., 1993)}$$

Solution:

Thus we get

$$x\left(\frac{\partial u}{\partial x}\right)+y\left(\frac{\partial u}{\partial y}\right)=\sin 2u \qquad ...(1)$$

Now differentiating (1) partially w.r.t. x and y respectively, we get

$$x\frac{\partial^2 u}{\partial x^2}+\frac{\partial u}{\partial x}+y\frac{\partial^2 u}{\partial x\,\partial y}=2\cos 2u.\frac{\partial u}{\partial x}, \qquad ...(2)$$

and $x\dfrac{\partial^2 u}{\partial x\,\partial y}+\dfrac{\partial u}{\partial y}+y\dfrac{\partial^2 u}{\partial y^2}=2\cos 2u\dfrac{\partial u}{\partial y}.$ ...(3)

Multiplying equations (2) by x, (3) by y and adding, we get

$$x^2\frac{\partial^2 u}{\partial x^2}+2xy\frac{\partial^2 u}{\partial x\,\partial y}+y^2\frac{\partial^2 u}{\partial y^2}+x\frac{\partial u}{\partial x}+y\frac{\partial u}{\partial y}=2\cos 2u\left(x\frac{\partial y}{\partial x}+y\frac{\partial u}{\partial y}\right)$$

or $x^2\dfrac{\partial^2 u}{\partial x^2}+2xy\dfrac{\partial^2 u}{\partial x\,\partial y}+y^2\dfrac{\partial^2 u}{\partial y^2}=\sin 2u=2\cos 2u\sin 2u,$ [by (1)]

or $x^2\dfrac{\partial^2 u}{\partial x^2}+2xy\dfrac{\partial^2 u}{\partial x\,\partial y}+y^2\dfrac{\partial^2 u}{\partial y^2}=\sin 4u-\sin 2u.$

Example 2:

If $z=\tan^{-1}\left(\dfrac{x+y}{\sqrt{x}+\sqrt{y}}\right)$ *then*

$x\dfrac{\partial z}{\partial x}+y\dfrac{\partial z}{\partial y}=\dfrac{1}{4}\sin 2z.$ **(D.U. B.Sc. (G), 1989, B.Sc. (H), 1992, 88)**

Solution:

$$z=\tan^{-1}\left(\frac{x+y}{\sqrt{x}+\sqrt{y}}\right)\Rightarrow \tan z=\frac{x+y}{\sqrt{x}+\sqrt{y}}.$$

Let u = tan z ...(1)

u is a homogeneous functions in x and y of degree $\frac{1}{2}$.

By Euler's Theorem $x\dfrac{\partial y}{\partial x}+y\dfrac{\partial y}{\partial y}=\dfrac{1}{2}.$

or $x\left(\sec^2 z\dfrac{\partial z}{\partial x}\right)+y\left(\sec^2 z\dfrac{\partial z}{\partial y}\right)=\dfrac{1}{2}\tan x$ by (1)

or $$x\frac{\partial z}{\partial x}+y\frac{\partial z}{\partial y}=\frac{1}{2}\frac{\tan z}{\sec^2 z}=\frac{1}{2}\sin z\cos z$$
$$=\frac{1}{4}\sin 2z.$$

Example 3:

If f(x, y, z) is a homogeneous function of the nth degree in x, y, z, prove that

$$x^2\frac{\partial^2 f}{\partial x^2}+y^2\frac{\partial^2 f}{\partial y^2}+z^2\frac{\partial^2 f}{\partial z^2}+2yz\frac{\partial^2 f}{\partial y\,\partial z}+2zx\frac{\partial^2 f}{\partial z\,\partial x}+2xy\frac{\partial^2 f}{\partial x\,\partial y}$$
$$= n(n-1)\,f(x, y, z).$$

Solution:

Here f(x, y, z) is a homogeneous function of the nth degree n, x, y, z. Therefore $\frac{\partial f}{\partial x}, \frac{\partial f}{\partial y}$ and $\frac{\partial f}{\partial z}$ are homogeneous functions f the (n – 1)th degree in x, y, z. So using Euler's theorem for $\frac{\partial f}{\partial x}$,

$$x\frac{\partial}{\partial x}\left(\frac{\partial f}{\partial x}\right)+y\frac{\partial}{\partial y}\left(\frac{\partial f}{\partial x}\right)+z\frac{\partial}{\partial z}\left(\frac{\partial f}{\partial x}\right)=(n-1)\frac{\partial f}{\partial x}$$

or $$x\frac{\partial^2 f}{\partial x^2}+y\frac{\partial^2 f}{\partial y\,\partial x}+z\frac{\partial^2 f}{\partial z\,\partial x}=(n-1)\frac{\partial f}{\partial x}. \quad ...(1)$$

Similarly, $$x\frac{\partial^2 f}{\partial x\,\partial y}+y\frac{\partial^2 f}{\partial y^2}+z\frac{\partial^2 f}{\partial y\,\partial z}=(n-1)\frac{\partial f}{\partial y} \quad ...(2)$$

and $$x\frac{\partial^2 f}{\partial x\,\partial y}+y\frac{\partial^2 f}{\partial z^2}=(n-1)\frac{\partial f}{\partial z} \quad ...(3)$$

Multiplying (1) by x, (2) by y and (3) by z and adding, we get

$$x^2\frac{\partial^2 f}{\partial x^2}+y^2\frac{\partial^2 f}{\partial y^2}+z^2\frac{\partial^2 f}{\partial z^2}+2yz\frac{\partial^2 f}{\partial y\,\partial z}+2zx\frac{\partial^2 f}{\partial z\,\partial x}+2xy\frac{\partial^2 f}{\partial x\,\partial y}$$
$$=(n-1)\left(x\frac{\partial f}{\partial y}+\frac{\partial f}{\partial y}z\frac{\partial f}{\partial z}\right)$$
$$= (n-1)\,nf(x, y, z) = n(n-1)\,f(x, y, z).$$

Example 4:

Verify Euler's theorem in the following cases:

(i) $u = x^{-4}\ 3x^3y + 5x^2y^2 + 4xy^3 - 2y^4.$

(ii) $u = \dfrac{x(x^3 - y^3)}{x^3 + y^3},$ **(Meerut, 1999P)**

(iii) $u = \dfrac{x^{1/4} + y^{1/4}}{x^{1/5} + y^{1/5}},$

(iv) $u = axy + byz = czx,$ **(Meerut, 1997)**

(v) $u = x^n \log\left(\dfrac{y}{x}\right),$ **(Meerut, 1999P)**

(vi) $u = 1/\sqrt{(x^2 + y^2)}.$ **(Meerut, 1999)**

Solution:

(i) We have $= x^4 - 3x^3y + 5x^2y^2 + 4xy^3 - 2y^4$. Obviously u is a homogeneous function of x and y of degree 4. So by Euler's Theorem, we must have

$$x\left(\frac{\partial u}{\partial x}\right) + y\left(\frac{\partial u}{\partial y}\right) = 4u. \text{ Let us verify it.}$$

We have $\left(\dfrac{\partial u}{\partial x}\right) = 4x^3 - 9x^2y + 10xy^2 + 4y^3,$

and $\left(\dfrac{\partial u}{\partial y}\right) = -3x^3 + 10x^2y + 12xy^2 - 8y^3.$

$$\therefore \quad x\frac{\partial u}{\partial x} + y\frac{\partial u}{\partial y} = x(4x^3 - 9x^2y + 10xy^2 + 4y^3)$$

$$+ y(-3x^3 + 10x^2y + 12xy^2 - 8y^3)$$

$= 4(x^4 - 3x^3y + 5x^2y^2 + 4xy^3 - 2y^4)$

$= 4u$. This verifies Euler's Theorem.

(ii) We have $u = \dfrac{x(x^3 - y^3)}{x^3 + y^3}$ which is obviously a homogeneous function of x and y of degree 4 – 3 *i.e.*, 1. Note that each term in the numerator is of degree 4 while each term in the denominator is of degree 3. In order to verify Euler's Theorem we are to know that

$$x\frac{\partial u}{\partial x}+y\frac{\partial u}{\partial y}=1.u=u.$$

Now $\log u = \log x + \log (x^3 - y^3) - \log(x^3 + y^3)$. ...(1)

Differentiating (1) partially w.r.t. x and y respectively, we get

$$\frac{1}{u}\frac{\partial u}{\partial x}=\frac{1}{x}+\frac{3x^2}{x^3-y^3}-\frac{3x^2}{x^3+y^3} \quad ...(2)$$

and $$\frac{1}{u}\frac{\partial u}{\partial y}=0-\frac{3y^2}{x^3-y^3}-\frac{3y^2}{x^3+y^3}. \quad ...(3)$$

Multiplying (2) by x and (3) by y and adding, we get

$$\frac{1}{u}\left(x\frac{\partial u}{\partial x}+y\frac{\partial u}{\partial y}\right)=1+\frac{3(x^3-y^3)}{x^3-y^3}-\frac{3(x^3+y^3)}{x^3+y^3}$$

$$= 1 + 3 - 3 = 1.$$

$\therefore\ x\frac{\partial u}{\partial x}+y\frac{\partial u}{\partial y}=u$. This verifies Euler's Theorem.

(iii) Here u is a homogeneous function of x and y of degree $\frac{1}{4}-\frac{1}{5}$ *i.e.*, $\frac{1}{20}$. So by Euler's Theorem we must have

$$x\left(\frac{\partial u}{\partial x}\right)+y\left(\frac{\partial u}{\partial y}\right)=\frac{1}{20}u.$$

Let us verify it. We have

$\log u = \log(x^{1/4} + y^{1/4}) - \log (x^{1/5} + y^{1/5})$.

$$\therefore\ \frac{1}{u}\frac{\partial u}{\partial x}=\frac{\frac{1}{4}x^{-3/4}}{x^{1/4}+y^{1/4}}-\frac{\frac{1}{5}x^{-4/5}}{x^{1/5}+y^{1/5}}$$

and $$\frac{1}{u}\frac{\partial u}{\partial y}=\frac{\frac{1}{4}y^{-3/4}}{x^{1/4}+y^{1/4}}-\frac{\frac{1}{5}y^{-4/5}}{x^{1/5}+y^{1/5}}.$$

$$\therefore\ \frac{1}{u}\left(x\frac{\partial u}{\partial x}+y\frac{\partial u}{\partial y}\right)=\frac{1}{4}\frac{x^{1/4}+y^{1/4}}{x^{1/4}+y^{1/4}}-\frac{1}{5}\frac{x^{1/5}+y^{1/5}}{x^{1/5}+y^{1/5}}=\frac{1}{4}-\frac{1}{5}=\frac{1}{20}.$$

$\therefore\ x\frac{\partial u}{\partial x}+y\frac{\partial u}{\partial y}=\frac{1}{20}u$. This verifies Euler's Theorem.

(iv) We have = axy + byz + czx, which is a homogeneous function of x, y and z of degree 2. So in order to verify Euler's Theorem, we must show that $x\frac{\partial u}{\partial x}+y\frac{\partial u}{\partial y}+z\frac{\partial u}{\partial z}=2u$.

Now $\frac{\partial u}{\partial x}=ay+cz$, $\frac{\partial u}{\partial y}=ax+bz$, and $\frac{\partial u}{\partial z}=by+cx$.

$\therefore\ x\frac{\partial u}{\partial x}+y\frac{\partial u}{\partial y}+z\frac{\partial u}{\partial z}$ = x (ay + cz) + y (ax + bz) + z (by + cx)

= 2 (axy + byz + czx) = 2u. This verifies Euler's Theorem.

(v) Here u is a homogeneous function of x and y of degree n. So by Euler's Theorem we must have

$$x\left(\frac{\partial u}{\partial x}\right)+y\left(\frac{\partial u}{\partial y}\right)=nu.$$

Now do the verification yourself.

(vi) Here $u=\frac{1}{\sqrt{(x^2+y^2)}}=\frac{1}{x\sqrt{[1+(y/x^2)]}}=x^{-1}.\frac{1}{\sqrt{\left[1+\sqrt{(y/x)^2}\right]}}$ is a homogeneous function of x and y of degree – 1. Now proceed yourself.

Example 5:

If $u=\tan^{-1}\left(\frac{x^3+y^3}{x+y}\right)$, *show that*

$$x\left(\frac{\partial u}{\partial x}\right)+y\left(\frac{\partial u}{\partial y}\right)=\sin 2u.$$

(Meerut, 1992, 93, 98 P, 90 S; Delhi, 93, 91; Lucknow, 90; Gorakphur, 99; Rohilkhand, 91)

Solution:

We have tan u = $(x^3 + y^3)/(x + y)$ = v, say. Then v is a homogeneous function of x and y of degree 3 – 1 *i.e.*, 2. Therefore by Euler's theorem.

We have $$x\left(\frac{\partial v}{\partial x}\right)+y\left(\frac{\partial v}{\partial y}\right)=2v. \quad ...(1)$$

Now v = tan u.

$\therefore\quad \frac{\partial v}{\partial x}=\sec^2 u\frac{\partial u}{\partial x}$ and $\frac{\partial v}{\partial y}=\sec^2 u\frac{\partial u}{\partial y}$.

Putting these values in (1), we get

$$x\sec^2 u\frac{\partial u}{\partial x}+y\sec^2 u\frac{\partial u}{\partial y}=2v$$

or $$x\frac{\partial u}{\partial x}+y\frac{\partial u}{\partial y}=\frac{2v}{\sec^2 u}=\frac{2\tan u}{\sec^2 u}=2\sin u\cos u=\sin 2u.$$

Example 6:

If $z = xy\, f\left(\frac{y}{x}\right)$, *show that* $x\left(\frac{\partial z}{\partial x}\right)+y\left(\frac{\partial z}{\partial y}\right)=2z$.

Show also that if z is a constant,

$$\frac{f'\left(\frac{y}{x}\right)}{f\left(\frac{y}{x}\right)}=\frac{x\left\{y+x\left(\frac{dy}{dx}\right)\right\}}{y\left\{y-x\left(\frac{dv}{dx}\right)\right\}}.$$

Solution:

We have, $z = x^2.\left(\frac{y}{x}\right)f\left(\frac{y}{x}\right)$, so that z is a homogeneous function of x and y of degree 2.

Hence by Euler's Theorem, we have

$$x\left(\frac{\partial z}{\partial x}\right)+y\left(\frac{\partial z}{\partial y}\right)=2z.$$

If z be a constant, then differentiating

$z = xy\ f\left(\frac{y}{x}\right)$ logarithmically, w.r.t. x, we get

$$0=\frac{1}{x}+\frac{1}{y}\frac{dy}{dx}+\frac{f'\left(\frac{y}{x}\right)}{f\left(\frac{y}{x}\right)}.\frac{x\left(\frac{dy}{dx}\right)-y}{x^2}$$

$$=\frac{y+x\left(\frac{dy}{dx}\right)}{xy}+\frac{f'\left(\frac{y}{x}\right)}{f\left(\frac{y}{x}\right)}.\frac{x\left(\frac{dy}{dx}\right)-y}{x^2}$$

Hence $$\frac{f'\left(\frac{y}{x}\right)}{f\left(\frac{y}{x}\right)}=\frac{x\left\{y+x\left(\frac{dy}{dx}\right)\right\}}{y\left\{y-x\left(\frac{dy}{dx}\right)\right\}}.$$

Example 7:

If $u = \log\dfrac{x^3+y^3}{x+y}$, *show that* $x\dfrac{\partial u}{\partial x}+y\dfrac{\partial u}{\partial y}=2$.

(Meerut, 1990 P, D.U. Maths (H), 2000)

Solution:

We have $e^u = \dfrac{x^3+y^3}{x+y} = v$, say.

Obviously $v = (x^3 + y^3)/(x + y)$ is a homogeneous function of x and y of degree 3 – 1 *i.e.*, 2. Therefore by Euler's theorem, we have

$$x\frac{\partial v}{\partial x}+y\frac{\partial v}{\partial y}=2.v=2v \qquad ...(1)$$

Now $v = e^u$.

$$\therefore \frac{\partial v}{\partial x}=e^u\frac{\partial u}{\partial x} \quad \text{and} \quad \frac{\partial v}{\partial y}=e^u\frac{\partial u}{\partial y}.$$

Putting these values in (1), we get

$$xe^u\frac{\partial u}{\partial x}+y\,e^u\frac{\partial u}{\partial y}=2e^u$$

$$\text{or } e^u\left(x\frac{\partial u}{\partial x}+y\frac{\partial u}{\partial y}\right)=2e^u \quad \text{or} \quad x\frac{\partial u}{\partial x}+y\frac{\partial u}{\partial y}=2.$$

Example 8:

If $u = x\phi\left(\dfrac{y}{x}\right)+\psi\left(\dfrac{y}{x}\right)$, *prove that*

$$x^2\frac{\partial^2 u}{\partial x^2}+2xy\frac{\partial^2 u}{\partial x\,\partial y}+y^2\frac{\partial^2 u}{\partial y^2}=0.$$ **(Rohilkhand, 1996; Agra, 95)**

Solution:

Let $u = z_1 + z_2$, where $z_1 = x\phi\left(\dfrac{y}{x}\right)$ and $z_2 = \psi\left(\dfrac{y}{x}\right)$. Obviously z_1 is a homogeneous function of x and y of degree 1 and z_2 is a homogeneous function of x and y of degree zero. Now

$$x\frac{\partial u}{\partial x}+y\frac{\partial u}{\partial y}=x\frac{\partial}{\partial x}(z_1+z_2)+y\frac{\partial}{\partial y}(z_1+z_2)$$

$$= \left(x\frac{\partial z_1}{\partial x} + y\frac{\partial z_1}{\partial y} \right) + \left(x\frac{\partial z_2}{\partial x} + y\frac{\partial z_2}{\partial y} \right) = 1.z_1 + 0.z_2. \text{ (by Euler's theorem).}$$

Thus, $x\left(\frac{\partial u}{\partial x}\right) + y\left(\frac{\partial u}{\partial y}\right) = z_1$ ...(1)

Differentiating equation (1) partially w.r.t. x and y respectively, we get

$$x\frac{\partial^2 u}{\partial x^2} + \frac{\partial u}{\partial x} + y\frac{\partial^2 u}{\partial x\, \partial y} = \frac{\partial z_1}{\partial x}, \quad ...(2)$$

and $x\frac{\partial^2 u}{\partial x\, \partial y} + \frac{\partial u}{\partial y} + y\frac{\partial^2 u}{\partial y^2} = \frac{\partial z_1}{\partial y}.$ ...(3)

Multiplying (2) by x and (3) by y and adding, we get

$$x^2\frac{\partial^2 u}{\partial x^2} + 2xy\frac{\partial^2 u}{\partial x\, \partial y} + y^2\frac{\partial^2 u}{\partial y^2} + x\frac{\partial u}{\partial x} + y\frac{\partial u}{\partial y} = x\frac{\partial z_1}{\partial x} + y\frac{\partial z_1}{\partial y}$$

or $x^2\frac{\partial^2 u}{\partial x^2} + 2xy\frac{\partial^2 u}{\partial x\, \partial y} + y^2\frac{\partial^2 u}{\partial y^2} + z_1 = 1.z_1,$

$$\left[\because x\left(\frac{\partial u}{\partial x}\right) + y\left(\frac{\partial u}{\partial y}\right) = z_1 \text{by (1), and } x\left(\frac{\partial z_1}{\partial x}\right) + y\left(\frac{\partial z_1}{\partial y}\right) = 1.z_1 \text{ by Euler's theorem} \right]$$

or $x^2\frac{\partial^2 u}{\partial x^2} + 2xy + y^2\frac{\partial^2 u}{\partial y^2} = 0.$

Example 9:

If $u = x^2y^2/(x + y)$, show that

$$x\left(\frac{\partial u}{\partial x}\right) + y\left(\frac{\partial u}{\partial y}\right) = 3u.$$

Solution:

We have $u = \frac{x^2y^2}{x + y^2} = \frac{x^3\left(\frac{y}{x}\right)^2}{\left[1 + \left(\frac{y}{x}\right)\right]} = x^3 f\left(\frac{y}{x}\right)$, say.

Thus u is a homogeneous function of x and y of degree 3.

Therefore by Euler's theorem, we have $x\frac{\partial u}{\partial x}+y\frac{\partial u}{\partial y}=3u$.

Example 10:

If $u = \sin^{-1}\left(\frac{x+y}{\sqrt{x}+\sqrt{y}}\right)$, *show that*

$x\left(\frac{\partial u}{\partial x}\right)+y\left(\frac{\partial u}{\partial y}\right)=\frac{1}{2}\tan u$. **(Rohilkhand, 1990; Agra, 92; Kanpur, 98)**

Solution:

We have sin u = $(x+y)/(\sqrt{x}+\sqrt{y})$ = v, say.

Then v is a homogeneous function of x and y of degree $\left(1-\frac{1}{2}\right)$ *i.e.*, $\frac{1}{2}$. Applying Euler's Theorem for v, we have

$$x\left(\frac{\partial v}{\partial x}\right)+y\left(\frac{\partial v}{\partial y}\right)=\frac{1}{2}v$$

or $x\frac{\partial}{\partial x}(\sin u)+y\frac{\partial}{\partial y}(\sin u)=\frac{1}{2}\sin u$, [∵ v = sin u]

or $x\cos u\left(\frac{\partial u}{\partial x}\right)+y\cos u\left(\frac{\partial u}{\partial y}\right)=\frac{1}{2}\sin u$

or $x\left(\frac{\partial u}{\partial x}\right)+y\left(\frac{\partial u}{\partial y}\right)=\frac{1}{2}\tan u$.

Example 11:

If u be a homogeneous function of x and y of degree n, show that

$$x\left(\frac{\partial^2 u}{\partial x^2}\right)+y\left(\frac{\partial^2 u}{\partial x\,\partial y}\right)=(n-1)\left(\frac{\partial u}{\partial x}\right),$$

and $x\left(\frac{\partial^2 u}{\partial x\,\partial y}\right)+y\left(\frac{\partial^2 u}{\partial y^2}\right)=(n-1)\left(\frac{\partial u}{\partial y}\right)$.

Hence deduce that

$$x^2 \frac{\partial^2 u}{\partial x^2} + 2xy \frac{\partial^2 u}{\partial x \, \partial y} + y^2 \frac{\partial^2 u}{\partial y^2} = n(n-1)\, u.$$

(Allahabad, 1991; Vikram, 93; Agra, 94; Gorakhpur, 93; I.C.S. 95; Jiwaji, 95; Kurukshetra, 98)

Solution:

By Euler's Theorem, we have

$$x\left(\frac{\partial u}{\partial x}\right) + y\left(\frac{\partial u}{\partial y}\right) = nu \qquad ...(1)$$

Differentiating (1) partially w.r.t. x, we get

$$x \frac{\partial^2 u}{\partial x^2} + \frac{\partial u}{\partial x} + y \frac{\partial^2 u}{\partial x \, \partial y} = n \frac{\partial u}{\partial x}$$

or $$x \frac{\partial^2 u}{\partial x^2} + y \frac{\partial^2 u}{\partial x \, \partial y} = n \frac{\partial u}{\partial x} - \frac{\partial u}{\partial x} = (n-1)\frac{\partial u}{\partial x} \qquad ...(2)$$

Proved.

Similarly, differentiating (1) partially w.r.t. 'y', we get

$$x \frac{\partial^2 u}{\partial x \, \partial y} + y \frac{\partial^2 u}{\partial y^2} = (n-1)\, \frac{\partial u}{\partial y}. \qquad ...(3)$$

Proved.

Now multiplying (2) by x and (3) by y and adding, we get

$$x^2 \frac{\partial^2 u}{\partial x^2} + 2xy \frac{\partial^2 u}{\partial x \, \partial y} + y^2 \frac{\partial^2 u}{\partial y^2}$$

$$= (n-1)\left[x \frac{\partial u}{\partial x} + y \frac{\partial u}{\partial y}\right] = (n-1)\, nu = n(n-1)u.$$

Example 12:

If $u = \sin^{-1}\left\{\left(\sqrt{x} - \sqrt{y}\right)/\left(\sqrt{x} + \sqrt{y}\right)\right\}$, *show that*

$$\frac{\partial u}{\partial x} = -\frac{y}{x}\frac{\partial u}{\partial y}.$$

(Rohilkhand, 1997; Kurukshetra, 93)

Solution:

We have $\sin u = \left(\sqrt{x} - \sqrt{y}\right)/\left(\sqrt{x} + \sqrt{y}\right) = v$, say. Then v is a homogeneous function of x and y of degree $\frac{1}{2} - \frac{1}{2}$ *i.e.*, 0.

Therefore by Euler's Theorem, we have $x\frac{\partial v}{\partial x}+y\frac{\partial v}{\partial y}=0$. v = 0.

Now v = sin u. ...(1)

$\therefore \frac{\partial v}{\partial x}=\cos u\frac{\partial u}{\partial x}, \frac{\partial v}{\partial y}=\cos u\frac{\partial u}{\partial y}$.

Putting these values in (1), we get

$$x\cos u\frac{\partial u}{\partial x}+y\cos u\frac{\partial u}{\partial y}=0 \text{ or } \cos u\left(x\frac{\partial u}{\partial x}+y\frac{\partial u}{\partial y}\right)=0$$

or $x\frac{\partial u}{\partial x}+y\frac{\partial u}{\partial y}$ = 0, since cos u ≠ 0. Hence $\frac{\partial u}{\partial x}=-\frac{y}{x}\frac{\partial u}{\partial y}$.

6.6 TOTAL DERIVATIVES

If u = f(x, y), where x = $\phi_1(t)$ and y = $\phi_2(t)$, then

$$\frac{du}{dt}=\frac{\partial u}{\partial x}.\frac{dx}{dt}+\frac{\partial u}{\partial y}.\frac{dy}{dt}.$$

Here $\frac{du}{dt}$ is called the **Total Differential Coefficient** of u with respect to t while $\frac{\partial u}{\partial x}$ and $\frac{\partial u}{\partial y}$ are partial derivatives of u.

Proof:

Let t be given a small increment δt, and let the corresponding changes in u, x and y be δu, δx and δy rexpectively. We have then

$$u = f(x, y), \quad ...(1)$$

and $u = \delta u = f(x + \delta x, y + \delta y)$. ...(2)

$\therefore \delta u = f(x + \delta x, y + \delta y) - f(x, y)$

$= [f(x + \delta x, y + \delta y) - f(x, y + \delta y)] + [f(x, y + \delta y) - f(x, y)]$. **(Note)**

$$\therefore \frac{\delta u}{\delta t}=\left\{\frac{f(x+\delta x,y+\delta y)-f(x,y+\delta y)}{\delta t}\right\}+\left\{\frac{f(x,y+\delta y)-f(x,y)}{\delta t}\right\}$$

$$=\left\{\frac{f(x+\delta x, y+\delta y)-f(x, y+\delta y)}{\delta x}\right\}.\frac{\delta x}{\delta t}$$

$$+\left\{\frac{f(x, y+\delta y)-f(x,y)}{\delta y}\right\}.\frac{\delta y}{\delta t} \quad ...(3)$$

Let $\delta t \to 0$ so that $\delta x \to 0$ and $\delta y \to 0$.

Now $\lim_{\delta t \to 0} \frac{\delta u}{\delta t} = \frac{du}{dt}$, $\lim_{\delta t \to 0} \frac{\delta x}{\delta t} = \frac{dx}{dt}$ and $\lim_{\delta t \to 0} \frac{\delta y}{\delta t} = \frac{dy}{dt}$.

Also $\lim_{\delta y \to 0} \frac{f(x, y+\delta y) - f(x, y)}{\delta y} = \frac{\partial f}{\partial y} = \frac{\partial u}{\partial y}$,

and $\lim_{\delta y \to 0} \frac{f(x+\delta x, y+\delta y) - f(x, y+\delta y)}{\delta x} = \frac{\partial f}{\partial x} = \frac{\partial u}{\partial x}$.

Since δx and δy tend to zero with δt and the functions involved are all supposed to be continuous, therefore the limit (3) becomes

$$\frac{du}{dt} = \frac{\partial u}{\partial x} \cdot \frac{dx}{dt} + \frac{\partial u}{\partial y} \cdot \frac{dy}{dt}.$$

In the same way if $u = f(x, y, z)$, where x, y, z are all functions of some variable t, then

$$\frac{du}{dt} = \frac{\partial u}{\partial x} \cdot \frac{dx}{dt} + \frac{\partial u}{\partial y} \cdot \frac{dy}{dt} + \frac{\partial u}{\partial z} \cdot \frac{dz}{dt}.$$

This result can be extended to any number of variables.

Cor. 1: *If u be a function of x and y, where y is a function of x, then*

$$\frac{du}{dx} = \frac{\partial u}{\partial x} + \frac{\partial u}{\partial y} \cdot \frac{dy}{dx}.$$

This result follows immediately by taking t = x in the formula of 6.6.

Cor. 2: *If* $u = f(x, y)$ *and* $x = f_1(t_1, t_2)$ *and* $y = f_2(t_1, t_2)$, *then*

$$\frac{\partial u}{\partial t_1} = \frac{\partial u}{\partial x} \cdot \frac{\partial x}{\partial t_1} + \frac{\partial u}{\partial y} \cdot \frac{\partial y}{\partial t_1} \text{ and } \frac{\partial u}{\partial t_2} = \frac{\partial u}{\partial x} \cdot \frac{\partial x}{\partial t_2} + \frac{\partial u}{\partial y} \cdot \frac{\partial y}{\partial t_2}.$$

Cor 3: *If x and y are connected by an equation of the form* $f(x, y) = 0$ *then*

$$\frac{dy}{dx} = -\frac{\partial f / \partial x}{\partial f / \partial y}.$$

Since $f(x, y) = 0$, therefore by cor. 1, we get

$$0 = \frac{\partial f}{\partial x} + \frac{\partial f}{\partial y} \cdot \frac{dy}{dx},$$

from which the required result follows.

Example 1:

Prove that $\frac{d^2y}{dx^2}+\frac{2a^2x^2}{y^5}=0$, *where* $y^3 - 3ax^2 + x^3 = 0.$

(Delhi, 1993)

Solution:

Let $f(x, y) \equiv y^3 - 3ax^2 + x^3 = 0.$

Then $p=\frac{\partial f}{\partial x}=-6ax+3x^2$, $q=\frac{\partial f}{\partial y}=3y^2$,

$$r=\frac{\partial^2 f}{\partial x^2}=-6a+6x,\ s=\frac{\partial^2 f}{\partial x\,\partial y}=0,\ t=\frac{\partial^2 f}{\partial y^2}=6y.$$

Now $\frac{d^2y}{dx^2}=-\frac{q^2r-2pqs+p^2t}{q^3}$,

$$=-\frac{(-6a+6x)\left(3y^2\right)^2+6y\left(-6x+3x^2\right)^2}{\left(3y^2\right)^3}$$

$$=-\frac{2(-a+x)y^3+2\left(-2ax+x^2\right)^2}{y^5}$$

$$=-\frac{2x\left(y^3+x^3-3ax^2\right)-2ay^3+8a^2x^2-2ax^3}{y^5}$$ **(Note)**

$$=-\frac{-2a\left(y^3+x^3\right)+8a^2x^2}{y^5},$$ $[\because\ y^3 + x^3 - 3ax^2 = 0]$

$$=-\frac{-2a\left(3ax^2\right)+8a^2x^2}{y^5}=-\frac{2a^2x^2}{y^5}.$$

Hence $\frac{d^2y}{dx^2}+\frac{2a^2x^2}{y^5}=0.$

Example 2:

If $(\tan x)^y + y^{\cot x} = a$, *find* $\frac{dy}{dx}$.

Solution:

Let $f(x, y) = (\tan x)^y + y^{\cot x} - a$. Then

$$\left(\frac{\partial f}{\partial x}\right) = y\ (\tan x)^{y-1}.\sec^2 x + y^{\cot x}.\log y.(-\ \text{cosec}^2 x),$$

and $$\left(\frac{\partial f}{\partial y}\right) = (\tan x)^y \log \tan x + (\cot x).y^{\cot x - 1}$$

Now we are given that f(x, y) = 0.

$$\therefore\ \frac{dy}{dx} = -\frac{\partial f/\partial x}{\partial f/\partial y} = -\frac{y(\tan x)^{y-1}\sec^2 x - y^{\cot x}.\log y.\text{cosec}^2 x}{(\tan x)^y \log \tan x + \cot x.y^{\cot x - 1}}.$$

Example 3:

If $x^3y^3 + 3x \sin y = e^y$, *find* $\frac{dy}{dx}$.

Solution:

Let $f(x, y) = x^3y^3 + 3x \sin y - e^x$. Then we have f(x, y) = 0.

$$\therefore\ \frac{dy}{dx} = -\frac{\partial f/\partial x}{\partial f/\partial y} = -\frac{\left(3x^2y^3 + 3 \sin y\right)}{3x^3y^2 + 3x \cos y - e^y}.$$

Example 4:

If $x^y + y^x = a^b$, *find* $\frac{dy}{dx}$. **(Meerut, 1991, 92)**

Solution:

Let $f(x, y) = x^y + y^x - a^b$. Then we have f(x, y) = 0.

$$\therefore\ \frac{dy}{dx} = -\frac{\partial f/\partial x}{\partial f/\partial y} = \frac{yx^{y-1} + y^x \log y}{x^y \log x + xy^{x-1}}.$$

Example 5:

If $\sqrt{\left(1-x^2\right)} + \sqrt{\left(1-x^2\right)} = a(x-y)$, *prove that*

$$\frac{dy}{dx} = \frac{\sqrt{\left(1-y^2\right)}}{\sqrt{\left(1-x^2\right)}}.$$ **(Meerut, 1997)**

Solution:

It is given that $\frac{\sqrt{\left(1-x^2\right)} + \sqrt{\left(1-y^2\right)}}{x-y} = a.$

Let $f(x, y) = \dfrac{\sqrt{(1-x^2)} + \sqrt{(1-y^2)}}{x-y} - a$. Then $f(x, y) = 0$.

$$\therefore \quad \frac{dy}{dx} = -\frac{\partial f/\partial x}{\partial f/\partial y}$$

$$= -\frac{\dfrac{\left\{\frac{1}{2}\left(1-x^2\right)^{-1/2}.(-2x)\right\}(x-y) - 1.\left\{\sqrt{(1-x)^2} + \sqrt{\left(1-y^2\right)}\right\}}{(x-y)^2}}{\dfrac{\left\{\frac{1}{2}(1-y)^{-1/2}.(-2y)\right\}(x-y) - (-1).\left\{\sqrt{\left(1-x^2\right)} + \sqrt{\left(1-y^2\right)}\right\}}{(x-y)^2}}$$

$$= -\frac{\dfrac{-x(x-y)}{\sqrt{\left(1-x^2\right)}} - \sqrt{\left(1-x^2\right)} - \sqrt{\left(1-y^2\right)}}{\dfrac{-y(x-y)}{\sqrt{\left(1-y^2\right)}} + \sqrt{\left(1-x^2\right)} + \sqrt{\left(1-y^2\right)}}$$

$$= -\frac{\sqrt{\left(1-y^2\right)}}{\sqrt{\left(1-x^2\right)}} . \frac{-x^2 + xy - \left(1-x^2\right) - \sqrt{\left(1-x^2\right)}\sqrt{\left(1-y^2\right)}}{-yx + y^2 + \sqrt{\left(1-x^2\right)}\,\sqrt{\left(1-y^2\right)} + 1 - y^2}$$

$$= -\frac{\sqrt{\left(1-y^2\right)}}{\sqrt{\left(1-x^2\right)}} . \frac{xy - 1 - \sqrt{\left(1-x^2\right)}\sqrt{\left(1-y^2\right)}}{-xy + 1 - \sqrt{\left(1-x^2\right)}\sqrt{\left(1-y^2\right)}}$$

$$= \frac{\sqrt{(1-y^2)}}{\sqrt{(1-r^2}}$$

Example 6:

If $u = \log\{(x^2 + y^2)/xy\}$, find du.

Solution:

We have $u = \log(x^2 + y^2) - \log x - \log y$.

$$\therefore \quad \frac{\partial u}{\partial x} = \frac{2x}{x^2+y^2} - \frac{1}{x} = \frac{2x^2 - x^2 - y^2}{x\left(x^2+y^2\right)} = \frac{x^2 - y^2}{x\left(x^2+y^2\right)},$$

and $\dfrac{\partial u}{\partial y}=\dfrac{2y}{x^2+y^2}-\dfrac{1}{y}=\dfrac{2y^2-x^2-y^2}{y\left(x^2+y^2\right)}=\dfrac{\left(y^2-x^2\right)}{y\left(x^2+y^2\right)}.$

Now $du=\dfrac{\partial u}{\partial x}dx+\dfrac{\partial u}{\partial y}dy=\dfrac{\left(x^2-y^2\right)}{x\left(x^2+y^2\right)}dx+\dfrac{\left(y^2-x^2\right)}{y\left(x^2+y^2\right)}dy$

$$=\frac{x^2-y^2}{xy\left(x^2+y^2\right)}\ (y\ dx-x\ dy).$$

Example 7:

If $u = \sin (x^2 + y^2)$, *where* $a^2x^2 + b^2y^2 = c^2$, *find* $\dfrac{du}{dx}$.

Solution:

We have $\dfrac{du}{dx}=\dfrac{\partial u}{\partial x}+\dfrac{\partial u}{\partial y}.\dfrac{dy}{dx}.$ ...(1)

Now $u = \sin (x^2 + y^2)$.

$\therefore\ \dfrac{\partial u}{\partial y} = 2x \cos (x^2 + y^2)$ and $\dfrac{\partial u}{\partial y} = 2y \cos (x^2 + y^2)$.

Since $a^2x^2 + b^2y^2 = c^2$,

Therefore $2a^2x + 2b^2y\left(\dfrac{dy}{dx}\right) = 0$ or $\dfrac{dy}{dx} = -\dfrac{\left(a^2x\right)}{\left(b^2y\right)}.$

$\therefore$ from (1), $\dfrac{du}{dx} = 2x \cos (x^2 + y^2) - [2y \cos (x^2 + y^2)]\left[\dfrac{\left(a^2x\right)}{\left(b^2y\right)}\right]$

$$= [2x \cos (x^2 + y^2)]\left[1-\left(\frac{a^2}{b^2}\right)\right].$$

Example 8:

If $ax^2 + 2xhy + by^2 + 2gx + 2fy + c = 0$, *find* $\dfrac{d^2y}{dx^2}$.

Solution:

Let $F(x, y) \equiv ax^2 + 3hxy + by^2 + 3gx + 2fy + c = 0$. Then with usual notation *i.e.* $\dfrac{\partial F}{\partial x}=p,\ \dfrac{\partial F}{\partial y}=q$, etc., we have

$p = 2(ax + hy + g)$, $q = 2(hx + by + f)$,

$r = 2a$, $s = 2h$ and $t = 2b$.

$$\frac{d^2y}{dx^2} = -8[(hx + by + f)^2 a - 2(ax + hy + g)(hx + by + f)h + b(ax + hy + g)^2]/\{8(hx + by + f)^3\}$$

$$= \frac{-\left[\left(ab - h^2\right)\left(ax^2 + 2hxy + by^2 + 2gx + 2fy\right) + af^2 + bg^2 - 2fgh\right]}{(hx + by + f)^3}$$

$$= \frac{-\left[\left(ab - h^2\right)(-c) + af^2 + bg^2 - 2f\,gh\right]}{(hx + by + f)^3}$$

$$= [abc + 2fgh - af^2 - bg^2 - ch^2]/(hx + by + f)^3.$$

Example 9:

Formula for the second differential coefficient of an implicit function.

If $f(x, y) = 0$ be an implicit function of x and y, find a formula for $\frac{d^2y}{dx^2}$.

Solution:

We have $\frac{dy}{dx} = -\frac{\partial f/\partial x}{\partial f/\partial y} = -\frac{p}{q}$, ...(1)

where $\frac{\partial f}{\partial x}$ and $\frac{\partial f}{\partial y}$ have been denoted by p and q respectively. Also using the notations

$r = \frac{\partial^2 f}{\partial x^2}$, $s = \frac{\partial^2 f}{\partial x\, \partial y}$ and $t = \frac{\partial^2 f}{\partial y^2}$, we have from (1),

$$\frac{d^2y}{dx^2} = \frac{d}{dx}\left(\frac{dy}{dx}\right) = \frac{d}{dx}\left(-\frac{p}{q}\right) = -\frac{q\left(\frac{dp}{dx}\right) - p\left(\frac{dq}{dx}\right)}{q^2} \qquad ...(2)$$

But $\frac{dp}{dx} = \frac{\partial q}{\partial x} + \frac{\partial q}{\partial y}.\frac{dy}{dx} = s + t\left(-\frac{p}{q}\right) = \frac{qs - pt}{q}$,

and $\frac{dq}{dx} = \frac{\partial q}{\partial x} + \frac{\partial q}{\partial y}.\frac{dy}{dx} = s + t\left(-\frac{p}{q}\right) = \frac{qs - pt}{q}$.

Substituting in (2) the values of $\frac{dp}{dx}$ and $\frac{dq}{dx}$, we get

$$\frac{d^2y}{dx^2} = -\left[q\left(\frac{qr-ps}{q}\right) - p\left(\frac{qs-pt}{q}\right)\right]\cdot\left[\frac{1}{q^2}\right]$$

$= -(q^2r - 2pqs + p^2t)/q^3.$ **(Remember)**

Example 10:

(a) *If f(x, y) = and f(y, z) = 0, show that*

$$\frac{\partial f}{\partial y}\cdot\frac{\partial \phi}{\partial z}\cdot\frac{dz}{dx} = \frac{\partial f}{\partial x}\cdot\frac{\partial \phi}{\partial y}.$$ **(Meerut, 1993; 97, Allahabad, 99, 98)**

(b) *If the curves f(x, y) = 0 and φ(x, y) = 0 touch, show that at point of contact*

$$\frac{\partial f}{\partial x}\cdot\frac{\partial \phi}{\partial y} = \frac{\partial f}{\partial y}\cdot\frac{\partial \phi}{\partial x}.$$

Solution:

(a) From f(x, y) = 0, we have $\frac{dy}{dx} = -\frac{\partial f/\partial x}{\partial f/\partial y}$. ...(1)

From φ(y, z) = 0, we have $\frac{dz}{dy} = -\frac{\partial \phi/\partial y}{\partial \phi/\partial z}$. ...(2)

Multiplying the respective sides of (1) and (2), we have

$$\frac{dy}{dx}\cdot\frac{dz}{dy} = \left(\frac{\partial f}{\partial x}\cdot\frac{\partial \phi}{\partial y}\right)\Big/\left(\frac{\partial f}{\partial y}\cdot\frac{\partial \phi}{\partial z}\right)$$

or $\frac{dz}{dx}\cdot\frac{\partial f}{\partial y}\cdot\frac{\partial \phi}{\partial z} = \frac{\partial f}{\partial x}\cdot\frac{\partial \phi}{\partial y}$.

(b) The curves will touch if at their common point they have the same value of $\frac{dy}{dx}$. Now for the curve f(x, y) = 0, we have

$\frac{dy}{dx} = -\frac{\partial f/\partial f}{\partial f/\partial y}$ and for the curve φ(x, y) = 0, we have

$$\frac{dy}{dx} = -\frac{\partial \phi/\partial x}{\partial \phi/\partial y}.$$

∴ the two curves touch if at their common point, we have

$$\frac{-\partial f/\partial x}{\partial f/\partial y} = \frac{-\partial \phi/\partial x}{\partial \phi/\partial y} \quad i.e., \quad \frac{\partial f}{\partial x}\cdot\frac{\partial \phi}{\partial y} = \frac{\partial f}{\partial y}\cdot\frac{\partial \phi}{\partial x}.$$

EXERCISES

1. If $x^x y^y z^z = c$, show that $x = y = z$

$$\frac{\partial^2 z}{\partial x\, \partial y} = -(x \log ex)^{-1}.$$

2. If u $\tan^{-1} \dfrac{xy}{\sqrt{1+x^2+y^2}}$ show that

$$\frac{\partial^2 u}{\partial x\, \partial y} = \frac{1}{(1+x^2+y^2)^{3/2}}.$$

3. If $u = e^{xyz}$ show that

$$\frac{\partial^2 u}{\partial x\, \partial y\, \partial z} = (1 + 3xyz + x^2y^2z^2)e^{xyz}.$$

4. Verify Euler's Theorem for

$$z = \tan^{-1}\left(\frac{y}{x}\right).$$

5. It $u = \sin^{-1} \dfrac{x^2+y^2}{x+y}$ show that

$$x\frac{\partial u}{\partial x} + y\frac{\partial u}{\partial y} = \tan u$$ (D.U. B.Sc. (H), 1994)

6. If $z = \tan^{-1}\left(\dfrac{x^2+y^2}{x+y}\right)$ find $\dfrac{\partial z}{\partial x} + \dfrac{\partial z}{\partial y}$ and $\dfrac{\partial^2 z}{\partial x\, \partial y}$.

7. Show that the function $z = \sin\left(\dfrac{y}{z}\right)$ $\dfrac{\partial^2 z}{\partial y\, \partial x} = \dfrac{\partial^2 z}{\partial x\, \partial y}$.

8. If $u = (x^2 + y^2 + z^2)^{-3/2}$ show that $x\dfrac{\partial u}{\partial x} + y\dfrac{\partial u}{\partial y} + z\dfrac{\partial u}{\partial z} = -4$.

9. If $z = (x + y) + (x + y)\, \phi\left(\dfrac{y}{x}\right)$ prove that

$$x\left(\frac{\partial^2 z}{\partial x^2} - \frac{\partial^2 z}{\partial y\, \partial x}\right) = y\left(\frac{\partial^2 z}{\partial y^2} - \frac{\partial^2 z}{\partial x\, \partial y}\right).$$

10. If $u = \sin^{-1}\left(\dfrac{\sqrt{x}-\sqrt{y}}{\sqrt{x}+\sqrt{y}}\right)$ show that $x\dfrac{\partial u}{\partial x} + y\dfrac{\partial u}{\partial y} = 0$.

11. If $z = x\left(\frac{y}{x}\right)$ prove that $x\frac{\partial z}{\partial x} + y\frac{\partial z}{\partial y} = z$.

12. If $u \tan^{-1}\left(\frac{x+y}{\sqrt{x}+\sqrt{y}}\right)$ prove that

$$x^2\frac{\partial^2 u}{\partial x^2} + 2xy\frac{\partial^2 y}{\partial x\,\partial y} + y^2\frac{\partial^2 y}{\partial y^2} = \frac{\sin 2u}{8}\left\{2\cos^2 u - 3\right\}.$$

13. If $u = At^{-1/2}\, e^{-x^2/4a^2t}$ prove that

$$\frac{\partial u}{\partial t} = a^2\frac{\partial^2 u}{\partial x^2}.$$

14. Verify $\frac{\partial^2 u}{\partial x\,\partial y} = \frac{\partial^2 u}{\partial y\,\partial x}$ for the function

$$u = x^2\tan^{-1}\frac{y}{x} - y^2\tan^{-1}\frac{x}{y}.$$ **(D.U. B.Sc. (G), 1998)**

15. If $u = e^{xyz}$ show that

$$\frac{\partial^2 u}{\partial x\,\partial y\,\partial z} = (1 + 3xyz + x^2y^2z^2)e^{xyz}.$$

7

INDETERMINATE FORMS

7.1 DEFINITION

The limit of $\phi(x)/\psi(x)$ as $x \to 0$ is in general, is equal to the limit of the numerator divided by the limit of the dinominator. But when these two limits are both zero the quotient reduces to the form 0/0 which is meaning less this does not imply that $\lim_{x\to a} [\phi(x)\ \psi(x)]$ is meaningless or that it does net exist. The only conclusion which we may draw is that the method adopted by us is unsuitable.

A fraction whose numerator and denominator both tend to zero as $x \to a$ is called the **Indeterminate Form 0/0.** It has no definite value. The other indeterminate forms are ∞/∞, $\infty - \infty$, $0 \times \infty$, 1^∞, 0^0 and ∞^0.

If a fraction $\psi(x)/\phi(x)$ takes the indeterminate form 0/0 when $x \to a$ it does not mean that $\lim_{x\to a} \dfrac{\psi(x)}{\phi(x)}$ will not exist. For example the fraction $\dfrac{x^2-a^2}{x-a}$ takes the indeterminate form $\dfrac{0}{0}$ when $x \to a$. But $\lim_{x\to a} \dfrac{x^2-a^2}{x-a} = \lim_{x\to a} \dfrac{(x-a)(x+a)}{x-a} = \lim_{x\to a}(x+a) = 2a$, showing that the limit exists in this case.

Thus, $\lim_{x\to a} \dfrac{\psi(x)}{\phi(x)}$ may exist even if the fraction $\dfrac{\psi(x)}{\phi(x)}$ takes the indeterminate form 0/0. In this chapter, we shall give methods to find the limits of the functions which take indeterminate forms.

7.2 THE FORM 0/0: L'HOSPITAL'S RULE

If $\phi(x)$ and $\phi(x)$ be two functions of x which can be expanded by Taylor's Theorem in the neighbourhood of $x = a$ and if $f(a) = \phi(a) = 0$, then

$$\lim_{x\to a}\frac{f(x)}{\phi(x)} = \lim_{x\to a}\frac{f'(x)}{\phi'(x)},$$

provided, the latter limit exists, finite or infinite.

(Gorakhpur, 1992; Delhi, 96; Agra, 94; Punjab, 96)

Proof:

By Taylor's theorem, we know that

$$\lim_{x\to a}\frac{f(x)}{\phi(x)}$$

$$= \lim_{x\to a}\frac{f(a)+(x-a)\,f'(a)+\frac{(x-a)^2}{2!}f''(a)+\ldots+R_1}{\phi(a)+(x-a)\,\phi'(a)+\frac{(x-a)^2}{2!}\phi''(a)+\ldots+R_2},$$

where $R_1 = \frac{(x-a)^n}{n!}f^n\{a+\theta_1(x-a)\}, 0<\theta_1<1$

and $R_2 = \frac{(x-a)^n}{n!}\phi^n\{a+\theta_2(x-a)\}, 0<\theta_2<1.$

Now, since f(a) = 0 and $\phi(a) = 0$, we get, after dividing both numerator and denominator by (x – a),

$$\lim_{x\to a}\frac{f(x)}{\phi(x)} = \lim_{x\to a}\frac{f'(a)+(x-a)\,\{(1/2!)f''(a)f''(a)+\ldots\}}{\phi'(a)+(x-a)\,\{(1/2!)\,\phi''(a)+\ldots\}}$$

$$= \frac{f'(a)}{\phi'(a)} = \lim_{x\to a}\frac{f'(x)}{\phi'(x)}.$$

This proves the theorem which i generally known as *L'Hospital's Rule.*

Note: If $f'(a) = f''(a) = \ldots = f^{n-1}(a) = 0$

and $\phi'(a) = \phi''(a) = \ldots = \phi^{n-1}(a) = 0$

but $\phi^n(a)$ and $\phi^n(a)$ are not both zero, then by repeated application of Hospital's rule, we have

$$\lim_{x\to a}\frac{f(x)}{\phi(x)} = \lim_{x\to a}\frac{f^n(x)}{\phi^n(x)}.$$

Rule: *If the limit of $f(x)/\phi(x)$ as $x \to a$ takes the form 0/0, differentiate the numerator and denominator separately w.r.t. x and obtain a new function $f'(x)/\phi'(x)$. Now as $x \to a$, if it again w.r.t. x and repeat the above process, till indeterminate form persists.*

Remember: *While applying L'Hospital's Rule we are not to differentiate $f(x)/\phi(x)$ as a fraction. The numerator and denominator must be differentiated separately.*

7.3 METHOD OF EXPANSION (ALGEBRAIC METHODS)

In many cases th limit of an indeterminate form can be easily obtained by using some well known Algebraic and Trigonometrical Expansions. We can also make use of some well-known limits in order to solve the problems or to shorten the work. The following expansions should be remembered. The student should note that differentiating the numerator or the denominator repeatedly amount realy to finding the coefficients in the expansion of the function concerned by Maclaurin's Theorem and if the expansion is already known, the work can be shortened.

$$(1+x)^n = 1 + nx + \frac{n(n-1)}{2!}x^2 + \frac{n(n-1)(n-2)}{3!}x^3 + \ldots$$

$$(1-x)^{-1} = 1 + x + x^2 + x^3 + \ldots, |x| < 1$$

$$a^x = 1 + x \log a + \frac{x^2}{2!}(\log a)^2 + \frac{x^3}{3!}(\log a)^3 + \ldots$$

$$e^x = 1 + x + \frac{x^2}{2!} + \frac{x^3}{3!} + \ldots$$

$$\sin x = x - \frac{x^3}{3!} + \frac{x^5}{5!} - \ldots, \cos x = 1 - \frac{x^2}{2!} + \frac{x^4}{4!} - \ldots$$

$$\tan x = x + \frac{x^3}{3} + \frac{2}{15}x^5 + \ldots$$

$$\log(1+x) = x - \frac{x^2}{2} + \frac{x^3}{3} - \frac{x^4}{4} + \ldots, |x| < 1$$

$$\log(1-x) = -\left(x + \frac{x^2}{2} + \frac{x^3}{3} + \frac{x^4}{4} + \ldots\right), |x| < 1$$

$$\sin^{-1} x = x + \frac{x^3}{6} + \frac{3x^5}{40} + \ldots; \tan^{-1} x = x - \frac{x^3}{3} + \frac{x^5}{5} - \ldots$$

$$\sinh x = x + \frac{x^3}{3!} + \frac{x^5}{5!} + \ldots; \cosh x = 1 + \frac{x^2}{2!} + \frac{x^4}{4!} + \ldots$$

Also remember that

log 1 = 0, log e = 1, log ∞ = ∞; log 0 = − ∞.

Sometimes the use of the following limits shortens the work:

(i) $\lim_{x \to 0} \frac{\sin x}{x} = 1$,

(ii) $\lim_{x \to 0} \cos x = 1$,

(iii) $\lim_{x \to 0} \frac{\tan x}{x} = 1$,

(iv) $\lim_{x \to 0} (1 + x)^{1/x} = e$,

(v) $\lim_{x \to 0} (1 + nx)^{1/x} = e^n$,

(vi) $\lim_{x \to \infty} \left(1 + \frac{1}{x}\right)^x = e$,

(vii) $\lim_{x \to \infty} \left(1 + \frac{a}{x}\right)^x = e^a$.

Example 1:

Evaulate $\lim_{x \to 0} \frac{1 - \cos x}{x \log(1 + x)}$.

Solution:

We have $\lim_{x \to 0} \frac{1 - \cos x}{x \log(1 + x)}$ [form 0/0]

$= \lim_{x \to 0} \frac{\sin x}{\log(1 + x) + \{x/(1 + x)\}}$, [form 0/0]

$= \lim_{x \to 0} \frac{\cos x}{\frac{1}{1 + x} + \frac{1 + x - x}{(1 + x)^2}} = \lim_{x \to 0} \frac{\cos x}{\frac{1}{1 + x} + \frac{1}{(1 + x)^2}} = \frac{1}{2}$. **Ans.**

Example 2:

Evaulate $\lim_{x \to 1} \frac{\log x}{x - 1}$. **(Agra, 1993; Meerut, 95, 98)**

Solution:

We have $\lim_{x \to 1} \frac{\log x}{x - 1}$ [form 0/0]

$$\lim_{x \to 1} \frac{1/x}{1} = 1.$$ **Ans.**

Example 3:

Evaulate $\displaystyle \lim_{x \to 0} \frac{x - \sin x}{\tan^3 x}.$

Solution:

We have $\displaystyle \lim_{x \to 0} \frac{x - \sin x}{\tan^3 x} = \lim_{x \to 0} \left\{ \frac{x - \sin x}{x^3} \cdot \left(\frac{x}{\tan x} \right)^3 \right\}$ **(Note)**

$$= \lim_{x \to 0} \frac{x - \sin x}{x^3} \cdot \lim_{x \to 0} \left(\frac{x}{\tan x} \right)^3$$

$$= \lim_{x \to 0} \frac{x - \sin x}{x^3}, \qquad \left[\because \lim_{x \to 0} \frac{x}{\tan x} = 1 \right]$$

$$= \frac{1}{6}.$$

Example 4:

Evaluate $\displaystyle \lim_{x \to 0} \frac{x - \tan x}{x^3}.$

(Delhi, 1995; Kanpur, 92; Kashmir, 92; Meerut, 93)

Solution:

We have $\displaystyle \lim_{x \to 0} \frac{x - \tan x}{x^3},$ [form 0/0]

$$= \lim_{x \to 0} \frac{1 - \sec^2 x}{3x^2}, \qquad \text{[form 0/0]}$$

$$= \lim_{x \to 0} \frac{-2 \sec x . \sec x \tan x}{6x}$$

$$= \lim_{x \to 0} - \frac{\sec^2 x}{3} \cdot \frac{\tan x}{x} = -\frac{1}{3} \times 1 = -\frac{1}{3}.$$ **Ans.**

Aliter: $\displaystyle \lim_{x \to 0} \frac{x - \tan x}{x^3}$

$$= \lim_{x \to 0} \frac{x - \left[x + (x^3/3) + (2x^5/15) + \ldots \right]}{x^3}$$

$$= \lim_{x\to 0}\left(-\frac{1}{3}-\frac{2}{15}x^2 - \ldots\right) = -\frac{1}{3}.$$

Example 5:

Evaluate

(a) $\lim_{x\to 0}\frac{\sin x}{x}$,

(b) $\lim_{x\to 0}\frac{\tan x}{x}$,

(c) $\lim_{x\to 0}\frac{\tan ax}{\sin bx}$.

Solution:

(a) We have, $\lim_{x\to 0}\frac{\sin x}{x}$ $\left[\text{form } \frac{0}{0}\right]$

$$= \lim_{x\to 0}\frac{\cos x}{1} = 1.$$ **Ans.**

(b) Proceeding as in part (a), we get

$$\lim_{x\to 0}\frac{\tan x}{x} = 1$$ **Ans.**

(c) $\lim_{x\to 0}\frac{\sin ax}{\sin bx}$, $\left[\text{form}\frac{0}{0}\right]$

$$= \lim_{x\to 0}\frac{a\cos ax}{b\cos bx} = \frac{a}{b}.$$ **Ans.**

Example 6:

Evaluate $\lim_{x\to 0}\frac{a^x - 1}{b^x - 1}$.

Solution:

We have $\lim_{x\to 0}\frac{a^x - 1}{b^x - 1}$ $\left[\text{form}\frac{0}{0}\right]$

$$= \lim_{x\to 0}\frac{a^x \log a}{b^x \log b} = \frac{\log a}{\log b} = \log_b a.$$ **Ans.**

Example 7:

Evaluate $\lim_{x\to 0}\frac{x - \sin x}{x^3}$. **(Rohilkhand, 1983; Meerut, 94)**

Solution:

We have $\lim\limits_{x\to 0}\dfrac{x-\sin x}{x^3}$ [form 0/0 so we shall apply L' Hospital's Rule)

$$= \lim_{x\to 0}\frac{1-\cos x}{3x^2} \qquad \left[\text{form again } \frac{0}{0}\right]$$

$$= \lim_{x\to 0}\frac{\sin x}{6x} \qquad \left[\text{form again } \frac{0}{0}\right]$$

$$= \lim_{x\to 0}\frac{\cos x}{6} = \frac{1}{6}.$$

Ans.

Aliter: $\lim\limits_{x\to 0}\dfrac{x-\sin x}{x^3} = \lim\limits_{x\to 0}\dfrac{x-\left[x-\left(\frac{x^3}{3!}\right)+\left(\frac{x^5}{5!}\right)-\ldots\right]}{x^3}$

$$= \lim_{x\to 0}\frac{\left(\frac{x^3}{6}\right)-\left(\frac{x^5}{120}\right)+\ldots}{x^3}$$

$$= \lim_{x\to 0}\left(\frac{1}{6}-\frac{x^2}{120}+\ldots\right) = \frac{1}{6}.$$

Ans.

Example 8:

Evaluate $\lim\limits_{x\to 0}\dfrac{1-\cos x}{3x^2}$. **(Meerut, 1998S)**

Solution:

We have $\lim\limits_{x\to 0}\dfrac{1-\cos x}{3x^2}$, $\left[\text{form } \frac{0}{0}\right]$

$$= \lim_{x\to 0}\frac{\sin x}{6x}. \qquad \left[\text{form } \frac{0}{0}\right]$$

$$= \lim_{x\to 0}\frac{\cos x}{6} = \frac{1}{6}.$$

Ans.

Example 9:

Evaluate $\lim\limits_{x\to 0}\dfrac{x^2+2\cos x-2}{x\sin^3 x}$. **(Meerut, 1991)**

Solution:

We have $\lim\limits_{x\to 0}\dfrac{x^2+2\cos x-2}{x\sin^3 x}$

$$= \lim_{x\to 0}\left[\frac{x^2+2\cos x-2}{x^4}.\frac{x^3}{\sin^3 x}\right]$$

$$= \lim_{x\to 0}\left(\frac{x}{\sin x}\right)^3.\lim_{x\to 0}\frac{x^2+2\cos x-2}{x^4}$$

$$= \lim_{x\to 0}\frac{x^2+2\cos x-2}{x^4}, \qquad \left[\because \lim_{x\to 0}\frac{x}{\sin x}=1\right]$$

$$= \lim_{x\to 0}\frac{2x-2\sin x}{4x^3} \text{ by L'Hospital's Rule for the form } \frac{0}{0}$$

$$= \lim_{x\to 0}\frac{2-2\cos x}{12x^2} \qquad \left[\text{form again } \frac{0}{0}\right]$$

$$= \lim_{x\to 0}\frac{2\sin x}{24x}=\frac{1}{12}\lim_{x\to 0}\frac{\sin x}{x}=\frac{1}{12}.1=\frac{1}{12}.$$ **Ans.**

Example 10:

Evaluate $\lim_{x\to 0}\frac{a\sin x-\sin ax}{x(\cos x-\cos ax)}$.

Solution:

We have $\lim_{x\to 0}\frac{a\sin x-\sin ax}{x(\cos x-\cos ax)}$, $\left[\text{form } \frac{0}{0}\right]$

$$= \lim_{x\to 0}\frac{a\cos x-a\cos ax}{(\cos x-\cos ax)+x(-\sin x+a\sin ax)}, \qquad \left[\text{form } \frac{0}{0}\right]$$

$$= \lim_{x\to 0}\frac{-a\sin x+a^2\sin ax}{(-\sin x+a\sin ax)+(-\sin x+a\sin ax)+x(-\cos x+a\cos ax)}$$

$$= \lim_{x\to 0}\frac{-a\sin x+a^2\sin ax}{2(a\sin ax-\sin x)+x(a^2\cos ax-\cos x)}, \qquad \left[\text{form } \frac{0}{0}\right]$$

$$= \lim_{x\to 0}\frac{-a\cos x+a^3\cos ax}{2(a^2\cos ax-\cos x)+(a^2\cos ax-\cos x)+x(-a^3\sin ax+\sin x)}$$

$$= \lim_{x\to 0}\frac{-a\cos x+a^3\cos ax}{3(a^2\cos ax-\cos x)+x(\sin x-a^3\sin ax)}$$

$$= \frac{-a+a^3}{3(a^2-1)}=\frac{a}{3}\frac{(a^2-1)}{(a^2-1)}=\frac{a}{3}.$$ **Ans.**

Example 11:

Evaluate $\lim\limits_{x\to 0}\dfrac{e^{x}-e^{x\cos x}}{x-\sin x}$.

Solution:

We have $\lim\limits_{x\to 0}\dfrac{e^{x}-e^{x\cos x}}{x-\sin x}$, $\left[\text{form }\dfrac{0}{0}\right]$

$$=\lim_{x\to 0}\frac{e^{x}-e^{x\cos x}(\cos x-x\sin x)}{1-\cos x}$$ (by L' Hospital's Rule)

$\left[\text{form again }\dfrac{0}{0}\right]$

$$=\lim_{x\to 0}.\frac{e^{x}-e^{x}\cos x(\cos x-x\sin x)^{2}-e^{x\cos x}(-2\sin x-x\cos x)}{\sin x}$$

by L' Hospital's Rule

$$=\lim_{x\to 0}\frac{e^{x}-e^{x\cos x}(\cos x-x\sin x)^{2}+e^{x\cos x}(2\sin x+x\cos x)}{\sin x},$$

$\left[\text{form }\dfrac{0}{0}\right]$

$$\lim_{x\to 0}=\frac{\begin{array}{r}e^{x}-e^{x}\cos x(\cos x-x\sin x)^{3}-2e^{x\cos x}(\cos x-x\sin x)\\ \times(-2\sin x-x\cos x)+e^{x\cos x}(\cos x-x\sin x)\\ .(2\sin x+x\cos x)+e^{x\cos x}(3\cos x-x\sin x)\end{array}}{\cos x}$$

$$=\frac{1-1(1-0)^{3}-2\times1\times1\times0+1\times1\times0+1\times(3-0)}{1}$$

$$=\frac{1-1+3}{1}=3.$$ **Ans.**

Example 12:

Evaluate $\lim\limits_{x\to 0}\dfrac{e^{ax}-e^{ax}}{\log(1+bx)}$.

Solution:

We have $\lim\limits_{x\to 0}\dfrac{e^{ax}-e^{-ax}}{\log(1+bx)}$, $\left[\text{form}\dfrac{0}{0}\right]$

$$= \lim_{x\to 0} \frac{ae^{ax} + ae^{-ax}}{\frac{b}{(1+bx)}} = \frac{a+a}{b} = \frac{2a}{b}.$$ **Ans.**

Example 13:

Evaluate $\lim_{x\to 0} \frac{\log(1+x^3)}{\sin^3 x}$.

Solution:

We have $\lim_{x\to 0} \frac{\log(1+x^3)}{\sin^3 x} = \lim_{x\to 0} \left\{ \frac{\log(1+x^3)}{x^3} \cdot \left(\frac{x}{\sin x}\right)^3 \right\}$

$$= \lim_{x\to 0} \frac{\log(1+x^3)}{x^3}, \text{ since } \lim_{x\to 0} \frac{x}{\sin x} = 1, \qquad \left[\text{form again } \frac{0}{0}\right]$$

$$= \lim_{x\to 0} \frac{\frac{3x^2}{(1+x^3)}}{3x^2} = \lim_{x\to 0} \frac{1}{1+x^3} = 1.$$ **Ans.**

Example 14:

Evaluate $\lim_{x\to 0} \frac{x\cos x - \log(1+x)}{x^2}$.

(Agra, 1992, Meerut, 92, 96 BP, 98; Vikram, 97)

Solution:

We have $\lim_{x\to 0} \frac{x\cos x - \log(1+x)}{x^2}$ $\left[\text{form} \frac{0}{0}\right]$

$$= \lim_{x\to 0} \frac{\cos x - x\sin x - \frac{1}{(1+x)}}{2x} \qquad \left[\text{form} \frac{0}{0}\right]$$

$$= \lim_{x\to 0} \frac{-\sin x - \sin x - x\cos x + \frac{1}{(1+x)^2}}{2} = \frac{1}{2}.$$

Example 14 (a):

Evaluate $\lim_{x\to 0} \frac{e^x - e^{-x} - 2x}{x^2 \sin x}$.

Solution:

$$\lim_{x\to 0}\frac{e^x-e^{-x}-2x}{x^2\sin x},\qquad\left[\text{form}\frac{0}{0}\right]$$

$$=\lim_{x\to 0}\left\{\frac{e^x-e^{-x}-2x}{x^3}.\left(\frac{x}{\sin x}\right)\right\},\qquad\textbf{(Note)}$$

$$=\lim_{x\to 0}\frac{e^x-e^{-x}-2x}{x^3},\qquad\left[\text{form}\frac{0}{0}\right]$$

$$=\lim_{x\to 0}\frac{e^x+e^{-x}-2}{3x^2},\qquad\left[\text{form}\frac{0}{0}\right]$$

$$=\lim_{x\to 0}\frac{e^x-e^{-x}}{6x},\qquad\left[\text{form}\frac{0}{0}\right]$$

$$=\lim_{x\to 0}\frac{e^x+e^{-x}}{6}=\frac{1+1}{6}=\frac{1}{3}.$$

Example 15:

Evaluate $\lim_{x\to 0}\frac{\sin x\,\sin^{-1}x}{x^2}$. **(Rohilkhand, 1999; Rajathan, 99)**

Solution:

$$\lim_{x\to 0}\frac{\sin x\,\sin^{-1}x}{x^2}=\lim_{x\to 0}\left(\frac{\sin^{-1}x}{x}.\frac{\sin x}{x}\right)$$

$$=\lim_{x\to 0}\frac{\sin^{-1}x}{x},\qquad\left[\text{form}\frac{0}{0}\right]$$

$$=\lim_{x\to 0}\frac{\dfrac{1}{\sqrt{(1-x^2)}}}{1}=1$$

Example 16

$$\lim_{x\to 0}\frac{\log(1+kx^2)}{1-\cos x}.$$

Solution:

We have $\lim_{x\to 0}\frac{\log(1+kx^2)}{1-\cos x}$, $\left[\text{form}\frac{0}{0}\right]$

$$= \lim_{x\to 0} \frac{\frac{2kx}{(1+kx^2)}}{\sin x} = \lim_{x\to 0}\left(\frac{2k}{1+kx^2}.\frac{x}{\sin x}\right)$$

$= 2k \times 1 = 2k.$ **Ans.**

Example 16:

Evaluate $\lim_{x\to\frac{1}{2}} \frac{\cos^2 \pi x}{e^{2x} - 2ex}$. **(Indore, 1994)**

Solution:

We have $\lim_{x\to\frac{1}{2}} \frac{\cos^2 \pi x}{e^{2x} - 2ex}$, $\left[\text{form}\frac{0}{0}\right]$

$$= \lim_{x\to\frac{1}{2}} \frac{2\cos \pi x.(-\pi \sin \pi x)}{2e^{2x} - 2e}$$

$$= \lim_{x\to\frac{1}{2}} \frac{-\pi \sin 2\pi x}{2e^{2x} - 2e}, \qquad \left[\text{form}\frac{0}{0}\right]$$

$$= \lim_{x\to\frac{1}{2}} \frac{-\pi^2 \cos 2\pi x}{4e^{2x}} = \frac{-2\pi^2(-1)}{4e} = \frac{\pi^2}{2e}.$$

Example 17

Evaluate $\lim_{\theta\to 0} \frac{\sin\theta - \theta\cos\theta}{\sin\theta - \theta}$.

Solution:

$$\lim_{\theta\to 0} \frac{\sin\theta - \theta \sin\theta}{\sin\theta - \theta}, \qquad \left[\text{form}\frac{0}{0}\right]$$

$$= \lim_{\theta\to 0} \frac{\cos\theta - \cos\theta + \theta\sin\theta}{\cos\theta - 1} = \lim_{\theta\to 0} \frac{\sin\theta}{\cos\theta - 1}, \qquad \left[\text{form}\frac{0}{0}\right]$$

$$= \lim_{\theta\to 0} \frac{\theta\cos\theta + \sin\theta}{-\sin\theta}, \qquad \left[\text{form}\frac{0}{0}\right]$$

$$= \lim_{\theta\to 0} \frac{\cos\theta - \theta\sin\theta + \cos\theta}{-\cos\theta} = \frac{1-0+1}{-1} = -2.$$ **Ans.**

Example 18:

Evaluate $\lim_{x\to 1} \frac{x^x - x}{1 - x + \log x}$. **(Kanpur, 1999)**

Solution:

$$\lim_{x\to 1}\frac{x^x - x}{1 - x + \log x}, \qquad \left[\text{form}\frac{0}{0}\right]$$

$$= \lim_{x\to 1}\frac{x^x(1+\log x)-1}{-1+\left(\frac{1}{x}\right)}, \qquad \left\{\because \frac{d}{dx}(x^x) = x^x(1+\log x)\right\}$$

$$\left[\text{form again}\frac{0}{0}\right]$$

$$= \lim_{x\to 1}\frac{x^x\left(\frac{1}{x}\right)+x^x(1+\log x).(1+\log x)}{\left(-\frac{1}{x^2}\right)}$$

$$= \frac{1+1(1+0)(1+0)}{-1} = -2.$$ **Ans.**

Example 19:

Evaluate $\lim_{x\to 0}\frac{e^x - e^{\sin x}}{x-\sin x}$. **(Kanpur, 1999; Meerut, 94P, 96P)**

Solution:

Here Nr. $= e^x - e^{\sin x} = e^x - e^{x-\frac{x^3}{3!}+\frac{x^5}{5!}-\ldots}$

$$= e^x - e^x . e^{-\frac{x^3}{3!}+\frac{x^5}{5!}-\ldots}$$

$$= e^x(1 - e^z), \text{ where } z = -\frac{x^3}{3!}+\frac{x^5}{5!}-\ldots$$

$$= e^x\left[1-\left(1+z+\frac{z^2}{2!}+\ldots\right)\right] = -e^x\left[z+\frac{z^2}{2!}+\ldots\right]$$

$$= -e^x\left[\left(-\frac{x^3}{3!}+\frac{x^5}{5!}-\ldots\right)+\frac{1}{2!}\left(-\frac{x^3}{3!}+\frac{x^5}{5!}-\ldots\right)^2+\ldots\right]$$

$$= -e^x\left[-\frac{x^3}{6}+\frac{x^5}{120}-\ldots\right] = x^3e^x\left(\frac{1}{6}-\frac{x^2}{120}+\ldots\right).$$

Also Denom. $= x - \sin x = x - \left(x-\frac{x^3}{3!}+\frac{x^5}{5!}-\ldots\right)$

$$= \frac{x^3}{6}-\frac{x^5}{120}+\ldots = x^3\left(\frac{1}{6}-\frac{x^2}{120}+\ldots\right).$$

$$\therefore \lim_{x\to 0}\frac{e^x - e^{\sin x}}{x-\sin x} = \lim_{x\to 0}\frac{x^3.e^x\left[\left(\frac{1}{6}\right)-\left(\frac{x^2}{120}\right)+\ldots\right]}{x^3\left[\left(\frac{1}{6}\right)-\left(\frac{x^2}{120}\right)+\ldots\right]}$$

$$= \lim_{x\to 0}\frac{e^x\left[\left(\frac{1}{6}\right)-\left(\frac{x^2}{120}\right)+\ldots\right]}{\left(\frac{1}{6}\right)-\left(\frac{x^2}{120}\right)+\ldots} = \frac{\frac{1}{6}}{\frac{1}{6}} = 1.$$

Ans.

Example 20:

Evaluate $\lim_{x\to 0}\frac{\left[e^x + log\{(1-x)/e\}\right]}{tan\,x - x}$.

(Rohilkhand, 1996, Gorakhpur, 92; Kanpur, 99; Lucknow, 95)

Solution:

$$\lim_{x\to 0}\frac{e^x + \log\left(\frac{1-x}{e}\right)}{\tan x - x} = \lim_{x\to 0}\frac{e^x + \log(1-x) - \log e}{\tan x - x}$$ **(Note)**

$$= \lim_{x\to 0}\frac{e^x + \log(1-x) - 1}{\tan x - x},$$ $\left[\text{form}\frac{0}{0}\right]$

$$= \lim_{x\to 0}\frac{1 + x + \frac{x^2}{2!} + \frac{x^3}{3!} + \ldots + \left(-x - \frac{x^2}{2} - \frac{x^3}{3} - \ldots\right) - 1}{\left(x + \frac{x^3}{3} + \frac{2x^5}{15} + \ldots\right) - x}$$

$$= \lim_{x\to 0}\frac{-\frac{1}{6}x^3(1 + \text{terms containing x and its higher powers})}{\frac{1}{3}x^3(1 + \text{terms containing x and its higher powers})}$$

$$= -\frac{1}{2}.$$

Ans.

Example 21:

Evaluate (a) $\lim_{x\to 0}\frac{(1+x)^n - 1}{x}$,

(b) $\lim_{x\to 0} \frac{1-\sqrt{(1-x^2)}}{x^2}$,

(c) $\lim_{x\to 0} \frac{xe^x - \log(1+x)}{x^2}$, **(Agra, 1996)**

(d) $\lim_{x\to 0} \frac{\log(1-x^2)}{\log \cos x}$, **(Meerut, 1997)**

(e) $\lim_{x\to 1} \frac{x^5 - 2x^3 - 4x^2 + 9x - 4}{x^4 - 2x^3 + 2x - 1}$. **(Meerut, 1993)**

Solution:

(a) We have

$$\lim_{x\to 0} \frac{(1+x)^n - 1}{x}, \qquad \left[\text{form} \frac{0}{0}\right]$$

$$= \lim_{x\to 0} \frac{n(1+x)^{n-1}}{1}, \qquad \text{[by L'Hospital's Rule]}$$

$= n.$

Aliter. We have

$$\lim_{x\to 0} \frac{(1+x)^n - 1}{x} = \lim_{x\to 0} \frac{1 + nx + \frac{n(n-1)}{2!}x^2 + \ldots - 1}{x}$$

$$= \lim_{x\to 0} \left\{ n + \frac{n(n-1)}{2!}x + \ldots \right\} = n.$$

(b) $\lim_{x\to 0} \frac{1-\sqrt{(1-x^2)}}{x^2}$ [form 0/0 so we shall apply L' Hospital's Rule]

$$= \lim_{x\to 0} \frac{0 = \left\{\frac{1}{2}(1-x^2)^{-1/2}(-2x)\right\}}{2x}$$

$$= \lim_{x\to 0} \frac{1-\sqrt{(1-x^2)}}{2x} = \lim_{x\to 0} \frac{1}{2\sqrt{(1-x^2)}} = \frac{1}{2}$$

(c) $\lim_{x\to 0} \frac{xe^x - \log(1+x)}{x^2}$, [form 0/0 so we shall apply Hospital's Rule]

$$= \lim_{x\to 0} \frac{xe^x + e^x - \left\{\frac{1}{(1+x)}\right\}}{2x}, \qquad \left[\text{form } \frac{0}{0}\right]$$

$$= \lim_{x\to 0} \frac{xe^x + e^x + e^x + \left\{\frac{1}{(1+x)^2}\right\}}{2} = \frac{0+1+1+1}{2} = \frac{3}{2}.$$

(d) $\lim_{x\to 0} \frac{\log(1-x^2)}{\log \cos x}$, [form 0/0 so we shall apply Hospital's Rule]

$$= \lim_{x\to 0} \frac{-\frac{2x}{(1-x^2)}}{-\frac{\sin x}{\cos x}} = \lim_{x\to 0}\left(\frac{2}{1-x^2}\cdot\frac{x}{\tan x}\right)$$

$$= \left(\lim_{x\to 0} \frac{2}{1-x^2}\right)\left(\lim_{x\to 0} \frac{x}{\tan x}\right) = 2\times 1 = 2.$$

(e) We have

$$= \lim_{x\to 1} \frac{x^5 - 2x^3 - 4x^2 + 9x - 4}{x^4 - 2x^3 + 2x - 1},$$

[form 0/0 so we shall apply L'Hospital's Rule]

$$= \lim_{x\to 1} \frac{5x^4 - 6x^2 - 8x + 9}{4x^3 - 6x^2 + 2}, \qquad \left[\text{form again } \frac{0}{0}\right]$$

$$= \lim_{x\to 1} \frac{20x^3 - 12x - 8}{12x^2 - 12x},$$ [form 0/0, ∴ again apply L'Hospital's rule]

$$= \lim_{x\to 1} \frac{60x^2 - 12}{24x - 12} = \frac{60-12}{24-12} = \frac{48}{12} = 4.$$

Example 22:

Evaluate $\lim_{x\to 0} \frac{\cosh x - \cos x}{x \sin x}$.

(Meerut, 1990; Agra, 95; Delhi, 96; Kashmir, 94; Punjab, 92)

Solution:

$$\lim_{x\to 0} \frac{\cosh - \cos x}{x \sin x}, \qquad \left[\text{form } \frac{0}{0}\right]$$

$$= \lim_{x\to 0}\left\{\frac{\cosh x - \cos x}{x^2}\cdot\left(\frac{x}{\sin x}\right)\right\},$$ **(Note)**

$$= \lim_{x\to 0}\frac{\cosh - \cos x}{x^2},$$ $\left[\text{form}\frac{0}{0}\right]$

$$= \lim_{x\to 0}\frac{\sinh x + \sin x}{2x},$$ $\left[\text{form}\frac{0}{0}\right]$

$$= \lim_{x\to 0}\frac{\cosh x + \cos x}{2} = \frac{1+1}{2} = 1.$$ **Ans.**

Example 23:

Evaluate $\lim_{x\to b}\frac{x^b - b^x}{x^x - b^b}$. **(Meerut, 1993, Rohilkhand, 97)**

Solution:

Here the form is $\frac{0}{0}$. To differentiate the Dr. we shall require the diff. coeff. of x^x. Let $y = x^x$; then $\log y = x \log x$.

Differentiating, $\frac{1}{y}\cdot\frac{dy}{dx} = x.\frac{1}{x} + \log x$

or $\frac{dy}{dx} = y(1 + \log x) = x^x(1 + \log x)$. $[\because y = x^x]$.

$\therefore \frac{d}{dx}(x^x) = x^x(1 + \log x)$.

$$\therefore \lim_{x\to b}\frac{x^b - b^x}{x^x - b^b} = \lim_{x\to b}\frac{bx^{b-1} - b^x \log b}{x^x(1+\log x) - 0},$$ [by Hospital's Rule]

$$= \frac{b.b^{b-1} - b^b \log b}{b^b(1+\log b)} = \frac{b^b(1-\log b)}{b^b(1+\log b)} = \frac{1-\log b}{1+\log b}.$$

Example 24:

Evaluate $\lim_{a\to b}\frac{a^b - b^a}{a^a - b^b}$

Solution:

Here $a \to b$. Therefore, while differentiating we shall regard a as variable and b as constant. **(Note)**

We have $\lim_{a\to b}\frac{a^b - b^a}{a^a - b^b}$, $\left[\text{form}\frac{0}{0}\right]$

$$= \lim_{a\to b} \frac{ba^{b-1} - b^a \log b}{a^a(1+\log a) - 0}, \qquad \left[\because \frac{d}{da}(a^a) = a^a(1+\log a)\right]$$

$$= \frac{b.b^{b-1} - b^b \log b}{b^b(1+\log b)} = \frac{b^b(1-\log b)}{b^b(1+\log b)} = \frac{1-\log b}{1+\log b}.$$ **Ans.**

Example 25:

Evaluate $\lim_{x\to 0} \frac{e^x - e^{-x} - 2\log(1+x)}{x \sin x}$.

Solution:

$$\lim_{x\to 0}\left\{\frac{e^x - e^{-x} - 2\log(1+x)}{x \sin x}\right\}$$

$$= \lim_{x\to 0}\left\{\frac{e^x - e^{-x} - 2\log(1+x)}{x^2}.\frac{x}{\sin x}\right\}$$ **(Note)**

$$= \lim_{x\to 0}\frac{e^x - e^{-x} - 2\log(1+x)}{x^2}, \qquad \left[\text{form } \frac{0}{0}\right]$$

$$= \lim_{x\to 0}\frac{e^x + e^{-x} - \frac{2}{(1+x)}}{2x}, \qquad \left[\text{form } \frac{0}{0}\right]$$

$$= \lim_{x\to 0}\frac{e^x - e^{-x} + \frac{2}{(1+x)^2}}{2} = \frac{1-1+2}{2} = 1.$$

Example 26:

Evaluate $\lim_{x\to 0}\frac{\sin 2x + 2\sin^2 x - 2\sin x}{\cos x - \cos^2 x}$.

Solution:

$$\lim_{x\to 0}\frac{\sin 2x + 2\sin^2 x - 2\sin x}{\cos x - \cos^2 x}, \qquad \left[\text{form } \frac{0}{0}\right]$$

$$= \lim_{x\to 0}\frac{2\cos 2x + 4\sin x \cos x - 2\cos x}{-\sin x - 2\cos x(-\sin x)}$$

$$= \lim_{x\to 0}\frac{2\cos 2x + 2\sin 2x - 2\cos x}{-\sin x + \sin 2x}, \qquad \left[\text{form } \frac{0}{0}\right]$$

$$= \lim_{x\to 0} -\frac{-4\sin 2x + 4\cos 2x + 2\sin x}{-\cos x + 2\cos 2x} = \frac{-0+4+0}{-1+2} = 4.$$ **Ans.**

Example 27:

Evaluate $\lim\limits_{x\to 0} \dfrac{e^x - \log(e+ex)}{x^2}$.

Solution:

$$\lim_{x\to 0} \frac{e^x - \log(e+ex)}{x^2} = \lim_{x\to 0} \frac{e^x - \log\{e(1+x)\}}{x^2}$$

$$= \lim_{x\to 0} \frac{e^x - \log e - \log(1+x)}{x^2}$$

$$= \lim_{x\to 0} \frac{e^x - 1 - \log(1+x)}{x^2}$$

$$= \lim_{x\to 0} \frac{e^x - 1 - \log(1+x)}{x^2}, \qquad \left[\text{form } \frac{0}{0}\right]$$

$$= \lim_{x\to 0} \frac{e^x - \left\{\dfrac{1}{(1+x)}\right\}}{2x}, \qquad \left[\text{form } \frac{0}{0}\right]$$

$$= \lim_{x\to 0} \frac{e^x + \left\{\dfrac{1}{(1+x)^2}\right\}}{2} = \frac{1+1}{2} = 1.$$

Example 28:

Evaluate $\lim\limits_{x\to 0} \dfrac{e^x - 2\cos x + e^{-x}}{x \sin x}$.

Solution:

$$\lim_{x\to 0} \frac{e^x - 2\cos x + e^{-x}}{x\sin x}$$

$$= \lim_{x\to 0} \left\{\frac{e^x - 2\cos x + e^{-x}}{x^2} \cdot \frac{x}{\sin x}\right\}, \qquad \textbf{(Note)}$$

$$= \lim_{x\to 0} \frac{e^x - 2\cos x + e^{-x}}{x^2}, \qquad \left[\text{form } \frac{0}{0}\right]$$

$$= \lim_{x \to 0} \frac{e^x + 2\sin x - e^{-x}}{2x} \qquad \left[\text{form } \frac{0}{0}\right]$$

$$= \lim_{x \to 0} \frac{e^x + 2\cos x + e^{-x}}{2} = \frac{1+2+1}{2} = 2. \qquad \textbf{Ans.}$$

Example 29:

Evaluate $\lim_{x \to +\infty} \dfrac{a^{1/x} - b^{1/x}}{\log\left\{\dfrac{x}{(x-1)}\right\}}$.

Solution:

We have $= \lim_{x \to +\infty} \dfrac{a^{1/x} - b^{1/x}}{-\log x/(x-1)}$

$$= \lim_{x \to +\infty} \frac{a^{1/x} - b^{1/x}}{\log\left[\dfrac{x}{x\left\{1-\left(\dfrac{1}{x}\right)\right\}}\right]} = \lim_{x \to +\infty} \frac{a^{1/x} - b^{1/x}}{\log\left[\dfrac{1}{\left\{1-\left(\dfrac{1}{x}\right)\right\}}\right]}$$

$$= \lim_{x \to +\infty} \frac{a^{1/x} - b^{1/x}}{\log 1 - \log (1-(1/x))}$$

$$= \lim_{x \to +\infty} \frac{a^{1/x} - b^{1/x}}{-\log\left\{1-\left(\dfrac{1}{x}\right)\right\}}, \qquad \left[\text{form } \frac{0}{0}\right]$$

$$= \lim_{x \to +\infty} \frac{\left(a^{1/x}\log a\right)\left(-\dfrac{1}{x^2}\right) - \left(b^{1/x}\log b\right)\left(-\dfrac{1}{x^2}\right)}{-\dfrac{1}{\left\{1-\left(\dfrac{1}{x}\right)\right\}\cdot\left(\dfrac{1}{x^2}\right)}} \quad \text{(by L' Hospital's Rule)}$$

$$= \lim_{x \to +\infty} \frac{b^{1/x}\log b - a^{1/x}\log a}{-1/\left\{1-\left(\dfrac{1}{x}\right)\right\}}, \text{ cancelling } \frac{1}{x^2} \text{ from the Nr. and the Dr.}$$

$$= \frac{b^0 \log b - a^0 \log a}{-\frac{1}{(1-0)}} = \frac{\log b - \log a}{-1} = \log a - \log b = \log\left(\frac{a}{b}\right).$$ **Ans.**

Example 30:

Evaluate $\lim\limits_{x \to 0} \frac{\tan x - \sin x}{x^3}$.

Solution:

We have $\lim\limits_{x \to 0} \frac{\tan x - \sin x}{x^3}$, $\left[\text{form } \frac{0}{0}\right]$

$$= \lim_{x \to 0} \frac{\left(x + \frac{x^3}{3} + \frac{2x^5}{15} + \ldots\right) - \left(x - \frac{x^3}{3!} + \frac{x^5}{5!} - \ldots\right)}{x^3}$$

$$= \lim_{x \to 0} \frac{\left(\frac{1}{3} + \frac{1}{3!}\right)x^3 + \left(\frac{2}{15} - \frac{1}{5!}\right)x^5 + \ldots}{x^3}$$

$$= \lim_{x \to 0} \frac{\frac{1}{2}x^3 + \left(\frac{1}{8}\right)x^5 + \ldots}{x^3} = \lim_{x \to 0}\left(\frac{1}{2} + \frac{1}{8}x^2 + \ldots\right) = \frac{1}{2}.$$ **Ans.**

Example 31:

Evaluate $\lim\limits_{x \to 0} \frac{e^x \sin x - x - x^2}{x^3}$ $\left[\textit{form } \frac{0}{0}\right]$.

Solution:

Required limit,

$$= \lim_{x \to 0} \frac{\left(1 + x + \frac{x^2}{2!} + \frac{x^4}{3!} + \frac{x^4}{4!} + \ldots\right)\left(x - \frac{x^3}{3!} + \frac{x^5}{5!} - \ldots\right) - x - x^2}{x^3}$$

$$= \lim_{x \to 0} \frac{\frac{1}{3}x^3 + \text{terms containing powers of x higher than}}{x^3}$$

$$= \lim_{x \to 0}\left(\frac{1}{3} + \text{terms containing x and its higher powers}\right) = \frac{1}{3}.$$ **Ans.**

Example 32:

Evaluate $\lim\limits_{x \to 0} \frac{e^x \sin x - x - x^2}{x^2 + x \log(1-x)}$. **(Gorakhpur, 1999)**

Solution:

$$\lim_{x\to 0}\frac{e^x \sin x - x - x^2}{x^2 + x\log(1-x)} \qquad \left[\text{form } \frac{0}{0}\right]$$

$$= \lim_{x\to 0}\frac{\left(1+x+\frac{x^2}{2!}+\frac{x^3}{3!}+\ldots\right)\left(x-\frac{x^3}{3!}+\frac{x^5}{5!}-\ldots\right)-x-x^2}{x^2+x\left(-x-\frac{x^2}{2}-\frac{x^3}{3}-\frac{x^4}{4}-\ldots\right)}$$

$$= \lim_{x\to 0}\frac{\left\{x+x^2-\frac{1}{6}x^3+\frac{1}{2}x^3+\text{terms containing powers of x higher than } 3\right\}-x-x^2}{-\frac{1}{2}x^3+\text{terms containing powers of x higher than } 3}$$

$$= \lim_{x\to 0}\frac{x^3\left[\frac{1}{3}+\text{terms containing x and its higher powers}\right]}{x^3\left[-\frac{1}{2}+\text{ terms containing x and its higher powers}\right]}$$

$$= \frac{\frac{1}{3}}{-\frac{1}{2}} = -\frac{2}{3}.$$ **Ans.**

Example 33:

Evaluate $\lim_{x\to 0}\frac{(1+x)^{1/x}-e}{x}$.

(Rohilkhand, 1990; Kanpur, 96; Gorakhpur, 97; Allahabad, 91; Delhi, 96; G.N.U., 94; Agra, 98; Bihar, 93; U.P.P.C.S., 97; Magadh, 91; Vikram, 96)

Solution:

Here $\lim_{x\to 0}\frac{(1+x)^{1/x}-e}{x}$ is of the form $\frac{0}{0}$ because $\lim_{x\to 0}(1+x)^{1/x}=e$. First we shall obtain an expansion for $(1 + x)^{1/x}$ in ascending powers of x. Let $y = (1 + x)^{1/x}$. Then

$$\log y = \frac{1}{x}\log(1+x) = \frac{1}{x}\left(x-\frac{x^2}{2}+\frac{x^3}{3}-\ldots\right) = 1-\frac{x}{2}+\frac{x^2}{3}-\ldots$$

$= 1 + z$, where $z = -\left(\frac{x}{2}\right)+\left(\frac{x^2}{3}\right) - ...$

$\therefore\ y = e^{1+z} = e.e^z = e.\left(1+\frac{z}{1!}+\frac{z^2}{2!}+...\right)$

$$= e\left[1+\left(-\frac{x}{2}+\frac{x^2}{3}-...\right)+\frac{1}{2}\left(-\frac{x}{2}+\frac{x^2}{3}-...\right)^2+...\right]$$

$$= e\left[1-\frac{x}{2}+\frac{x^2}{3}+\frac{1}{8}x^2 + \text{terms containing powers of x higher than } 3\right]$$

$$= e\left[1-\frac{1}{2}x+\frac{11}{24}x^2+...\right].$$

Now $= e\left[1-\frac{1}{2}x+\frac{11}{24}x^2+...\right]$

$$\lim_{x\to 0}\frac{(1+x)^{1/x}-e}{x} = \lim_{x\to 0}\frac{e\left[1-\frac{1}{2}x+\frac{11}{24}x^2+...\right]-e}{x}$$

$$= \lim_{x\to 0}\frac{e\left[-\frac{1}{2}x+\frac{11}{24}x^2+...\right]}{x}$$

$$= \lim_{x\to 0} e\left[-\frac{1}{2}+\frac{11}{24}x+...\right] = -\frac{1}{2}e.$$ **Ans.**

Example 34:

Find the values of a and b in order that

$\lim_{x\to 0}\frac{x(1-a\cos x)+b\sin x}{x^3}$ *may be equal to* $\frac{1}{3}$. **(Delhi, 1997)**

Solution:

The answer is $a = \frac{1}{2}$ and $b = -\frac{1}{2}$.

Example 35:

Find the values of a, b, c so that

$\lim_{x\to 0}\frac{ae^x - b\cos x + ce^{-x}}{x\sin x} = 2$. **(Meerut, 1992, 96)**

Solution:

Here the given limit

$$= \lim_{x \to 0} \frac{ae^x - b\cos x + ce^{-x}}{x \sin x}, \qquad \left[\text{form } \frac{a-b+c}{0}\right].$$

$\therefore$ For the given limit to be equal to 2, we must have

$$a - b + c = 0. \qquad ...(1)$$

Now, applying L'Hospital's rule for the form $\frac{0}{0}$, we have the given limit

$$= \lim_{x \to 0} \frac{ae^x + b\sin x - ce^{-x}}{\sin x + x\cos x}, \qquad \left[\text{form } \frac{a-c}{0}\right].$$

$\therefore$ for the given limit to be equal to 2, we must have

$$a - c = 0. \qquad ...(2)$$

Now, again applying L'Hospital's Rule for the form $\frac{0}{0}$, we have the given limit

$$= \lim_{x \to 0} \frac{ae^x + b\cos x + ce^{-x}}{\cos x + \cos x - x\sin x} = \frac{a+b+c}{2}.$$

$\therefore$ for the given limit to be equal to 2, we must have

$$(a + b + c)/2 = 2 \text{ i.e., } a + b + c = 4. \qquad ...(3)$$

Solving (1), (2) and (3), we get $a = 1$, $b = 2$, $c = 1$.

Example 36:

Evaluate $\lim_{x \to 0} \frac{\sin x - x + \frac{1}{6}x^3}{x^5}$.

(Meerut, 1997, 96 BP; Rohilkhand, 98)

Solution:

$$\lim_{x \to 0} \frac{\sin x - x + \frac{1}{x}x^3}{x^5}, \qquad \left[\text{form } \frac{0}{0}\right]$$

$$= \lim_{x \to 0} \frac{\left(x - \frac{x^3}{3!} + \frac{x^5}{5!} - \frac{x^7}{7!} + \ldots\right) - x + \frac{1}{6}x^3}{x^5}$$

$$= \lim_{x \to 0} \frac{\frac{x^5}{5!} - \frac{x^7}{7!} + \ldots}{x^5} = \lim_{x \to 0}\left(\frac{1}{5!} - \frac{x^2}{7!} + \ldots\right) = \frac{1}{5!} = \frac{1}{120}.$$

Example 37:

Evaluate $\lim\limits_{x\to 0} \dfrac{a^x - 1 - x \log a}{x^2}$.

Solution:

$$\lim_{x\to 0} \frac{a^x - 1 - x\log a}{x^2}, \qquad \left[\text{form } \frac{0}{0}\right]$$

$$= \lim_{x\to 0} \frac{a^x \log a - \log a}{2x}, \qquad \left[\text{form } \frac{0}{0}\right]$$

$$= \lim_{x\to 0} \frac{a^x (\log a)^2}{2} = \frac{1}{2}(\log a)^2.$$ **Ans.**

Example 38:

Evaluate $\lim\limits_{x\to 0} \dfrac{(1+x)^{1/x} - e + \frac{1}{2}ex}{x^2}$. **(Meerut, 1992, 93, 95, 98, 94P)**

Solution:

Here $\lim\limits_{x\to 0} \dfrac{(1+x)^{1/x} - e + \frac{1}{2}ex}{x^2}$ $\left[\text{form } \dfrac{0}{0}\right]$

$$= \lim_{x\to 0} \frac{e\left[1 - \frac{1}{2}x + \frac{11}{24}x^2 + \ldots\right] - e + \frac{1}{2}ex}{x^2}$$

$$= \lim_{x\to 0} \frac{e\left[\left(\frac{11}{24}\right)x^2 + \text{terms containing higher powers of x}\right]}{x^2}$$

$$= \lim_{x\to 0} e\left[\left(\frac{11}{24}\right) + \text{terms containing x and its higher powers}\right]$$

$$= \left(\frac{11}{24}\right)e.$$

Example 39:

Evaluate $\lim\limits_{x\to 0} \dfrac{x^{1/2}\tan x}{(e^x - 1)^{3/2}}$.

(Kumaun, 1993; Agra, 91; Bundelkhand, 98; U.P.P.C.S., 94)

Solution:

$$\lim_{x\to 0} \frac{x^{1/2}\tan x}{(e^x - 1)^{3/2}}, \qquad \left[\text{form } \frac{0}{0}\right]$$

$$= \lim_{x\to 0} \frac{x^{1/2} \tan x}{\left[\left\{1 + x + \left(\frac{x^2}{2!}\right) + \ldots\right\} - 1\right]^{3/2}}$$

$$= \lim_{x\to 0} \frac{x^{1/2} \tan x}{\left[x + \left(\frac{x^2}{2}\right) + \ldots\right]^{3/2}}$$

$$= \lim_{x\to 0} \frac{x^{1/2} \tan x}{x^{3/2}\left[1 + \left(\frac{x}{2}\right) + \ldots\right]^{3/2}}$$

$$= \lim_{x\to 0} \frac{\tan x}{x\left[1 + \left(\frac{x}{2}\right) + \ldots\right]^{3/2}}$$

$$= \lim_{x\to 0} \left[\frac{1}{\left[1 + \left(\frac{x}{2}\right) + \ldots\right]^{3/2}} \cdot \frac{\tan x}{x}\right]$$

$= 1 \cdot 1 = 1.$ **Ans.**

Example 40:

Evaluate $\lim_{x\to 0}\left\{\frac{a^x - b^x}{x}\right\}$.

(Rohilkhand, 1991; Kashmir, 94; Ranchi, 94; Vikram, 98; Meerut, 94)

Solution:

We have $\lim_{x\to 0}\left\{\frac{(a^x - b^x)}{x}\right\}$, $\left[\text{form } \frac{0}{0}\right]$

$= \lim_{x\to 0} \frac{a^x \log a - b^x \log b}{1}$, by Hospital's rule

$= \log a - \log b = \log\left(\frac{a}{b}\right)$. **Ans.**

Example 41:

Evaluate $\lim_{x\to 0} \frac{\tan x - x}{x^2 \tan x}$. **(Meerut, 1990P, 91S)**

Solution:

$$\lim_{x\to 0} \frac{\tan x - x}{x^2 \tan x} = \lim_{x\to 0}\left[\frac{\tan x - x}{x^3}.\frac{x}{\tan x}\right], \quad \textbf{(Note)}$$

$$= \lim_{x\to 0} \frac{\tan x - x}{x^3}, \quad \left[\text{form } \frac{0}{0}\right]$$

$$= \lim_{x\to 0} \frac{\sec^2 x - 1}{3x^2}, \quad \left[\text{form } \frac{0}{0}\right]$$

$$= \lim_{x\to 0} \frac{2\sec x.\sec x \tan x}{6x}$$

$$= \lim_{x\to 0}\left(\frac{1}{3}.\sec^2 x.\frac{\tan x}{x}\right) = \frac{1}{3}\cdot 1.1 = \frac{1}{3}.$$ **Ans.**

Example 42:

Find the values of a and b in order that

$\lim_{x\to 0} \frac{x(1 + a\cos x) - b\sin x}{x^3}$ *may be equal to 1.*

(Agra, 1990; Meerut, 1991, 93, 95, 90S, 92; Garhwal, 97; G.N.U. 92; Kashmir, 91; Kanpur, 97; Raj., 93)

Solution:

We have $\lim_{x\to 0} \frac{x(1 + a\cos x) - b\sin x}{x^3}$,

$\left[\text{form } \frac{0}{0} \text{ so we shall apply Hospital's rule}\right]$

$$= \lim_{x\to 0} \frac{1 + a\cos x - ax\sin x - b\cos x}{3x^2}. \quad \text{...(1)}$$

Now, the denominator of (1) $\to 0$ as $x \to 0$. Therefore, if the numerator of (1) does not tend to 0 as $x \to 0$, then the given limit cannot be equal to 1. Hence for the given limit to be equal to 1 the numerator of (1) must also $\to 0$ as $x \to 0$.

$\therefore \quad 1 + a - b = 0$ or $a - b = -1$. ...(2)

Now, if $1 + a - b = 0$, then (1) takes the form 0/0. Hence by applying L'Hospital's rule to (1), the given limit is equal to

$$\lim_{x\to 0}\frac{-a\sin x-a\sin x-ax\cos x+b\sin x}{6x}, \qquad \left[\text{form } \frac{0}{0}\right]$$

$$=\lim_{x\to 0}\frac{-a\cos x+ax\sin x+(b-2a)\cos x}{6}, \qquad \text{[by L'Hospital's Rule]}$$

$$=\frac{-a+b-2a}{6}=\frac{b-3a}{6}=1, \text{ (as given).}$$

$\therefore\ b - 3a = 6.$

Adding (2) and (3), we have $-2a = 5$ or $a = -\dfrac{5}{2}$. **Ans.**

$$\therefore\ b = a + 1 = \left(-\frac{5}{2}\right)+1=-\frac{3}{2}$$

Hence $a = -\dfrac{5}{2}$, $b = -\dfrac{3}{2}$.

Example 43:

Evaluate $\displaystyle\lim_{x\to 0}\frac{5\sin x-7\sin 2x+3\sin 3x}{\tan x-x}$.

(Meerut, 1993 S; Lucknow, 94)

Solution:

$$\lim_{x\to 0}\frac{5\sin x-7\sin 2x+3\sin 3x}{\tan x-x}, \qquad \left[\text{form } \frac{0}{0}\right]$$

$$=\lim_{x\to 0}\frac{5\left(x-\frac{x^3}{3!}+\ldots\right)-7\left\{(2x)-\frac{(2x)^3}{3!}+\ldots\right\}+3\left\{(3x)-\frac{(3x)^3}{3!}+\ldots\right\}}{\left(x+\frac{1}{3}x^3+\frac{2}{15}x^5+\ldots\right)-x}$$

$$=\lim_{x\to 0}\frac{x^3\left(-\frac{5}{6}+\frac{28}{3}-\frac{27}{2}+\text{higher powers of } x\right)}{\left(\frac{1}{3}x^3+\frac{2}{15}x^5+\ldots\right)}$$

$$=\lim_{x\to 0}\frac{-5+\text{terms containing } x \text{ and its hihger powers}}{\left(\frac{1}{3}+\frac{2}{15}x^2+\ldots\right)}$$

$$=-\frac{5}{1/3}=-15.$$ **Ans.**

Example 44:

If $\lim\limits_{x\to 0} \dfrac{\sin 2x + a \sin x}{x^3}$ *be finite, find the value of 'a' and the limit.*

Solution:

We have $\lim\limits_{x\to 0} \dfrac{\sin 2x + a \sin x}{x^3}$, $\left[\text{form } \dfrac{0}{0}\right]$

$$= \lim_{x\to 0} \frac{2\cos 2x + a \cos x}{3x^2} \quad ...(1)$$

Now, the denominator of (1) $\to 0$ as $x \to 0$.

But if the numerator of (1) does not tend to zero as $x \to 0$, then the given limit becomes infinite. Therefore for the given limit to be finite the numerator of (1) must $\to 0$ as $x \to 0$.

Hence $2 + a = 0$ or $a = -2$.

With this value of a, we get from (1), the given limit

$$= \lim_{x\to 0} \frac{2\cos 2x - 2\cos x}{3x^2}, \quad \left[\text{form } \frac{0}{0}\right]$$

$$= \lim_{x\to 0} \frac{-4\sin 2x + 2\sin x}{6x}, \quad \left[\text{form } \frac{0}{0}\right]$$

$$= \lim_{x\to 0} \frac{-8\cos 2x + 2\cos x}{6} = \frac{-8+2}{6} = -\frac{6}{6} = -1.$$

Form II: $\dfrac{\infty}{\infty}$.

If f(x) and $\phi(x)$ be two functions such that

$\lim\limits_{x\to a} f(x) = \infty$ and $\lim\limits_{x\to a} \phi(x) = \infty$, then

$\lim\limits_{x\to a} \dfrac{f(x)}{\phi(x)} = \lim\limits_{x\to 0} \dfrac{f'(x)}{\phi'(x)}$, provided the limit exists.

Proof:

We have, $\lim\limits_{x\to a} \dfrac{f(x)}{\phi(x)}$

$$= \lim_{x\to a} \left[\frac{1}{\phi(x)}\right] / \left[\frac{1}{f(x)}\right], \quad \left[\text{form } \frac{0}{0}\right]$$

$$= \lim_{x\to a} \left[-\frac{1}{\{\phi(x)\}^2}\phi'(x)\right] / \left[-\frac{1}{\{f(x)\}^2}f'(x)\right], \quad \text{[By L'Hospital's Rule]}$$

$$= \left[\lim_{x\to a} \frac{\phi'(x)}{f'(x)}\right]\left[\lim_{x\to a} \frac{f(x)}{\phi(x)}\right]^2 \qquad \text{...(1)}$$

Now three cases arise:

Case I: *When* $\lim_{x\to a} f(x)/\phi(x)$ *is neither zero nor infinite.*

In this case dividing both sides of (1) by

$$\left[\lim_{x\to 0}\left\{\frac{f(x)}{\phi(x)}\right\}\right]^2, \text{ we get}$$

$$\left[\lim_{x\to 0}\left\{\frac{f(x)}{\phi(x)}\right\}\right]^{-1} = \lim_{x\to a} \frac{\phi'(x)}{f'(x)}$$

or $$\lim_{x\to a} \frac{f(x)}{\phi(x)} = \lim_{x\to a} \frac{f'(x)}{\phi'(x)}.$$

Case II: *When* $\lim_{x\to a}\left\{\frac{f(x)}{\phi(x)}\right\} = 0$.

In this case, we have

$$1 + \lim_{x\to a} \frac{f(x)}{\phi(x)} = \lim_{x\to a}\left[1 + \frac{f(x)}{\phi(\ )}\right]$$

$$= \lim_{x\to a} \frac{\phi(x) + f(x)}{\phi(x)}, \qquad \left[\text{form } \frac{\infty}{\infty}\right]$$

$$= \lim_{x\to a} \frac{\phi'(x) + f'(x)}{\phi'(x)},$$

[by case I, since the form is ∞/∞ and the limit is equal to 1 + 0 i.e., 1 which is neither zero nor infinite]

$$= \lim_{x\to a}\left[1 + \frac{f'(x)}{\phi'(x)}\right] = 1 + \lim_{x\to a} \frac{f'(x)}{\phi'(x)}. \qquad \text{...(2)}$$

Subtracting 1 form both sides of (2), we get

$$\lim_{x\to a} \frac{f(x)}{\phi(x)} = \lim_{x\to a} \frac{f'(x)}{\phi'(x)}.$$

Case III: When $\lim_{x\to a}\left\{\frac{f(x)}{\phi(x)}\right\}$ is infinite. In this case,

$$\lim_{x\to a}\frac{\phi(x)}{f(x)}=0=\lim_{x\to a}\frac{\phi'(x)}{f'(x)}$$ [by case II].

$$\therefore \quad \lim_{x\to a}\frac{f(x)}{\phi(x)}=\lim_{x\to a}\frac{f'(x)}{\phi'(x)}.$$

Note: The above proposition is also true when $x \to \infty$ or $-\infty$ in place of a.

Important: It is interesting to note that in both cases whe the form is $\frac{\infty}{\infty}$ or $\frac{0}{0}$ the rule of evaluating th limit by differentiating the numerator and denominator separately holds good. But while evaluating $\lim_{x\to a}\left\{\frac{f(x)}{\phi(x)}\right\}$, when it is of the form $\frac{\infty}{\infty}$, it is sometimes necessary to change it into the form $\frac{0}{0}$, otherwise the process of differentiating the numerator and the denominator will never end.

Remember: $\log 0 = -\infty$, and $\log \infty = \infty$.

SOME MORE SOLVED EXAMPLES

Example 1:

Evaluate $\lim_{x\to\infty}\frac{x^2+2x}{5-3x^2}$.

Solution:

We have, $\lim_{x\to\infty}\frac{x^2+2x}{5-3x^2}$ $\left[\text{form } \frac{\infty}{\infty}\right]$

$$=\lim_{x\to\infty}\frac{2x+2}{-6x},$$ $\left[\text{form } \frac{\infty}{\infty}\right]$

$$\lim_{x\to\infty}\frac{2}{-6}=-\frac{1}{3}.$$

Aliter: $$\lim_{x\to\infty}\frac{x^2+2x}{5-3x^2}=\lim_{x\to\infty}\frac{x^2\left\{1+\left(\frac{2}{x}\right)\right\}}{x^2\left\{\left(\frac{5}{x^2}\right)-3\right\}}$$

$$=\lim_{x\to\infty}\frac{1+\left(\frac{2}{x}\right)}{\left(\frac{5}{x^2}\right)-3}=-\frac{1}{3}.$$

Example 2:

Evaluate $\lim_{x\to 0} \frac{\log x}{\cot x}$. **(Kanpur, 1998; Meerut, 98)**

Solution:

We have, $\lim_{x\to 0} \frac{\log x}{\cot x}$, $\left[\text{form } \frac{\infty}{\infty}\right]$

$= \lim_{x\to 0} \frac{\frac{1}{x}}{-\text{cosec}^2 x}$, $\left[\text{form } \frac{\infty}{\infty}\right]$

$= \lim_{x\to 0} \frac{-\sin^2 x}{x}$, $\left[\text{form } \frac{0}{0}\right]$

$$= \lim_{x\to 0} \frac{-2\sin x \cos x}{1} = \frac{-2\times 0\times 1}{1} = 0.$$

Example 3:

Evaluate $\lim_{x\to 0} \frac{\log \sin x}{\log x}$.

Solution:

We have $\lim_{x\to 0} \frac{\log \sin x}{\log x}$, $\left[\text{form } \frac{\infty}{\infty}\right]$

$$= \lim_{x\to 0} \frac{\frac{\cos x}{\sin x}}{\frac{1}{x}} = \lim_{x\to 0} x \cot x = \lim_{x\to 0} \frac{x}{\tan x} = 1.$$

Example 4:

Evaluate $\lim_{x\to 0} \frac{\log x^2}{\cot x^2}$.

Solution:

We have $\lim_{x\to 0} \frac{\log x^2}{\cot x^2}$ $\left[\text{form } \frac{\infty}{\infty}\right]$

$$= \lim_{x\to 0} \frac{\left(\frac{1}{x^2}\right).2x}{\left(-\text{cosec}^2 x\right)(2x)} = \lim_{x\to 0}\left(-\frac{\sin^2 x^2}{x^2}\right)$$

$= \lim_{x\to 0} \left\{\left(\frac{\sin x^2}{x^2}\right)\left(-\sin x^2\right)\right\}$ **(Note)**

$$= \lim_{x\to 0}\left(-\sin x^2\right) = 0. \qquad \left[\because \lim_{x\to 0}\frac{\sin x^2}{x^2} = 1\right]$$

Example 5:

Evaluate $\lim_{x\to 0}\dfrac{\log \tan x}{\log x}$.

Solution:

We have $\lim_{x\to 0}\dfrac{\log \tan x}{\log x}$ $\qquad \left[\text{form } \dfrac{\infty}{\infty}\right]$

$$= \lim_{x\to 0}\frac{\dfrac{\sec^2 x}{\tan x}}{\dfrac{1}{x}} = \lim_{x\to 0}\left\{\left(\frac{x}{\tan x}\right)\left(\frac{1}{\cos^2 x}\right)\right\} = 1\times 1 = 1.$$

Example 6:

Evaluate $\lim_{x\to \pi/2}\dfrac{\tan 5x}{\tan x}$. **(Agra, 1998)**

Solution:

$\lim_{x\to\pi/2}\dfrac{\tan 5x}{\tan x}$ $\qquad \left[\text{form } \dfrac{\infty}{\infty}\right]$

$$= \lim_{x\to\pi/2}\left(\frac{\sin 5x}{\sin x}.\frac{\cos x}{\cos 5x}\right)$$

$$= \lim_{x\to\pi/2}\frac{\sin 5x}{\sin x}.\lim_{x\to\pi/2}\frac{\cos x}{\cos 5x}$$

$$= 1 + \lim_{x\to\pi/2}\frac{\cos x}{\cos 2x}, \qquad \left[\text{form } \frac{0}{0}\right]$$

$$= \lim_{x\to\pi/2}\frac{-\sin x}{-5\sin 5x}, \qquad \text{(by L'Hospital's Rule)}$$

$$= \frac{-1}{-5} = \frac{1}{5}.$$

Example 7:

Evaluate $\lim_{x\to 0}\dfrac{\log \sin 2x}{\log \sin x}$. **(Meerut, 1994)**

Solution:

We have $\lim_{x\to 0} \frac{\log \sin 2x}{\log \sin x}$, $\left[\text{form } \frac{\infty}{\infty}\right]$

$$= \lim_{x\to 0} \frac{\left(\frac{1}{\sin 2x}\right)(2\cos 2x)}{\left(\frac{1}{\sin x}\right)\cos x} = \lim_{x\to 0} \frac{2\cot 2x}{\cot x}, \qquad \left[\text{form } \frac{\infty}{\infty}\right]$$

$$= \lim_{x\to 0} \frac{2\tan x}{\tan 2x}, \qquad \left[\text{form } \frac{0}{0}\right]$$

$$= \lim_{x\to 0} \frac{2\sec^2 x}{2\sec^2 2x}, \qquad \text{(By L'Hospital's Rule)}$$

$= 1.$

Example 7 (a):

Evaluate $\lim_{x\to\infty}\left(\frac{\log x}{x}\right).$

Solution:

We have $\lim_{x\to\infty} \frac{\log x}{x}$, $\left[\text{form } \frac{\infty}{\infty}\right]$

$$= \lim_{x\to\infty} \frac{\frac{1}{x}}{1} = \lim_{x\to\infty}\left(\frac{1}{x}\right) = 0 \cdot$$

Example 8:

Evaluate $\lim_{x\to 0} \frac{\log \log \left(1-x^2\right)}{\log \log \cos x}.$ **(Meerut, 1994)**

Solution:

We have, $\lim_{x\to 0} \frac{\log \log \left(1-x^2\right)}{\log \log \cos x}$, $\left[\text{form } \frac{\infty}{\infty}\right]$

$$= \lim_{x\to 0} \frac{\frac{1}{\log\left(1-x^2\right)} \cdot \frac{1}{1-x^2} \cdot (-2x)}{\frac{1}{\log \cos x} \cdot \frac{1}{\cos x} \cdot (-\sin 2x)}$$

$$= 2 \lim_{x\to 0} \frac{x \cos x \log \cos x}{\sin x.(1-x^2)\log(1-x^2)}$$

$$= 2 \lim_{x\to 0} \frac{x}{\sin x}. \lim_{x\to 0} \frac{\cos x}{1-x^2}. \lim_{x\to 0} \frac{\log \cos x}{\log(1-x^2)}$$

$$= 2\times 1\times 1\times \lim_{x\to 0} \frac{\log \cos x}{\log(1-x^2)}, \qquad \left[\text{form } \frac{0}{0}\right]$$

$$= 2 \lim_{x\to 0} \frac{\frac{1}{\cos x}.(-\sin x)}{\frac{1}{1-x^2}.(-2x)} = 2\times\frac{1}{2}. \lim_{x\to 0}\left(\frac{\sin x}{x}.\frac{1-x^2}{\cos x}\right)$$

$= 1.$

Example 9 (a):

Evaluate $\lim\limits_{x\to\infty} x^m e^{-x}$. **(K.U., 1997)**

Solution:

We have $\lim\limits_{x\to\infty} x^m e^{-x} = \lim\limits_{x\to\infty} \frac{x^m}{e^x}$, $\left[\text{form } \frac{\infty}{\infty}\right]$

$$= \lim_{x\to\infty} \frac{m x^{m-1}}{e^x}, \qquad \left[\text{form } \frac{\infty}{\infty}\right]$$

$$= \lim_{x\to\infty} \frac{m(m-1)x^{m-2}}{e^x}, \qquad \left[\text{form } \frac{\infty}{\infty}\right]$$

$$= \lim_{x\to\infty} \frac{m(m-1)(m-2)\ldots 3\cdot 2\cdot 1}{e^x} = \lim_{x\to\infty} \frac{m!}{e^x}$$

$= 0,$ $[\because\ e^x \to \infty \text{ when } x \to \infty]$.

Example 9 (b):

Evaluate $\lim\limits_{x\to 0+} \frac{\operatorname{cosec} x}{\log x}$. **(Agra, 1998)**

Solution:

$\lim\limits_{x\to 0+} \frac{\operatorname{cosec} x}{\log x}$, $\left[\text{form } \frac{\infty}{\infty}\right]$

$$= \lim_{x\to 0+} \frac{-\operatorname{cosec} x \cot x}{\frac{1}{x}} = \lim_{x\to 0+} -\left[\left(\frac{x}{\tan x}\right)\operatorname{cosec} x\right]$$

$$= \lim_{x\to 0} + \left(-\operatorname{cosec} x\right) = -\infty.$$

Example 10 (a):

Evaluate $\lim\limits_{x\to 0} \dfrac{\log \sin x}{\cot x}$.

Solution:

$$\lim_{x\to 0} \frac{\log \sin x}{\cot x}, \qquad \left[\text{form } \frac{\infty}{\infty}\right]$$

$$= \lim_{x\to 0} \frac{\dfrac{\cos x}{\sin x}}{-\operatorname{cosec}^2 x} = \lim_{x\to 0}\left(-\frac{\cos x}{\sin x}\sin^2 x\right)$$

$$= \lim_{x\to 0}\left(-\sin x \cos x\right) = 0.$$

Example 10 (b):

Evaluate $\lim\limits_{x\to \pi/2} \dfrac{\log\left(x-\frac{1}{2}\pi\right)}{\tan x}$. **(Meerut, 1999P, 97)**

Solution:

We have $\lim\limits_{x\to 1/2\pi} \dfrac{\log\left(x-\frac{1}{2}\pi\right)}{\tan x}$ $\left[\text{form } \dfrac{\infty}{\infty}\right]$

$$= \lim_{x\to \pi/2} \frac{\dfrac{1}{\left(x-\frac{1}{2}\pi\right)}}{\sec^2 x} = \lim_{x\to \pi/2} \frac{\cos^2 x}{x-\frac{1}{2}\pi}, \qquad \left[\text{form } \frac{0}{0}\right]$$

$$= \lim_{x\to \pi/2} \frac{-2\cos x \sin x}{1} = \lim_{x\to \pi/2}\left(-\sin 2x\right) = 0.$$

Example 11:

Evaluate $\lim\limits_{x\to\infty} \dfrac{3x+4}{\sqrt{\left(2x^2+5\right)}}$. **(K.U., 1995)**

Solution:

We have $\lim\limits_{x\to\infty} \dfrac{3x+4}{\sqrt{\left(2x^2+5\right)}}$, $\left[\text{form } \dfrac{\infty}{\infty}\right]$

$$= \lim_{x\to\infty} \frac{x\left[3+\left(\frac{4}{x}\right)\right]}{x\left[\sqrt{\left\{2+\left(\frac{5}{x^2}\right)\right\}}\right]} = \lim_{x\to\infty} \frac{3+\left(\frac{4}{x}\right)}{\sqrt{\left\{2+\left(\frac{5}{x^2}\right)\right\}}} = \frac{3}{\sqrt{2}}.$$

Example 12:

Evaluate $\lim_{x\to\infty} \frac{\log(x-a)}{\log(e^x-e^a)}$.

Solution:

$$\lim_{x\to a} \frac{\log(x-a)}{\log(e^x-e^a)}, \qquad \left[\text{form } \frac{\infty}{\infty}\right]$$

$$= \lim_{x\to a} \frac{\frac{1}{(x-a)}}{\left\{\frac{1}{(e^x-e^a)}\right\}e^x} = \lim_{x\to a} \frac{e^x-e^a}{e^x(x-a)}, \qquad \left[\text{form } \frac{0}{0}\right]$$

$$= \lim_{x\to a} \frac{e^x}{e^x(x-a)+e^x} = \lim_{x\to a} \frac{e^x}{e^x[(x-a)+1]}$$

$$= \lim_{x\to a} \frac{1}{x-a} = 1.$$

Example 13:

Evaluate $\lim_{x\to 0} \frac{\log(\tan^2 2x)}{\log(\tan^2 x)}$. **(Magadh, 1992)**

Solution:

We have $\lim_{x\to 0} \frac{\log(\tan^2 2x)}{\log(\tan^2 x)}$

$$= \lim_{x\to 0} \frac{2\log(\tan 2x)}{2\log(\tan x)}, \qquad \left[\text{form } \frac{\infty}{\infty}\right]$$

$$= \lim_{x\to 0} \frac{\left(\frac{1}{\tan 2x}\right).2\sec^2 2x}{\left(\frac{1}{\tan x}\right)\sec^2 x} = \lim_{x\to 0} \frac{2\tan x \cos^2 x}{\tan 2x \cos^2 2x}$$

$$= \lim_{x \to 0} \frac{2 \sin x \cos x}{\sin 2x \cos 2x} = \lim_{x \to 0} \frac{\sin 2x}{\sin 2x \cos 2x}$$

$$= \lim_{x \to 0} \frac{1}{\cos 2x} = \frac{1}{1} = 1.$$

7.4 FORM III $0 \times \infty$

This form can be easily reduced to the form $\frac{0}{0}$ or to the form $\frac{\infty}{\infty}$.

Let $\lim_{x \to a} f(x) = 0$ and $\lim_{x \to a} \phi(x) = \infty$.

Then we can write $\lim_{x \to a} f(x).\phi(x)$

$$= \lim_{x \to a} \frac{f(x)}{\frac{1}{\phi(x)}}, \qquad \left[\text{form } \frac{0}{0}\right]$$

$$= \lim_{x \to a} \frac{\phi(x)}{\frac{1}{f(x)}}. \qquad \left[\text{form } \frac{\infty}{\infty}\right]$$

Thus, $\lim_{x \to a} f(x).\phi(x)$ is reduced to the form $\frac{0}{0}$ or $\frac{\infty}{\infty}$ which can now be evaluated by L'Hospital's Rule or otherwise.

Example 1:

Evaluate $\lim_{x \to 0} x^m (\log x)^n$, *where m, n are positive integers.*

(Allahabad, 1991)

Solution:

We have $\lim_{x \to 0} x^m (\log x)^n$, $\qquad$ [form $0 \times \infty$]

$$= \lim_{x \to 0} \frac{(\log x)^n}{x^{-m}}, \qquad \left[\text{form } \frac{\infty}{\infty}\right]$$

$$= \lim_{x \to 0} \frac{n(\log x)^{n-1}\left(\frac{1}{x}\right)}{-mx^{-m-1}} = \lim_{x \to 0} \left[-\frac{n}{m}.\frac{(\log x)^{n-1}}{x^{-m}}\right], \qquad \left[\text{form } \frac{\infty}{\infty} \text{ if } n > 1\right]$$

$$= \lim_{x \to 0} \left(-\frac{n}{m}\right).\frac{(n-1)(\log x)^{n-2}.\left(\frac{1}{x}\right)}{-mx^{-m-1}}$$

$$= \lim_{x\to 0} (-1)^2 \frac{n(n-1)}{m^2} \cdot \frac{(\log x)^{n-2}}{x^{-m}}, \qquad \left[\text{form } \frac{\infty}{\infty} \text{ if } n>2\right]$$

$$= \lim_{x\to 0} (-1)^n \cdot \frac{n(n-1)(n-2)\ldots\text{upto n factors}}{m^n} \cdot \frac{(\log x)^{n-n}}{x^{-m}},$$

[by repeated application of the above process]

$$= \lim_{x\to 0} (-1)^n \frac{n!}{m^n} . x^m = (-1)^n \frac{n!}{m^n} . \lim_{x\to 0} x^m = 0.$$

Example 2:

Evaluate $\lim_{x\to 0}\left(\frac{1}{x^2} - \cot^2 x\right)$.

Solution:

We have $\lim_{x\to 0}\left(\frac{1}{x^2} - \cot^2 x\right)$, [form $\infty - \infty$]

$$= \lim_{x\to 0}\left(\frac{1}{x^2} - \frac{\cos^2 x}{\sin^2 x}\right)$$

$$= \lim_{x\to 0} \frac{\sin^2 x - x^2 \cos^2 x}{x^2 \sin^2 x}, \qquad \left[\text{form } \frac{0}{0}\right]$$

$$= \lim_{x\to 0}\left(\frac{\sin^2 x - x^2 \cos^2 x}{x^4} \cdot \frac{x^2}{\sin^2 x}\right)$$

$$= \lim_{x\to 0} \frac{\sin^2 x - x^2 \cos^2 x}{x^4}$$

$$= \lim_{x\to 0} \frac{\left[x - \left(\frac{x^3}{3!}\right) + \ldots\right]^2 - x^2\left[1 - \left(\frac{x^2}{2!}\right) + \left(\frac{x^4}{4!}\right) - \ldots\right]^2}{x^4}$$

$$= \lim_{x\to 0} \frac{\left[x^2 - \left(\frac{x^3}{3!}\right) + \ldots\right] - x^2\left(1 - x^2 + \ldots\right)}{x^4}$$

$$= \lim_{x\to 0} \frac{x^2 - \left(\frac{x^4}{3}\right) + \ldots - x^2 + x^4 + \ldots}{x^4}$$

$$= \lim_{x\to 0} \frac{\left(\frac{2}{3}\right)x^4 + \text{terms containing higher powers of x}}{x^4} = \frac{2}{3}.$$

Example 3:

Evaluate $\lim_{x\to\infty}\left\{x\tan\left(\frac{1}{x}\right)\right\}.$ **(Meerut, 1992S)**

Solution:

We have $\lim_{x\to\infty}\left[\tan\left(\frac{1}{x}\right)\right].x,$ [form $0 \times \infty$]

$= \lim_{x\to\infty}\left[\tan\left(\frac{1}{x}\right)\right]\Big/\left[\frac{1}{x}\right],$ $\left[\text{form } \frac{0}{0}\right]$

$$= \lim_{x\to\infty}\frac{\left[\left(-\frac{1}{x^2}\right)\sec^2\left(\frac{1}{x}\right)\right]}{\left[-\frac{1}{x^2}\right]} = \lim_{x\to\infty}\sec^2\left(\frac{1}{x}\right) = 1.$$

Example 4:

Evaluate $\lim_{x\to 1}(1-x).\tan\frac{\pi x}{2}.$ **(Meerut, 1993, 97; Agra, 95)**

Solution:

We have $\lim_{x\to 1}(1-x).\tan\frac{\pi x}{2},$ [form $0 \times \infty$]

$= \lim_{x\to 1}\frac{1-x}{\cot\left(\frac{\pi x}{2}\right)},$ $\left[\text{form } \frac{0}{0}\right]$

$$= \lim_{x\to 1}\frac{-1}{-\frac{\pi}{2}.\text{cosec}^2\frac{\pi x}{2}} = \lim_{x\to 1}\frac{2}{\pi}\sin^2\frac{\pi x}{2}.\frac{2}{\pi}.$$

Example 5:

Evaluate $\lim_{x\to\infty} 2^x \sin\left(\frac{2}{2^x}\right).$

Solution:

We have $\lim_{x\to\infty} 2^x\sin\left(\frac{2}{2^x}\right),$ [form $\infty \times 0$]

$$= \lim_{x\to\infty} \frac{\sin\left(a.2^{-x}\right)}{2^{-x}}, \qquad \left[\text{form } \frac{0}{0}\right]$$

$$= \lim_{x\to\infty} \frac{\left\{\cos\left(a.2^{-x}\right)\right\}.a.2^{-x}(\log 2).(-1)}{2^{-x}(\log 2).(-1)}$$

$$= \lim_{x\to\infty} a\cos\left(\frac{a}{2^x}\right) = a\cos 0 = a.1 = a.$$

Example 6:

Evaluate $\lim_{x\to 0} x\log x$. **(Kanpur, 1995, Agra, 96; Meerut, 99, 95)**

Solution:

$$\lim_{x\to 0} x\log x, \qquad [\text{form } 0\times\infty]$$

$$\lim_{x\to 0} \frac{\log x}{\frac{1}{x}} \qquad \left[\text{form } \frac{\infty}{\infty}\right]$$

$$= \lim_{x\to 0} \frac{\frac{1}{x}}{-\left(\frac{1}{x^2}\right)} = \lim_{x\to 0}(-x) = 0.$$

Example 7:

Evaluate $\lim_{x\to 1}\left(\sec\frac{\pi}{2x}\right).\log x$.

Solution:

We have $\lim_{x\to 1}\left(\sec\frac{\pi}{2x}\right).\log x$ $\qquad [\text{form } \infty\times 0]$

$$= \lim_{x\to 1} \frac{\log x}{\cos\left(\frac{\pi}{2x}\right)} \qquad \left[\text{form } \frac{0}{0}\right]$$

$$= \lim_{x\to 1} \frac{\frac{1}{x}}{-\left[\sin\left(\frac{\pi}{2x}\right)\right].\left(-\frac{\pi}{2x^2}\right)} = \lim_{x\to 1} \frac{2x}{\pi}\operatorname{cosec}\frac{\pi}{2x} = \frac{2}{\pi}.$$

Example 8:

Evaluate $\lim\limits_{x\to\infty}\left(a^{1/x}-1\right)x$. **(Meerut, 1990, 98P)**

Solution:

We have $\lim\limits_{x\to\infty}\left(a^{1/x}-1\right)x$ [form $0\times\infty$]

$$=\lim_{x\to\infty}\frac{a^{1/x}-1}{\frac{1}{x}} \qquad \left[\text{form } \frac{0}{0}\right]$$

$$=\lim_{x\to\infty}\frac{a^{1/x}.\log a.\left(-\frac{1}{x^2}\right)}{-\frac{1}{x^2}}$$

$$=\lim_{x\to\infty}a^{1/x}\log a=a^0\log a=\log a.$$

7.5 FORM IV: [$\infty-\infty$]

When $\lim\limits_{x\to a}f(x)=\infty$ and $\lim\limits_{x\to a}\phi(x)=\infty$, we have

$$=\lim_{x\to a}\left[f(x)-\phi(x)\right], \qquad [\text{form } \infty-\infty]$$

$$=\lim_{x\to a}\left[\frac{1}{\frac{1}{f(x)}}-\frac{1}{\frac{1}{\phi(x)}}\right]$$

$$=\lim_{x\to a}\frac{\left\{\frac{1}{\phi(x)}\right\}-\left\{\frac{1}{f(x)}\right\}}{\left\{\frac{1}{f(x)}\right\}.\left\{\frac{1}{\phi(x)}\right\}} \qquad \left[\text{form } \frac{0}{0}\right]$$

Now, this can be evaluated by applying L'Hospital's Rule.

Example 1:

Evaluate $\lim\limits_{x\to 0}\dfrac{\cot x-\left(\frac{1}{x}\right)}{x}$. **(Kanpur, 1996; Gorakhpur, 99)**

Solution:

We have $\lim\limits_{x\to 0} \dfrac{\cot x - \left(\dfrac{1}{x}\right)}{x}$

$$= \lim_{x\to 0} \frac{x \cos x - \sin x}{x^2 \sin x} \qquad \left[\text{form } \frac{0}{0}\right]$$

$$= \lim_{x\to 0} \left(\frac{x \cos x - \sin x}{x^3} \cdot \frac{x}{\sin x}\right)$$

$$= \lim_{x\to 0} \frac{x \cos x - \sin x}{x^3} \qquad \left[\text{form } \frac{0}{0}\right]$$

$$= \lim_{x\to 0} \frac{\cos x - x \sin x - \cos x}{3x^2}$$

$$= \lim_{x\to 0} \left(-\frac{1}{3} \cdot \frac{\sin x}{x}\right) = -\frac{1}{3} \cdot 1 = -\frac{1}{3}.$$

Example 2:

Evaluate $\lim\limits_{x\to 0} \left(\dfrac{1}{x^2} - \dfrac{1}{\sin^2 x}\right)$.

(Agra, 1997; Raj., 97; Gurunanak, 93; Ranchi, 96; Meerut, 96, 98P, 96; Rohilkhand, 97; Vikram, 95; Delhi, 95)

Solution:

We have $\lim\limits_{x\to 0} \left(\dfrac{1}{x^2} - \dfrac{1}{\sin^2 x}\right)$ [form $\infty - \infty$]

$$= \lim_{x\to 0} \frac{\sin^2 x - x^2}{x^2 \sin^2 x}, \qquad \left[\text{form } \frac{0}{0}\right]$$

$$= \left(\lim_{x\to 0} \frac{\sin^2 x - x^2}{x^4}\right) \cdot \left(\lim_{x\to 0} \frac{x^2}{\sin^2 x}\right), \qquad \textbf{(Note)}$$

$$= \lim_{x\to 0} \frac{\sin^2 x - x^2}{x^4}, \qquad \left[\because \lim_{x\to 0} \frac{x^2}{\sin^2 x} = 1\right]$$

$$= \lim_{x\to 0} \frac{1}{x^4}\left[\left(x - \frac{x^3}{3!} + \dots\right)^2 - x^2\right]$$

$$= \lim_{x\to 0} \frac{1}{x^4}\left[\left\{x^2 - \frac{1}{3}x^4 + \dots\right\} - x^2\right]$$

$$= \lim_{x\to 0}\left[-\frac{1}{3} + \text{terms containing x and its higher powers}\right]$$

$$= -\frac{1}{3}.$$

Example 3:

Evaluate $\lim_{x\to \pi/2}(\sec x - \tan x)$. **(Vikram, 1990; Indore, 90)**

Solution:

We have $\lim_{x\to \pi/2}(\sec x - \tan x)$, [form $\infty - \infty$]

$$= \lim_{x\to \pi/2}\left(\frac{1}{\cos x} - \frac{\sin x}{\cos x}\right) = \lim_{x\to \pi/2}\frac{1-\sin x}{\cos x}, \qquad \left[\text{form } \frac{0}{0}\right]$$

$$= \lim_{x\to \pi/2}\frac{-\cos x}{-\sin x} = \lim_{x\to \pi/2}\cot x = 0.$$

Example 4:

Evaluate $\lim_{x\to 0}\left[\frac{1}{e^x - 1} - \frac{1}{x}\right]$.

Solution:

We have $\lim_{x\to 0}\left[\frac{1}{e^x - 1} - \frac{1}{x}\right]$, [form $\infty - \infty$]

$$= \lim_{x\to 0}\frac{x - e^x + 1}{x(e^x - 1)} \qquad \left[\text{form } \frac{0}{0}\right]$$

$$= \lim_{x\to 0}\frac{1 - e^x}{e^x - 1 + xe^x} \qquad \left[\text{form } \frac{0}{0}\right]$$

$$= \lim_{x\to 0}\frac{-e^x}{e^x + e^x + xe^x} = -\frac{1}{2}.$$

Example 5:

Evaluate $\lim_{x\to 0}\left(\operatorname{cosec}^3 x - \frac{1}{x^3}\right)$.

Solution:

We have $\lim_{x\to 0}\left(\operatorname{cosec}^3 x - \frac{1}{x^3}\right)$, [form $\infty - \infty$]

$$= \lim_{x\to 0}\left(\frac{1}{\sin^3 x} - \frac{1}{x^3}\right) = \lim_{x\to 0}\frac{x^3 - \sin^3 x}{x^3 \sin^3 x}$$

$$= \lim_{x\to 0}\left\{\left(\frac{x^3 - \sin^3 x}{x^6}\right).\left(\frac{x}{\sin x}\right)^3\right\}$$

$$= \lim_{x\to 0}\frac{x^3 - \sin^3 x}{x^6} = \lim_{x\to 0}\frac{x^3 - \left\{x - \left(\frac{x^3}{3!}\right) + \left(\frac{x^5}{5!}\right) - \ldots\right\}^3}{x^6}$$

$$= \lim_{x\to 0}\frac{x^3 - \left[x + \left\{-\left(\frac{x^3}{3!}\right) + \left(\frac{x^5}{5!}\right) - \ldots\right\}\right]^3}{x^6}$$

$$= \lim_{x\to 0}\frac{x^3 - \left[x^3 + 3x^2\left\{-\left(\frac{x^3}{3!}\right) + \left(\frac{x^5}{5!}\right)\right\} + \ldots\right]}{x^6}$$

$$= \lim_{x\to 0}\frac{\frac{1}{2}x^5 + \text{terms containing powers of x higher than } 5}{x^6}$$

$$= \lim_{x\to 0}\frac{x^5\left[\frac{1}{2} + \text{terms containing x and its higher powers}\right]}{x^6}$$

$$= \lim_{x\to 0}\frac{\frac{1}{2} + \text{terms containing x and its higher powers}}{x} = \infty.$$

Example 6:

Evaluate $\lim_{x\to 0}\left(\frac{1}{x} - \cot x\right)$. **(Agra, 1999; Utkal, 92)**

Solution:

We have $\lim_{x\to 0}\left(\frac{1}{x} - \cot x\right)$, [form $\infty - \infty$]

$$= \lim_{x\to 0}\left(\frac{1}{x} - \frac{\cos x}{\sin x}\right) = \lim_{x\to 0}\frac{\sin x - x\cos x}{x \sin x}$$

$$= \lim_{x\to 0}\left(\frac{\sin x - x\cos x}{x^2}\cdot\frac{x}{\sin x}\right)$$

$$= \lim_{x\to 0}\frac{\sin x - x\cos x}{x^2}, \qquad \left[\text{form } \frac{0}{0}\right]$$

$$= \lim_{x\to 0}\frac{\cos x - \cos x + x\sin x}{2x} = \lim_{x\to 0}\left(\frac{1}{2}\sin x\right) = 0.$$

Example 7:

Evaluate $\lim_{x\to 1}\left[\frac{x}{x-1} - \frac{1}{\log x}\right]$.

Solution:

We have $\lim_{x\to 1}\left[\frac{x}{x-1} - \frac{1}{\log x}\right]$, $\qquad$ [form $\infty - \infty$]

$$= \lim_{x\to 1}\left[\frac{x\log x - x + 1}{(x-1)\log x}\right], \qquad \left[\text{form } \frac{0}{0}\right]$$

$$= \lim_{x\to 1}\left[\frac{x.\left(\frac{1}{x}\right) + \log x - 1}{(x-1)\left(\frac{1}{x}\right) + \log x}\right]$$

$$= \lim_{x\to 1}\frac{\log x}{1 - \left(\frac{1}{x^2}\right) + \left(\frac{1}{x}\right)} = \frac{1}{1+1} = \frac{1}{2}.$$

Example 8:

Evaluate $\lim_{x\to 0}\left[\frac{1}{x(1+x)} - \frac{\log(1+x)}{x^2}\right]$.

Solution:

$\lim_{x\to 0}\left[\frac{1}{x(1+x)} - \frac{\log(1+x)}{x^2}\right]$, $\qquad$ [form $\infty - \infty$]

$$= \lim_{x\to 0}\left[\frac{x - (1+x)\log(1+x)}{x^2(1+x)}\right], \qquad \left[\text{form } \frac{0}{0}\right]$$

$$= \lim_{x\to 0} \frac{1-(1+x).\left\{\frac{1}{(1+x)}\right\}-\log(1+x)}{2x+3x^2}$$

$$= \lim_{x\to 0} \frac{-\log(1+x)2x+3x^2}{2x+3x^2}, \qquad \left[\text{form } \frac{0}{0}\right]$$

$$= \lim_{x\to 0} \frac{-\left\{\frac{1}{(1+x)}\right\}}{2+6x} = -\frac{1}{2}.$$

Example 9:

Evaluate $\lim_{x\to 0}\left[\frac{1}{x}-\frac{1}{x^2}\log(1+x)\right]$. **(Gorakhpur, 1991)**

Solution:

We have $\lim_{x\to 0}\left[\frac{1}{x}-\frac{1}{x^2}\log(1+x)\right]$

$$= \lim_{x\to 0} \frac{x-\log(1+x)}{x^2}, \qquad \left[\text{form } \frac{0}{0}\right]$$

$$= \lim_{x\to 0} \frac{1-\left\{\frac{1}{(1+x)}\right\}}{2x} = \lim_{x\to 0} \frac{1+x-1}{2x(1+x)}$$

$$= \lim_{x\to 0} \frac{1}{2(1+x)} = \frac{1}{2}.$$

Example 10:

Evaluate $\lim_{x\to \pi/2}\left(\sec x-\frac{1}{1-\sin x}\right)$. **(Vikram, 1998; Meerut, 1999S)**

Solution:

We have $\lim_{x\to \pi/2}\left(\sec x-\frac{1}{1-\sin x}\right)$, [form ∞ − ∞]

$$= \lim_{x\to \pi/2}\left(\frac{1}{\cos x}-\frac{1}{1-\sin x}\right) = \lim_{x\to \pi/2}\left[\frac{1-\sin x-\cos x}{\cos x(1-\sin x)}\right], \qquad \left[\text{form } \frac{0}{0}\right]$$

$$= \lim_{x\to\pi/2}\left[\frac{-\cos x+\sin x}{-\sin x(1-\sin x)+\cos x(-\cos x)}\right]$$

$$= \lim_{x\to\pi/2}\left(\frac{\sin x-\cos x}{-\sin x+\sin^2 x-\cos^2 x}\right)-\frac{1}{-1+1}=\infty.$$

Example 11:

Evaluate $\lim_{x\to 0}(\cos x)^{\cot^2 x}$.

(Allahabad, 1997; Rohilkhand, 98, 99; Gorakhpur, 96)

Solution:

Let $y=\lim_{x\to 0}(\cos x)^{\cot^2 x}$. [form 1^∞]

$\therefore\ \log y = \lim_{x\to 0}(\cot^2 x).(\log\cos x)$, [form $\infty\times 0$]

$= \lim_{x\to 0}\frac{\log\cos x}{\tan^2 x}$, $\left[\text{form } \frac{0}{0}\right]$

$= \lim_{x\to 0}\frac{\left\{\left(\frac{1}{\cos x}\right).(-\sin x)\right\}}{2\tan x.\sec^2 x}$, [by L'Hospital's rule]

$$= \lim_{x\to 0}\frac{-\tan x}{2\tan x.\sec^2 x}=\lim_{x\to 0}\frac{-1}{2\sec^2 x}=-\frac{1}{2}.$$

$\therefore\ y=e^{-1/2}$ i.e., $\lim_{x\to 0}(\cos x)^{\cot^2 x}=e^{-1/2}$.

Example 12:

Evaluate $\lim_{x\to\pi/2}(\sin x)^{\tan x}$.

(Kanpur, 1989; Agra, 94, 97; Gorakhpur, 99; Meerut, 98S)

Solution:

Let $y=\lim_{x\to\pi/2}(\sin x)^{\tan x}$ [form 1^∞]

$\therefore\ \log y=\lim_{x\to\pi/2}\tan x.\log\sin x$, [form $\infty\times 0$]

$= \lim_{x\to\pi/2}\frac{\log\sin x}{\cot x}$, $\left[\text{form } \frac{0}{0}\right]$

$$= \lim_{x \to \pi/2} \frac{\left(\frac{1}{\sin x}\right)\cos x}{-\operatorname{cosec}^2 x}, \qquad \text{[by L'Hospital's Rule]}$$

$$= \lim_{x \to \pi/2} (-2 \sin x \cos x) = 0.$$

$\therefore\ y = e^0 = 1.$

Example 13:

Evaluate $\lim_{x \to \infty} \left(1 + \frac{a}{x}\right)^x$. **(Kanpur, 1998; Rohilkhand, 92; Allahabad, 92; Gorakhpur, 94; Meerut, 97, 99S)**

Solution:

Let $y = \lim_{x \to \infty} \left(1 + \frac{a}{x}\right)^x$. [form 1^∞]

$\therefore\ \log y = \lim_{x \to \infty} \left\{x \log\left(1 + \frac{a}{x}\right)\right\}$, [form $\infty \times 0$]

$$= \lim_{x \to \infty} \frac{\log\left\{1 + \left(\frac{a}{x}\right)\right\}}{\frac{1}{x}}, \qquad \left[\text{form } \frac{0}{0}\right]$$

$$= \lim_{x \to \infty} \frac{\left[\frac{1}{\left\{1 + \left(\frac{a}{x}\right)\right\}}\right] \cdot \left(-\frac{a}{x^2}\right)}{-\left(\frac{1}{x^2}\right)} = \lim_{x \to \infty} \frac{a}{1 + \left(\frac{a}{x}\right)} = a.$$

$\therefore\ y = e^a$ or $\lim_{x \to \infty} \left(1 + \frac{a}{x}\right)^x = e^a$.

Example 14:

Evaluate $\lim_{x \to 0} (1 + x)^{1/x}$. **(Kanpur, 1997; Gorakhpur, 97)**

Solution:

Let $y = \lim_{x \to 0} (1 + x)^{1/x}$, [form 1^∞]

$$\therefore \quad \log y = \lim_{x\to 0} \frac{\log(1+x)}{x}, \qquad \left[\text{form } \frac{0}{0}\right]$$

$$= \lim_{x\to 0} \frac{\frac{1}{(1+x)}}{1} = \lim_{x\to 0} \frac{1}{1+x} = 1\cdot$$

$$\therefore \quad y = e^1 = e \text{ or } \lim_{x\to 0}(1+x)^{1/x} = e.$$

Example 15:

Evaluate $\lim\limits_{x\to\infty}\left(1+x^{-2}\right)^x$.

Solution:

$$\text{Let } y = \lim_{x\to\infty}\left(1+\frac{1}{x^2}\right)^x \qquad [\text{form } 1^\infty]$$

$$\therefore \log y = \lim_{x\to\infty} x \log\left(1+\frac{1}{x^2}\right) \qquad [\text{form } 1^\infty]$$

$$= \lim_{x\to\infty} x\left[\frac{1}{x^2}-\frac{1}{2x^4}+\ldots\right] \qquad \left[\text{Expanding } \log\left\{1+\left(\frac{1}{x^2}\right)\right\}\right]$$

$$= \lim_{x\to\infty}\left[\frac{1}{x}-\frac{1}{2x^3}+\ldots\right] = 0.$$

$$\therefore y = e^0 = 1.$$

7.6 THE FORMS 0^0, 1^∞, ∞^0

Suppose $\lim\limits_{x\to a}\left[f(x)\right]^{\phi(x)}$ takes any one of these three forms.

Then let $y = \lim\limits_{x\to a}\left[f(x)\right]^{\phi(x)}$.

Taking logarithm of both sides, we get

$$\log y = \lim_{x\to a} \phi(x).\log f(x).$$

Now in any of the above three cases log y takes the form $0 \times \infty$ which is changed to the form $\left[\frac{0}{0}\right]$ or $\left[\frac{\infty}{\infty}\right]$ whichever is convenient and then its limit is evaluated by L'Hospital's Rule or by using standard Expansions.

Example 1:

Evaluate $\lim_{x \to 0} \left(\frac{\tan x}{x} \right)^{1/x^2}$.

(Meerut, 1991, 95, 97; Avadh, 97; Kanpur, 97; Rohilkhand, 91; Gorakhpur, 90)

Solution:

We have

$$\log y = \lim_{x \to 0} \frac{1}{x^2} \left[\frac{x^2}{3} + \frac{7}{90} x^4 + \ldots \right]$$

$$= \lim_{x \to 0} \left[\frac{1}{3} + \frac{7}{90} x^2 + \ldots \right] = \frac{1}{3}.$$

$\therefore\ y = e^{1/3}$.

Example 1 (a):

Evaluate $\lim_{x \to 0} + \left(\frac{\tan x}{x} \right)^{1/x^3}$. **(U.P.P.C.S., 1995; Meerut, 99)**

Solution:

We have

$$\log y = \lim_{x \to 0} + \frac{1}{x^3} \left[\frac{x^2}{3} + \frac{7}{90} x^4 + \ldots \right]$$

$$= \lim_{x \to 0} + \left[\frac{1}{3x} + \frac{7}{90} x + \ldots \right] = +\infty.$$

$\therefore\ y = e^{\infty} = \infty$.

Example 2:

Evaluate $\lim_{x \to 0} (\cos x)^{\cot^2 x}$.

(Allahabad, 1997; Rohilkhand, 98, 99; Gorakhpur, 96)

Solution:

Let $y = \lim_{x \to 0} (\cos x)^{\cot^2 x}$. [form 1^∞]

$\therefore\ \log y = \lim_{x \to 0} (\cot^2 x).(\log \cos x)$, [form $\infty \times 0$]

$= \lim_{x \to 0} \frac{\log \cos x}{\tan^2 x}$ $\left[\text{form } \frac{0}{0}\right]$

$$= \lim_{x\to 0} \frac{\left\{\left(\frac{1}{\cos x}\right).(-\text{sub } x)\right\}}{2\tan x.\sec^2 x},$$ [by L'Hospital's Rule]

$$= \lim_{x\to 0} \frac{-\tan x}{2\tan x.\sec^2 x} = \lim_{x\to 0} \frac{-1}{2\sec^2 x} = -\frac{1}{2}.$$

$\therefore$ $y = e^{-1/2}$ i.e., $\lim_{x\to 0}(\cos x)^{\cot^2 x} = e^{-1/2}$.

Example 3:

Evaluate $\lim_{x\to\infty}\left(\frac{1}{x}\right)^{1/x}$. **(Agra, 1994)**

Solution:

Let $y = \lim_{x\to\infty}\left(\frac{1}{x}\right)^{1/x}$ [form 0^0]

$\therefore$ $\log y = \lim_{x\to 0} \frac{1}{x}\log\left(\frac{1}{x}\right)$, [form $0 \times \infty$]

$$= \lim_{x\to\infty} \frac{-\log x}{x}$$ $\left[\because \log\left(\frac{1}{x}\right) = \log 1 - \log x = -\log x\right]$

$$= -\lim_{x\to\infty} \frac{\left(\frac{1}{x}\right)}{1} = 0.$$

$\therefore$ $y = e^0 = 1$.

Example 4:

Evaluate $\lim_{x\to\infty}\left(\frac{1}{2}\pi - \tan^{-1}x\right)^{1/x}$.

Solution:

Let $y = \lim_{x\to\infty}\left(\frac{1}{2}\pi - \tan^{-1}x\right)^{1/x}$ [form 0^0]

$\therefore$ $\log y = \lim_{x\to\infty}\left(\frac{1}{x}\right)\log\left(\frac{1}{2}\pi - \tan^{-1}x\right)$

$$= \lim_{x\to\infty} \frac{\log\left(\frac{1}{2}\pi - \tan^{-1}x\right)}{x}$$ $\left[\text{form } \frac{\infty}{\infty}\right]$

$$= \lim_{x\to\infty} \frac{\left\{\frac{1}{\left(\frac{1}{2}\pi - \tan^{-1} x\right)}\right\}.\left\{-\frac{1}{(1+x^2)}\right\}}{1}$$

$$= \lim_{x\to\infty} \left\{\frac{-\frac{1}{(1+x^2)}}{\left(\frac{1}{2}\pi - \tan^{-1} x\right)}\right\} \qquad \left[\text{form } \frac{0}{0}\right]$$

$$= \lim_{x\to\infty} \frac{\left\{\frac{1}{(1+x^2)^2}\right\}.2x}{-\frac{1}{(1+x^2)}}, \qquad \text{[by L'Hospital's Rule]}$$

$$= \lim_{x\to\infty} \left\{\frac{-2x}{1+x^2}\right\} \qquad \left[\text{form } \frac{\infty}{\infty}\right]$$

$$= \lim_{x\to\infty} \left\{\frac{-2}{2x}\right\} = 0.$$

$\therefore\ y = e^0 = 1.$

Example 5:

Evaluate $\lim_{x\to \pi/2} (\sin x)^{\tan x}$.

(Kanpur, 1999; Agra, 94, 97; Gorakhpur, 99; Meerut, 98S)

Solution:

Let $y = \lim_{x\to\pi/2} \tan x.\log \sin x$, [form 1^∞]

$\therefore\ \log y = \lim_{x\to\pi/2} \tan x.\log \sin x$, [form $\infty \times 0$]

$$= \lim_{x\to\pi/2} \frac{\log \sin x}{\cot x}, \qquad \left[\text{form } \frac{0}{0}\right]$$

$$= \lim_{x\to\pi/2} \frac{\left(\frac{1}{\sin x}\right)\cos x}{-\text{cosec}^2 x}, \qquad \text{[by L'Hospital's Rule]}$$

$= \lim_{x \to \pi/2} (-\sin x \cos x) = 0$.

$\therefore\ y = e^0 = 1.$

Example 6:

Evaluate $\lim_{x \to \infty} \left(1 + x^{-2}\right)^x$.

Solution:

$$\text{Let } y = \lim_{x \to \infty} \left(1 + \frac{1}{x^2}\right)^x, \qquad [\text{form } 1^\infty]$$

$$\therefore\ \log y = \lim_{x \to \infty} x \log\left(1 + \frac{1}{x^2}\right), \qquad [\text{form } \infty \times 0]$$

$$= \lim_{x \to \infty} x\left[\frac{1}{x^2} - \frac{2}{x^4} + \ldots\right], \qquad \left[\text{Expanding } \log\left\{1 + \left(\frac{1}{x^2}\right)\right\}\right]$$

$$= \lim_{x \to \infty} \left[\frac{1}{x} - \frac{1}{2x^3} + \ldots\right] = 0.$$

$\therefore\ y = e^0 = 1.$

Example 7:

Evaluate $\lim_{x \to 0} \left(2 - \frac{x}{a}\right)^{\tan(\pi x/2a)}$. **(G.N.U., 1997; Delhi, 90)**

Solution:

$$\text{Let } y = \lim_{x \to 0} \left(2 - \frac{x}{a}\right)^{\tan(\pi x/2a)} \qquad [\text{form } 1^\infty]$$

$$\therefore\ \log y = \lim_{x \to a} \tan\left(\frac{\pi x}{2a}\right)\left[\log\left(2 - \frac{x}{a}\right)\right], \qquad [\text{form } \infty \times 0]$$

$$= \lim_{x \to a} \left[\left\{\log\left(2 - \frac{x}{a}\right)\right\} / \cot\left(\frac{\pi x}{2a}\right)\right], \qquad \left[\text{form } \frac{0}{0}\right]$$

$$= \lim_{x \to a} \left[\left(-\frac{1}{a}\right)\left(2 - \frac{x}{a}\right)^{-1}\right] / \left[\left\{-\operatorname{cosec}^2\left(\frac{\pi x}{2a}\right)\right\} \frac{\pi}{2a}\right]$$

$$= \lim_{x \to a} \frac{1}{a} . \frac{2a}{\pi} . \frac{1}{\left\{2 - \left(\frac{x}{a}\right)\right\}} \sin^2\left(\frac{\pi x}{2a}\right) = \frac{2}{\pi}.$$

$\therefore\ y = e^{2/\pi}.$

Example 8:

Evaluate $\lim_{x \to \pi/4} (\tan x)^{\tan 2x}$. **(Rohilkhand, 1998; Raj. 98)**

Solution:

Let $y = \lim_{x \to \pi/4} \left[(\tan x)^{\tan 2x}\right]$ [form 1^∞]

$\therefore$ log $\log y = \lim_{x \to \pi/4} \left[(\tan 2x).(\log \tan x)\right]$ [form $\infty \times 0$]

$$= \lim_{x \to \pi/4} \left[\frac{\log \tan x}{\cot 2x}\right], \qquad \left[\text{form } \frac{0}{0}\right]$$

$$= \lim_{x \to \pi/4} \left[\frac{\left(\frac{1}{\tan x}\right).\sec^2 x}{-2\operatorname{cosec}^2(2x)}\right] = \lim_{x \to \pi/4} \left[\frac{\sec^2 x.\sin^2(2x)}{-2\tan x}\right]$$

$$= \frac{\left(\sqrt{2}\right)^2.(1)^2}{-2.1} = -1.$$

$$= y = e^{-1} = \frac{1}{e}.$$

Example 9:

Evaluate $\lim_{x \to 0} \left(\frac{a^x + b^x}{2}\right)^{1/x}$ **(Rohilkhand, 1992, 96)**

Solution:

Let $y = \lim_{x \to 0} \left(\frac{a^x + b^x}{2}\right)^{1/x}$ [form 1^∞]

$$\therefore \log y = \lim_{x \to 0} \left[\frac{1}{x} \log\left\{\frac{a^x + b^x}{2}\right\}\right]$$

$$= \lim_{x \to 0} \frac{\log\left\{\left(a^x + b^x\right)/2\right\}}{x} \qquad \left[\text{form } \frac{0}{0}\right]$$

$$= \lim_{x \to 0} \frac{\log\left(a^x + b^x\right) - \log 2}{x} \qquad \left[\text{form } \frac{0}{0}\right]$$

$$= \lim_{x \to 0} \frac{\dfrac{a^x \log x + b^x \log b}{a^x + b^x}}{1} = \lim_{x \to 0} \frac{a^x \log a + b^x \log b}{a^x + b^x}$$

$$= \frac{\log a + \log b}{1+1} = \frac{1}{2} \log (ab) = \log \sqrt{(ab)}\,.$$

$$\therefore\ y = \sqrt{(ab)}\,.$$

Example 10:

Evaluate $\lim_{x \to 0} (\operatorname{cosec} x)^{1/\log x}$.

(Agra, 1998, 95; Rajasthan, 97; Indore, 92)

Solution:

Let $y = \lim_{x \to 0} (\operatorname{cosec} x)^{1/\log x}$. [form ∞^0]

$$\therefore\ \log y = \lim_{x \to 0} \frac{1}{\log x} (\log \operatorname{cosec} x), \qquad \left[\text{form } \frac{\infty}{\infty}\right]$$

$$= \lim_{x \to 0} \frac{\left(\dfrac{1}{\operatorname{cosec} x}\right)(-\operatorname{cosec} x \cot x)}{\dfrac{1}{x}}$$

$$= \lim_{x \to 0} \left(\frac{-x}{\tan x}\right), \qquad \left[\text{form } \frac{0}{0}\right]$$

$$= \lim_{x \to 0} \left(\frac{-1}{\sec^2 x}\right) = -1.$$

$$\therefore\ y = e^{-1} = \frac{1}{e}.$$

Example 11:

Evaluate $\lim_{x \to \infty} \left(1 + \dfrac{a}{x}\right)^x$. **(Kanpur, 1998; Rohilkhand, 92; Allahabad, 92; Gorakhpur, 94; Meerut, 97, 99S)**

Solution:

Let $y = \lim_{x \to \infty} \left(1 + \dfrac{a}{x}\right)^x$. [form 1^∞]

$\therefore\ \log y = \lim_{x\to\infty}\left\{x\log\left(1+\frac{a}{x}\right)\right\},$ [form $\infty \times 0$]

$$= \lim_{x\to\infty}\frac{\log\left\{1+\left(\frac{a}{x}\right)\right\}}{\frac{1}{x}}, \qquad \left[\text{form } \frac{0}{0}\right]$$

$$= \lim_{x\to\infty}\frac{\left[\frac{1}{\left\{1+\left(\frac{a}{x}\right)\right\}}\right]\cdot\left(-\frac{a}{x^2}\right)}{-\left(\frac{1}{x^2}\right)} = \lim_{x\to\infty}\frac{a}{1+\left(\frac{a}{x}\right)} = a$$

$\therefore\ y = e^a$ or $\lim_{x\to\infty}\left(1+\frac{a}{x}\right)^x = e^a.$

Example 12:

Evaluate $\lim_{x\to 0}(1+x)^{1/x}$ **(Kanpur, 1997; Gorakhpur, 97)**

Solution:

Let $y = \lim_{x\to 0}(1+x)^{1/x}$ [form 1^∞].

$$\therefore\ \log y = \lim_{x\to 0}\frac{\log(1+x)}{x} \qquad \left[\text{form } \frac{0}{0}\right]$$

$$= \lim_{x\to 0}\frac{\frac{1}{(1+x)}}{1} = \lim_{x\to 0}\frac{1}{1+x} = 1\cdot$$

$\therefore\ y = e^1 = e$ or $\lim_{x\to 0}(1+x)^{1/x} = e.$

Example 13:

Evaluate $\lim_{x\to 0}(\cos x)^{1/x^2}.$ **(Meerut, 1991 Kanpur, 1996)**

Solution:

Let $y = \lim_{x\to 0}(\cos x)^{1/x^2}$ [form 1^∞].

$$\therefore \log y = \lim_{x\to 0}\left(\frac{1}{x^2}\right)(\log \cos x), \qquad [\text{form } \infty \times 0]$$

$$= \lim_{x\to 0}\frac{\log \cos x}{x^2}, \qquad \left[\text{form } \frac{0}{0}\right]$$

$$= \lim_{x\to 0}\frac{\left(\frac{1}{\cos x}\right)(-\sin x)}{2x}, \qquad [\text{by L'Hospital's Rule}]$$

$$= \lim_{x\to 0}\left(\frac{-\tan x}{2x}\right), \qquad \left[\text{form } \frac{0}{0}\right]$$

$$= \lim_{x\to 0}\left(\frac{-\sec^2 x}{2}\right), \qquad [\text{by L'Hospital's Rule}]$$

$$= -\frac{1}{2}. \qquad \therefore y = e^{-1/2}.$$

Example 14:

Evaluate $\lim_{x\to 0}(\cot x)^{\sin x}$ **(Allahabad, 1994; Andhra, 91)**

Solution:

Let $y = \lim_{x\to 0}(\cot x)^{\sin x}$ $[\text{form } \infty^0]$

$$\therefore \log y = \lim_{x\to 0}(\sin x).(\log \cot x) \qquad [\text{form } 0 \times \infty]$$

$$= \lim_{x\to 0}\left(\frac{\log \cot x}{\operatorname{cosec} x}\right), \qquad \left[\text{form } \frac{\infty}{\infty}\right]$$

$$= \lim_{x\to 0}\frac{\left(\frac{1}{\cot x}\right).(-\operatorname{cosec}^2 x)}{-\operatorname{cosec} x \cot x} = \lim_{x\to 0}\left(\frac{\operatorname{cosec} x}{\cot^2 x}\right)$$

$$= \lim_{x\to 0}\left(\frac{\sin x}{\cos^2 x}\right) = 0.$$

$\therefore y = e^0 = 1.$

Example 15:

Evaluate $\lim_{x\to 1/2\pi}(\sec x)^{\cot x}$. **(Gorakhpur, 1991)**

Solution:

Let $y = \lim_{x \to 1/2\pi} (\sec x)^{\cot x}$. [form ∞^0]

$\therefore \log y = \lim_{x \to \pi/2} (\cot x).(\log \sec x)$, [form $0 \times \infty$]

$= \lim_{x \to \pi/2} \frac{\log \sec x}{\tan x}$ $\left[\text{form } \frac{\infty}{\infty}\right]$

$$= \lim_{x \to \pi/2} \frac{\left(\frac{1}{\sec x}\right).\sec x \tan x}{\sec^2 x}$$

$$= \lim_{x \to \pi/2} \frac{\tan x}{\sec^2 x} = \lim_{x \to \pi/2} (\sin x \cos x) = 0.$$

$\therefore y = e^0 = 1.$

Example 16:

Evaluate $\lim_{x \to 0} \left(\frac{\tan x}{x}\right)^{1/x}$ **(Allahabad, 1997; Gurunanak, 93; Magadh, 96; Bihar, 92; Kanpur, 95; Meerut, 94P, 96P)**

Solution:

Let $y = \lim_{x \to 0} \left(\frac{\tan x}{x}\right)^{1/x}$ [form 1^∞]

$\therefore \log y = \lim_{x \to 0} \frac{1}{x} \log \frac{\tan x}{x}$

$$= \lim_{x \to 0} \left[\frac{1}{x} \log \left\{\frac{x + \left(\frac{x^3}{3}\right) + \left(\frac{2x^5}{15}\right) + \ldots}{x}\right\}\right],$$

(writing the expansion for tan x)

$$= \lim_{x \to 0} \frac{1}{x} \log \left[1 + \frac{x^2}{3} + \frac{2x^4}{15} + \ldots\right]$$

$$= \lim_{x \to 0} \frac{1}{x} \log(1 + z), \text{ where } z = \frac{x^2}{3} + \frac{2x^4}{15} + \ldots$$

$= \lim_{x \to 0} \frac{1}{x}\left[z - \frac{z^2}{2} + \ldots\right]$, [Expanding log (1 + z)]

$$= \lim_{x \to 0} \frac{1}{x}\left[\left(\frac{x^2}{3} + \frac{2x^4}{15} + \dots\right) - \frac{1}{2}\left(\frac{x^2}{13} + \frac{2x^4}{15} + \dots\right)^2 + \dots\right]$$

$$= \lim_{x \to 0} \frac{1}{x}\left[\frac{x^2}{3} + \left(\frac{2}{15} - \frac{1}{18}\right)x^4 + \dots\right]$$

$$= \lim_{x \to 0} \frac{1}{x}\left[\frac{x^2}{3} + \frac{7}{90}x^4 + \dots\right]$$

$$= \lim_{x \to 0}\left[\frac{x}{3} + \frac{7}{90}x^3 + \dots\right] = 0.$$

$\therefore\ y = e^0 = 1.$

Example 17:

Evaluate $\lim\limits_{x \to 0} (\cos x)^{1/x}$. **(Magadh, 1990)**

Solution:

Let $y = \lim\limits_{x \to 0} (\cos x)^{1/x}$. [form 1^∞]

$\therefore\ \log y = \lim\limits_{x \to 0} \left(\frac{1}{x}\right)(\log \cos x),$ [form $\infty \times 0$]

$= \lim\limits_{x \to 0} \frac{\log \cos x}{x}$ $\left[\text{form } \frac{0}{0}\right]$

$$= \lim_{x \to 0} \frac{\left(\frac{1}{\cos x}\right)(-\sin x)}{1} = \lim_{x \to 0} (-\tan x) = 0\cdot$$

$\therefore\ y = e^0 = 1.$

Example 18:

Evaluate $\lim\limits_{x \to 1} (x)^{1/(x-1)}$. **(Vikram, 1997)**

Solution:

Let $y = \lim\limits_{x \to 1} x^{1/(x-1)}$. [form 1^∞]

$\therefore\ \log y = \lim\limits_{x \to 1} \frac{1}{(x-1)} \log x = \lim\limits_{x \to 1} \frac{\log x}{(x-1)},$ $\left[\text{form } \frac{0}{0}\right]$

$$= \lim_{x \to 1} \frac{\frac{1}{x}}{1} = 1. \text{ Hence } y = e^1 = e.$$

Example 19:

Evaluate $\lim\limits_{x \to 0} \left(\dfrac{\sinh x}{x} \right)^{1/x^2}$. **(Garhwal, 1993; Gorakhpur, 93)**

Solution:

We have $\log y = \lim\limits_{x \to 0} \dfrac{1}{x^2} \left[\dfrac{x^2}{6} - \dfrac{1}{180} x^4 + \ldots \right]$

$$= \lim_{x \to 0} \left[\frac{1}{6} - \frac{1}{180} x^2 + \ldots \right] = \frac{1}{6}.$$

$\therefore\ y = e^{1/6}$.

Example 20:

Evaluate $\lim\limits_{x \to 0} \left[\dfrac{2(\cosh x - 1)}{x^2} \right]^{1/x^2}$ **(Meerut, 1994)**

Solution:

Let $y = \lim\limits_{x \to 0} \left[\dfrac{2(\cosh x - 1)}{x^2} \right]^{1/x^2}$, [form 1^∞].

$$\therefore\ \log y = \lim_{x \to 0} \frac{1}{x^2} \log \left\{ \frac{2(\cosh x - 1)}{x^2} \right\}$$

$$= \lim_{x \to 0} \frac{1}{x^2} \log \left\{ \frac{2}{x^2} \left(1 + \frac{x^2}{2!} + \frac{x^4}{4!} + \ldots - 1 \right) \right\}$$

$$= \lim_{x \to 0} \frac{1}{x^2} \log \left\{ \frac{2}{x^2} \left(\frac{x^2}{2} + \frac{x^4}{24} + \ldots \right) \right\}$$

$$= \lim_{x \to 0} \frac{1}{x^2} \log \left[1 + \left(\frac{x^2}{12} + \ldots \right) \right]$$

$$= \lim_{x \to 0} \frac{1}{x^2} \left[\left(\frac{x^2}{12} + \ldots \right) - \frac{1}{2} \left(\frac{x^2}{12} + \ldots \right)^2 + \ldots \right]$$

$$= \lim_{x\to 0} \frac{1}{x^2}\left[\frac{x^2}{12} + \text{higher powers of x}\right]$$

$$= \lim_{x\to 0}\left[\frac{1}{12} + \text{terms containing x and its higher powers}\right] = \frac{1}{12}.$$

$\therefore\ y = e^{1/12}.$

Example 21:

Evaluate $\lim\limits_{x\to\infty} \dfrac{Ax^n + Bx^{n-1} + Cx^{n-2} + \dots}{ax^m + bx^{m-1} + cx^{m-2} + \dots}.$

Solution:

$$\text{Given limit} = \lim_{x\to\infty} \frac{x^n\left[A + \left(\frac{B}{x}\right) + \left(\frac{C}{x^2}\right) + \dots\right]}{x^m\left[a + \left(\frac{b}{x}\right) + \left(\frac{c}{x^2}\right) + \dots\right]}$$

$$= \lim_{x\to\infty}\left(x^{n-m}.\frac{A}{a}\right).$$

Now, if $n > m$, $\lim\limits_{x\to\infty}\left(\dfrac{A}{a}x^{n-m}\right) = \infty;$

if $n = m$, $\lim\limits_{x\to\infty} \dfrac{A}{a}x^{n-m} = \lim\limits_{x\to\infty} \dfrac{A}{a} = \dfrac{A}{a};$

if $n < m$, $\lim\limits_{x\to\infty} \dfrac{A}{a}x^{n-m} = \lim\limits_{x\to\infty}\left(\dfrac{A}{a}.\dfrac{1}{x^{m-n}}\right) = 0.$

Example 22:

Evaluate $\lim\limits_{x\to 0} x^x.$

Solution:

Let $y = \lim\limits_{x\to 0} x^x.$ [firn 0^0]

$\therefore\ \log y = \lim\limits_{x\to 0} x \log x,$ [form $0 \times \infty$]

$= \lim\limits_{x\to 0} \dfrac{\log x}{\left(\frac{1}{x}\right)},$ $\left[\text{form } \dfrac{\infty}{\infty}\right]$

$$= \lim_{x\to 0} \frac{\frac{1}{x}}{\left(-\frac{1}{x^2}\right)} = \lim_{x\to 0}(-x) = 0.$$

$\therefore\ y = e^0 = 1.$

Example 23:

Evaluate $\lim\limits_{x\to 0} \dfrac{x^x - 1}{x}$.

Solution:

$$\lim_{x\to 0} \frac{x^x - 1}{x} \qquad \left[\text{form } \frac{0}{0}, \because \lim_{x\to 0} x^x = 1\right]$$

$$= \lim_{x\to 0} \frac{\frac{d}{dx}\left(x^x\right)}{1} \qquad \text{[by L'Hospital's Rule]}$$

$$= \lim_{x\to 0} \left\{\frac{d}{dx}\left(x^x\right)\right\}$$

Now let $y = x^x$; then $\log y = x \log x$. Now differentiating,

$$\frac{1}{y}\frac{dy}{dx} = x\left(\frac{1}{x}\right) + 1.\log x = (1 + \log x).$$

$$\therefore\ \frac{dy}{dx} = y\,(1 + \log x). \quad \text{Thus, } \frac{d}{dx}\left(x^x\right) = x^x(1 + \log x).$$

$$\text{Hence } \lim_{x\to 0} \frac{x^x - 1}{x} = \lim_{x\to 0}\left[x^x(1 + \log x)\right]$$

$$= 1(1 - \infty) = -\infty. \qquad \left[\because \lim_{x\to 0} x^x = 1\right]$$

Example 24:

Evaluate $\lim\limits_{x\to 0}\left(a^x + x\right)^{1/x}$. **(Kanpur, 1990)**

Solution:

$$\text{Let } y = \lim_{x\to 0}\left(a^x + x\right)^{1/x} \qquad [\text{form } 1^\infty]$$

$$\therefore\ \log y = \lim_{x\to 0}\left[\frac{1}{x}\log\left(a^x + x\right)\right]$$

$$= \lim_{x\to 0} \frac{\log\left(a^x + x\right)}{x} \qquad \left[\text{form } \frac{0}{0}\right]$$

$$= \lim_{x\to 0} \frac{\left\{\frac{1}{\left(a^x + x\right)}\right\}\left(a^x \log a + 1\right)}{1}$$

$$= \lim_{x\to 0} \frac{a^x \log a + 1}{a^x + x} = \frac{\log a + 1}{1+0} = \log a + 1$$

$= \log a + \log e = \log (ae),$ $\qquad [\because 1 = \log e]$

$\therefore\ y = ae.$

Example 25:

Evaluate $\lim_{x\to 0} (\sin x)^{\tan x}$.

Solution:

Let $y = \lim_{x\to 0} (\sin x)^{\tan x}$ $\qquad$ [form 0^0]

$\therefore\ \log y = \lim_{x\to 0} (\tan x).(\log \sin x),$ $\qquad$ [form $0 \times \infty$]

$$= \lim_{x\to 0} \frac{\log \sin x}{\cot x}, \qquad \left[\text{form } \frac{\infty}{\infty}\right]$$

$$= \lim_{x\to 0} \frac{\left(\frac{1}{\sin x}\right)\cos x}{-\operatorname{cosec}^2 x}, \qquad \left[\text{form } \frac{\infty}{\infty}\right]$$

$$= \lim_{x\to 0} \left[-\left(\frac{\sin^2 x}{\tan x}\right)\right], \qquad \left[\text{form } \frac{0}{0}\right]$$

$$= \lim_{x\to 0} \left[\frac{-2\sin x \cos x}{\sec^2 x}\right] = 0.$$

$\therefore\ y = e^0 = 1.$

Example 26:

Evaluate $\lim_{x\to 0} \left(\frac{\sin x}{x}\right)^{1/x}$ $\qquad$ **(K.U., 1997)**

Solution:

Let $y = \lim_{x\to 0} \left(\frac{\sin x}{x}\right)^{1/x}$ $\qquad \left[\text{form } 1^\infty, \text{since } \lim_{x\to 0} \frac{\sin x}{x} = 1\right].$

$$\therefore \log y = \lim_{x\to 0}\left[\frac{1}{x}\log\frac{\sin x}{x}\right]$$

$$= \lim_{x\to 0}\frac{1}{x}.\log\left\{\frac{x-\left(\frac{x^3}{3!}\right)+\left(\frac{x^5}{5!}\right)-...}{x}\right\}$$

$$= \lim_{x\to 0}\frac{1}{x}.\log\left(1-\frac{x^2}{3!}+\frac{x^4}{5!}-...\right)$$

$$= \lim_{x\to 0}\frac{1}{x}\log\left[1-\left(\frac{x^2}{6}-\frac{x^4}{120}+...\right)\right]$$

$$= \lim_{x\to 0}\frac{1}{x}\log(1-z),\text{ where } z=\frac{x^2}{6}-\frac{x^4}{120}+...$$

$$= \lim_{x\to 0}\frac{1}{x}\left(-z-\frac{z^2}{2}-...\right)$$

$$= \lim_{x\to 0}\frac{1}{x}\left[-\left(\frac{x^2}{6}-\frac{x^4}{120}+...\right)-\frac{1}{2}\left(\frac{x^2}{6}-\frac{x^4}{120}+...\right)^2-...\right]$$

$$= \lim_{x\to 0}\frac{1}{x}\left[-\frac{x^2}{6}+\left(\frac{x^4}{120}-\frac{x^4}{72}\right)+...\right]$$

$$= \lim_{x\to 0}\frac{1}{x}\left[-\frac{x^2}{6}-\frac{x^4}{180}+...\right]$$

$$= \lim_{x\to 0}\left[-\frac{x}{6}-\frac{x^3}{180}+...\right]=0.$$

$\therefore$ y = e^0 = 1.

Example 27:

Evaluate $\lim_{x\to 0}\left(\frac{\sin x}{x}\right)^{1/x^2}$.

(Kashmir, 1995; Ranchi, 95; Meerut, 99; Andhra, 91)

Solution:

We have

$$\log y = \lim_{x\to 0} \frac{1}{x^2}\left[-\frac{x^2}{6} - \frac{x^4}{180} + \ldots\right]$$

$$= \lim_{x\to 0}\left[-\frac{1}{6} - \frac{x^2}{180} + \ldots\right] = -\frac{1}{6}.$$

Example 28:

Evaluate $\lim_{x\to 0} + \left(\frac{\sin x}{x}\right)^{1/x^3}$

Solution:

We have

$$\log y = \lim_{x\to 0} + \frac{1}{x^3}.\left[-\frac{x^2}{6} - \frac{x^4}{180} + \ldots\right] = -\infty.$$

$y = e^{-\infty} = 0.$

Example 29:

Evaluate $\lim_{x\to 0}\left(\frac{\sinh x}{x}\right)^{1/x}.$

Solution:

Let $y = \lim_{x\to 0}\left(\frac{\sinh x}{x}\right)^{1/x}$ [form 1^∞]

$$\therefore \log y = \lim_{x\to 0}\left[\frac{1}{x}\log\left\{\frac{\sinh x}{x}\right\}\right].$$

Now $\sinh x = \frac{1}{2}\left(e^x - e^{-x}\right) = x + \frac{x^3}{3!} + \frac{x^5}{5!} + \ldots$

$$\therefore \log y = \lim_{x\to 0}\frac{1}{x}\log\left\{\frac{x + \left(\frac{x^3}{3!}\right) + \left(\frac{x^5}{5!}\right) + \ldots}{x}\right\}$$

$$= \lim_{x\to 0}\frac{1}{x}\log\left[1 + \frac{x^2}{6} + \frac{x^4}{120} + \ldots\right]$$

$$= \lim_{x\to 0}\frac{1}{x}\log(1+z), \text{ where } z = \frac{x^2}{6} + \frac{x^4}{120} + \ldots$$

$$= \lim_{x\to 0} \frac{1}{x}\left[z - \frac{z^2}{2} + \dots\right]$$

$$= \lim_{x\to 0} \frac{1}{x}\left[\left(\frac{x^2}{6} + \frac{x^4}{120} + \dots\right) - \frac{1}{2}\left(\frac{x^2}{6} + \frac{x^4}{120} + \dots\right)^2 + \dots\right]$$

$$= \lim_{x\to 0} \frac{1}{x}\left[\frac{x^2}{6} + \left(\frac{1}{120} - \frac{1}{72}\right)x^4 + \dots\right]$$

$$= \lim_{x\to 0} \frac{1}{x}\left[\frac{x^2}{6} - \frac{1}{80}x^4 + \dots\right]$$

$$= \lim_{x\to 0}\left[\frac{x}{6} - \frac{1}{180}x^3 + \dots\right] = 0.$$

$\therefore\ y = e^0 = 1.$

Example 30:

Evaluate $\lim_{x\to 0}\left(\dfrac{a^x + b^x}{2}\right)^{1/x}$ **(Kanpur, 1994, 97)**

Solution:

Let $y = \lim_{x\to 0}\left(\dfrac{a^x + b^x}{2}\right)^{1/x}$ [form 1^∞]

$$\therefore\ \log y = \lim_{x\to 0}\left[\frac{1}{x}\log\left\{\frac{a^x + b^x}{2}\right\}\right]$$

$$= \lim_{x\to 0} \frac{\log\{(a^x + b^x)/2\}}{x} \qquad \left[\text{form } \frac{0}{0}\right]$$

$$= \lim_{x\to 0} \frac{\log(a^x + b^x) - \log 2}{x} \qquad \left[\text{form } \frac{0}{0}\right]$$

$$= \lim_{x\to 0} \frac{\dfrac{a^x \log x + b^x \log b}{a^x + b^x}}{1} = \lim_{x\to 0} \frac{a^x \log a + b^x \log b}{a^x + b^x}$$

$$= \frac{\log a + \log b}{1+1} = \frac{1}{2}\log(ab) = \log\sqrt{(ab)}.$$

$\therefore\ y = \sqrt{(ab)}.$

Example 31:

Evaluate $\lim_{x \to 0} \frac{\log\left(\tan^2 2x\right)}{\log\left(\tan^2 x\right)}$. **(Allahabad, 1993)**

Solution:

We have $\lim_{x \to 0} \frac{\log\left(\tan^2 2x\right)}{\log\left(\tan^2 x\right)}$

$$= \lim_{x \to 0} \frac{2\log(\tan 2x)}{2\log(\tan x)}, \qquad \left[\text{form } \frac{\infty}{\infty}\right]$$

$$= \lim_{x \to 0} \frac{\left(\frac{1}{\tan 2x}\right).2\sec^2 2x}{\left(\frac{1}{\tan x}\right)\sec^2 x} = \lim_{x \to 0} \frac{2\tan x \cos^2 x}{\tan 2x \cos^2 2x}$$

$$= \lim_{x \to 0} \frac{2\sin x \cos x}{\sin 2x \cos 2x} = \lim_{x \to 0} \frac{\sin 2x}{\sin 2x \cos 2x}$$

$$= \lim_{x \to 0} \frac{1}{\cos 2x} = \frac{1}{1} = 1.$$

8

TANGENTS AND NORMALS

8.1 TANGENT

Definition: *The tangent to a plane curve at the point P(x, y) on it is defined as the limiting position of the chord PQ as the point Q(x + δx, y + δy) approaches the point P, provided such a limiting position exists.*

Remark: The term "tends to" has been defined so far only in connection with variables which takes up numerical values. Here this term use for position.

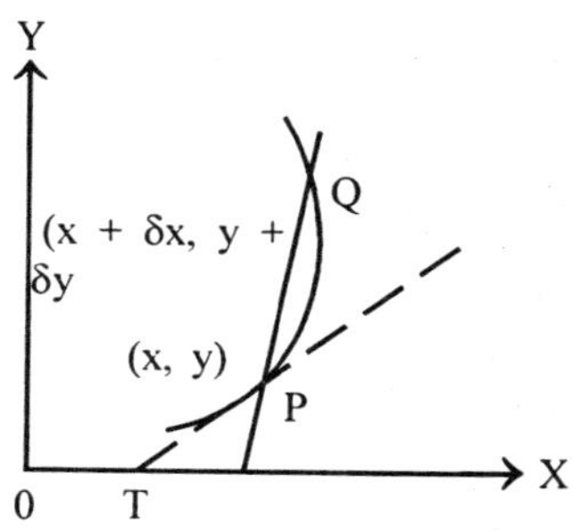

Fig. 8.1

Equation of the Tangent: Let P be any point (x, y) on the curve y = f(x), and Q a neighbouring point (x + δx, y + δy) on it. The point Q may be taken on either side of P. The equation of the chord PQ is

$$Y - y = \frac{(y+\delta y)-y}{(x+\delta x)}(X-x),$$ [Here (X, Y) are the current coordinates of any point on the chord]

or $$Y-y=\left(\frac{\delta y}{\delta x}\right)(X-x). \quad ...(1)$$

Now as $Q \to P$, $\delta x \to 0$, $\frac{\delta y}{\delta x} \to \frac{dy}{dx}$, and the chord PQ tends to the tangent at P. Therefore taking limit of (1) as $Q \to P$, we see that the equation of the tangent to the curve y = f(x) at the point (x, y) is

$$Y - y = \left(\frac{dy}{dx}\right)(X - x). \qquad ...(2)$$

Geometrical Meaning of $\frac{dy}{dx}$: The equation (2) of the tangent at any point P(x, y) on the curve y = f(x) may be written as

$$Y = \left(\frac{dy}{dx}\right)X + \left(y - x\frac{dy}{dx}\right),$$

which is of the form Y = mX + c. On comparing the two equations, we have $m = \frac{dy}{dx}$. Therefore if ψ is the angle which the tangent to the curve y = f(x) at the point (x, y) on it makes with the positive direction of x-axis then

$$\frac{dy}{dx} = \tan\psi.$$

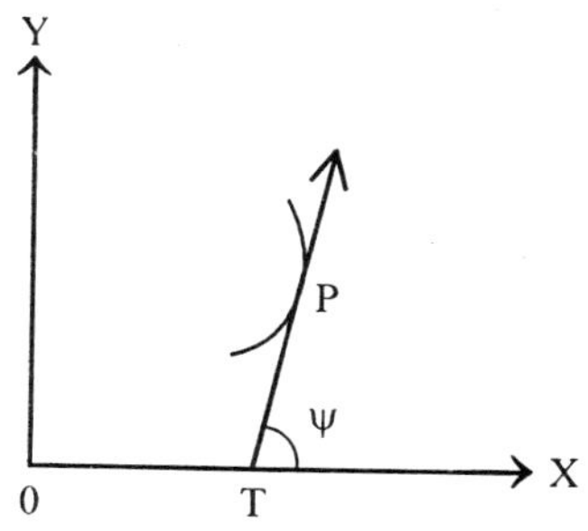

Fig. 8.2

Remember: *The differential coefficient $\frac{dy}{dx}$ at any point (x, y) on the curve y = f(x) is equal to tan ψ, where ψ is the angle which the positive direction of the tangent at P to the curve makes with the positive direction of the axis of x.*

The tangent of the angle which the tangent at P to the curve makes with the x-axis is usually called the gradient of the curve at the point P.

Note 1: The equation of the tangent at any point (x_1, y_1) on the curve $y = f(x)$ is $y - y_1 = \left(\frac{dy}{dx}\right)_{(x_1, y_1)}(x - x_1)$. Here (x, y) are the current coordinates of any poin on the tangent.

Note 2: If the equations of the curve be given in the parametric form, say, x = f(t) and y = ϕ(t), then

$$\frac{dy}{dx} = \frac{\frac{dy}{dt}}{\frac{dx}{dt}} = \frac{\phi'(t)}{f'(t)}$$

Therefore, the equation of the tangent at any point 't' on the curve is given by

$$Y - \phi(t) = \frac{\phi'(t)}{f'(t)}\left[X - f(t)\right]$$

Note 3: If the equation of a curve is given in the form f(x, y) = 0, then

$$\frac{dy}{dx} = -\frac{\frac{\partial f}{\partial x}}{\frac{\partial f}{\partial y}}$$

8.2 TANGENTS PARALLEL AND PERPENDICULAR TO THE X-AXIS

The tangent at any point is *parallel* to the x-axis if and only if tan $\psi = 0$ *i.e.*, if any only if $\frac{dy}{dx} = 0$. Again the tangent at any point is *perpendicular* to the x-axis if any only if $\frac{dy}{dx}$ is infinite or if and only if $\frac{dy}{dy} = 0$.

8.3 NORMAL

Definition: *The normal to a curve at any point P on it is the straight line which passes through P and which is perpendicular to the tangent to the curve at P.*

Equation of the Normal: Let P be any given point (x, y) on the curve y = f(x). The gradient of the tangent at (x, y) to the curve $= \frac{dy}{dx}$. Therefore the gradient of the normal at (x, y) to the curve

$$-\frac{-1}{\frac{dy}{dx}} = -\frac{dx}{dy}.$$

Hence, the equation of the normal at (x, y) to the curve is

$$Y - y = -\left(\frac{dx}{dy}\right)(X - x)$$

or $$\left(\frac{dy}{dx}\right)(Y - y) + (X - x) = 0.$$

8.4 ANGLE OF INTERSECTION OF TWO CURVES

The angle of intersection of two curves is defined as the angle between their tangents at their point of intersection.

Let the equations of the two curves be

$$f(x, y) = 0 \qquad ...(1)$$

and $$f(x, y) = 0. \qquad ...(2)$$

Let $\left(\frac{dy}{dx}\right)_1$ stand for the $\frac{dy}{dx}$ of the curve (1) and $\left(\frac{dy}{dx}\right)_2$ for the $\frac{dy}{dx}$ of the curve (2).

Suppose (x_1, x_2) is a point of intersection of (1) and (2). Let m_1 and m_2 be the gradients (or slopes) of the tangents at the point (x_1, y_1) to the curves (1) and (2) respectively. Then

$$m_1 = \left(\frac{dy}{dx}\right)_1 \text{ at the point } (x_1, y_1)$$

and $$m_2 = \left(\frac{dy}{dx}\right)_2 \text{ at the point } (x_1, y_1).$$

Suppose θ is the angle of intersection of (1) and (2) at the point (x_1, y_1).

If $m_1 = m_2$, the angle of intersection θ is 0°. In this case the two curves have the same tangent at th point (x_1, y_1) and thus the two curves touch each other at the point (x_1, y_1).

If $m_1 = \infty$, $m_2 = 0$ or if $m_1 = 0$, $m_2 = \infty$, the angle of intersection θ is 90°.

If $m_1 m_2 = -1$, again the angle of intersection is 90°.

In all other cases, the acute angle θ between the tangents is given by

$$\theta = \tan^{-1}\left|\frac{m_1 - m_2}{1 + m_1 m_2}\right|.$$

If the angle of intersection of two curves is 90°, we say that the curves *intersect orthogonally.*

Important: *The curves (1) and (2) intersect orthogonally at the point (x_1, y_1) if*

$$\left(\frac{dy}{dx}\right)_1 \cdot \left(\frac{dy}{dx}\right)_2 = -1 \text{ at the point } (x_1, y_1).$$

Note: If equatioins (1) and (2) intersect at more than one points, we should find their angle of intersection at each point.

Example 1:

Prove that $\frac{x}{a} + \frac{y}{b} = 1$ touches the curve $y = b\, e^{-x a}$ at the point where the curve crosses the axis of y. **(Kashmir, 1993; Delhi, 96)**

Solution:

Differentiating the equation of the curve $y = b\, e^{-x/a}$, we get

$$\frac{dy}{dx} = be^{-x/a}\left(-\frac{1}{a}\right) - \left(\frac{b}{a}\right)e^{-x/a}.$$

The given curve crosses the axis of y where x = 0. So putting x = 0 in the equation of the curve, we get $y = b,\ e^0 = b.\ 1 = b$.

Thus, the given curve crosses the axis of y at the point (0, b).

Now $$\left(\frac{dy}{dx}\right)_{(0,\ b)} = -\left(\frac{b}{a}\right)e^0 = -\frac{b}{a}$$

Hence, the equation of the tangent to the curve at the point (0, b) is

$y - b = \left(-\frac{b}{a}\right)(x - 0)$ or $ay - ab = -bx$ or $ay + bx = ab$ or $\frac{x}{a} + \frac{y}{b} = 1$, dividing each side by ab.

Example 2:

Find the equation of the tangent at the point (x_1, y_1) to the ellipse

$$\frac{x^2}{a^2} + \frac{y^2}{b^2} = 1.$$ **(Meerut, 1995, 94P)**

Solution:

Differentiating the equation of the given curve, we get

$$\frac{2x}{a^2} + \frac{2y}{b^2}\frac{dy}{dx} = 0 \text{ or } \frac{dy}{dx} = -\frac{b^2x}{a^2y}.$$

$$\therefore \quad \left(\frac{dy}{dx}\right)_{(x_1,\ y_1)} = -\frac{b^2x_1}{a^2y_1}.$$

Hence the required equation of the tangent at the point (x_1, y_1) is

$$y - y_1 = -\frac{b^2x_1}{a^2y_1}(x - x_1) \text{ or } a^2y_1(y - y_1) = -b^2x_1(x - x_1)$$

or $$b^2xx_1 + a^2yy_1 = b^2x_1^2 + a^2y_1^2$$

or $$\frac{xx_1}{a^2} + \frac{yy_1}{b^2} = \frac{x_1^2}{a^2} + \frac{y_1^2}{b^2}$$ [dividing each side by a^2b^2]

or $$\left(\frac{xx_1}{a^2}\right) + \left(\frac{yy_1}{b^2}\right) = 1, \left[\because \text{the point } (x_1, y_1) \text{ lies on } \left(\frac{x^2}{a_2}\right) + \left(\frac{y^2}{b^2}\right) = 1\right].$$

Example 3:

Show that the normal to the curve $5x^5 + 10x^3 + x + 2y + 6 = 0$ at $(0, -3)$ is also a tangent at two other points of the curve.

Solution:

The given curve is

$$5x^5 - 10x^3 + x + 2y + 6 = 0. \qquad ...(1)$$

Differentiating (1), we get

$$25x^4 - 30x^2 + 1 + 2\left(\frac{dy}{dx}\right) = 0. \qquad ...(2)$$

Putting x = 0 and y = – 3 in (2), we get $1 + 2\left(\frac{dy}{dx}\right)_{(0,\,-3)} = 0$ *i.e.*, $\left(\frac{dy}{dx}\right)_{(0,\,-3)} = -\frac{1}{2}$. Therefore the gradient of the normal to (1) at the point (0, – 3) is 2 and the equation of this normal is

$$y + 3 = 2(x - 0) \text{ or } y = 2x - 3. \qquad ...(3)$$

Now, if the normal (3) is to be a tangent to (1) at two other points of (1), it should meet (1) at two pairs of coincident points. Solving (1) and (3), we get $5x^5 - 10x^3 + x + 2(2x - 3) + 6 = 0$ or $x(x - 1)^2 (x + 1)^2 = 0$. Thus, x = 1 and x = – 1 give two pairs of coincident points at which (3) meets (1). Hence (3) touches (1) at the points where x = 1 and – 1.

Example 4:

Find the equation of the normal at the point (x_1, y_1) to the ellipse

$$\left(\frac{x^2}{a^2}\right)+\left(\frac{y^2}{b^2}\right)=1.$$

Solution:

We have $$\frac{dy}{dx}=-\left(\frac{b^2x}{a^2y}\right).$$

Therefore, $$\left(\frac{dy}{dx}\right)_{(x_1,y_1)}=-\frac{b^2x_1}{a^2y_1}.$$

Hence, the equation of the normal at the point (x_1, y_1) is

$$-\left(\frac{b^2x_1}{a^2y_1}\right)(y-y_1)+(x-x_1)=0$$

or $$(x - x_1) = \left(\frac{b^2 x_1}{a^2 y_1}\right)(y - y_1)$$

or $$\frac{x - x_1}{\frac{x_1}{a^2}} = \frac{y - y_1}{\frac{y_1}{b^2}}.$$

Example 5:

Find the equation of the tangent at the point 't' to the curve $x = a \sin^3 t$, $y = b \cos^3 t$. **(Meerut, 1992, 97, 94 P)**

Solution:

We have $$\frac{dx}{dt} = 3a \sin^2 t \cos t,$$

and $$\frac{dy}{dt} = -3b \cos^2 t \sin t.$$

$$\therefore \quad \frac{dy}{dx} = \frac{\frac{dy}{dt}}{\frac{dx}{dt}} = -\frac{3b \cos^2 t \sin t}{3a \sin^2 t \cos t} = -\frac{b \cos t}{a \sin t}.$$

Hence the required equation of the tangent at the point 't' is

$$y - b \cos^3 t = -\frac{b \cos t}{a \sin t}(x - a \sin^3 t)$$

or $$bx \cos t + ay \sin t = ab \sin t \cos t (\sin^2 t + \cos^2 t)$$
$$= ab \sin t \cos t.$$

Dividing each side by ab sin t cos t, we get

$$\frac{x}{a \sin t} + \frac{y}{b \cos t} = 1 \quad \text{or} \quad \left(\frac{x}{a}\right).\text{cosec } t + \left(\frac{y}{b}\right).\sec t = 1.$$

Example 6:

If the normal to the curve $x^{2/3} + y^{2/3}$ *makes an angle* ϕ *with the axis of x, show that its equation is*

$y \cos f - x \sin f = a \cos 2f.$ **(Meerut, 1990, 93, 96 BP, 97)**

Solution:

The given curve is $x^{2/3} + y^{2/3} = a^{2/3}$. ...(1)

Differentiating (1), we get

$$\frac{2}{3}x^{-1/3}+\frac{2}{3}y^{-1/3}\frac{dy}{dx}=0 \text{ or } \frac{dy}{dx}=-\left(\frac{y}{x}\right)^{1/3}.$$

Hence the gradient of the normal to (1) at the point (x, y)

$$=-\frac{dx}{dy}=\left(\frac{x}{y}\right)^{1/3}=\tan\phi \text{ (given).}$$

$$\therefore \qquad x^{1/3}\frac{x^{1/3}}{y^{1/3}}=\frac{\sin\phi}{\cos\phi}$$

$$\text{or} \qquad \cdot\frac{x^{1/3}}{\sin f}=\frac{y^{1/3}}{\cos\phi}=\frac{\sqrt{\left(x^{2/3}+y^{2/3}\right)}}{\sqrt{\left(\sin^2\phi+\cos^2\phi\right)}}=\frac{\sqrt{\left(a^2/3\right)}}{1}, \text{ from (1).}$$

$$\therefore \qquad x^{1/3}=a^{1/3}\sin\phi \text{ or } x=a\sin^3\phi.$$

Similarly, we have $y = a\cos^3\phi$.

Thus, $(a\sin^3\phi, a\cos^3\phi)$ is the point on (1) the normal at which makes an angle ϕ with the axis of x. Hence the required equation of the normal is

$y - a\cos^3\phi = \tan\phi\,(x - a\sin^3\phi)$

or $y\cos\phi - a\cos^4\phi = x\sin\phi - a\sin^4\phi$

or $y\cos\phi - x\sin\phi = a(\cos^4\phi - \sin^4\phi)$

$= a(\cos^2\phi + \sin^2\phi)(\cos^2\phi - \sin^2\phi)$

$= a\cos 2\phi$.

Example 7:

Prove that curve $\left(\frac{x}{a}\right)^n+\left(\frac{y}{b}\right)^n=2$ *touches the straight line* $\left(\frac{x}{a}\right)+\left(\frac{y}{b}\right)=2$ *at the point (a, b) whatever be the value of n.*

(Meerut, 1991, 92, 93, 94)

Solution:

Differentiating the equation of the curve, we get

$$n\left(\frac{x}{a}\right)^{n-1}.\frac{1}{a}+n\left(\frac{y}{b}\right)^{n-1}.\frac{1}{b}\frac{dy}{dx}=0 \text{ or } \frac{dy}{dx}=-\frac{b}{a}\left(\frac{bx}{ay}\right)^{n-1}.$$

$$\therefore \qquad \left(\frac{dy}{dx}\right)_{at(a,\,b)}=-\frac{b}{a}\left(\frac{ba}{ab}\right)^{n-1}=-\frac{b}{a}.$$

Hence equation of the tangent to the curve at the point (a, b) is

$$y - b = \left(-\frac{b}{a}\right)(x - a) \text{ or } bx + ay = 2ab \text{ or } \left(\frac{x}{a}\right) + \left(\frac{y}{b}\right) = 2,$$

which is free from n. Hence the required result follows.

Example 8:

Find the equation of the normal at the point 't' on the curve $x = a \sin^3 t$, $y = b \cos^3 t$. **(Meerut, 1994P, 97)**

Solution:

We get $$\frac{dy}{dx} = -\frac{(b \cos t)}{(a \sin t)}.$$

Therefore the graidient of the normal at the point 't'.

$$= -\left(\frac{dx}{dy}\right) = \frac{(a \sin t)}{(b \cos t)}.$$

Hence the required equation of the normal at the point 't' is

$$y - b \cos^3 t = \frac{a \sin t}{b \cos t}\left(x - a \sin^3 t\right)$$

or $$by \cos t - b^2 \cot^4 t = ax \sin t - a^2 \sin^4 t$$

or $$ax \sin t - by \cos t = a^2 \sin^4 t - b^2 \cos^4 t.$$

Example 9:

Find the equation of the tangent and the normal at the point 't' on the curve $x = a (t + \sin t)$, $y = a (1 - \cos t)$. **(Meerut, 1995)**

Solution:

We have $$\frac{dx}{dt} = a(1 + \cos t), \frac{dy}{dt} = a \sin t.$$

$$\therefore \quad \frac{dy}{dx} = \frac{\frac{dy}{dt}}{\frac{dx}{dt}} = \frac{a \sin t}{a(1 + \cos t)} = \frac{2 \sin \frac{1}{2}t \cos \frac{1}{2}t}{2 \cos^2 \frac{1}{2}t} = \tan \frac{1}{2}t,$$

$\therefore$ the equation of the tangent at the point 't' is

$$y - a(1 - \cos t) = \left(\tan \frac{1}{2}t\right)\{x - a(t + \sin t)\}$$

or $$y - 2a \sin^2 \frac{1}{2}t = \left(\tan \frac{1}{2}t\right)(x - at) - a \sin t \tan \frac{1}{2}t$$

or $$y - 2a\sin^2\frac{1}{2}t = (x - at)\tan\frac{1}{2}t - 2a\sin^2\frac{1}{2}t$$

or $$y = (x - at)\tan\frac{1}{2}t.$$

Again the equation of the normal at the point 't' is

$$y - a(1 - \cos t) = -\frac{1}{\tan\frac{1}{2}t}\{x - a(t + \sin t)\}$$

or $$\left(y - 2a\sin^2\frac{1}{2}t\right)\tan^2\frac{1}{2}t = -x + a(t + \sin t)$$

or $$x + y\tan\frac{1}{2}t = a\left(t + \sin t + 2\sin^2\frac{1}{2}t\tan\frac{1}{2}t\right).$$

Example 10:

Prove that the condition that $x\cos\alpha + y\sin\alpha = p$ touches the curve $x^m y^n = a^{m+n}$ is

$$p^{m+n} m^m n^n = (m + n)^{m+n} a^{m+n} \cos^m\alpha \sin^n\alpha.$$

(Delhi, 1990; Meerut, 97, 94; Lucknow, 97)

Solution:

The given curve is $x^m y^n = a^{m+n}$. ...(1)

Taking logarithm of both sides of (1), we have

$$m\log x + n\log y = (m + n)\log a.$$

Differentiating it, we get

$$\frac{m}{x} + \frac{n}{y}\frac{dy}{dx} = 0 \text{ or } \frac{dy}{dx} = -\left(\frac{m}{n}\right)\left(\frac{y}{x}\right).$$

Suppose the straight line

$$x\cos\alpha + y\sin\alpha = p \qquad ...(2)$$

touches the curve (1) at the point (x_1, y_1).

The equation of the tangent to (1) at the point (x_1, y_1) is

$$y - y_1 = -\frac{my_1}{nx_1}(x - x_1)$$

or $$\left(\frac{mx}{x_1}\right) + \left(\frac{ny}{y_1}\right) = (m + n). \qquad ...(3)$$

Since the equations (2) and (3) represent the same straight line, therefore comparing the coefficients, we get

$$\frac{\cos\alpha}{\left(\frac{m}{x_1}\right)} = \frac{\sin\alpha}{\left(\frac{n}{y_1}\right)} = \frac{p}{(m+n)}$$

i.e., $x_1 = \dfrac{mp}{(m+n)\cos\alpha}$ and $y_1 = \dfrac{np}{(m+n)\sin\alpha}$. ...(4)

Now the point (x_1, y_1) lies on (1). Therefore we have

or $$\left(\frac{mp}{(m+n)\cos\alpha}\right)^m \cdot \left(\frac{np}{(m+n)\sin\alpha}\right)^n = a^{m+n},$$ from (4)

or $$m^m n^n p^{m+n} = a^{m+n} (m+n)^{m+n} \cos^m\alpha \sin^n\alpha.$$

Example 11:

Find the coordinates of the points on the curve $y = x^2 + 3x + 3$ the tangent at which passes through the origin.

Solution:

From the equation of the curve, we have $\dfrac{dy}{dx} = 2x + 3$.

$\therefore$ the equation of the tangent to the given curve at the point (x_1, y_1) is $y - y_1 = (2x_1 + 3)(x - x_1)$. If this tangent passes through the origin (0, 0), we have $0 - y_1 = (2x_1 + 3)(0 - x_1)$

i.e., $$y_1 = 2x_1^2 + 3x_1. \quad ...(1)$$

Since the point (x_1, y_1) lies on the given curve, therefore

$$y_1 = x_1^2 + 3x_1 + 4. \quad ...(2)$$

From (1) and (2), we have $2x_1^2 + 3x_1 = x_1^2 + 3x_1 + 4$ or $x_1^2 = 4$ or $x_1 = \pm 2$. Putting $x_1 = 2$ and -2 in (1), we get $y_1 = 14$ and 2 respectively. Hence the required points are (2, 14) and (– 2, 2).

Example 12:

(a) *Find the tangents to the curve $y = x^3 - 2x^2 + x - 2$ which are (i) parallel to the axis of x, (ii) parallel to the straight line $y = x$ bisecting the angle between the coordinate axes.*

(b) *Show that the abscissae of the points on the curve $y = x(x - 2)(x - 4)$ where the tangents are parallel to the axis of x are given by $x = \pm\left(2/\sqrt{3}\right)$.*

(c) *Find the equations of the tangent and normal to the curve y(x – 2) (x – 3) – x + 7 = 0 at the point where it cuts the axis of x.*

(Meerut, 1995, 96, 99; Avadh, 97)

Solution:

(a) The given curve is

$$y = x^3 - 2x^2 + x - 2. \qquad ...(1)$$

Differentiating (1), we $\frac{dy}{dx} = 3x^2 - 4x + 1$.

(i) If the tangent to (1) at the point (x, y) is parallel to x-axis we have $\frac{dy}{dx} = 0$ *i.e.*, $3x^2 - 4x + 1 = 0$ *i.e.*, $x = 1, \frac{1}{3}$.

Putting $x = 1, \frac{1}{3}$ in (1), we get respectively $y = -2, -\frac{50}{27}$.

(ii) If the tangent to (1) at the point (x, y) is parallel to the line y = x whose gradient is 1, we have $\frac{dy}{dx} = 1$ *i.e.*, $3x^2 - 4x + 1 = 1$ *i.e.*, $(3x - 4) = 0$ *i.e.*, $x = 0, \frac{4}{3}$. From (1), x = 0 gives y = –2 and $x = \frac{4}{3}$ gives $y = -\frac{50}{27}$. Thus the tangents to the curve (2) at the points (0, – 2) and $\left(\frac{4}{3}, -\frac{50}{27}\right)$ are parallel to the line y = x. The equations of these tangents are $y - (-2) = 1(x - 0)$ and $y - \left(-\frac{50}{27}\right) = 1\left(x - \frac{4}{3}\right)$ *i.e.*, $x - y = 2$ and $x - y = \frac{86}{27}$.

(b) The given curve is $y = x(x - 2)(x - 4) = x^3 - 6x^2 + 8x$. We have $\frac{dy}{dx} = 3x^2 - 12x + 8$. The tangent to the given curve at the point (x, y) is parallel to the axis of x if $\frac{dy}{dx} = 0$ *i.e.*, $3x^2 - 12x + 8 = 0$ *i.e.*, $x = \frac{\{12 \pm \sqrt{(144 - 96)}\}}{6} = 2 \pm \left(\frac{2}{\sqrt{3}}\right)$. Hence the abscissae of the required points are $2 \pm \left(\frac{2}{\sqrt{3}}\right)$.

(c) The given curve cuts the axis of x at the point (7, 0).
Differentiating the equation of the curve, we get

$$\frac{dy}{dx}(x^2 - 5x + 6) + y(2x - 5) - 1 = 0.$$

Putting x = 7 and y = 0 in it we have

$$\left(\frac{dy}{dx}\right)_{(7,\,0)} = \frac{1}{20}.$$

$\therefore$ equation of tangent to the given curve at the point (7, 0) is

$$y - 0 = \left(\frac{1}{20}\right)(x - 7) \text{ i.e., } 20y - x + 7 = 0.$$

Also the equation of normal is

$$y - 0 = -20(x - 7) \text{ i.e., } y + 20x - 140 = 0.$$

Example 13:

Show that the condition that the curves $ax^2 + by^2 = 1$ and $a'x^2 + b'y^2 = 1$ should intersect orthogonally is that

$$\frac{1}{a} - \frac{1}{b} = \frac{1}{a'} - \frac{1}{b'}.$$

(Kumaun, 1993; Meerut, 98, 99S, 91P, 92; Lucknow, 92; Avadh, 97)

Solution:

The given curves are

$$ax^2 + by^2 = 1, \quad ...(1)$$

and

$$a'x^2 + b'y^2 = 1. \quad ...(2)$$

Let (x_1, y_1) be a point of intersection of (1) and (2). Then (x_1, y_1) satisfies both (1) and (2). Therefore

$$ax_1^2 + by_1^2 - 1 = 0 \quad ...(3)$$

and

$$a'x_1^2 + b'y_1^2 - 1 = 0 \quad ...(4)$$

Solving (3) and (4), we get

$$\frac{x_1^2}{-b+b'} = \frac{y_1^2}{-a'+a} = \frac{1}{ab'-a'b}. \quad ...(5)$$

Differentiating (1), we get

$$2ax + 2by\left(\frac{dy}{dx}\right) = 0$$

or

$$\frac{dy}{dx} = -\frac{ax}{(by)}.$$

Again differentiating (2), we get $\dfrac{dy}{dx} = -\dfrac{a'x}{(b'y)}.$

Now equation (1) and (2) intersect orthogonally if at their point of intersection (x_1, y_1), we have

$$\left(\frac{dy}{dx}\right)_1 \cdot \left(\frac{dy}{dx}\right)_2 = 1$$

or $$\left(-\frac{a x_1}{b y_1}\right)\left(-\frac{a' x_1}{b' y_1}\right) = -1$$

Putting the values or x_1^2 and y_1^2 from (5) in (6), we get

$$aa'\frac{(-b+b')}{ab'-a'b} + bb'\frac{(a-a')}{ab'-a'b} = 0$$

or $$aa'(b' - b) = bb'(a' - a)$$

or $$\frac{b'-b}{bb'} = \frac{a'-a}{aa'} \text{ or } \frac{1}{b} - \frac{1}{b'} = \frac{1}{a} - \frac{1}{a'}$$

as the required condition.

Example 14:

Find the points on the curve $y = x/(1 - x^2)$, where the tangent is inclined at an angle $\frac{\pi}{4}$ to the x-axis.

Solution:

The given curve is $y = \frac{x}{(1-x^2)}$. ...(1)

Differentiating (1), we get

$$\frac{dy}{dx} = \frac{(1-x^2) - x(-2x)}{(1-x^2)^2} = \frac{1+x^2}{(1-x^2)^2}.$$

If the tangent to (1) at the point (x, y) makes an angle $\frac{\pi}{4}$ with the axis of x, we have $\frac{dy}{dx} = \tan\frac{\pi}{4} = 1$.

$$\therefore \quad \frac{(1+x^2)}{(1-x^2)^2} = 1,$$

or $$x^4 - 3x^2 = 0$$

or $$x^2(x^2 - 3) = 0.$$

$\therefore$ x = 0, $\pm\sqrt{3}$. Putting x = 0, $\sqrt{3}$ and $-\sqrt{3}$ in (1), we get respectively y = 0, $-\frac{\sqrt{3}}{2}$. Hence the required points are (0, 0), $\left(\sqrt{3}, -\frac{\sqrt{3}}{2}\right)$ and $\left(-\sqrt{3}, \frac{\sqrt{3}}{2}\right)$.

Example 15:

Prove that all points of the curve $y^2 = 4a\{x + a\sin(x/a)\}$ at which the tangent is parallel to the axis of x lie on a parabola.

Solution:

The given curve is

$$y^2 = 4a\left\{x + a\sin\left(\frac{x}{a}\right)\right\}. \quad ...(1)$$

Differentiating (1), we get

$$2y\left(\frac{dy}{dx}\right) = 4a\left\{1 + \cos\left(\frac{x}{a}\right)\right\}$$

or $$\frac{dy}{dx} = \left(\frac{2a}{y}\right)\left\{1 + \cos\left(\frac{x}{a}\right)\right\}.$$

Suppose that the tangent to (1) at the point (x_1, y_1) is parallel to x-axis. Then $\frac{dy}{dx}$ at $(x_1, y_1) = 0$

i.e., $$\left(\frac{2a}{y_1}\right)\left\{1 + \cos\left(\frac{x_1}{a}\right)\right\} = 0$$

i.e., $$\cos\left(\frac{x_1}{a}\right) = -1. \quad ...(2)$$

Also (x_1, y_1) lies on (1). Therefore we have

$$y_1^2 = 4a\left\{x_1 + a\sin\left(\frac{x_1}{a}\right)\right\} = 4a(x_1 + 0).$$

$$\left[\because \text{ from (2), } \cos\left(\frac{x_1}{a}\right) = -1 \Rightarrow \sin\left(\frac{x_1}{a}\right) = 0\right]$$

or $$y_1^2 = 4ax_1.$$

Hence, the required locus of (x_1, y_1) is $y^2 = 4ax$ which is a parabola.

Example 16:

If the curve $x^m y^n = a^{m+n}$, prove that the portion of the tangent intercepted between the axes is divided at its point of contact into segments which are in a constant ratio.

Solution:

The given curve is $x^m y^n = a^{m+n}$. ...(1)

Differentiating (1) after taking logarithm of both sides, we get

$$\left(\frac{m}{x}\right)+\left(\frac{n}{y}\right)\left(\frac{dy}{dx}\right)=0$$

or $$\frac{dy}{dx}=\frac{(-my)}{nx}.$$

$\therefore$ the equation of the tangent to (1) at the point (x_1, y_1) is

$$y - y_1 = -\left(\frac{my_1}{nx_1}\right)(x - x_1)$$

or $$nx_1y - nx_1y_1 = -my_1x + mx_1y_1$$

or $$my_1x + nx_1y = (m+n)x_1y_1$$

or $$\frac{x}{\frac{\{(m+n)x_1\}}{m}}+\frac{y}{\frac{\{(m+n)y_1\}}{n}} = 1. \qquad ...(2)$$

The tangent (2) meets the x-axis at the point $\left(\frac{m+n}{m}x_1, 0\right)$ and the y-axis at the point $\left(0, \frac{m+n}{n}y_1\right)$. Let the point of contact (x_1, y_1) divide the line joining these points in the ratio k: 1. Then

$$x_1 = \frac{1.\left(\frac{m+n}{m}x_1\right)+k.0}{k+1} = \frac{(m+n)x_1}{m(k+1)}.$$

Since $x_1 \neq 0$, therefore $1=\frac{m+n}{m(m+1)}$

or $mk + m = m + n$

or $k = \frac{n}{m}$ = constant. Hence the result.

Example 17:

Tangents are drawn from the origin to the curve $y = \sin x$. Prove that their points of contact lie on $x^2y^2 = x^2 - y^2$.

(Meerut, 1990 P, 91S; Lucknow, 93; Kanpur, 91)

Solution:

Differentiating the equation of the given curve $y = \sin x$, we get $\frac{dy}{dx} = \cos x$.

Let (x_1, y_1) be the point of contact of a tangent drawn from the origin to the curve $y = \sin x$. We have $\frac{dy}{dx}$ at $(x_1, y_1) = \cos x_1$. Therefore the equation of the tangent to the curve $y = \sin x$ at the point (x_1, y_1) on it is

$$y - y_1 = \cos x_1.(x - x_1).$$

Since this tangent passes through the origin *i.e.*, the point (0, 0), therefore

$$0 - y_1 = \cos x_1.(0 - x_1) \text{ or } y_1 = x_1 \cos x_1$$

or $$\left(\frac{y_1}{x_1}\right) = \cos x_1 \quad ...(1)$$

Again the point (x_1, y_1) lies on the curve $y = \sin x$. Therefore,

$$y_1 = \sin x_1. \quad ...(2)$$

Squaring and adding (1) and (2), we get

$$\left(\frac{y_1}{x_1}\right)^2 + y_1^2 = 1 \text{ or } x_1^2 y_1^2 = x_1^2 - y_1^2.$$

Hence generalising we get the locus of (x_1, y_1) as

$$x^2y^2 = x^2 - y^2.$$

Example 18:

Prove that the curves $y = e^{-ax} \sin bx$, $y = e^{-ax}$ touch at the points for which $bx = 2m\pi + \frac{1}{2}\pi$, where m is an integer.

Solution:

The given curves are

$$y = e^{-ax} \sin bx \quad ...(1)$$

and $$y = e^{-ax} \quad ...(2)$$

Solving the equations (1) and (2) for x, we get

$$e^{-ax} \sin bx = e^{-ax}$$

or $\sin bx = 1$. [$\because e^{-ax} \neq 0$ for any real number x]

Now, the general solution of the equation $\sin bx = 1$ is given by $bx = 2m\pi + \frac{1}{2}\pi$, where m is any integer.

$\therefore$ at the points where the curves (1) and (2) meet, we have

$$bx = 2m\pi + \frac{1}{2}\pi.$$

Now $$\left(\frac{dy}{dx}\right)_1 = -ae^{-ax}\sin bx + be^{-ax}\cos bx$$

and $$\left(\frac{dy}{dx}\right)_2 = -ae^{-ax}.$$

At the points where the curves (1) and (2) meet, we have $\sin bx = 1$ and so $\cos bx = 0$.

$\therefore$ at the points where the two curves meet, we have

$$\left(\frac{dy}{dx}\right)_1 = -ae^{-ax} = \left(\frac{dy}{dx}\right)_2$$

i.e., the two curves have the same tangent at these points.

Hence the two curves touch at the points for which $bx = 2mx + \frac{1}{2}\pi$.

Example 19:

Find the points at which the tangent to the curve

$$ax^2 + 2hxy + by^2 = 1$$

is (i) parallel to and (ii) perpendicular to the axis of x.

Solution:

Differentiating the equation of the curve, we get

$$2ax + 2h\left(x\frac{dy}{dx} + y\right) + 2by\frac{dy}{dx} = 0 \text{ or } \frac{dy}{dx} = -\left(\frac{ax + hy}{hx + by}\right).$$

(i) If the tangent is parallel to x-axis , we have $\frac{dy}{dx} = 0$

i.e., $ax + hy = 0$. ...(1)

(ii) IF the tangent is $\perp$ to x-axis, we have $\frac{dx}{dy} = 0$

i.e., $hx + by = 0$. ...(2)

Hence the required points are obtained by solving

$$ax^2 + 2hxy + by^2 = 1 \text{ with (1) and (2) respectively.}$$

Example 20:

Prove that normal of the curve $y^2 = 4ax$ touches the curve

$$27ay^2 = 4(x - 2a)^3.$$ **(Kanpur, 1999)**

Solution:

The equation of the normal at any point $(am^2, -2am)$ of the parabola

$$y^2 = 4ax \text{ is } y = mx - 2am - am^3. \qquad ...(1)$$

The parametric equations of the curve $27ay^2 = 4(x - 2a)^3$ can be written as

$$y = 2at^3,\ x = 2a + 3at^2.$$

$$\therefore \quad \frac{dy}{dx} = \frac{\frac{dy}{dt}}{\frac{dx}{dt}} = \frac{6at^2}{6at} = t.$$

$\therefore$ the equation of tangent at the point 't' is

$$y - 2at^3 = t(x - 2a - 3at^2)$$

$$y = tx - 2at - at^3.$$

Example 21:

If $p = x \cos \alpha + y \sin \alpha$ touches the curve

$$\left(\frac{x}{a}\right)^{n/(n-1)} + \left(\frac{y}{b}\right)^{n/(n-1)} = 1,$$

prove that $p^n = (a \cos \alpha)^n + (b \sin \alpha)^n$.

(Delhi, 1993, 91, 99; Allahabad, 90; Meerut, 96P, 91P)

Solution:

The equation of the curve is

$$\left(\frac{x}{a}\right)^{n/(n-1)} + \left(\frac{y}{b}\right)^{n/(n-1)} = 1. \qquad ...(1)$$

Differentiating (1), we get

$$\frac{1}{a^{n/(n-1)}}\cdot\left(\frac{n}{n-1}\right)x^{1/(n-1)} + \frac{1}{b^{n/(n-1)}}\left(\frac{n}{n-1}\right)y^{1/(n-1)}\cdot\frac{dy}{dx} = 0$$

or

$$\frac{dy}{dx} = -\frac{b^{n/(n-1)}}{a^{n/(n-1)}}\cdot\frac{x^{1/(n-1)}}{y^{1/(n-1)}}.$$

Hence the equation of th tangent at (x, y) to the curve (1) is

$$Y - y = -\frac{b^{n/(n-1)}}{a^{n/(n-1)}} \cdot \frac{x^{1/(n-1)}}{y^{1/(n-1)}}(X - x)$$

or $$\frac{Xx^{1/(n-1)}}{a^{n/(n-1)}} + \frac{Yy^{1/(n-1)}}{b^{n/(n-1)}} = \left(\frac{x}{a}\right)^{n/(n-1)} + \left(\frac{y}{b}\right)^{n/(n-1)}$$

$= 1$, form (1). ...(2)

Now suppose the straight line

$$X \cos \alpha + Y \sin \alpha = p, \quad ...(3)$$

touches the curve equation (1) at the point (x, y). Then (2) and (3) are the equations of the same straight line. Therefore comparing the coefficients, we get

$$\frac{\cos \alpha}{\left(\dfrac{x^{1/(n-1)}}{a^{n/(n-1)}}\right)} = \frac{\sin \alpha}{\left(\dfrac{y^{1/(n-1)}}{b^{n/(n-1)}}\right)} = \frac{p}{1}$$

or $$a \cos \alpha = p\left(\frac{x}{a}\right)^{1/(n-1)}$$

and $$b \sin \alpha = p\left(\frac{y}{b}\right)^{1/(n-1)}$$

Raising both sides to the power n and adding, we get

$$(a \cos \alpha)^n + (b \sin \alpha)^n = p^n\left\{\left(\frac{x}{a}\right)^{n/(n-1)} + \left(\frac{y}{b}\right)^{n/(n-1)}\right\}$$

$= p^n.1,$ [$\because$ (x, y) lies on (1)]

p^n, which is the required condition.

Example 22:

If $p = x \cos \alpha + y \sin \alpha$ *touches* $\left(\dfrac{x}{a}\right)^m + \left(\dfrac{y}{b}\right)^m = 1$, *prove that* $p^{m/(m-1)} = (a \cos \alpha)^{m/(m-1)} + (b \sin \alpha)^{m/(m-1)}$.

Solution:

The equation of the curve is $\left(\dfrac{x}{a}\right)^m + \left(\dfrac{y}{b}\right)^m = 1.$ ...(1)

Differentiating (1), we get

$$m\left(\frac{x}{a}\right)^{m-1}\left(\frac{1}{a}\right)+m\left(\frac{y}{b}\right)^{m-1}\left(\frac{1}{b}\right)\left(\frac{dy}{dx}\right)=0$$

or
$$\frac{dy}{dx}=-\frac{\left(\frac{1}{a}\right)\left(\frac{x}{a}\right)^{m-1}}{\left(\frac{1}{b}\right)\left(\frac{y}{b}\right)^{m-1}}.$$

Hence, the equation of the tangent at (x, y) to (1) is

$$Y-y=-\frac{\left(\frac{1}{a}\right)\left(\frac{x}{a}\right)^{m-1}}{\left(\frac{1}{b}\right)\left(\frac{y}{b}\right)^{m-1}}(X-x)$$

or
$$\frac{X}{a}\left(\frac{x}{a}\right)^{m-1}+\frac{Y}{b}\left(\frac{y}{b}\right)^{m-1}=\left(\frac{x}{a}\right)^{m}+\left(\frac{y}{b}\right)^{m}=1, \text{ from (1).} \qquad ...(2)$$

Now, suppose the straight line

$$X\cos\alpha+Y\sin\alpha=p, \qquad ...(3)$$

touches the curve (1) at the point (x, y). Then (2) and (3) are the equations of the same straight line and so comparing the coefficients, we get

$$\frac{\cos a}{\left(\frac{1}{a}\right)\left(\frac{x}{a}\right)^{m-1}}=\frac{\sin a}{\left(\frac{1}{b}\right)\left(\frac{y}{b}\right)^{m-1}}=\frac{p}{1}.$$

$$\therefore \qquad \left(\frac{x}{a}\right)^{m-1}=\frac{(a\cos\alpha)}{p} \text{ and } \left(\frac{y}{b}\right)^{m-1}=\frac{(b\sin\alpha)}{p}.$$

Hence $\left(\frac{a\cos\alpha}{p}\right)^{m/(m-1)}+\left(\frac{b\sin\alpha}{p}\right)^{m/(m-1)}=\left(\frac{x}{a}\right)^{m}+\left(\frac{y}{b}\right)^{m}=1,$

[$\because$ (x, y) lies on (1)].

Thus, we get the required condition.

Example 23:

Find the angle of intersection of the curves

$x^2-y^2=a^2$ *and* $x^2+y^2=a^2\sqrt{2}$. **(Delhi, 1999)**

Solution:

The given curves are

$$x^2 - y^2 = a^2, \qquad ...(1)$$

and $$x^2 + y^2 = a^2\sqrt{2}. \qquad ...(2)$$

Solving equation (1) and (2) to get their points of intersection, we have

$$x^2 = \frac{a^2(\sqrt{2}+1)}{2} \text{ and } y^2 = \frac{a^2(\sqrt{2}-1)}{2}.$$

Hence $x^2y^2 = \frac{a^4(2-1)}{4} = \frac{a^4}{4}$.

Differentiating the equation (1), we get $\frac{dy}{dx} = \frac{(2x)}{(2y)} = \frac{x}{y}$ and differentiating the equation (2), we get $\frac{dy}{dx} = -\frac{x}{y}$. If θ is the angle of intersection of (1) and (2), at their point of intersection (x, y), then

$$\theta = \tan^{-1}\left|\frac{\frac{x}{y} - \left(-\frac{x}{y}\right)}{1 + \left(\frac{x}{y}\right)\left(-\frac{x}{y}\right)}\right| = \tan^{-1}\left|\frac{2xy}{y^2 - x^2}\right|$$

$$= \tan^{-1}\left|\frac{\sqrt{(4x^2y^2)}}{y^2 - x^2}\right| = \tan^{-1}\left|\frac{\sqrt{a^4}}{-a^2}\right|,$$ [$\because$ at the point of intersection (x, y), we have $x^2 - y^2 = a^2$ and $4x^2y^2 = a^4$]

$$= \tan^{-1} |-1| = \tan^{-1} 1 = \frac{\pi}{4}.$$

Example 24:

Find the angles of intersection of the parabolas

$$y^2 = 4ax \text{ and } x^2 = 4by.$$ **(Meerut, 1990, 91)**

Solution:

The given curves are $y^2 = 4ax$, ...(i)

and $x^2 = 4by$. ...(ii)

Solving (i) and (ii), we get on eliminating y

$$x^4 = 64ab^2x \text{ or } x(x^3 - 64ab^2) = 0.$$

$\therefore$ $x = 0$ and $4a^{1/3}\, b^{2/3}$.

Substituting these values of x in (ii), we get

$$y = 0 \text{ for } x = 0$$

and $\quad y = 4a^{2/3}\ b^{1/3}$ for $x = 4a^{1/3}\ b^{2/3}$.

Therefore, (0, 0) and $(4a^{1/3}\ b^{2/3})$, $4a^{2/3}\ b^{1/3})$ are the two points of intersection of equation (i) and (ii).

Differentiating (i), we get

$$2y\frac{dy}{dx} = 4a \quad i.e., \quad \frac{dy}{dx} = \frac{2a}{y}.$$

Differentiating (ii), we have

$$2x = 4b\frac{dy}{dx} \quad i.e., \quad \frac{dy}{dx} = \frac{x}{2b}.$$

Angle of intersection at (0, 0).

$$\frac{dy}{dx} \text{ of (i) at } (0, 0) = \infty$$

and
$$\left(\frac{dy}{dx}\right) \text{ of (ii) at } (0, 0) = 0.$$

$\therefore$ the angle of intersection at (0, 0) is 90°.

Angle of intersection at $(4a^{1/3}\ b^{2/3}, 4a^{2/3}\ b^{1/3})$.

$$\frac{dy}{dx} \text{ or (i) at } (4a^{1/3}\ b^{2/3}, 4a^{2/3}\ b^{1/3}) = \frac{a^{1/3}}{2b^{1/3}},$$

and
$$\frac{dy}{dx} \text{ or (ii) at } (4a^{1/3}\ b^{2/3}, 4a^{2/3}\ b^{1/3}) = \frac{2a^{1/3}}{b^{1/3}}.$$

Therefore, if θ is the acute angle between the tangents to the two curves at the point $(4a^{1/3}\ b^{2/3}, 4a^{2/3}\ b^{1/3})$, then

$$\theta = \tan^{-1}\left|\frac{\frac{2a^{1/3}}{b^{1/3}} - \frac{a^{1/3}}{2b^{1/3}}}{1 + \frac{2a^{1/3}}{b^{1/3}}\cdot\frac{a^{1/3}}{2b^{1/3}}}\right| = \tan^{-1}\frac{3a^{1/3}b^{1/3}}{2\left(a^{2/3} + b^{2/3}\right)}.$$

8.5 LENGTH OF CARTESIA TANGENT, NORMAL, SUBTANGENT AND SUBNORMAL

Let P(x, y) be any point on the curve y = f(x). Let the tangent and normal at P on the curve meet the x-axis in T and N respectively and let PS be th ordinate of the point P. Then ST is called the **Subtangent** and NS the **Subnormal** at the point P of the curve. If ψ be the angle which the tangent at P makes with x-axis, then

$\angle PTS = \angle SPN = y$

and $\tan \psi = \frac{dy}{dx}$.

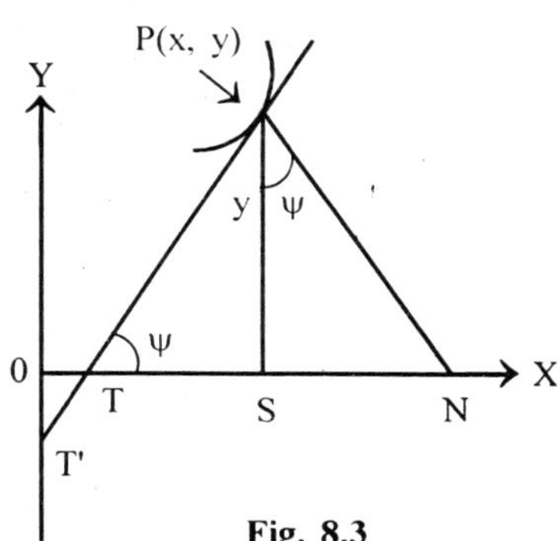

Fig. 8.3

Now at the point P of the curve, we get

length of tangent

$= PT = y \operatorname{cosec} \psi$

$= \psi\sqrt{(1+\cot^2 \psi)}$

$= y\sqrt{\left\{1+\left(\frac{dx}{dy}\right)^2\right\}}.$

Length of Subtangent

$$= TS = y \cot \psi = y\frac{dx}{dy} = \frac{y}{\left(\frac{dy}{dx}\right)}.$$

Length of Normal $= PN = \psi \sec y = \sqrt{(1+\tan^2 \psi)}$

$$= y\sqrt{\left\{1+\left(\frac{dy}{dx}\right)\right\}},$$

and **Length of Subnormal**

$$= SN = y \tan \psi = y\left(\frac{dy}{dx}\right).$$

Intercepts Made by the Tangent on the Coordinates Axes: The equation of the tangent at the point P(x, y) is $Y-y = \left(\frac{dy}{dx}\right)(X - x)$.

It meets the x-axis where Y = 0 *i.e.*, where $0 - y = \left(\frac{dy}{dx}\right)(X - x)$

or X *i.e.*, OT $= \frac{x - \frac{y}{dy}}{dx}$. Again the tangent meets the y-axis, where X = 0 *i.e.* where $Y - y = \left(\frac{dy}{dx}\right)(0 - x)$ or Y *i.e.*, $OT' = y - x\left(\frac{dy}{dx}\right)$. Hence the **intercept cut off by the tangent on x-axis** $= x - \dfrac{y}{\dfrac{dy}{dx}} = x - y\dfrac{dx}{dy}$, and

the intercept cut off by the tangent on y-axis $= y - x\left(\dfrac{dy}{dx}\right)$.

Example 1:

Prove that for the catenary $y = c \cosh\left(\dfrac{x}{c}\right)$, *the perpendicular dropped from the foot of the ordinate upon the tangent is of constant length.*

(Delhi, 1990; Gorakhpur, 1990)

Solution:

Differentiating the given equation equation of the curve, we get

$$\frac{dy}{dx} = \sinh\left(\frac{x}{c}\right).$$

$$\therefore \qquad \tan\psi = \frac{dy}{dx} = \sinh\left(\frac{x}{c}\right).$$

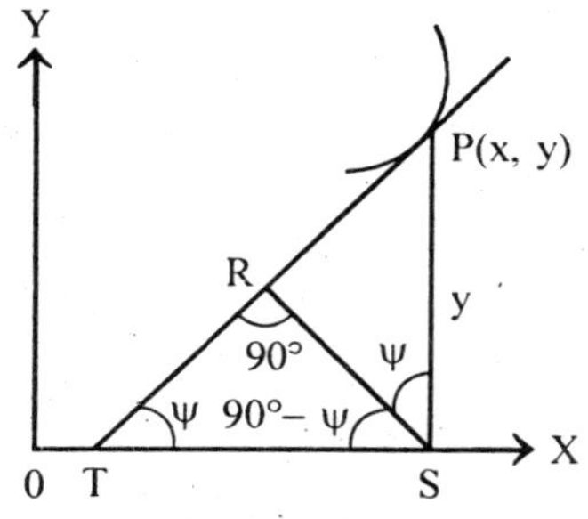

Fig. 8.4

Let P(x, y) be any point on the catenary $y = c \cosh\left(\dfrac{x}{c}\right)$.

Let the tangent at P meet the x-axis the T and let PS be the ordinate of the point *i.e.*, PS = y. Also let R be the foot of the perpendicular drawn from S on the tangent at P. If ψ be the angle which the tangent at P makes with the x-axis, then

$$\angle PTS = \angle PSR = \psi.$$

We are to find the length of SR.

We have, $SR = PS \cos\psi = y \cos\psi = \dfrac{y}{\sec\psi} = \dfrac{y}{\sqrt{(1+\tan^2\psi)}}$

$$= \frac{c\cosh\left(\frac{x}{c}\right)}{\sqrt{\left\{1+\sinh^2\left(\frac{x}{c}\right)\right\}}} \qquad \text{[putting for y and tan ψ]}$$

$$= \frac{c\cosh\left(\frac{x}{c}\right)}{\cosh\left(\frac{x}{c}\right)} = c = \text{constant}.$$

Example 2:

Show that the subtangent at any point of the curve $x^m y^n = a^{m+n}$ varies as the abscissa. **(Delhi, 1991; Meerut, 99)**

Solution:

The given curve is $x^m y^n = a^{m+n}$ ...(1)

Taking logarithm of both sides of (1), we get

$$m \log x + n \log y = (m + n) \log a.$$

Differentiating, we have

$$\frac{m}{x} + \frac{n}{y}\frac{dy}{dx} = 0 \text{ or } \frac{dy}{dx} = -\frac{my}{nx}.$$

$$\text{Now, subtangent } = \frac{y}{\frac{dy}{dx}} = y\left(\frac{-nx}{my}\right) = -\frac{nx}{m}.$$

$$= (\text{some constant}).x.$$

$\therefore$ Subtangent varies as the abscissa x.

Example 2(a):

Show that in the case of the curve $\beta y^2 = (x + \alpha)^3$, the square of the subtangent varies as the subnormal. **(Delhi, 1990; Meerut, 97S)**

Solution:

$$\text{Here} \quad \frac{dy}{dx} = \frac{3(x+\alpha)^2}{2\beta y}.$$

$$\text{Also subtangent } = \frac{y}{\frac{dy}{dx}} = \frac{2\beta y^2}{3(x+\alpha)^2} = \frac{2(x+\alpha)^3}{3(x+\alpha)^2} = \frac{2}{3}(x+\alpha)$$

$$\text{and subnormal } = y.\frac{dy}{dx} = \frac{3(x+\alpha)^2}{2\beta}.$$

$$\therefore \quad \frac{(\text{subtangent})^2}{\text{subnormal}} = \frac{\frac{4(x+\alpha)^2}{9}}{\frac{3(x+\alpha)^2}{2\beta}} = \frac{8\beta}{27} = \text{constant}.$$

Thus, $(\text{subtangent})^2 \propto$ subnormal.

Example 3:

Find the subtangent, subnormal, normal, tangent and the intercepts on the axes at the point 't' on the cycloid $x = a(t + \sin t)$, $y = a(1 - \cos t)$.

Solution:

We have $\dfrac{dy}{dx}=\dfrac{\dfrac{dy}{dt}}{\dfrac{dx}{dt}}=\dfrac{a\sin t}{a(1+\cos t)}$

$$=\frac{2a\sin\frac{1}{2}t\cos\frac{1}{2}r}{2a\cos^2\frac{1}{2}t}=\tan\frac{1}{2}t.$$

$\therefore$ $\tan\psi=\dfrac{dy}{dx}=\tan\dfrac{1}{2}t$ *i.e.*, $\psi=\dfrac{1}{2}t$.

Now, **subtangent** $=\dfrac{y}{\dfrac{dy}{dx}}=\dfrac{a(1-\cos t)}{\tan\frac{1}{2}t}=\dfrac{2a\sin^2\frac{1}{2}t}{\tan\frac{1}{2}t}$

$$=2a\sin\frac{1}{2}t\cos\frac{1}{2}t=a\sin t;$$

subnormal $=y.\left(\dfrac{dy}{dx}\right)=a(1-\cos t)\tan\dfrac{1}{2}t=2a\sin^2\dfrac{1}{2}t\tan\dfrac{1}{2}t;$

length of tangent $=y\operatorname{cosec}\psi=a(1-\cos t)\operatorname{cosec}\dfrac{1}{2}t,\left[\because\psi=\dfrac{1}{2}t\right]$

$$=2a\sin^2\frac{1}{2}t\sec\frac{1}{2}t=2a\sin\frac{1}{2}t\tan\frac{1}{2}t;$$

intercept on x-axis $=x-\dfrac{y}{\dfrac{dy}{dx}}=a(t+\sin t)-\dfrac{a(1-\cos t)}{\tan\frac{1}{2}t}$

$$=a(t+\sin t)-2a\sin^2\frac{1}{2}t\cot\frac{1}{2}t$$

$$=a(t+\sin t)-a\sin t=at$$

and **intercept on y-axis** $=y-x\left(\dfrac{dy}{dx}\right)$

$$=a(1-\cos t)-a(t+\sin t)\tan\frac{1}{2}t$$

$$=2a\sin^2\frac{1}{2}t-at\tan\frac{1}{2}t-a\sin t\tan\frac{1}{2}t$$

$$=2a\sin^2\frac{1}{2}t-at\tan\frac{1}{2}t-2a\sin\frac{1}{2}t\cos\frac{1}{2}t\tan\frac{1}{2}t$$

$$= 2a \sin^2 \frac{1}{2}t - at \tan\frac{1}{2}t - 2a \sin^2 \frac{1}{2}t$$

$$= -at \tan\frac{1}{2}t.$$

Example 4:

If x_1, y_1 be the parts of the axes of x and y intercepted by the tangent at any point (x, y) on the curve $\left(\frac{x}{a}\right)^{2/3} + \left(\frac{y}{b}\right)^{2/3} = 1$, show that

$$\left(\frac{x_1^2}{a^2}\right) + \left(\frac{y_1^2}{b^2}\right) = 1.$$

(Meerut, 1994, 97)

Solution:

The given curve is $\left(\frac{x}{a}\right)^{2/3} + \left(\frac{y}{b}\right)^{2/3} = 1.$ ...(1)

The coordinates of any point (x, y) on (1) may be taken as $x = a \cos^3 t$, $y = b \sin^3 t$, where t is the parameter. **(Note)**

Now $$\frac{dy}{dx} = \frac{\frac{dy}{dt}}{\frac{dx}{dt}} = \frac{3b \sin^2 t \cos t}{-3a \cos^2 t \sin t} = -\frac{b \sin t}{a \cos t}.$$

The equation of the tangent at the point 't' to the curve (1) is

$$y - b \sin^3 t = -\frac{b \sin t}{a \cos t}\left(x - a\cos^3 t\right)$$

or $$bx \sin + ay \cos t = ab \sin t (\cos^2 t + \sin^2 t)$$
$$= ab \sin t \cos t$$

or $$\frac{bx \sin t}{ab \sin t \cos t} + \frac{ay \cos t}{ab \sin t \cos} = 1,$$

[dividing each side by ab sin t cos t]

or $$\frac{x}{a \cos t} + \frac{y}{b \sin t} = 1.$$ ...(2)

If x_1 and y_1 are the intercepts made by the straight line (2) on the axes of x and y respectively, then

$$x_1 = a \cos t \text{ and } y_1 = b \sin t.$$

[Note that we have compared (2) with the equation of a straight line in intercepts form x/a + y/b = 1].

Now $\frac{x_1^2}{a^2}+\frac{y_1^2}{b^2}=\left(a^2\cos^2 t\right)\left(\frac{1}{a^2}\right)+\left(b^2\sin^2 t\right)\left(\frac{1}{b^2}\right)$

$= \cos^2 t + \sin^2 t = 1.$

Example 5 (a):

Show that the subnormal at any point of a parabola is of constant length and the subtangent varies as the abscissa of the point of contact.

(Garhwal, 1993; Meerut, 93)

Solution:

Equation of the parabola is $y^2 = 4ax$. Differentiating it, we have

$$2y\left(\frac{dy}{dx}\right)=4a \text{ or } \frac{dy}{dx}=\frac{2a}{y}.$$

Now subtangent $=\frac{y}{\frac{dy}{dx}}=\frac{y}{\frac{2a}{y}}=\frac{y^2}{2a}=\frac{4ax}{2a}=2x.$

Therefore subtangent $\propto x$.

Again subnormal $=y\left(\frac{dy}{dx}\right)=y\left(\frac{2a}{y}\right)=2a$ which is a constant.

Example 5 (b):

Prove that for the curve $y = be^{x/a}$, the tangent is of constant length and the sub-normal varies as the square of the ordinate.

(Meerut, 1993, 96P)

Solution:

The given curve is $y = be^{x/a}$.

Differentiating it, we have

$$\frac{dy}{dx}=\left(\frac{b}{a}\right)e^{x/a}=\frac{y}{a}. \qquad \textbf{(Note)}$$

$\therefore$ subtangent $=\frac{y}{\frac{dy}{dx}}=\frac{y}{\frac{y}{a}}=a$ = constant.

Also subnormal $=y.\frac{dy}{dx}=y.\frac{y}{a}=\frac{1}{a}y^2.$

$\therefore$ subnormal $\propto y^2$ *i.e.*, square of the ordinate.

Example 6:

Prove that in the ellipse $\frac{x^2}{a^2}+\frac{y^2}{b^2}=1$, *the length of the normal varies inversely as the perpendicular from the origin on the tangent.*

(Delhi, 1992)

Solution:

Differentiating the equation of the given ellipse, we get

$$\frac{dy}{dx}=\frac{\left(-b^2x\right)}{\left(a^2y\right)}.$$

Therefore, the tangent to the ellipse at the point (x, y) is

$$Y-y=\left(-\frac{b^2x}{a^2y}\right)(X-x)$$

or $$b^2xX + a^2yY = b^2x^2 + a^2y^2 = a^2b^2. \qquad ...(1)$$

[$\because$ from the equation of the ellipse, we have $b^2x^2 + a^2y^2 = a^2b^2$].

Now, p = the length of the perpendicular drawn from (0, 0) to the tangent (1)

$$=\frac{a^2b^2}{\sqrt{\left(b^4x^2+a^4y^2\right)}}.$$

Also the length of the normal at the point (x, y) = y sec ψ

$$=y\sqrt{\left\{1+\left(\frac{dy}{dx}\right)^2\right\}}=y\sqrt{\left(1+\frac{b^2x^2}{d^4y^2}\right)}=\frac{\sqrt{\left(a^4y^2+b^4x^2\right)}}{a^2}$$

We have to prove that the length of the normal is proportional to $\frac{1}{p}$ *i.e.*,

(the length of the normal) = (some constant).$\left(\frac{1}{p}\right)$

or (the length of the normal).p = some constant.

Now, (the length of the normal).p

$$=\frac{\sqrt{\left(a^4y^2+b^4x^2\right)}}{a^2}.\frac{a^2b^2}{\sqrt{\left(b^4x^2+a^4x^2\right)}}=b^2 = \text{constant}.$$

Hence the required result follows.

Example 7:

In the tractrix $= x\ a^{x} = a\left(\cos t + \log \tan\frac{1}{2}t\right)$, $y = a \sin t$ *prove that the portion of the tangent intercepted between the curve and x-axis is of constant length.*

(Meerut, 1993; Gorakhpur, 96; Delhi, 94; Allahabad, 91; Agra, 90)

Solution:

Differentiating the given parametric equation ofthe curve w.r.t. t, we get

$$\frac{dx}{dt} = a\left\{-\sin t + \frac{1}{\tan\frac{1}{2}t}.\left(\sec^2\frac{1}{2}t\right).\frac{1}{2}\right\}$$

$$= a\left(-\sin t + \frac{1}{2\sin\frac{1}{2}t\cos\frac{1}{2}t}\right)$$

$$= a\left(-\sin t + \frac{1}{\sin t}\right) = \frac{a\left(1-\sin^2 t\right)}{\sin t} = \frac{a\cos^2 t}{\sin t}$$

and $$\frac{dy}{dt} = a \cos t.$$

$$\therefore \qquad \frac{dy}{dx} = \frac{\frac{dy}{dt}}{\frac{dx}{dt}} = (a \cos t).\frac{\sin t}{a\cos^2 t} = \tan t$$

Hence $$\tan\psi = \frac{dy}{dx} = \tan t, \text{ or } \psi = t.$$

Now the length of the portion of the tangent intercepted between the curve and x-axis = the length of the tangent

$= y \operatorname{cosec}\psi = a \sin t \operatorname{cosec} t,$ $[\because \psi = t, \text{ and } y = a \sin t]$

$= a$ *i.e.*, constant.

Example 8:

Find the abscissa of the point on the curve $ay^2 = x^3$, *the normal at which cuts off equal intercepts from the axes.*

Solution:

Differentiating the equation of the given curve, we get $\frac{dy}{dx} = \frac{3x^2}{(2y)}$. If the normal to the curve at the point (x, y) cuts off equal intercepts from the coordinates axes, then it makes an angle of 45° or 135° with the axis of x. Therefore the gradient of the normal is equal to tan 45° or tan 135°. Hence we have

$$-\frac{dx}{dy} = 1 \text{ or } -1, \qquad \left[\because \text{gradient of the normal} = -\frac{dx}{dy}\right]$$

i.e., $$\frac{dy}{dx} = -1 \text{ or } 1.$$

Squaring we get $\left(\frac{dy}{dx}\right)^2 = 1$

or $$\left\{\frac{3x^2}{(2ay)}\right\}^2 = 1 \text{ or } 4a^2y^2 = 9x^4.$$

Solving this with the equation of the curve $ay^2 = x^3$, we get

$$4ax^3 = 9x^4 \text{ or } x = 0, \frac{4a}{9}.$$

But at x = 0, we have y = 0 and the normal to the curve at (0, 0) does not cut off any intercept from the coordinate axes. Hence, the abscissa of the required point $= \frac{4a}{9}$.

Example 9:

Prove that for the curve $x^{2/3} + y^{2/3} = a^{2/3}$, the portion of the tangent intercepted between the axes is constant length.

(Meerut, 1993; Gorakhpur, 92; Delhi, 92, 97)

Solution:

The equation of the give curve is

$$x^{2/3} + y^{2/3} = a^{2/3}. \qquad ...(1)$$

Differentiating (1) w.r.t. x, we get

$$\left(\frac{2}{3}\right)x^{-1/3} + \left(\frac{2}{3}\right)y^{-1/3}\left(\frac{dy}{dx}\right) = 0$$

or $$\frac{dy}{dx}=\frac{\left(-x^{-1/3}\right)}{\left(y^{-1/3}\right)}.$$

Hence the equation of the tangent to (1) at the point (x, y) is

$$Y-y=-\frac{x^{-1/3}}{y^{-1/3}}(X-x)$$

or $$Yy^{-1/3}+Xx^{-1/3}=x^{2/3}+y^{2/3}=a^{2/3}, \text{ from (1)}$$

or $$\frac{X}{x^{1/3}a^{2/3}}+\frac{y}{y^{1/3}a^{2/3}}=1. \qquad ...(2)$$

If the tangent (2) cuts off the intercepts OP and OQ from the axes of x and y respectively, then

$$OP = x^{1/3}a^{2/3}, \text{ and } OQ = y^{1/3}a^{2/3}.$$ **(Meerut, 1998P)**

Now the length of the tangent intercepted between the axes

$$=PQ=\sqrt{\left(OP^2+OQ^2\right)}=\sqrt{\left(x^{2/3}a^{4/3}+y^{2/3}a^{4/3}\right)}$$

$$=a^{2/3}\sqrt{\left(x^{2/3}+y^{2/3}\right)}=a^{2/3}\sqrt{a^{2/3}},$$ [∵ the point (x, y) lies on (1)]

$$=a^{2/3}a^{1/3}=a=\text{constant}.$$

Example 10:

In the catenary $y = a\cosh\left(\frac{x}{a}\right)$, prove that the length of the portion of the normal intercepted between the curve and the x-axis is $\frac{y^2}{a}$.

(Jiwaji, 1990; Magadh, 96; Utkal, 90)

Solution:

Differentiating the equation of the given catenary, we get

$$\frac{dy}{dx}=a\left\{\sinh\left(\frac{x}{a}\right)\right\}.\left(\frac{1}{a}\right)=\sinh\left(\frac{x}{a}\right).$$

Now, the length ofthe portion ofthe normal intercepted between the curve and the x-axis = the length of the normal

$$=y\sec\psi=y\sqrt{\left(1+\tan^2\psi\right)}=y\sqrt{\left\{1+\left(\frac{dy}{dx}\right)^2\right\}}$$

$$= y\sqrt{\left\{1+\sinh^2\left(\frac{x}{a}\right)\right\}} = y\cosh\left(\frac{x}{a}\right)$$

$$= y.\left(\frac{y}{a}\right) = \frac{y^2}{a}. \qquad \left[\because y = a\cosh\left(\frac{x}{a}\right)\right]$$

Example 10 (a):

In the curve $x^{m+n} = a^{m-n} y^{2n}$, *prove that the mth power of the subtangent varies as the nth power of the subnormal.*

Solution:

The given curve is $x^{m+n} = a^{m-n} y^{2n}$. ...(1)

Taking logarithm of both sides of (1), we get

$(m + n) \log x = (m - n) \log a + 2n \log y.$

Now differentiating w.r.t. x, we get

$$\frac{m+n}{x} = 0 + \frac{2n}{y}.\frac{dy}{dx} \text{ i.e. } \frac{dy}{dx} = \left(\frac{m+n}{2n}\right).\frac{y}{x}. \qquad ...(2)$$

Now, we want to prove that $(\text{subtangent})^m \propto (\text{subnormal})^n$

or $(\text{subtangent})^m/(\text{subnormal})^n$ = constant.

$$= \frac{[y/(dy/dx)]^m}{[y(dy/dx)]^n} = \frac{y^{m-n}}{\left(\frac{dy}{dx}\right)^{m+n}} = y^{m-n}.\frac{(2n)^{m+n} x^{m+n}}{(m+n)^{m+n} y^{m+n}} \qquad \text{[from (2))]}$$

$$= \frac{(2n)^{m+n}}{(m+n)^{m+n}}.\frac{x^{m+n}}{y^{2n}} = \frac{(2n)^{m+n}}{(m+n)} a^{m-n}, \qquad \text{from (1)}$$

= constant.

Hence the required result follows.

Example 11:

If the tangent to the curve $x^{1/2} + y^{1/2} = a^{1/2}$ *at any point on it cuts the axes OX, OY at P, Q respectively prove that*

OP + OQ = a. **(Delhi, 96, 1999; Kanpur, 96; Agra, 95)**

Solution:

The given curve is $\left(\frac{x}{a}\right)^{1/2} + \left(\frac{y}{a}\right)^{1/2} = 1.$ ...(1)

We see that the point $x = a\cos^4 t$, $y = a\sin^4 t$ satisfies the equation (1) for all values of t. Therefore, the coordinates of any point (x, y) on (1) may be taken as

$x = a\cos^4 t$, $y = a\sin^4 t$, where t is the parameter.

Now $$\frac{dy}{dx} = \frac{\frac{dy}{dt}}{\frac{dx}{dt}} = \frac{4a\sin^3 t\cos t}{-4a\cos^3 t\sin t} = -\frac{\sin^2 t}{\cos^2 t}.$$

The equation of the tangent at the point 't' to (1) is

$$y - a\sin^4 t = -\frac{\sin^2 t}{\cos^2 t}\left(x - a\cos^4 t\right)$$

or $$x\sin^2 t + y\cos^2 t = a\sin^2 t\cos^2 t(\cos^2 t + \sin^2 t)$$
$$= a\sin^2 t\cos^2 t$$

or $$\frac{x\sin^2 t}{a\sin^2 t\cos^2 t} + \frac{y\cos^2 t}{a\sin^2 t\cos^2 t} = 1, \text{ d}$$

Dividing each side by $a\sin^2 t\cos^2 t$

or $$\frac{x}{a\cos^2 t} + \frac{y}{a\sin^2 t} = 1. \qquad ...(2)$$

If OP and OQ are the intercepts made by the straight line (2) on the axes of x and y respectively then

$$OP = a\cos^2 t \text{ and } OQ = a\sin^2 t.$$

Hence $OP + OQ = a\cos^2 t + a\sin^2 t = a$.

8.6 POLAR COORDINATES

If P be any point on the xy-plane then its position can be indicated by stating:

(i) its distance r from a fixed point O, and

(ii) the indication θ of the line OP to a fixed straight line OX called the **Initial Line.**

The fixed point O is called the **Pole** and (r, θ) are called the **Polar Coordinates** of the point P.

The distance OP = r is called the **Radius Vector** of P, and ∠XOP = θ is called the **Vectorial Angle** of P.

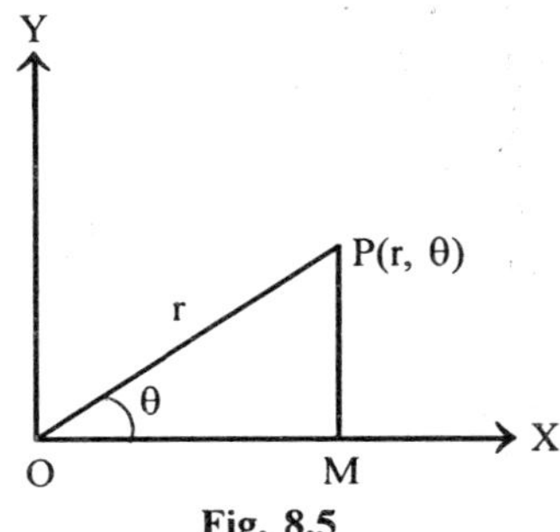

Fig. 8.5

If r and θ are given, there is one and only one point which will have the coordinates (r, θ). If (x, y) be the cartesian coordinates of P, then the formulae of conversion are

$$x = r \cos \theta, \; y = r \sin \theta.$$

Also, we have $x^2 + y^2 = r^2$ and $\theta = \tan^{-1}\left(\frac{y}{x}\right)$.

8.7 ANGLE BETWEEN RADIUS VECTOR AND TANGENT

(Meerut, 1990, 96 BP; Lucknow, 91, 97; Kanpur, 94, 92; Delhi, 96, 94; Allahabad, 99)

Let P(r, θ) and Q(r + δr, θ + δθ) be two neighbouring points on the curve r = f(θ). The line TPT' is tangent to this curve at P. Also let φ be the angle between the tangent at P and the radius vector OP. We have to find φ.

Draw QM perpendicular to OP. As Q → P, we have δr → 0, δθ → 0, the chord PQ tends to the tangent at P and the

$$\angle QPM \to \phi.$$

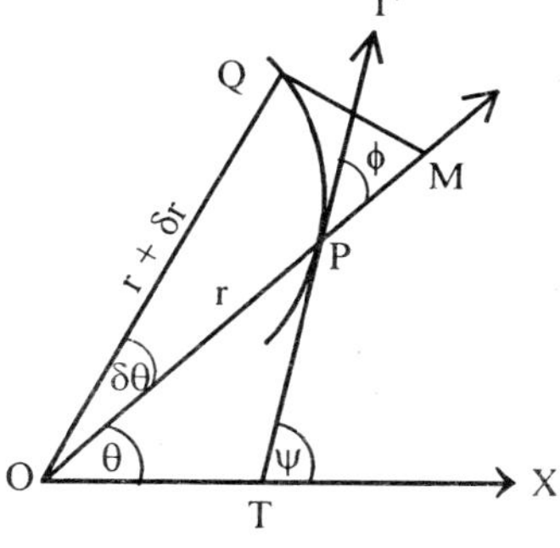

Fig. 8.6

Now $\tan \phi = \lim_{\delta\theta \to 0} \tan \angle QPM$

$$\lim_{\delta\theta\to 0} \frac{QM}{PM} = \lim_{\delta\theta\to 0} \frac{QM}{OM - OP}$$

$$= \lim_{\delta\theta\to 0} \frac{(r+\delta r)\sin \delta\theta}{(r+\delta r)\cos \delta\theta - r}$$

$$= \lim_{\delta\theta\to 0} \frac{(r+\delta r)\left\{(\delta\theta) - (d\theta)^3/3! + \ldots\right\}}{(r+\delta r)\left\{1 - (\delta\theta)^2/2! + \ldots\right\} - r}$$

$= \lim_{\delta\theta\to 0} \frac{r\delta\theta}{\delta r}$, neglecting small quantities of the second and higher order

$$= r\left(\frac{d\theta}{dr}\right).$$

Hence $\tan \phi = r\frac{d\theta}{dr}$ or $\cot \phi = \frac{1}{r}\frac{dr}{d\theta}$.

Note: From the Figure, we have an important relation ψ = θ + φ.

Important: If the equation of a curve is of the form r = f(θ), then differentiating w.r.t. θ after taking logarithm of both sides, we at once get cot φ.

8.8 ANGLE OF INTERSECTION OF TWO POLAR CURVES

Let the two curves $r = f(\theta)$ and $r = F(\theta)$ intersect at P and let the values of ϕ at the point P fo the two curves be ϕ_1 and ϕ_2 respectively. Then the angle of intersection of the two curves at P(*i.e.*, the angle between the tangents to the two curves at P) is evidently equal to $|\phi_1 - \phi_2|$ or $\phi_1 \sim \phi_2$.

If α is the acute angle of intersection of the two curves at P, we have

$$\tan \alpha = |\tan (\phi_1 - \phi_2)| = \left|\frac{\tan \phi_1 - \tan \phi_2}{1 + \tan \phi_1 \tan \phi_2}\right|.$$

$$\therefore \qquad \alpha = \tan^{-1} \left|\frac{\tan \phi_1 - \tan \phi_2}{1 + \tan \phi_1 \tan \phi_2}\right|.$$

In particular, the two curves cut orthogonally *i.e.*, at right angles if $\tan \phi_1 \tan \phi_2 = -1$ or if $\cot \phi_1 \cot \phi_2 = -1$.

8.9 POLAR SUBTANGENT AND POLAR SUBNORMAL

Let the tangent and normal at any point $P(r, \theta)$ on the curve meet the straight line through the pole perpendicular to the radius vector OP in T and N respectively. Then OT is called the **Polar Subtangent** and ON is called the **Polar Subnormal** at P.

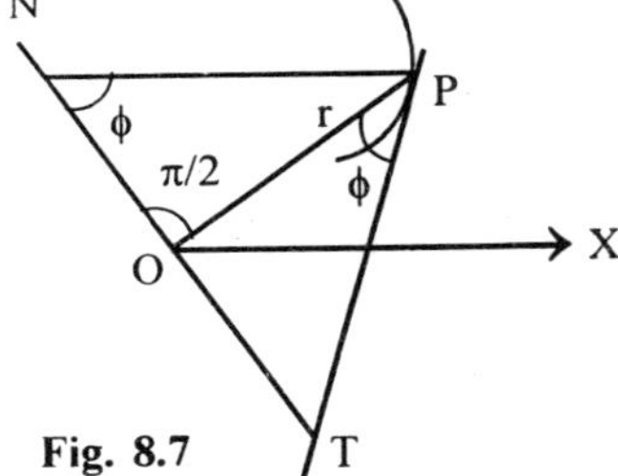

Fig. 8.7

Thus, **polar subtangent**

$$= OT = OP \tan \phi$$

$$= r \tan \phi = r\left(r\frac{d\theta}{dr}\right) = r^2 \frac{d\theta}{dr},$$ **(Agra, 1998)**

and **polar subnormal**

$$= ON = OP \cot \phi = r \cot \phi = r\frac{1}{r}\frac{dr}{d\theta} = \frac{dr}{d\theta}.$$

Als **length of polar tangent**

$$= PT = OP \sec \phi = OP\sqrt{(1 + \tan^2 \phi)}$$

$$= r\sqrt{\left[1 + r^2\left(\frac{d\theta}{dr}\right)^2\right]},$$

and **length of polar normal**

$$= PN = OP \operatorname{cosec} \phi = OP\sqrt{(1 + \cot^2 \phi)}$$

$$= r\sqrt{\left[1+\frac{1}{r^2}\left(\frac{dr}{d\theta}\right)^2\right]} = \sqrt{\left[r^2+\left(\frac{dr}{d\theta}\right)^2\right]}.$$

8.10 LENGTH OF THE PERPENDICULAR FROM POLE TO TANGENT

Let p be the length of perpendicular OT drawn from the pole O to tangent at any point P(r, θ) on the curve

$$r = f(\theta).$$

We have ∠OPT = φ. From the right angled triangle OPT, we have

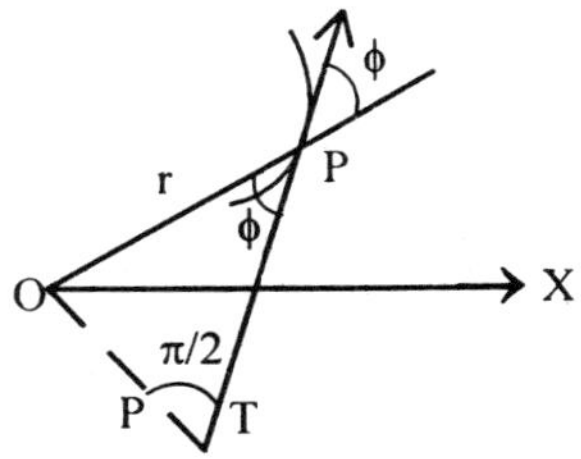

$$\frac{p}{r} = \sin\phi$$

or $$p = r\sin\phi$$

From (1), we have

$$\frac{1}{p^2} = \left(\frac{1}{r^2}\right)\text{cosec}^2\phi$$

$$= \left(\frac{1}{r^2}\right)\left(1+\cot^2\phi\right)$$

Fig. 8.8

$$= \frac{1}{r^2}\left[1+\frac{1}{r^2}\left(\frac{dr}{d\theta}\right)^2\right] = \frac{1}{r^2}+\frac{1}{r^4}\left(\frac{dr}{d\theta}\right)^2.$$

Thus, $$\frac{1}{p^2} = \frac{1}{r^2}+\frac{1}{r^4}\left(\frac{dr}{d\theta}\right)^2.$$ **(Kanpur, 1995)**

Let $u = \frac{1}{r}$. Then $\frac{du}{d\theta} = -\frac{1}{r^2}\frac{dr}{d\theta}$.

∴ $$\frac{1}{p^2} = u^2 = \left(\frac{du}{d\theta}\right)^2$$

8.11 PEDAL EQUATION

The relation between p and r for a given curve is called its pedal equation where r is the radius vector of any point on the curve and p is the length of perpendicular from the pole to the tangent at that point.

The pedal equation of a curve is usually writtena s p = f(r) or r = f(p) or f(p, r) = 0.

Case I: To form the pedal equation of a curve whose cartesian equation is given. Let

$$f(x, y) = 0, \qquad ...(1)$$

be the cartesian equation of the given curve.

The equation of the tangent at any point P(x, y) on the curve (1) is

$$Y - y = \left(\frac{dy}{dx}\right)(X - x)$$

or

$$Y - \left(\frac{dy}{dx}\right).X + \left\{x.\left(\frac{dy}{dx}\right) - y\right\} = 0.$$

$\therefore$

$$p = \left[x\frac{dy}{dx} - y\right] / \sqrt{\left[1 + \left(\frac{dy}{dx}\right)^2\right]} \qquad ...(2)$$

Also $\quad r^2 = OP^2 = x^2 + y^2. \qquad ...(3)$

Eliminating x and y from the equations (1), (2) and (3), we obtain the required pedal equation of the curve.

Case II: The form the pedal equation of a curve whose polar equation is given.

Let $\quad f(r, \theta) = 0, \qquad ...(1)$

be the given polar equation of a curve.

We have $\quad p = r \sin \phi, \qquad ...(2)$

and

$$\cot \phi = \frac{1}{r}\frac{dr}{d\theta}. \qquad ...(3).$$

Eliminating θ and φ between equations (1), (2) and (3), we get the required pedal equation of the curve.

Important: Sometimes we do not get the value of φ from equation (3) in a convenient form. Then instead of using the relations (2) and (3), we can use the single relation

$$\frac{1}{p^2} = \frac{1}{r^2} + \frac{1}{r^4}\left(\frac{dr}{d\theta}\right)^2. \qquad ...(4)$$

Now, eliminating θ between (1) and (4), we get the required pedal equation.

Example 1:

Prove that the spirals $r^n = a^n \cos nq$ and $r^n = b^n \sin n\theta$ intersect orthogonally. **(Garhwal, 1993; Rohilkhand, 96; Kanpur, 95)**

Solution:

The given curves are

$$r^n = a^n \cos n\theta \quad \text{...(1)}$$

and $$r^n = b^n \sin n\theta. \quad \text{...(2)}$$

From the equation of the curve (1), on taking logarithm, we get

$$n \log r = n \log a + \log \cos n\theta.$$

$$\therefore \quad \left(\frac{1}{r}\right)\left(\frac{dr}{d\theta}\right) = -\tan n\theta \text{ or } \cot \phi_1 = -\tan n\theta.$$

Again, from equation (2), we get $n \log r = n \log b + \log \sin n\theta$.

Differentiating w.r.t. θ, we get

$$\frac{n}{r}\frac{dr}{d\theta} = \frac{n \cos n\theta}{\sin n\theta} = n \cot n\theta.$$

$$\therefore \quad \frac{n}{r}\frac{dr}{d\theta} = \frac{n \cos n\theta}{\sin n\theta} = n \cot n\theta \text{ or } \cot \phi_2 = \cot n\theta.$$

Now at the point of intersection (r, θ) of equation (1) and (2), we have $\cot \phi_1 . \cot \phi_2 = (-\tan n\theta).(\cot n\theta) = -1$.

Hence, the curves equations (1) and (2) intersect orthogonally.

Example 2:

Find the angle between the tangent and the radius vector in the case of the curve $r^n = a^n \sec (n\theta + \alpha)$, and prove that this curve is intersected by the curve $r^n = b^n \sec (n\theta + \beta)$ at an angle which is independent of a and b. **(Rohilkhand, 1997; Kanpur, 99; Meerut, 96)**

Solution:

The given curves are

$$r^n = a^n \sec (n\theta + \alpha) \quad \text{...(1)}$$

and $$r^n = b^n \sec (n\theta + \beta) \quad \text{...(2)}$$

Taking logarithm of both sides of (1), we get

$$n \log r = n \log a + \log \sec (n\theta + \alpha).$$

Differentiating with respect to θ, we get

$$\frac{n}{r}\frac{dr}{d\theta} = 0 + \frac{1}{\sec (n\theta + \alpha)} . n \sec(n\theta + \alpha) \tan(n\theta + \alpha)$$

or $$\frac{1}{r}\frac{dr}{d\theta} = \tan (n\theta + \alpha)$$

or $$\cot \phi = \cot\left[\frac{1}{2}\pi - (n\theta + \alpha)\right]$$

or $$\phi = \frac{1}{2}\pi - (n\theta + \alpha),$$

which gives the angle between the tangent and the radius vector in the case of the curve (1).

Let $$\phi_1 = \frac{1}{2}\pi - (n\theta - \alpha).$$

If ϕ_2 denotes the angle between the tangent and the radius vector in the case of the curve (2), then proceeding as above, we have

$$\phi_2 = \frac{1}{2}\pi - (n\theta + \beta).$$

$\therefore$ the angle of intersection of (1) and (2) = $\phi_1 - \phi_2 = \beta - \alpha$ which is independent of a and b.

Example 3:

Show that the pedal equation of the ellipse

$$\frac{x^2}{a} + \frac{y^2}{b^2} = 1 \text{ is } \frac{1}{p^2} = \left(\frac{1}{a^2}\right) + \left(\frac{1}{b^2}\right) - \left(\frac{r^2}{a^2 b^2}\right).$$

(Jhansi, 1999; Kanpur, 1996; Gorakhpur, 93, 98; Lucknow, 99; Meerut, 97)

Solution:

The equation of the curve is $\left(\frac{x^2}{a^2}\right) + \left(\frac{y^2}{b^2}\right) = 1.$ ...(1)

The co-ordinates (x, y) of any point P on (1) may be taken as x = a cos t, y = b sin t, where t is the parameter.

These give $$\frac{dy}{dx} = \frac{\frac{dy}{dt}}{\frac{dx}{dt}} = \frac{b \cos t}{a \sin t}$$

Hence the equation of the tangent to (1) at the point 't' is

$$y - b \sin t = -\frac{b \cos t}{a \sin t}(x - a \cos t)$$

or $$bx \cos t + ay \sin t - ab (\sin^2 t + \cos^2 t) = 0$$

or $$ab - bx \cos t - ay \sin t = 0. \qquad ...(2)$$

$\therefore$ p = the length of perpendicular from (0, 0) to (2)

$$= \frac{ab}{\sqrt{(a^2 \sin^2 t + b^2 \cos^2 t)}}.$$

$\therefore \qquad \dfrac{1}{p^2}=\dfrac{\left(a^2\sin^2 t+b^2\cos^2 t\right)}{a^2b^2}$ ...(3)

Also $\qquad r^2 = x^2 + y^2 = a^2 \cos^2 t + b^2 \sin^2 t$

$= a^2(1 - \sin^2 t) + b^2 (1 - \cos^2 t)$

$= a^2 + b^2 - a^2\sin^2 t - b^2 \cos^2 t$ ...(4)

Eliminating t between (3) and (4), we obtain the pedal equation of (1). From (4), we get

$$a^2 \sin^2 t + b^2 \cos^2 t = (a^2 + b^2) - r^2.$$

Substituting this value in (3), we get

$$\frac{1}{p^2}=\frac{\left\{\left(a^2+b^2\right)-r^2\right\}}{a^2b^2}$$

or $\qquad \dfrac{1}{p^2}=\left(\dfrac{1}{a^2}\right)+\left(\dfrac{1}{b^2}\right)-\left(\dfrac{r^2}{a^2b^2}\right),$

which is the required pedal equation of the ellipse (1).

Example 4:

Find the angle at which the radius vector cuts curve

$r = a(1 - \cos\theta)$. **(Meerut, 1993)**

Solution:

The given curve is

$r = a\ (1 - \cos\theta).$ ...(1)

Taking logarithm of both sides of (1), we get

$$\log r = \log a + \log (1 - \cos\theta).$$

Now differentiating w.r.t. θ, we get

$$\frac{1}{r}\frac{dr}{d\theta}=0+\frac{-(-\sin\theta)}{1-\cos\theta}=\frac{\sin\theta}{1-\cos\theta}$$

$$=\frac{2\sin\frac{1}{2}\theta\cos\frac{1}{2}\theta}{2\sin^2\frac{1}{2}\theta}=\cot\frac{1}{2}\theta.$$

$\therefore \qquad \cot\phi = \cot\frac{1}{2}\theta$ or $\phi = \dfrac{\theta}{2}.$

Example 5:

Find the length of the polar tangent and polar normal for the curve $r = a(1 + \cos\theta)$.

Solution:

The equation of the curve is

$$r = a(1 + \cos\theta). \qquad ...(1)$$

Taking logarithm of both sides of (1), we get

$$\log r = \log a + \log(1 + \cos\theta).$$

Now differentiating w.r.t. θ, we get

$$\cot\phi = \frac{1}{r}\frac{dr}{d\theta} = 0 + \frac{-\sin\theta}{1+\cos\theta} = \frac{-2\sin\frac{1}{2}\theta\cos\frac{1}{2}\theta}{2\cos^2\frac{1}{2}\theta}$$

$$= -\tan\frac{1}{2}\theta = \cot\left(\frac{\pi}{2}+\frac{\theta}{2}\right).$$

$$\therefore \qquad \phi = \frac{\pi}{2}+\frac{\theta}{2}.$$

Now the length of polar tangent

$$= r\sec\phi = r\sec\left(\frac{\pi}{2}+\frac{\theta}{2}\right)$$

$$= r\ \text{cosec}\frac{1}{2}\theta = a(1+\cos\theta)\ \text{cosec}\frac{1}{2}\theta.$$

Again the length of polar normal

$$= r\ \text{cosec}\,\phi = r\ \text{cosec}\left(\frac{\pi}{2}+\frac{\theta}{2}\right)$$

$$= r\sec\frac{\theta}{2} = a(1+\cos\theta)\sec\frac{1}{2}\theta$$

$$= 2a\cos^2\frac{1}{2}\theta\sec\frac{1}{2}\theta = 2a\cos\frac{1}{2}\theta.$$

Example 6:

Find the angle at which the radius vector cuts the curve
$\frac{l}{r} = 1 + e\cos\theta.$ **(Meerut, 1994, 95, 97)**

Solution:

Taking logarithm of both sides of the equation of the curve, we get

$$\log l - \log r = \log(1 + e\cos\theta).$$

Differentiating w.r.t. θ, we get

$$-\frac{1}{r}\frac{dr}{d\theta}=\frac{1}{(1+e\cos\theta)}(-e\sin\theta).$$

$$\therefore \qquad \cot\phi=\frac{1}{r}\frac{dr}{d\theta}=\frac{e\sin\theta}{1+\cos\theta},$$

or $$\tan\phi=\frac{1+e\cos\theta}{e\sin\theta}$$

or $$\phi=\tan^{-1}\left[\frac{(1+e\cos\theta)}{e\sin\theta}\right].$$

Example 7:

Show that in the equiangluar spiral $r = ae^{\theta\cot\alpha}$, the tangent is inclined at a constant angle α to the radius vector.

(Rohilkhand, 1999; Meerut, 94, 99; Gorakhpur, 95)

Solution:

The curve is

$$r = ae^{\theta\cot\alpha}. \qquad ...(1)$$

$$\therefore \qquad \frac{dr}{d\theta}=ae^{\theta\cot\alpha}.\cot\alpha=r\cot\alpha. \qquad [\because \text{ from (1), } r = ae^{\theta\cot\alpha}]$$

Now $$\tan\phi=r\frac{d\theta}{dr}=\frac{r}{\frac{dr}{d\theta}}=\frac{r}{r\cot\alpha}=\tan\alpha.$$

$$\therefore \qquad \phi = \alpha = \text{constant}.$$

Example 8:

Find the angle f for the curve

$$aq = (r^2 - a^2)^{1/2} - a\cos^{-1}\left(\frac{a}{r}\right).$$

(Agra, 1993; Kanpur, 90, 99; Meerut, 91, 95, 97)

Solution:

The given curve is

$$a\theta = (r^2 - a^2)^{1/2} - a\cos^{-1}\left(\frac{a}{2}\right). \qquad ...(1)$$

Differentiating w.r.t. r, we get

$$a\frac{d\theta}{dr}=\frac{1}{2}\left(r^2-a^2\right)^{1/2}.2r+\frac{a}{\sqrt{\left\{1-\left(\frac{a}{r}\right)^2\right\}}}.\left(-\frac{a}{r^2}\right)$$

$$=\frac{r}{\sqrt{\left(r^2-a^2\right)}}-\frac{a^2}{r\sqrt{\left(r^2-a^2\right)}}=\frac{r^2-a^2}{r\sqrt{\left(r^2-a^2\right)}}=\frac{\sqrt{\left(r^2-a^2\right)}}{r}$$

$$\therefore\ r\frac{d\theta}{dr}=\frac{\sqrt{\left(r^2-a^2\right)}}{a}\ \text{or}\ \tan\phi=\frac{\sqrt{\left(r^2-a^2\right)}}{a}.$$

Now $\cos\phi = (\sec\phi)^{-1} = \{1 + \tan^2\phi\}^{-1/2}$

$$=\left\{1-\frac{r^2-a^2}{a^2}\right\}^{-1/2}=\left(\frac{r^2}{a^2}\right)^{-1/2}=\frac{a}{r}.$$

$$\therefore\qquad \phi=\cos^{-1}\left(\frac{a}{r}\right).$$

Note: If not required otherwise, we can also write

$$\phi=\tan^{-1}\left\{\sqrt{\frac{\left(r^2-a^2\right)}{a}}\right\}.$$

Example 9:

Find the polar subtangent for the following curves:

(a) $\frac{l}{r}=1+e\cos\theta$ **(Meerut, 1998)**

(b) $r = ae^{\theta\cot\alpha}$

(c) $r = a(1 + \cos\theta)$ **(Agra, 1996)**

(d) $r = a(1 - \cos\theta)$ **(Meerut, 1991; Agra, 94; Rohilkhand, 91)**

(e) $\frac{2a}{r}=1-\cos\theta.$ **(Agra, 1998)**

Solution:

(a) Differentiating w.r.t. θ, we have

$$\left(-\frac{l}{r^2}\right)\left(\frac{dr}{d\theta}\right)=-e\sin\theta.$$

$$\therefore \qquad \frac{1}{r^2}\frac{dr}{d\theta} = \frac{e \sin\theta}{l}.$$

Now polar subtangent $= r^2\left(\dfrac{d\theta}{dr}\right) = \dfrac{l}{(e \sin\theta)}$.

(b) We have $\dfrac{dr}{d\theta} = ae^{\theta \cot\alpha}.\cot\alpha = r\cot\alpha.$

$\therefore$ polar subtangent

$$= r^2\frac{d\theta}{dr} = r^2.\frac{1}{\frac{dr}{d\theta}} = r^2.\frac{1}{r\cot\alpha} = r\tan\alpha.$$

(c) We have $\dfrac{dr}{d\theta} = -a\sin\theta.$

$\therefore$ polar subtangent

$$= r^2\frac{d\theta}{dr} = a^2(1+\cos\theta)^2.\left(-\frac{1}{a\sin\theta}\right) = -a\frac{\left(2\cos^2\frac{1}{2}\theta\right)^2}{2\sin\frac{1}{2}\theta\cos\frac{1}{2}\theta}$$

$$= -2a\cos^2\frac{1}{2}\theta\cot\frac{1}{2}\theta, \text{ or } = 2a\cos^2\frac{1}{2}\theta\cot\frac{1}{2}\theta$$

on neglecting the negative sign because the polar subtangent is a length.

(d) We have $\dfrac{dr}{d\theta} = a\sin\theta.$

$\therefore$ polar subtangent

$$= r^2\frac{d\theta}{dr} = a^2(1-\cos\theta)^2.\frac{1}{a\sin\theta} = \frac{a^2\left(2\sin^2\frac{1}{2}\theta\right)^2}{2a\sin\frac{1}{2}\theta\cos\frac{1}{2}\theta}$$

$$= 2a\sin^2\frac{1}{2}\theta\tan\frac{1}{2}\theta.$$

(e) Differentiating w.r.t. θ, we get

$$\left(-\frac{2a}{r^2}\right)\left(\frac{dr}{d\theta}\right) = \sin\theta.$$

$\therefore \quad \dfrac{1}{r^2}\dfrac{dr}{d\theta} = -\dfrac{1}{2a}\sin\theta.$ Now polar subtangent

$$= r^2\left(\frac{d\theta}{dr}\right) = -\frac{2a}{\sin\theta} = -2a\,\text{cosec}\,\theta.$$

Since the polar subtangent is a length, therefore neglecting the –ive sign, we get polar subtangent = 2a cosec θ.

Example 10:

Find the polar subtangent and polar subnormal for the curve $r = a\theta$.

(Lucknow, 1990; Kanpur, 98)

Solution:

The equation of the curve is r = aθ. Therefore,

$$\frac{dr}{d\theta} = a. \text{ Also } \frac{d\theta}{dr} = \frac{1}{a}.$$

Now, polar subtangent $= r^2\left(\frac{d\theta}{dr}\right) = \frac{a^2\theta^2}{a} = a\theta^2 = \frac{r^2}{a}$, and polar subnormal $= \frac{dr}{d\theta} = a =$ constant.

Example 11:

Find the angle of intersection of the curves

(a) $r = a\ (1 + \cos\theta)$, $r = b\ (1 - \cos\theta)$.

(Rohilkhand, 1999; Delhi, 98; Meerut, 98S; Kanpur, 98)

(b) $r = a\cos\theta$, $2r = a$.

Solution:

(a) The given curves are $r = a(1 + \cos\theta)$ ...(1)

and $r = b\ (1 - \cos\theta)$. ...(2)

Differentiating equation (1) logarithmically, we get

$$\frac{1}{r}\frac{dr}{d\theta} = -\frac{\sin\theta}{1+\cos\theta} = -\frac{2\sin\frac{1}{2}\theta\cos\frac{1}{2}\theta}{2\cos^2\frac{1}{2}\theta}$$

$$\therefore \quad \cot\phi_1 = \left(\frac{1}{r}\right)\left(\frac{dr}{d\theta}\right) = -\tan\frac{1}{2}\theta = \cot\left(\frac{1}{2}\pi + \frac{1}{2}\theta\right);$$

so that $\phi_1 = \frac{1}{2}\pi + \frac{1}{2}\theta$.

Again differentiating (2) logarithmically, we get

$$\frac{1}{r}\frac{dr}{d\theta} = \frac{\sin\theta}{1-\cos\theta}.$$

$$\therefore \quad \cot \phi_2 = \frac{1}{r}\frac{dr}{d\theta} = \frac{2\sin\frac{1}{2}\theta\cos\frac{1}{2}\theta}{2\sin^2\frac{1}{2}\theta} = \cot\frac{1}{2}\theta \text{; so } \phi_2 = \frac{1}{2}\theta.$$

Now the angle of intersection of (1) and (2) = $|\phi_1 - \phi_2|$

$$= \frac{1}{2}\pi + \frac{1}{2}\theta - \frac{1}{2}\theta = \frac{1}{2}\pi.$$

(b) For the first curve, $\frac{dr}{d\theta} = -a\sin\theta$.

$$\therefore \quad \tan\phi_1 = r\frac{d\theta}{dr} = a\cos\theta.\frac{1}{(-a\sin\theta)} = -\cot\theta = \tan\left(\frac{1}{2}\pi + \theta\right)$$

so that $\phi_1 = \frac{1}{2}\pi + \theta$.

For the second curve, $\frac{dr}{d\theta} = 0$.

$$\therefore \quad \tan\phi_2 = r\frac{d\theta}{dr} = \frac{r}{\frac{dr}{d\theta}} = \frac{\frac{a}{2}}{0} = \infty$$

so that $\quad \phi_2 = \frac{1}{2}\pi$.

Now the angle of intersection of the two curves

$$= \phi_1 \sim \phi_2 = \left(\frac{1}{2}\pi + \theta\right) - \frac{1}{2}\pi = \theta,$$

where θ is to be found at the point of intersection of the two curves. Solving the equations of the two curves for θ, we get

$$\cos\theta = \frac{1}{2} \text{ or } \theta = \frac{1}{3}\pi.$$

Thus, at the point of intersection, we have $\theta = \frac{1}{3}\pi$. Hence at the point of intersection $\theta = \frac{1}{3}\pi$, the required angle of intersection $= \frac{\pi}{3}$.

Example 12:

Find the angle of intersection between the pair of curves $r = 6\cos\theta$ and $r = 2(1 + \cos\theta)$. **(Meerut, 1992)**

Solution:

The given curves are

$$r = 6 \cos \theta \qquad ...(1)$$

and $$r = 2(1 + \cos \theta) \qquad ...(2)$$

From (1), on taking logarithm, we get

$$\log r = \log 6 + \log \cos \theta.$$

$$\therefore \qquad \cos \phi_1 = \frac{1}{r}\frac{dr}{d\theta} = \frac{-\sin \theta}{\cos \theta} = -\tan \theta$$

$$= \cot\left(\frac{1}{2}\pi + \theta\right). \text{ Thus } \phi_1 = \frac{1}{2}\pi + \theta.$$

Again from (2), on taking logarithm, we get

$$\log r = \log 2 + \log (1 + \cos \theta).$$

$$\therefore \quad \cot \phi_2 = \frac{1}{r}\frac{dr}{d\theta} = \frac{-\sin \theta}{1 + \cos \theta} = \frac{-2\sin\frac{1}{2}\theta \cos\frac{1}{2}\theta}{2\cos^2\frac{1}{2}\theta}$$

$$= -\tan\frac{1}{2}\theta = \cot\left(\frac{1}{2}\pi + \frac{1}{2}\theta\right).$$

Thus, $\phi_2 = \frac{1}{2}\pi + \frac{1}{2}\theta.$

Now the angle of intersection of (1) and (2)

$$= \phi_1 \sim \phi_2 = \left(\frac{1}{2}\pi + \theta\right) - \left(\frac{1}{2}\pi + \frac{1}{2}\theta\right) = \frac{1}{2}\theta,$$

where θ is the vectorial angle of the point of intersection of (1) and (2).

Now, to get θ for the point of intersection of (1) and (2), we have on eliminating r between (2) and (2).

$6 \cos \theta = 2(1 + \cos \theta)$ or $1 + \cos \theta = 2 \cos \theta$

or $2 \cos \theta = 1$ or $\cos \theta = \frac{1}{2}$ or $\theta = \frac{1}{3}\pi$.

$\therefore$ the required angle of intersection $= \frac{1}{2}\left(\frac{1}{3}\pi\right) = \frac{\pi}{6}$.

Example 13:

Show that the curves r = a(1 + sin θ) and r = a(1 − sin θ) cut orthogonally. **(Delhi, 1996; Meerut, 97)**

Solution:

The given curves are

$$r = a(1 + \sin \theta) \qquad ...(1)$$

and $$r = a(1 - \sin \theta). \qquad ...(2)$$

From (1), on taking logarithm, we get

$$\log r = \log a + \log (1 - \sin \theta).$$

Now differentiating w.r.t. θ, we get

$$\frac{1}{r}\frac{dr}{d\theta} = 0 + \frac{\cos\theta}{1+\sin\theta} = \frac{\cos\theta}{1+\sin\theta}.$$

$$\therefore \qquad \cot \phi_1 = \frac{\cos\theta}{1+\sin\theta}.$$

Again, from equation (2), we get

$$\log r = \log a + \log (1 - \sin \theta).$$

Differentiating w.r.t. θ, we get

$$\frac{1}{r}\frac{dr}{d\theta} = \frac{-\cos\theta}{1-\sin\theta} \text{ or } \cot\phi_2 = \frac{-\cos\theta}{1-\sin\theta}.$$

$$\text{Now } \cot\phi_1.\cot\phi_2 = \left(\frac{\cos\theta}{1+\sin\theta}\right).\left(\frac{-\cos\theta}{1-\sin\theta}\right)$$

$$= \left(\frac{-\cos^2\theta}{1-\sin^2\theta}\right) = \left(\frac{-\cos^2\theta}{\cos^2\theta}\right) = -1.$$

Hence the curves (1) and (2) intersect orthogonally.

Example 14:

Find the pedal equation of the ellipse

$$\frac{l}{r} = 1 + e \cos \theta.$$

(Rohilkhand, 1992; Agra, 92, 97; Kanpur, 93; Meerut, 98)

Solution:

The given curve is

$$\frac{l}{r} = 1 + e \cos \theta \qquad ...(1)$$

Differentiating (1) w.r.t. θ, we get

$$\frac{-l\, dr}{d\theta} = -e \sin\theta \text{ or } \frac{1}{r^2}\frac{d}{d\theta} = \frac{e\sin\theta}{l} \qquad ...(2)$$

Now $$\frac{1}{p^2}=\left(\frac{1}{p^2}\right)+\left(\frac{1}{r^4}\right)\left(\frac{dr}{d\theta}\right)^2.$$

$$\therefore \quad \frac{1}{p^2}=\frac{1}{r^2}+\left(\frac{1}{r^2}\frac{dr}{d\theta}\right)^2=\frac{1}{r^2}+\frac{e^2\sin^2\theta}{l^2}, \qquad \text{[from (2)]}$$

$$=\frac{1}{r^2}+\frac{e^2}{l^2}\left(1-\cos^2\theta\right)=\frac{1}{r^2}+\frac{e^2}{l^2}-\frac{e^2}{l^2}\cos^2\theta.$$

But from (1), $e\cos\theta=\dfrac{(l-r)}{r}$.

$$\therefore \quad \frac{1}{p^2}=\frac{1}{r^2}+\frac{e^2}{l^2}-\frac{1}{l^2}\left(\frac{l-r}{r}\right)$$

$$=\frac{1}{r^2}+\frac{e^2}{l^2}-\frac{1}{l^2r^2}\left(l^2-2lr+r^2\right)=\frac{1}{r^2}+\frac{e^2}{l^2}-\frac{1}{r^2}+\frac{2l}{l^2r}-\frac{1}{l^2}.$$

Hence $$\frac{1}{p^2}=\frac{1}{l^2}\left(e^2+\frac{2l}{r}-1\right),$$

which is the required pedal equation.

Example 15:

Find the angle of intersection of the parabolas

$$r=\frac{a}{(1+\cos\theta)} \text{ and } r=\frac{b}{(1-\cos\theta)}.$$

(Kanpur, 1998; Rohilkhand, 96; Meerut, 97, 96P)

Solution:

The given curves are

$$r=\frac{a}{(1+\cos\theta)} \qquad ...(1)$$

and $$r=\frac{b}{(1-\cos\theta)}. \qquad ...(2)$$

From (1), on taking logarithm, we get

$$\log r=\log a-\log(1+\cos\theta).$$

Differentiating both sides w.r.t. θ, we get

$$\frac{1}{r}\frac{dr}{d\theta}=-\frac{(-\sin\theta)}{1+\cos\theta}=\frac{2\sin\frac{1}{2}\theta\cos\frac{1}{2}\theta}{2\cos^2\frac{1}{2}\theta}=\tan\frac{1}{2}\theta.$$

$\therefore \qquad \cot\phi_1 = \tan\frac{1}{2}\theta = \cot\left(\frac{1}{2}\pi - \frac{1}{2}\theta\right);$

or $\qquad \phi_1 = \frac{\pi}{2} - \frac{\theta}{2}.$

Again taking larithm of both sides of (2), we get

$$\log r = \log b - \log(1 - \cos\theta).$$

Differentiating w.r.t. θ, we get

$$\frac{1}{r}\frac{dr}{d\theta} = -\frac{(-\sin\theta)}{1-\cos\theta} = \frac{-\sin\theta}{1-\cos\theta}$$

$$= \frac{-2\sin\frac{1}{2}\theta\cos\frac{1}{2}\theta}{2\sin^2\frac{1}{2}\theta} = -\cot\frac{1}{2}\theta.$$

$\therefore \qquad \cot\phi_2 = -\cot\frac{1}{2}\theta = \cot\left(\pi - \frac{1}{2}\theta\right);$

or $\qquad \phi_2 = \pi - \frac{1}{2}\theta.$

Now the angle of intersection of (1) and (2)

$$= \phi_1 \sim \phi_2 = \left(\pi - \frac{1}{2}\theta\right) - \left(\frac{1}{2}\pi - \frac{1}{2}\theta\right) = \frac{\pi}{2}.$$

Hence (1) and (2) intersect orthogonally.

Note: We can also prove the result by showing that

$$\cot\phi_1 . \cot\phi_2 = -1.$$

Example 16:

Find the angle of intersection of the curves

$r^2 = 16\sin 2\theta$ and $r^2\sin 2\theta = 4$. **(Meerut, 1990)**

Solution:

The given curves are

$$r^2 = 16\sin 2\theta, \qquad ...(1)$$

and $\qquad r^2\sin 2\theta = 4. \qquad ...(2)$

From (1), $2\log r = \log 16 + \log\sin 2\theta.$

Therefore $\left(\frac{2}{r}\right)\left(\frac{dr}{d\theta}\right) = 2\cos\frac{2\theta}{\sin 2\theta}$

or $\qquad \cot\phi_1 = \left(\frac{1}{r}\right)\left(\frac{dr}{d\theta}\right) = \cot 2\theta$. Thus, $\phi_1 = 2\theta$.

From (2), $2 \log r + \log \sin 2\theta = \log 4$.

Therefore $\frac{2}{r}\frac{dr}{d\theta} + \frac{2\cos 2\theta}{\sin 2\theta} = 0$

or $\qquad \cot\phi_2 = \frac{1}{r}\frac{dr}{d\theta} = -\cot 2\theta = \cot(\pi - 2\theta)$.

Thus, $\qquad \phi_2 = \pi - 2\theta$.

Now the angle of intersection of (1) and (2) $= \phi_1 \sim \phi_2 = (\pi - 2\theta) - 2\theta = \pi - 4\theta$, where θ is to be found at the point where (1) and (2) intersect.

Eliminating r between (1) and (2), we get $\sin^2 2\theta = \frac{1}{4}$. Therefore, $\sin 2\theta = \pm\frac{1}{2}$. But $\sin 2\theta = -\frac{1}{2}$ is inadmissible because it gives imaginary values of r from (1) and (2). Now $\sin 2\theta = \frac{1}{2}$ gives $2\theta = \frac{1}{6}\pi$ or $\theta = \frac{\pi}{12}$.

Hence the angle of intersection of (1) and (2) at the point $\theta = \frac{\pi}{12}$ $\pi - 4\left(\frac{\pi}{12}\right)$ *i.e.*, $\frac{2\pi}{3}$.

Example 17:

If ϕ be the angle between tangent to a curve and the radius vector drawn from the origin of coordinates to the point of contact, prove that

$$\tan\phi = \frac{x\left(\frac{dy}{dx}\right) - y}{x + y\left(\frac{dy}{dx}\right)}.$$

(Meerut, 1992, 98)

Solution:

We have $\psi = \theta + \phi$. $\therefore \phi = \psi - \theta$.

$$\therefore \tan\phi = \tan(\psi - \theta) = \frac{\tan\psi - \tan\theta}{1 + \tan\psi\tan\theta}$$

$$= \frac{\left(\frac{dy}{dx}\right) - \left(\frac{y}{x}\right)}{1 + \left(\frac{dy}{dx}\right)\left(\frac{y}{x}\right)} = \frac{x\left(\frac{dy}{dx}\right) - y}{1 + y\left(\frac{dy}{dx}\right)}$$

Example 18:

Prove that the normal at any point (r, θ) to the curve $r^n = a^n \cos n\theta$ makes an angle $(n + 1)\,\theta$ with the initial line. **(Kanpur, 1997)**

Solution:

The given curve is

$$r^n = a^n \cos n\theta. \qquad \text{...(1)}$$

Taking logarithm of both sides of (1), we get

$$\frac{n}{r}\frac{dr}{d\theta} = 0 + \frac{1}{\cos n\theta}.(-n \sin n\theta) = -n \tan n\theta$$

or $$\cot \phi = \frac{1}{r}\frac{dr}{d\theta} = -\tan n\theta = \cot\left(\frac{1}{2}\pi + n\theta\right).$$

$\therefore$ $$\phi = \frac{1}{2}\pi + n\theta.$$

If ψ is the angle which the tangent at any point (r, θ) to the curve (1) makes with the initial line, then

$$\psi = \theta + \phi = \theta + \frac{1}{2}\pi + n\theta = \frac{1}{2}\pi + (n + 1)\theta.$$

The slope of the tangent at (r, θ)

$$= \tan \psi = \tan\left[\frac{1}{2}\pi + (n+1)\theta\right] = -\cot (n + 1)\,\theta.$$

$\therefore$ the slope of the normal to (1) at the point (r, θ)

$$= -\frac{1}{-\cot (n+1)\,\theta} = \tan (n+1)\,\theta.$$

Hence the normal to (1) at the point (r, θ) makes in angle $(n + 1)\theta$ with the axis of x *i.e.*, with the initial line.

Example 19:

Find the pedal equations of the curves

(i) $r^n = a^n \sin n\theta$ **(Rohilkhand, 1990, 93)**

(ii) $r^n = \cos n\theta$. **(Meerut, 1990S, 97; Delhi, 91; Agra, 98)**

Solution:

(i) The given curve is $r^n = a^n \sin n\theta$. ...(1)

Taking logarithm of both sides of (1), we get

$$n = \log r = n \log a + \log \sin n\theta.$$

Differentiating w.r.t. θ, we get

$$\frac{n}{r}\frac{dr}{d\theta} = \frac{n \cos n\theta}{\sin n\theta} = n \cot n\theta.$$

$$\therefore \quad \cot \phi = \left(\frac{1}{r}\right)\frac{dr}{d\theta} = \cot n\theta, \text{ or } \phi = n\theta.$$

Now $\pi = r \sin \phi = r \sin n\theta$. ...(2)

Eliminating θ between (1) and (2), we get the pedal equation of (1).

From (2), $\sin n\theta = \frac{p}{r}$. Substituting this value in (1), we get $r^n = a^n \left(\frac{p}{r}\right)$ or $r^{n+1} = pa^n$, which is the required pedal equation.

(ii) The given curve is $r^n = a^n \cos n\theta$. ...(1)

Taking logarithm, we get $n \log r = n \log a + \log \cos n\theta$.

Differentiating w.r.t. θ, we get

$$\frac{n}{r}\frac{dr}{d\theta} = \frac{-n \sin n\theta}{\cos n\theta} = -n \tan n\theta.$$

$$\therefore \cot \phi = \frac{1}{r}\frac{dr}{d\theta} = -\tan n\theta = \cot\left(\frac{1}{2}\pi + n\theta\right); \text{ or } \phi = \frac{1}{2}\pi + n\theta.$$

Now $\pi = r \sin \phi = r \sin \left(\frac{1}{2}\pi + n\theta\right) = r \cos n\theta$. ...(2)

From (2), we have $\cos n\theta = \frac{p}{r}$. Substituting this value in (1), we get $r^n = a^n\left(\frac{p}{r}\right)$ or $r^{n+1} = pa^n$, which is the required pedal equation.

Important Note: The find the pedal equation of a polar curve it is often convenient to find the angle φ and then to use the relation $\pi = r \sin \phi$. But if the angle φ cannot be obtained in a convenient form, we use the relation

$$\frac{1}{p^2} = \frac{1}{r^2} + \frac{1}{r^4}\left(\frac{dr}{d\theta}\right)^2.$$

Example 20:

Prove that the locus of the extremity of the polar subnormal of the curve $r = f(\theta)$ is $r = f'\left(\theta - \frac{\pi}{2}\right)$. Hence show that the locus of the extremity of the polar subnormal of the equiangular spiral $r = ae^{m\theta}$ is another equiangular spiral. **(Kanpur, 1999; Meerut, 98)**

Solution:

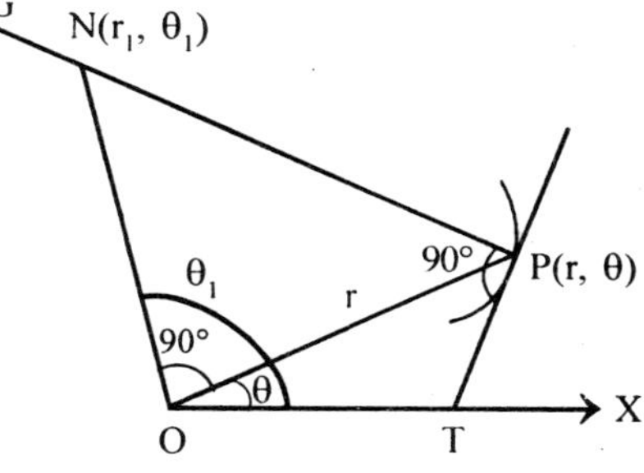

Fig. 8.9

First Part: Let P(r, θ) be any point on the curve $r = f(\theta)$ and PT and PG be the tangent and normal respectively at P. Let the straight line perpendicular to OP through O meet PG in $N(r_1, \theta_1)$.

Now, r_1 = ON = Polar subnormal of the curve $r = f(\theta)$ at the point P

$$= \frac{dr}{d\theta} = f'(\theta). \qquad ...(1)$$

Also $\theta_1 = \theta + \frac{\pi}{2}$; $\therefore\ \theta = \theta_1 - \frac{\pi}{2}$. ...(2)

Eliminating θ between (1) and (2), we get

$$r_1 = f'\left(\theta_1 - \frac{\pi}{2}\right).$$

∴ the locus of the point (r_1, θ_1) is

$$r = f'\left(\theta - \frac{\pi}{2}\right). \qquad ...(3)$$

Second Part: Comapring the equation of the curve $r = ae^{m\theta}$ with the equation $r = f(\theta)$, we have $f(\theta) = ae^{m\theta}$.

$$\therefore \qquad f'(\theta) = ame^{m\theta}.$$

$$\therefore \qquad f'\left(\theta - \frac{\pi}{2}\right) = ame^{m(\theta - \pi/2)} = ame^{m\theta}.e^{-m\pi/2}$$

$$= \lambda e^{m\theta}, \text{ where } \lambda\ ame^{-m\pi/2}.$$

From (3), the required locus is $r = \lambda e^{m\theta}$, which is also an equilangular spiral.

Example 21:

Prove that for the parabola $\frac{2a}{r} = (1 - \cos\theta)$,

(a) $f = \pi - \frac{1}{2}\theta$,

(b) $p = a \operatorname{cosec}\frac{1}{2}\theta$,

(c) $p^2 = ar$. **(Kanpur, 1999; Agra, 96, Jhansi, 98)**

Solution:

The given curve is $\frac{2a}{r} = 1 - \cos\theta$. ...(1)

Taking logarithm of both sides of (1), we get

$$\log 2a - \log r = \log(1 - \cos\theta).$$

Differentiating w.r.t. θ, we get

$$-\frac{1}{r}\frac{dr}{d\theta} = \frac{-(-\sin\theta)}{1-\cos\theta} = \frac{2\sin\frac{1}{2}\theta\cos\frac{1}{2}\theta}{2\sin^2\frac{1}{2}\theta} = \cot\frac{1}{2}\theta.$$

$$\therefore\ \cot\phi = \frac{1}{r}\frac{dr}{d\theta} - \cot\frac{1}{2}\theta = \cot\left(\pi - \frac{1}{2}\right).$$

Hence $\phi = \pi - \frac{1}{2}\theta$. **[The result (a) proved]**

Now $p = r\sin\phi = r\sin\left(\pi - \frac{1}{2}\theta\right) = r\sin\frac{1}{2}\theta$.

From (1), we have

$$r = \frac{2a}{(1-\cos\theta)} = \frac{2a}{\left(2\sin^2\frac{1}{2}\theta\right)} = \frac{a}{\left(\sin\frac{1}{2}\theta\right)} = a\ \text{cosec}\frac{1}{2}\theta.$$

The proves the result (b).

Now $p = r\sin\frac{1}{2}\theta$ and $r = \frac{a}{\left(\sin^2\frac{1}{2}\theta\right)}$. Eliminating θ between these we get

$$r = \frac{a}{\frac{p^2}{r^2}} = \frac{ar^2}{p^2} \qquad \therefore\ p^2 = \frac{ar^2}{r} = ar.$$

This proves the result (c).

Example 22:

Find the pedal equation of the parabola

$y^2 = 4a(x + a)$. **(Kanpur, 1999; Lucknow, 1991; Agra, 99)**

Solution:

The given curve is

$$y^2 = 4a(x + a) \qquad ...(1)$$

$$\frac{2y\,dy}{dx} = 4a$$

or $$\frac{dy}{dx}=\frac{2a}{y}.$$

Therefore, the equation of the tangent at (x, y) to (1) is

$$Y-y=\left(\frac{2a}{y}\right)(X-x)$$

or $$\left(\frac{2a}{y}\right)X-Y+y-\left(\frac{2a}{y}\right)x=0. \qquad ...(2)$$

$\therefore$ p = the length of the perpendicular from (0, 0) on (2)

$$=\left[y-\frac{2ax}{y}\right]\Big/\left(1+\frac{4a^2}{y^2}\right)^{1/2}=\frac{(y^2-2ax)}{\sqrt{(y^2+4a^2)}}$$

$$=\frac{4a(x+a)-2ax}{\sqrt{[4a(x+a)+4a^2]}}, \qquad [\because \text{ from (1), } y^2=4a(x+a)]$$

$$=\frac{2ax+4a^2}{\sqrt{[4a(x+2a)]}}=\frac{2a(x+2a)}{\sqrt{[4a(x+2a)]}}=\sqrt{[a(x+2a)]}.$$

Also $r^2 = x^2 + y^2 = x^2 + 4a(x + a) = (x + 2a)^2$; $r = (x + 2a)$.

Now $p^2 = a(x + 2a) = ar$. Hence $p^2 = ar$ is the required pedal equation of (1).

Example 23:

Find the pedal equation of the curve

$r = ae^{\theta \cot \alpha}$. **(Meerut, 1993; Utkal, 90)**

Solution:

The given curve is $r = ae^{\theta \cot \alpha}$ ...(1)

Taking logarithm of both sides of (1), we get

$\log r = \log a + \theta \cot \alpha \log e = \log a + \theta \cot \alpha$. $[\because \log e = 1]$

Differentiating w.r.t. θ, we get

$$\left(\frac{1}{r}\right)\frac{dr}{d\theta}=0+\cot\alpha=\cot\alpha.$$

$$\therefore \qquad \cot\phi=\left(\frac{1}{r}\right)\frac{dr}{d\theta}=\cot\alpha; \text{ or } \phi=\alpha.$$

Now $p = r\sin\phi = r\sin\alpha$. Hence $p = r\sin\alpha$ is the required pedal equation.

Example 24:

For the cardioid $r = a(1 - \cos\theta)$, prove that

(a) $\phi = \frac{1}{2}\theta$, **(Meerut, 1991, 93, 94, 96; Rohilkhand, 91)**

(b) $2ap^2 = r^3$. **(Garhwal, 1993; Meerut, 1990, 96; Agra, 95)**

Solution:

The tiven curve is

$$r = a(1 - \cos\theta). \qquad ...(1)$$

$$\frac{dr}{d\theta} = a \sin\theta.$$

(a) We have

$$\tan\phi = r\frac{d\theta}{dr} = \frac{a(1-\cos\theta)}{a\sin\theta} = \frac{2a\sin^2\frac{1}{2}\theta}{2a\sin\frac{1}{2}\theta\cos\frac{1}{2}\theta} = \tan\frac{\theta}{2}$$

$$\therefore\ \phi = \frac{\theta}{2}.$$

(b) We have

$$p = r\sin\phi = r\sin\frac{\theta}{2}, \qquad \left[\because \phi = \frac{\theta}{2}\right]$$

Now from (1), $r = 2a\sin^2\frac{1}{2}\theta = 2a\left(\frac{p^2}{r^2}\right)$, $\qquad \left(\because p = r\sin\frac{1}{2}\theta\right)$

$$\therefore\ 2ap^2 = r^3.$$

Example 25:

Find the pedal equation of the following curves:

(a) $x^2 + y^2 = 2ax$.

(b) $x^{2/3} + y^{2/3} = a^{2/3}$. **(Meerut, 1994)**

(c) $x^2 - y^2 = a^2$.

(d) $r^2 = a^2 \cos 2\theta$.

(e) $r^2 \cos 2\theta = a^2$. **(Delhi, 1992; Rohilkhand, 95)**

(f) $r = \text{sech}\, n\theta$.

(g) $r = a(1 + \cos\theta)$ **(Rohilkhand, 1992)**

(h) $r^m \cos m\theta = a^m$. **(Lucknow, 1997; Indore, 90)**

Solution:

(a) The given curve is $x^2 + y^2 = 2ax$. Changing the equation to polar co-ordinates by putting $x = r \cos \theta$ and $y = r \sin \theta$, we get

$$r^2 = 2ar \cos \theta \text{ or } r = 2a \cos \theta. \qquad ...(1)$$

$$\therefore \quad \frac{dr}{d\theta} = -2a \sin \theta.$$

Now $\tan \phi = r\dfrac{d\theta}{dr} = \dfrac{2a \cos \theta}{-2a \sin \theta} = -\cot \theta = \tan\left(\dfrac{1}{2}\pi + \theta\right);$

so that $\phi = \dfrac{1}{2}\pi + \theta.$

We have $p = r \sin \phi = r \sin\left(\dfrac{1}{2}\pi + \theta\right) = r \cos \theta = r\left(\dfrac{r}{2a}\right)$, from (1).

Hence $2ap = r^2$ is the required pedal equation.

(b) The given curve is $x^{2/3} + y^{2/3} = a^{2/3}$. ...(1)

The co-ordinates (x, y) of any point P on (1) may be taken as

$$x = a \cos^3 t, \; y = a \sin^3 t,$$

where t is the parameter. These give

$$\frac{dy}{dx} = \frac{\frac{dy}{dt}}{\frac{dx}{dt}} = \frac{3a \sin^2 t \cos t}{-3a \cos^2 t \sin t} = -\frac{\sin t}{\cos t}.$$

Hence the equation of the tangent to (1) at the point 't' is

$$y - a \sin^3 t = -\frac{\sin t}{\cos t}\left(x - a \cos^3 t\right)$$

or $\quad x \sin t + y \cos t = a \sin t \cos t (\cos^2 t + \sin^2 t)$

$$= a \sin t \cos t. \qquad ...(2)$$

$\therefore$ p = the length of the perpendicular from (0, 0) to (2)

$$= \frac{a \sin t \cos t}{\sqrt{\left(\sin^2 t + \cos^t\right)}} = a \sin t \cos t \qquad ...(3)$$

Now $r^2 = x^2 + y^2 = a^2 \cos^6 t + a^2 \sin^6 t = a^2 [(\cos^2 t)^3 + (\sin^2 t)^3]$
$= a^2[(\cos^2 t + \sin^2 t)^3 - 3 \cos^2 t \sin^2 t (\cos^2 t + \sin^2 t)]$

$$= a^2\left[1 - 3\left(\frac{p^2}{a^2}\right).1\right], \qquad \left[\because \text{ from (3), } \cos^2 t \sin^2 t = \frac{p^2}{a^2}\right]$$

$= a^2 - 3p^2.$

Hence the required pedal equation is $r^2 = a^2 - 3p^2$.

(c) The given curve is $x^2 - y^2 = a^2$. Changing to polar co-ordinate, the equation becomes

$r^2 \cos^2 \theta - r^2 \sin^2 \theta = a^2$ or $r^2 \cos^2 \theta = a^2$ or $r^2 \cos 2\theta = a^2$ ...(1)

Taking logarithm of (1), we get

$$2 \log r + \log \cos 2\theta = 2 \log a.$$

Differentiating w.r.t. θ, we have

$$\frac{2}{r}\frac{dr}{d\theta} - 2\frac{\sin 2\theta}{\cos 2\theta} = 0 \text{ or } \frac{1}{r}\frac{dr}{d\theta} = \tan 2\theta.$$

$$\therefore \cos \cot \phi = \left(\frac{1}{r}\right)\left(\frac{dr}{d\theta}\right) = \tan 2\theta = \cot\left(\frac{1}{2}\pi - 2\theta\right);$$

so that $\phi = \frac{1}{2}\pi - 2\theta$.

Now $p = r \sin \phi = r \sin\left(\frac{1}{2}\pi - 2\theta\right) = r \cos 2\theta$

$$= r\left(\frac{a^2}{r^2}\right) = \frac{a^2}{r}, \qquad \left[\because \text{ from (1)}, \cos 2\theta = \frac{a^2}{r^2}\right]$$

Hence the required pedal equation is $pr = a^2$.

(d) The given curve is $r^2 = a^2 \cos 2\theta$. ...(1)

Taking logarithm of (1), we get

$$2 \log r = 2 \log a + \log \cos 2\theta.$$

Differentiating w.r.t. θ, we get

$$\frac{2}{r}\frac{dr}{d\theta} = 0 - 2\frac{\sin 2\theta}{\cos 2\theta} \text{ or } \frac{1}{r}\frac{dr}{d\theta} = -\tan 2\theta.$$

$$\therefore \cot \phi = \frac{1}{r}\frac{dr}{d\theta} - \tan 2\theta = \cot\left(\frac{1}{2}\pi + 2\theta\right);$$

so that $\phi = \frac{1}{2}\pi + 2\theta$.

Now $p = r \sin \phi = r \sin\left(\frac{1}{2}\pi + 2\theta\right) = r \cos 2\theta = r\left(\frac{r^2}{a^2}\right)$, from (1).

Hence the required pedal equation is $pa^2 = r^3$.

(e) See part (c) of this Example.

(f) The given curve is $r = a \operatorname{sech} n\theta$ ...(1)

Differentiating (1), $\frac{dr}{d\theta} = -an \operatorname{sech} n\theta \tanh n\theta$.

Now $\frac{1}{p^2}=\frac{1}{r^2}+\frac{1}{r^4}\left(\frac{dr}{d\theta}\right)^2=\frac{1}{r^2}+\frac{1}{r^4}a^2n^2\operatorname{sech}^2 n\theta\tanh^2 n\theta$

$=\frac{1}{r^2}+\frac{n^2}{r^4}\left(a^2\operatorname{sech}^2 n\theta\right)\left(1-\operatorname{sech}^2 n\theta\right)[\because \operatorname{sech}^2 n\theta = 1 - \tanh^2 n\theta]$

$=\frac{1}{r^2}+\frac{n^2}{r^4}.r^2\left(1-\frac{r^2}{a^2}\right),$ [by (1)]

$=\frac{1}{r^2}+\frac{n^2}{r^2}-\frac{n^2}{a^2}=\left(n^2+1\right)\frac{1}{r^2}-\frac{n^2}{a^2}$

$\therefore$ the pedal equation is $\frac{1}{p^2}=\frac{\left(n^2+1\right)}{r^2}-\frac{n^2}{a^2}$ which is of the form

$\frac{1}{p^2}=\left(\frac{A}{r^2}\right)+B.$

(g) The given curve is $r = a(1 + \cos\theta)$. ...(1)

Taking logarithm, we get $\log r = \log a + \log(1 + \cos\theta)$.

Differentiating w.r.t. θ, we have

$$\frac{1}{r}\frac{dr}{d\theta}=\frac{-\sin\theta}{1+\cos\theta}=\frac{-2\sin\frac{1}{2}\theta\cos\frac{1}{2}\theta}{2\cos^2\frac{1}{2}\theta}=-\tan\frac{1}{2}\theta.$$

$$\therefore \cot\phi=\left(\frac{1}{r}\right)\left(\frac{dr}{d\theta}\right)=-\tan\frac{1}{2}\theta=\cot\left(\frac{1}{2}\pi+\frac{1}{2}\theta\right);$$

so that $\phi=\frac{1}{2}\pi+\frac{1}{2}\theta$.

Now $p = r\sin\phi = r\sin\left(\frac{1}{2}\pi+\frac{1}{2}\theta\right)=r\cos\frac{1}{2}\theta$. ...(2)

Eliminating θ between (1) and (2), we get the required pedal equation.

From (1), we have $r = 2a\cos^2\frac{1}{2}\theta=2a\left(\frac{p}{r}\right)^2$,

$\left[\because \text{from (2)}, \cos\frac{1}{2}\theta=\frac{p}{r}\right]$.

Hence the required pedal equation is $r^3 = 2ap^2$.

(h) The given curve is $r^m\cos m\theta = a^m$ ...(1)

Taking log, we get $m\log r + \log\cos m\theta + m\log a$.

Differentiating w.r.t. θ we have $\frac{m}{r}\frac{dr}{d\theta} = \frac{m \sin m\theta}{\cos m\theta} = 0$.

$$\therefore \cot\phi = \left(\frac{1}{r}\right)\left(\frac{dr}{d\theta}\right) = \tan m\theta = \cot\left(\frac{1}{2}\pi = m\theta\right);$$

so that $\phi = \frac{1}{2}\pi - m\theta$.

Now $p = r \sin\phi = r \sin\left(\frac{1}{2}\pi = m\theta\right) = r \cos m\theta = r\left(\frac{a^m}{r^m}\right)$ [by (1)]

Hence the required pedal equation is

$$p = a^m/r^{m-1} \text{ or } pr^{m-1} = a^m.$$

8.12. DIFFERENTIAL COEFFICIENT OF ARE LENGTH CARTESIAN CO-ORDINATES

Let P(x, y) and Q (x + δx, y + δy) be two neighbouring points on the curve y = f(x). Let the arc length AP = s, measured from a fixed point A on the curve. Then

(i) $\frac{dx}{dx} = \sqrt{\left\{1+\left(\frac{dy}{dx}\right)^2\right\}}$,

(ii) $\frac{ds}{dy} = \sqrt{\left\{1+\left(\frac{dx}{dy}\right)^2\right\}}$,

(iii) $\frac{ds}{dt} = \sqrt{\left\{\left(\frac{dx}{dt}\right)^2+\left(\frac{dy}{dt}\right)^2\right\}}$,

(iv) $\sin\psi = \frac{dy}{dx}$,

(v) $\cos\psi = \frac{dx}{ds}$,

(vi) $\left(\frac{dx}{ds}\right)^2+\left(\frac{dy}{ds}\right)^2 = 1$.

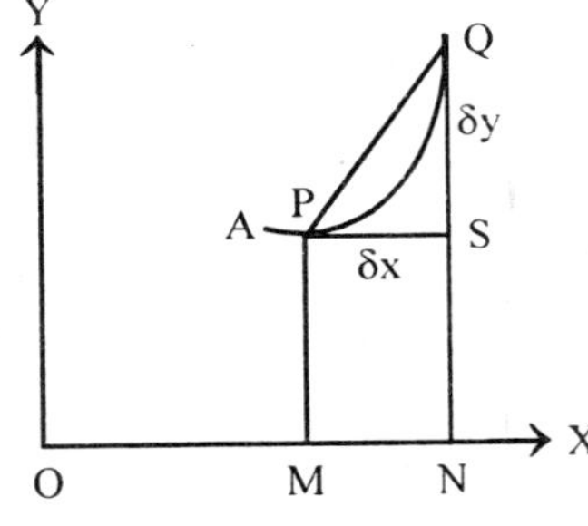

Fig. 8.10

Proof:

Let s + δs denote the length of the arc AQ, so that arc PQ = δs. Also let PM = y and QN = y + δy be the ordinates of P and Q. Then if PS be the perpendicular from P on QN, we have PS = ds and QS = δy. Join PQ. We have

$$(\text{Chord PQ})^2 = PS^2 + SQ^2 = (\delta x)^2 + (\delta y)^2 \qquad ...(1)$$

Dividing (1) throughout by $(dx)^2$, we get

$$\left(\frac{\text{chord PQ}}{\delta x}\right)^2 = 1 + \left(\frac{\delta y}{\delta x}\right)^2$$

or $$\left(\frac{\text{chord PQ}}{\text{arc PQ}} \cdot \frac{\text{arc PQ}}{\delta x}\right)^2 = 1 + \left(\frac{\delta y}{\delta x}\right)^2$$

or $$\left(\frac{\text{chord PQ}}{\text{arc PQ}}\right)^2 \cdot \left(\frac{\delta s}{\delta x}\right)^2 = 1 + \left(\frac{\delta y}{\delta x}\right)^2. \qquad [\because \text{arc PQ} = \delta s]$$

Taking limit of both sides as Q → P, *i.e.*, dx → 0, we get

$$\lim_{Q \to P}\left(\frac{\text{chord PQ}}{\text{arc PQ}}\right)^2 \cdot \lim_{\delta x \to 0}\left(\frac{\delta s}{\delta x}\right)^2 = \lim_{\delta x \to 0}\left[1 + \left(\frac{\delta y}{\delta x}\right)^2\right]$$

or $$\left(\frac{ds}{dx}\right)^2 = 1 + \left(\frac{dy}{dx}\right)^2,$$

$$\left[\because \lim_{Q \to P} \frac{\text{chord PQ}}{\text{arc PQ}} = 1, \lim_{\delta x \to 0} \frac{\delta y}{\delta x} = \frac{dy}{dx}, \text{etc.}\right]$$

Thus, $$\frac{ds}{dx} = \pm\sqrt{\left\{1 + \left(\frac{dy}{dx}\right)^2\right\}},$$

where positive or negative sign is to be taken before the radical sign according as s increases as x increases or decreases. Hence if s increases as x increases, we have

$$\frac{ds}{dx} = \sqrt{\left\{\frac{ds}{dx} = \sqrt{\left(1 + \left(\frac{dy}{dx}\right)^2\right)}\right\}}. \qquad \textbf{(Lucknow, 1997)}$$

Cor. 1: Dividing (1) throughout by $(\delta y)^2$ and proceeding to limits, we get

$$\frac{ds}{dy} = \pm\sqrt{\left\{1 + \left(\frac{dx}{dy}\right)^2\right\}}$$

where + ive or – ive sign is to be taken before the radical sign according as s increases as y increases or decreases.

Cor. 2: If the equations of the curve be given in the parametric form $x = f(t)$, $y = \phi(t)$, then s is obviously a function of t. In this case dividing (1) throughout by $(\delta t)^2$ and proceeding to limits, we get

$$\frac{ds}{dt} = \pm\sqrt{\left\{\left(\frac{dx}{dt}\right)^2 + \left(\frac{dy}{dt}\right)^2\right\}},$$

where + ive or – ive sign is to be taken before the radical sign according as s increases as t increases or decreases.

Cor. 3: We have

$$\cos\psi = \frac{1}{\sec\psi} = \frac{1}{\sqrt{(1+\tan^2\psi)}} = \frac{1}{\sqrt{\left\{1+\left(\frac{dy}{dx}\right)^2\right\}}} = \frac{1}{\frac{ds}{dx}} = \frac{dx}{ds}.$$

Again $\sin\psi$

$$= \frac{1}{\operatorname{cosec}\psi} = \frac{1}{\sqrt{(1+\cot^2\psi)}} = \frac{1}{\sqrt{\left\{1+\left(\frac{dx}{dy}\right)^2\right\}}} = \frac{1}{\frac{ds}{dy}} = \frac{dy}{ds}.$$

Now $\cos^2\psi + \sin^2\psi = 1$.

Therefore $\left(\frac{dx}{ds}\right)^2 + \left(\frac{dy}{ds}\right)^2 = 1.$

Note: It is easy to remember all the above results with the help of the adjoining hypothetical figure.

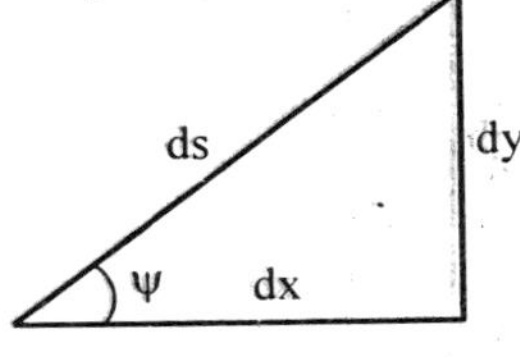

Fig. 8.11

8.13 DIFFERENTIAL COEFFICIENT OF ARC LENGTH. POLAR CO-ORDINATES

For the curve $r = f(\theta)$, we have

(i) $\frac{ds}{d\theta} = \sqrt{\left\{r^2 + \left(\frac{dr}{d\theta}\right)^2\right\}}$, **(Meerut 1993; Allahabad 92; Kanpur 96)**

(ii) $\frac{ds}{dr} = \sqrt{\left\{1 + r^2\left(\frac{d\theta}{dr}\right)^2\right\}}$,

(iii) $\cos\phi = \frac{dr}{ds}$,

(iv) $\sin\phi = r\frac{d\theta}{ds}$,

(v) $\left(\frac{dr}{ds}\right)^2 + \left(r\frac{d\theta}{ds}\right)^2 = 1.$

Proof:

Let s be the length of the arc AP of the curve r = f(θ). Here A is a fixed point on the curve and P is any point (r, θ). Take a point Q(r + δr, θ + δθ) on the curve in the neighbourhood of P such that arc AQ = s + δs. Then arc PQ = δs. Also δθ → 0 and δr → 0, as Q → P. Join PQ.

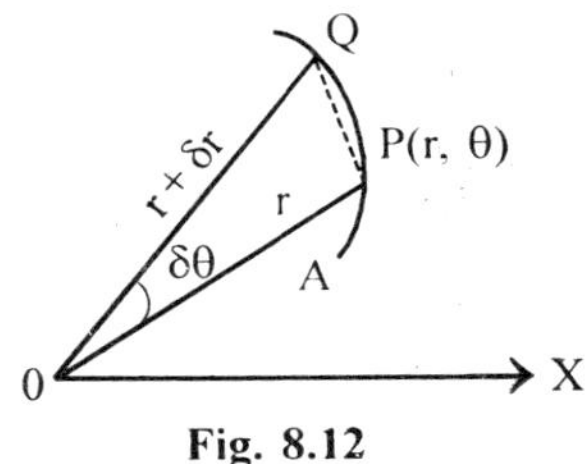

Fig. 8.12

From the ΔOPQ, we have

$$(\text{chord PQ})^2 = OP^2 + OQ^2 - 2OP.OQ \cos \angle QOP$$

$$= r^2 + (r + \delta r)^2 - 2r(r + \delta r) \cos \delta\theta$$

$$= (\delta r)^2 + 2r\delta r(1 - \cos \delta\theta) + 2r^2(1 - \cos \delta\theta)$$

$$= (\delta r)^2 + 4r.\delta r \sin^2 \frac{1}{2}\delta\theta + 4r^2 \sin^2 \frac{1}{2}\delta\theta.$$

Dividing by $(\delta\theta)^2$, we get

$$\left(\frac{\text{chord PQ}}{\text{arc PQ}}\right)^2 \cdot \left(\frac{\delta s}{\delta\theta}\right)^2 = \left(\frac{\delta r}{\delta\theta}\right)^2 + r.\left(\frac{\sin\frac{1}{2}\delta\theta}{\frac{1}{2}\delta\theta}\right)^2 \delta r + r^2\left(\frac{\sin^2\frac{1}{2}\delta\theta}{\frac{1}{2}\delta\theta}\right)^2.$$

Taking limit of both sides when Q → P, we get

$$\left(\frac{ds}{d\theta}\right)^2 = r^2 + \left(\frac{dr}{d\theta}\right)^2, \qquad \left[\because \lim_{Q\to P} \frac{\text{chord PQ}}{\text{arc PQ}} = 1\right]$$

Hence $\dfrac{ds}{d\theta}\sqrt{\left\{r^2 + \left(\dfrac{dr}{d\theta}\right)^2\right\}}$

Cor. 1: $\dfrac{ds}{dr} = \dfrac{ds}{d\theta}.\dfrac{d\theta}{dr} = \sqrt{\left\{r^2 + \left(\dfrac{dr}{d\theta}\right)^2\right\}}.\dfrac{d\theta}{dr}$

$$= \sqrt{\left\{1 + \left(r\frac{d\theta}{dr}\right)^2\right\}}.$$

Cor. 2: $\sin\phi = \dfrac{1}{\operatorname{cosec}\phi} = \dfrac{1}{\sqrt{(1+\cot^2\phi)}} = \dfrac{\tan\phi}{\sqrt{(1+\tan^2\phi)}}$

$$= \frac{\tan \phi}{\sqrt{\left\{1 + r^2 \left(\frac{d\theta}{dr}\right)^2\right\}}} = \frac{r\left(\frac{d\theta}{dr}\right)}{\frac{ds}{dr}} = r\frac{d\theta}{ds}.$$

Cor. 3: $\cos \phi = \frac{1}{\sec \phi} = \frac{1}{\sqrt{(1 + \tan^2 \phi)}} = \frac{1}{\sqrt{\left\{1 + r^2 \left(\frac{d\theta}{dr}\right)^2\right\}}}$

$$= \frac{1}{\frac{ds}{dr}} = \frac{dr}{ds}.$$

Cor. 4: We have $\cos^2 \phi + \sin^2 \phi = 1$.

$$\therefore \qquad \left(\frac{dr}{ds}\right)^2 + r^2 \left(\frac{d\theta}{ds}\right)^2 = 1.$$

Note: All the above results can be easily remembered from the adjoining hypothetical figure.

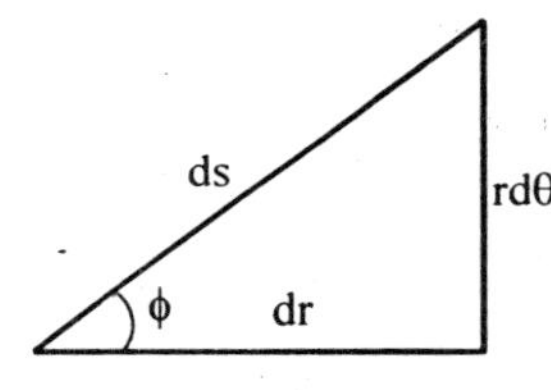

Fig. 8.13

Example 1:

From the curve $r = ae^{\theta \cot \alpha}$, *prove that* $\frac{s}{r}$ = *constant, s being measured from the origin.* **(Lucknow, 1992; Meerut, 95)**

Solution:

We have $r = ae^{\theta \cot \alpha}$.

$$\therefore \qquad \frac{dr}{d\theta}\; ae^{\theta \cot \alpha}.\cot \alpha = r \cot \alpha.$$

Now $\qquad \frac{ds}{dr} = \sqrt{\left\{1 + \left(r\frac{d\theta}{dr}\right)^2\right\}} = \sqrt{\left\{1 + r^2 . \frac{1}{(r \cot \alpha)^2}\right\}}$

$$= \sqrt{(1 + \tan^2 \alpha)} = \sec \alpha.$$

Integrating w.r.t. r, we get s = r sec α + C, where C is constant of integration. Since s is measured from the origin, therefore s = 0 when r = 0. This gives 0 = 0 + C *i.e.*, C = 0.

Hence s = r sec a or $\frac{s}{r}$ = sec α = constant.

Example 2:

Prove that for any curve,

$$\sin^2 \phi\left(\frac{d\phi}{d\theta}\right)+r\left(\frac{d^2r}{ds^2}\right) = 0. \qquad \textbf{(Meerut, 1998)}$$

Solution:

We know that $\frac{dr}{ds} = \cos\phi$. Differentiating both sides w.r.t. s, we get

$$\frac{d^2r}{ds^2} = -\sin\phi\left(\frac{d\phi}{ds}\right) = -\sin\phi\left(\frac{d\phi}{d\theta}\right).\left(\frac{d\theta}{ds}\right).$$

Multiplying both sides by r, we get

$$r\left(\frac{d^2r}{ds^2}\right) = -\sin\phi\left(\frac{d\phi}{d\theta}\right).r\left(\frac{d\theta}{ds}\right)$$

or $$r\left(\frac{d^2r}{ds^2}\right) = -\sin\phi\left(\frac{d\phi}{d\theta}\right)\sin\phi, \qquad \left[\because r\left(\frac{d\theta}{ds}\right) = \sin\phi\right]$$

or $$r\left(\frac{d^2r}{ds^2}\right) + \sin^2\phi\left(\frac{d\phi}{d\theta}\right) = 0.$$

Example 3:

For the curve $r^m = a^n \cos m\theta$, prove that

(a) $\frac{ds}{d\theta} = a(\sec m\theta)^{(m+1)/m}$, **(Vikram, 1990)**

(b) $\frac{ds}{d\theta}$ *varies inversely as $(m - 1)^{thg}$ power of r.*

(c) $a^{2m}\frac{d^2r}{ds^2} + mr^{2m-1} = 0.$ **(Lucknow, 1990; Kanpur, 96)**

Solution:

(a) We get

$$\frac{ds}{d\theta} = r\sec m\theta = (a^m\cos m\theta)^{1/m}.\sec m\theta$$

$$= a(\cos m\theta)^{1/m}.\sec m\theta = a(\sec m\theta)^{-1/m}\sec m\theta$$

$$= a(\sec m\theta)^{1+(-1/m)} = a(\sec m\theta)^{(m-1)/m}.$$

(b) We get

$$\frac{ds}{d\theta} = \frac{a^m}{r^{m-1}} = \text{(some constant)}.\frac{1}{r^{m-1}}.$$

Therefore $\frac{ds}{d\theta}$ varies inversely as (m – 1)th power of r.

(c) We get $\frac{1}{r}\cdot\frac{dr}{d\theta} = -\tan m\theta$.

Therefore $\cot\phi = \frac{1}{r}\frac{dr}{d\theta} = -\tan m\theta = \cot\left(\frac{p}{2} + m\theta\right)$.

Hence $\phi = \frac{1}{2}\pi + m\theta$. Now we know that $\frac{dr}{ds} = \cos\phi$.

Differentiating w.r.t. s, we get

$$\frac{d^2r}{ds^2} = -\sin\phi\frac{d\phi}{ds} = -\sin\phi\frac{d\phi}{d\theta}\frac{d\theta}{ds}$$

$$= -\sin\phi\frac{d\phi}{d\theta}\frac{\sin\phi}{r}, \qquad \left[\because r\left(\frac{d\theta}{ds}\right) = \sin\phi\right]$$

$$= -\frac{\sin^2\phi}{r}\frac{d\phi}{d\theta} = \frac{\sin^2\left(\frac{1}{2}\pi + m\theta\right)}{r}.m$$

$$\left[\because \phi = \frac{1}{2}\pi + m\theta, \text{ and so } \frac{d\phi}{d\theta} = m\right]$$

$$= -\frac{m\cos^2 m\theta}{r} = -\frac{m}{r}\cdot\left(\frac{r^m}{a^m}\right)^2, \qquad [\because = r^m = a^m \cos m\theta]$$

$$= -\frac{mr^{2m-1}}{a^{2m}}.$$

$$\therefore\ a^{2m}\frac{d^2r}{ds^2} + mr^{2m-1} = 0.$$

Example 4:

Show that for the curve

$$\theta = \cos^{-1}\frac{r}{k} - \frac{\sqrt{(k^2 - r^2)}}{r},\ r\frac{ds}{dr} \textit{ is constant.}$$

Solution:

Let r = k cos t.

Then $\theta = \cos^{-1}(\cos t) - \frac{\sqrt{(k^2 - k^2\cos^2 t)}}{k\cos t} = t - \tan t$.

These give $\frac{dr}{dt} = -k\sin t$ and $\frac{d\theta}{dt} = 1 - \sec^2 t$.

Now $$r\frac{d\theta}{dr} = r\frac{\frac{d\theta}{dt}}{\frac{dr}{dt}} = k\cos t.\frac{1-\sec^2 t}{-k\sin t}$$

$$= -\cot t.(-\tan^2 t) = \tan t.$$

$\therefore$ $$\frac{ds}{dr} = \sqrt{\left\{1+\left(r\frac{d\theta}{dr}\right)^2\right\}} = \sqrt{\left(1+\tan^2 t\right)} = \sec t$$

$$= \frac{1}{\cos t} = \frac{1}{\frac{r}{k}} = \frac{k}{r}.$$

Here $$r\left(\frac{ds}{dr}\right) = k = \text{constant.}$$

Example 5:

For the cycloid $x = a(1 - \cos t)$, $y = a(t + \sin t)$, find $\frac{ds}{dt}, \frac{ds}{dx}$ and $\frac{ds}{dy}$. **(Meerut, 1994)**

Solution:

Differentiating w.r.t. t, we have

$$\frac{dx}{dt} = a\sin t,\ \frac{dy}{dt} = a(1+\cos t).$$

Now $$\frac{ds}{dt} = \sqrt{\left\{\left(\frac{dx}{dt}\right)^2+\left(\frac{dy}{dt}\right)^2\right\}} = \sqrt{\left\{(a\sin t)^2 + a^2(1+\cos t)^2\right\}}$$

$$= a\sqrt{\left(2\sin\frac{1}{2}t+\cos\frac{1}{2}t\right)^2+\left(2a\cos\frac{1}{2}t\right)}.$$

$$= 2a\cos\frac{1}{2}t\sqrt{\left(\sin^2\frac{1}{2}t+\cos^2\frac{1}{2}t\right)} = 2a\cos\frac{1}{2}t$$

Also $$\frac{ds}{dx} = \frac{ds}{dt}.\frac{dt}{dx} = \frac{2a\cos\frac{1}{2}t}{\frac{dx}{dt}} = \frac{2a\cos\frac{1}{2}t}{a\sin t} = \frac{2a\cos\frac{1}{2}t}{2a\sin\frac{1}{2}t\cos\frac{1}{2}t}$$

$$= \operatorname{cosec}\frac{1}{2}t,$$

and $\dfrac{ds}{dy} = \dfrac{ds}{dt} \cdot \dfrac{dt}{dy} = \dfrac{2a\cos\frac{1}{2}t}{\frac{dy}{dt}} = \dfrac{2a\cos\frac{1}{2}t}{a(1+\cos t)} = \dfrac{2a\cos\frac{1}{2}t}{2a\cos^2\frac{1}{2}t}$

$= \sec\frac{1}{2}t.$

Example 6:

For any curve, prove that

(a) $\dfrac{ds}{d\theta} = \dfrac{r^2}{p},$ **(Kanpur, 1998)**

(b) $\dfrac{ds}{dr} = \dfrac{r}{\sqrt{(r^2 - p^2)}}.$ **(Kanpur, 1995)**

Solution:

(a) We know that $r\dfrac{d\theta}{ds} = \sin\phi$ or $\dfrac{d\theta}{ds} = \dfrac{\sin\phi}{r}.$

$\therefore \dfrac{ds}{d\theta} = \dfrac{r}{\sin\phi} = \dfrac{r}{\frac{p}{r}} = \dfrac{r^2}{p},$ $[\because p = r\sin\phi]$

(b) We know that $\dfrac{dr}{ds} = \cos\phi$

$\therefore \dfrac{ds}{dr} = \dfrac{1}{\cos\phi} = \dfrac{1}{\sqrt{(1-\sin^2\phi)}} = \dfrac{1}{\sqrt{\left\{1-\left(\frac{p}{2}\right)^2\right\}}} = \dfrac{1}{\sqrt{(r^2-p^2)}}.$

Example 7:

For the ellipse x = cos t, y = b sin t, prove that

$\dfrac{ds}{dt} = a\sqrt{(1-e^2\cos^2 t)}.$ **(Meerut, 1993)**

Solution:

Here $\dfrac{dx}{dt} = -a\sin t$ and $\dfrac{dy}{dt} = b\cos t.$

We have

$$\frac{ds}{dt} = \sqrt{\left\{\left(\frac{dx}{dt}\right)^2 + \left(\frac{dy}{dt}\right)^2\right\}} = \sqrt{(a^2\sin^2 t + b^2\cos^2 t)}$$

$$= \sqrt{\left\{a^2 \sin^2 t + a^2\left(1-e^2\right)\cos^2 t\right\}},$$

[$\because$ for the given ellipse, $b^2 = a^2\ (1 - a^2)$]

$$= a\sqrt{\left\{\left(\sin^2 t + \cos^2 t\right) - e^2 \cos^2 t\right\}} = a\sqrt{\left(1 - e^2 \cos^2 t\right)}.$$

Example 8:

If $r^m = a^m \cos mq$, prove that $\dfrac{ds}{d\theta} = \dfrac{a^m}{r^{m-1}}$.

Solution:

We have $r^m = a^m \cos m\theta$. ...(1)

Taking logarithm of both sides of (1), we get

$$m \log r = m \log a + 1 \log \cos m\theta.$$

Differentiating w.r.t. θ, we get

$$\frac{m}{r}\frac{dr}{d\theta} = 0 - \frac{m \sin m\theta}{\cos m\theta},\ \frac{dr}{d\theta} = -r \tan m\theta.$$

Now $\dfrac{ds}{d\theta} = \sqrt{\left\{r^2 + \left(\dfrac{dr}{d\theta}\right)^2\right\}} = \sqrt{\left(r^2 + r^2 \tan^2 m\theta\right)} = r \sec m\theta$

$$= \frac{r}{\cos m\theta} = \frac{r}{\dfrac{r^m}{a^m}}, \qquad \left[\because \text{from } (1), \cos m\theta = \frac{r^m}{a^m}\right]$$

$$= r.\frac{a^m}{r^m} = \frac{a^m}{r^{m-1}}.$$

EXERCISES

1. Trangents are drawn from the origin to the curve $y = \sin x$. Prove that their point of contact lie on $x^2y^2 = x^2 - y^2$.
2. If $p = x \cos \alpha + y \sin x$ touches the curve

   $$\left(\frac{x}{a}\right)^{n/n-1} + \left(\frac{y}{b}\right)^{n/n-1} = 1.$$

   Prove that $p^n = (a \cos \alpha)^n + (b \sin \alpha)^b$.
3. Prove that $\dfrac{x}{a} + \dfrac{y}{b} = 1$ touches the curve $y = be^{-x/a}$ at the point where the curve crosses the axis of y.

4. Prove that all points on the curve

$$y^2 = 4a\left[x + a\sin\left(\frac{x}{a}\right)\right]$$

at which the tangent is parallel to the axis of a lie on a parabola.

5. Show that the condition that the curves

$$ax^2 + by^2 = 1 \text{ and } a'x^2 + b'y^2 = 1$$

should intersect orthogonally is that

$$\frac{1}{a} - \frac{1}{b} = \frac{1}{a'} - \frac{1}{b'}.$$

6. Prove that the curve $\left(\frac{x}{a}\right)^n + \left(\frac{y}{b}\right)^n = 2$ touch the straight line $\frac{x}{a} + \frac{y}{b} = 2$ at the point (a, b) what ever be the value of n

7. In the curve $x^m y^b = a^{m+n}$ prove that the portion of the tangent intercepted between the axis is divided at its point of contact into segments which are in a constant ratio.

8. Prove that in the ellips $\frac{x^2}{a^2} + \frac{y^2}{b^2} = 1$ the length of the normal varies inversely as the perpendicular from the origin on the tangent.

9. Show that the exponential curve $y = be^{x/a}$ the subtangent is of constant length and the subnormal varies as the square of the ordinates.

10. In the curve $x^{m+n} = a^{m-n} y^{2n}$ prove that the mth power of the subtangent varies as the nth power of the subnormal.

11. Show that the subtangent at any point of the curve $x^m y^n$ varies as the abscissa.

12. Find the length of the normal and subnormal of the curve

$$y = \frac{1}{2}a\left[e^{x/a} + e^{-x/a}\right].$$

13. Show that the pedal equation of $r = a(1 - \cos\alpha)$ is $r^3 = 2ap^2$.

14. Show that the pedal equation of the lemniscate $r^2 = a^2 \cos 2\theta$ is $r^2 = a^2 p$.

15. Prove that the length of the perpendicular from the pole on the tangent to the ellipse $\frac{l}{r} = 1 + e\cos x$) is given by $\frac{1}{p^2} = \frac{1}{l^2}\left(\frac{2l}{r} - l + e^2\right)$.

16. Show that the pedal equation of the ellipse is $\frac{x^2}{a^2} + \frac{y^2}{b^2} = 1$

is $\frac{1}{p^2} = \frac{1}{a^2} + \frac{1}{b^2} - \frac{r^2}{a^2b^2}$.

17. Show that the equiangular spiral $r = ae^{\theta \cot \alpha}$ the tangent is inclined at a constant angle to the radius vector.

18. Show that in the curve $r = a\theta$ the polar subnormal is constant and in the curve $n\theta = a$. The polar subtangent is constant.

19. Prove that the condition that $x \cos \alpha + y \sin \alpha = p$ should touch

$x^m y^n = a^{m+n}$. is $p^{m+n}.m^m.n^n = (m + n)^{m+n} a^{m+n}.\cos^m \alpha.\sin^n \alpha$.

20. Show that in the curve $y = x \log (x^2 = a^2)$ the sum of the tangent and the subtangent varies as the product of the co-ordinates of the point.

21. Show that the abscissa of the points on the curve $y = x(x - 2)(x - a)$ where the tangent are parallel to the axis of x are given by $x = 2 \pm 2\sqrt{3}$.

22. Prove that the curves $y = e^{-ax} \sin bx$, $y = e^{-ax}$ touch at the point for which $bx = 2m\pi + \frac{\pi}{2}$ where m is an integer.

23. Show that the pedal equation of the spiral $r = a \operatorname{sech} n\theta$ is of the form $\frac{1}{p^2} = \frac{A}{r^2} + B$ and that the pedal equations of the curves $r = ae^{m\theta}$, $r\theta = a$, $r \sin n\theta = a$ and $r \cosh \theta = a$ are all of the same form.

9

CURVATURE

9.1 INTRODUCTION

Of the two curves shown here, one bend more sharply then the other in other words of P one has a greater curvature than the other but in order to get a quantitative estimate of curvature we shall first have to give a careful definition.

9.2 DEFINITION

Let P and Q be two neighbouring points on a given curve, such that the are PQ is concave towards its chord. Let the normals at P and Q intersect in N.

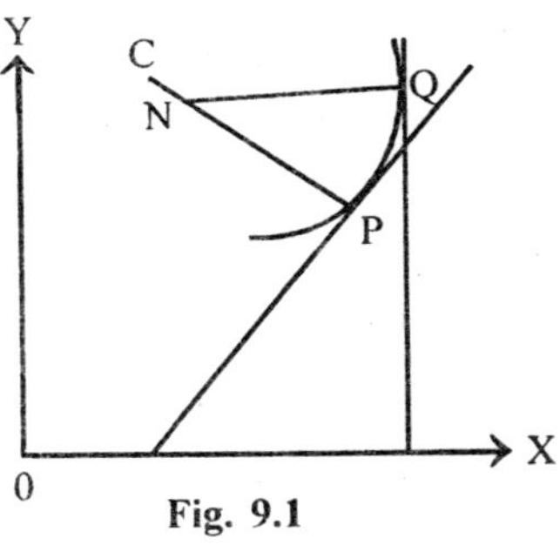

Fig. 9.1

If as Q $\to$ P, N tends to the definite position C, then C is called the **Centre of Curvature** of the curve at the point P.

The distance CP is called the radius of curvature of the curve at P and is usually denoted by ρ.

The circle with its centre at C and radius CP is called the **Circle of Curvature** of the curve at P.

Any chord of this circle drawn through the point P is called a **Chord of Curvature.**

9.3 RADIUS OF CURVATURE OF INTRINSIC CURVES

(Gorakhpur, 1999; Kanpur, 96; Indore, 97; Vikram, 92)

Let P and Q be two neighbouring points on the curve APQ and the tangents at these two points make angles ψ and $\psi + \delta\psi$ respectively with a fixed line, say, x-axis. Also let arc AP = s and arc AQ = s + δs an let these points meets at R.

Let the tangents at P and Q intersect at R, and N be the point of intersection of the normals at these two points. We have

$\angle$ PNQ = $\angle$ TRT' = $\delta\psi$.

Suppose N $\rightarrow$ C as Q $\rightarrow$ P. Then the radius of curvature at P = $\rho = \lim\limits_{Q \to P} PN$.

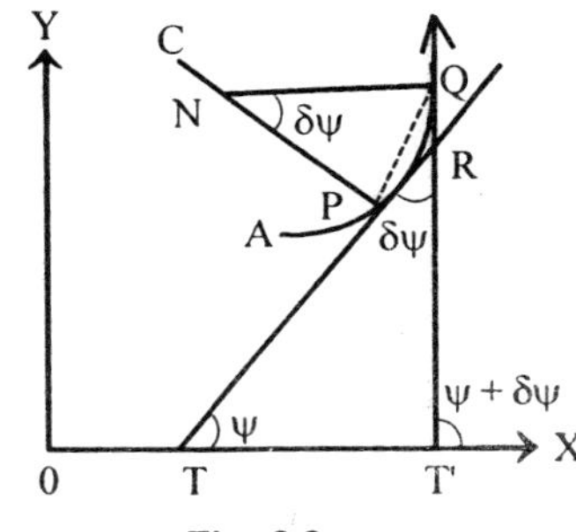

Fig. 9.2

From Δ PNQ, we have

$$\frac{PN}{\sin NQP} = \frac{\text{chord } PQ}{\sin PNQ}$$

$$= \frac{\text{chord } PQ}{\sin \delta\psi}$$

$$\therefore \quad PN = \frac{\text{chord } PQ}{\sin \delta\psi} = NQP$$

$$= \frac{\text{chord } PQ}{\delta s} \cdot \frac{\delta s}{\delta\psi} \cdot \frac{\delta\psi}{\sin \delta\psi} . \sin NQP$$

As Q $\rightarrow$ P, $\delta\psi \rightarrow 0$, $\delta s \rightarrow 0$, chord PQ $\rightarrow$ tangent at P, QN $\rightarrow$ normal at P and consequently $\angle NQP \rightarrow \frac{1}{2}\pi$.

$$\therefore \; \rho = \lim_{Q \to P} PN$$

$$= \left(\lim_{Q \to P} \frac{\text{chord } PQ}{\text{arc } PQ}\right) \cdot \left(\lim_{\delta\psi \to 0} \frac{\delta\psi}{\sin \delta\psi}\right) \left(\lim_{\delta\psi \to 0} \frac{\delta\psi}{\sin \delta\psi}\right) \left(\lim_{Q \to P} \sin NQP\right)$$

$$= 1 . \frac{ds}{d\psi} . 1 . \sin\frac{\pi}{2} = \frac{ds}{d\psi} . \text{ Hence } \rho = \frac{ds}{d\psi}.$$

Note: The relation between s and ψ is called the intrinsic equation of a curve. Therefore, $\rho = \dfrac{ds}{d\psi}$ is known as intrinsic formula for radius of curvature.

Cor.: The curvature of the curve at P is defined as the reciprocal of the radius of curvature at P. Hence the curvature at $P = \frac{1}{\rho} = \frac{d\psi}{ds}$.

9.4 CARTESIAN FORMULA FOR RADIUS OF CURVATURE

(Garhwal, 1993; Meerut, 91S, 91, 97; Indore, 93; Gorakhpur, 92 96; Magadh, 96; Rohilkhand, 92; Bundelkhand, 97; Lucknow, 92, 97, 95; Agra, 95)

Let y = f(x) be the equation of the curve. The inclination ψ of the tangent at any point to the axis of x is given by $\tan\psi = \frac{dy}{dx}$.

Differentiating w.r.t. s, we get

$$\sec^2\psi\frac{d\psi}{ds} = \frac{d^2y}{dx^2}.\frac{dx}{ds} = \frac{d^2y}{dx^2}\cos\psi. \qquad \left[\because \frac{dx}{ds} = \cos\psi\right]$$

$$\therefore \frac{ds}{d\psi} = \frac{\sec^2\psi}{\left(\frac{d^2y}{dx^2}\right).\cos\psi} = \frac{\sec^3\psi}{\frac{d^2y}{dx^2}}$$

$$= \frac{\{1+\tan^2\psi\}^{3/2}}{\frac{d^2y}{dx^2}} = \frac{\left\{1+\left(\frac{dy}{dx}^2\right)^2\right\}^{3/2}}{\frac{d^2y}{dx^2}}.$$

Hence $\rho = \frac{ds}{d\psi} = \frac{\left\{1+\left(\frac{dy}{dx}\right)^2\right\}^{3/2}}{\frac{d^2y}{dx^2}}$.

Note 1: The value of ρ is + ive or – ive according as $\frac{d^2y}{dx^2}$ is positive or negative. However in numerical problems we shall be required to find only the length of the radius of curvature and we shall not be concerned with its sign. So we should ignore the negative sign whenever we get a negative value for ρ.

Note 2: From the definition, it is obvious the value of ρ depends on the curve and not on the co-ordinate axes chosen. Therefore, interchanging x and y in the above formula for ρ, we obtain

$$\rho = \frac{\left\{1+\left(\frac{dx}{dy}\right)^2\right\}^{3/2}}{\frac{d^2x}{dy^2}}.$$

This formula is specially useful when $\frac{dy}{dx}$ is infinite, *i.e.*, when the tangent is parallel to the x-axis.

Definition: *If at any point of a curve* $\frac{d^2y}{dx^2}=0$, *the point is called* a **Point of Inflexion.**

Example 1:

Find the radius of curvature at the point $\left(\frac{1}{4},\frac{1}{4}\right)$ *of the curve* $\sqrt{x}+\sqrt{y}=1$. **(Gorakhpur, 1992; Kanpur, 95)**

Solution:

Here a = 1 and instead of (x_1, y_1) the point is $\left(\frac{1}{4},\frac{1}{4}\right)$. The answer is $\frac{1}{\sqrt{2}}$.

Example 1 (a):

In the curve $y = ae^{x/a}$, *prove that* $r = a\sec^2\theta\,\mathrm{cosec}\,\theta$, *where* $\theta = \tan^{-1}\left(\frac{y}{a}\right)$.

Solution:

We have $\frac{dy}{dx}=a\left(e^{x/a}\right)\left(\frac{1}{a}\right)=e^{x/a}=\frac{y}{a}$;

and $\frac{d^2y}{dx^2}=\left(\frac{1}{a}\right)\left(\frac{dy}{dx}\right)=\left(\frac{1}{a}\right)\left(\frac{y}{a}\right)=\frac{y}{a^2}$.

$$\therefore \qquad \rho=\frac{\left[1+\left(\frac{y^2}{a^2}\right)\right]^{3/2}}{\frac{y}{a^2}}=\frac{\left(a^2+y^2\right)^{3/2}}{ay}=\frac{\left(a^2+a^2\tan^2\theta\right)^{3/2}}{a.a\tan\theta}$$

[$\because$ given $y = a\tan\theta$]

$= a\sec^3\theta\cot\theta = a\sec^2\theta\,\mathrm{cosec}\,\theta$. **Ans.**

Example 2:

In the ellipse $\left(\frac{x^2}{a^2}\right)+\left(\frac{y^2}{b^2}\right)$ *= 1, show that the radius of curvature at an end of the major axis is equal to the semi-lotus rectum of the ellipse.*

(Garhwal, 1997)

Solution:

The ellipse is $\left(\frac{x^2}{a^2}\right)+\left(\frac{y^2}{b^2}\right)=1$. ...(1)

Differentiating, w.r.t. 'x' $\frac{2x}{a^2}+\frac{2y}{b^2}\frac{dy}{dx}=0$

$$\Rightarrow \qquad \frac{dy}{dx}=-\frac{b^2}{a^2}\left(\frac{x}{y}\right).$$

Differentiating again, w.r.t. 'x' $\frac{d^2y}{dx^2}=-\frac{b^2}{a^2}\left[\frac{y.1-x\left(\frac{dy}{dx}\right)}{y^2}\right]$

$$=-\frac{b^2}{a^2y^2}\left[y-x\left(-\frac{b^2}{a^2y}\right)\right]$$

$$=-\frac{b^2}{a^2y^2}\left(\frac{a^2y^2+b^2x^2}{a^2y}\right)=-\frac{b^2}{a^2y^2}\left(\frac{a^2b^2}{a^2y}\right)$$

[$\because$ from (1), $a^2y^2 + b^2x^2 = a^2b^2$]

$$=-\frac{b^4}{a^2y^3}.$$

$$\therefore \rho \text{ at the point } (x, y) = \frac{\left[1+\left(\frac{dy}{dx}\right)^2\right]^{3/2}}{\frac{d^2y}{dx^2}}$$

$$=\frac{\left[1+\left(-\frac{b^2x}{a^2y}\right)^2\right]^{3/2}}{\frac{b^4}{(a^2y^3)}}, \text{ neglecting the } - \text{ ive sign}$$

$$=\frac{\left(a^4y^2+b^4x^2\right)^{3/2}}{a^4b^4}.$$

Now the co-ordinates of one-end of major axis are (a, 0).

$$\therefore\ [\rho]_{\text{at }(a,\,0)} = \frac{\left[a^4.0 + b^4a^2\right]^{3/2}}{a^4b^4} = \frac{b^6a^3}{a^4b^4} = \frac{b^2}{a}$$

= semi latus-rectum of the ellipse. **Hence proved.**

Example 3:

Find the radius of curvature at (x, y) on the curve

$a^2y = x^3 - a^3.$ **(Meerut, 1997, 94S)**

Solution:

We have $a^2y = x^3 - a^3$.

$$\therefore\ \frac{dy}{dx} = \frac{1}{a^2}\left(3x^2\right);\ \frac{d^2y}{dx^2} = \frac{6x}{a^2}.$$

$$\therefore\ \rho = \frac{\left[1+\left(\frac{dy}{dx}\right)^2\right]^{3/2}}{\frac{d^2y}{dx^2}} = \left[\frac{1+\left(\frac{3x^2}{a^2}\right)^2}{\frac{6x}{a^2}}\right]^{3/2}$$

$$= \frac{\left(a^4+9x^4\right)^{3/2}}{a^6}.\frac{a^2}{6x} = \frac{\left(a^4+9x^4\right)^{3/2}}{6a^4x}.$$ **Ans.**

Example 4:

Find the radius of curvature at (x, y) of the curve

$y = \frac{1}{2}.c.(e^{x\,c} + e^{-x\,c}) = c\cosh\left(\frac{x}{c}\right).$ **(Meerut, 1997 S)**

Solution:

We have $y = c\cosh\left(\frac{x}{c}\right)$.

$$\therefore\ \frac{dy}{dx} = c\left(\sinh\frac{x}{c}\right).\frac{1}{c} = \sinh\left(\frac{x}{c}\right),$$

and $\frac{d^2y}{dx^2} = \left(\frac{1}{c}\right)\cosh\left(\frac{x}{c}\right).$

$$\therefore\ \rho = \frac{\left[1+\left(\frac{dy}{dx}\right)^2\right]^{3/2}}{\frac{d^2y}{dx^2}} = \frac{\left[1+\sinh^2\left(\frac{x}{c}\right)\right]^{3/2}}{\left(\frac{1}{c}\right)\cosh\left(\frac{x}{c}\right)} = \frac{c\left[\cosh^2\left(\frac{x}{c}\right)\right]^{3/2}}{\cosh\left(\frac{x}{c}\right)}$$

$$= c \cosh^2\left(\frac{x}{c}\right) = c\left(\frac{y}{c}\right)^2 = \frac{y^2}{c}.$$

Example 5:

If CP, CD be a pair of conjugate semi-diameters of an ellipse, prove that the radius of curvature at P is $\frac{CD^3}{ab}$, *a and b being the lengths of the semi-axes of the ellipse.* **(Meerut, 1992, 94, 95)**

Solution:

If ρ be the radius of curvature of at the point P, we get

$$\rho = \frac{\left(a^2 \sin^2 t + b^2 \cos^2 t\right)^{3/2}}{ab}.$$

Now the eccentric angle of the point D is $'t + \frac{1}{2}\pi'$. Therefore the co-ordinates of D are $\left\{a \cos\left(\frac{1}{2}\pi + t\right), b \sin\left(\frac{1}{2}\pi + t\right)\right\}$, *i.e.*,

(– a sin t, b cos t). Also C is the point (0, 0). Therefore

$CD = (a^2 \sin^2 t + b^2 \cos^2 t)^{1/2}$ or $CD^3 = (a^2 \sin^2 t + b^2 \cos^2 t)^{3/2}$.

Hence $\rho = \frac{CD^3}{ab}$.

Note: Taking the co-ordinates of any point P(x, y) on the given ellipse as x = a cos t, y = b sin t. Then as proved we have ρ at the point 't'

$$= (a^2 \sin^2 t + b^2 \cos^2 t)^{3/2} (ab).$$

Now at an extremity of the major axis, we have t = 0. Therefore at an extremity of the major axis, we have

$$\rho = \frac{\left(a^2 \sin^2 0 + b^2 \cos^2 t\right)^{3/2}}{ab} = \frac{b^3}{ab} = \frac{b^2}{a}$$

= the length of semi-latus rectum. **Hence proved.**

Example 6:

Find the curvature at the point $\left(\frac{3a}{2}, \frac{3a}{2}\right)$ *of the curve* $x^3 + y^3 = 3axy$. **(Meerut, 1999, 93S, 99 P)**

Solution:

The curve is $x^3 + y^3 = 3axy$. ...(1)

Differentiating both sides (1) w.r.t. x, we have

$$3x^2+3y^2\left(\frac{dy}{dx}\right)=3ay+3ax\left(\frac{dy}{dx}\right)$$

or $$x^2+y^2\left(\frac{dy}{dx}\right)=ay+ax\left(\frac{dy}{dx}\right). \qquad ...(2)$$

$\therefore$ $$\frac{dy}{dx}=\frac{x^2-ay}{ax-y^2}, \text{ so that } \left(\frac{dy}{dx}\right)_{\left(\frac{3}{2}a,\frac{3}{2}a\right)}=-1.$$

Again, differentiating both sides of (2) w.r.t. x we have

$$2x+2y\left(\frac{dy}{dx}\right)^2+y^2\frac{d^2y}{dx^2}=a\frac{dy}{dx}+a\frac{dy}{dx}+a\frac{dy}{dx}+a\frac{dy}{dx}+ax\frac{d^2y}{dx^2}$$

or $$\left(ax-y^2\right)\frac{d^2y}{dx^2}=2x+2y\left(\frac{dy}{dx}\right)^2-2a\frac{dy}{dx}. \qquad ...(3)$$

Putting $x=\frac{3a}{2}, y=\frac{3a}{2}$ and $\left(\frac{dy}{dx}\right)_{\left(\frac{3a}{2}.\frac{3a}{2}\right)}=-1$ in (3), we obtain

$$\left(\frac{d^2y}{dx^2}\right)_{\left(\frac{3a}{2},\frac{3a}{2}\right)}=-\frac{32}{3}.\frac{1}{a}=\frac{32}{(3a)}.$$

Hence the radius of curvature ρ at $\left(\frac{3a}{2},\frac{3a}{2}\right)$ is

$$=\left[\frac{\left\{1+\left(\frac{dy}{dx}\right)^2\right\}^{3/2}}{\frac{d^2y}{dx^2}}\right]_{\left(\frac{3a}{2},\frac{3a}{2}\right)}$$

$$=\frac{(1+1)^{3/2}}{\left(-\frac{32}{3a}\right)}=-\frac{3a}{8\sqrt{2}}.$$

$\therefore$ Curvature at $\left(\frac{3a}{2},\frac{3a}{2}\right)=\frac{1}{\rho}=-\frac{8\sqrt{2}}{3a}$.

Neglecting the negative sign, the value of curvature at

$$\left(\frac{3a}{2},\frac{3a}{2}\right)=\frac{8\sqrt{2}}{3a}.$$

Ans.

Example 7:

Find the radius of curvature at the point (s, ψ) on the following curves:

(i) $s = 4a \sin \psi$,

(ii) $s = c \tan \psi$,

(iii) $s = 8a \sin^2 \frac{1}{6}\psi$.

Solution:

(i) We have $s = 4a \sin \psi$.

$$\therefore \ r = \frac{ds}{d\psi} = 4a \cos \psi \, .$$

(ii) We have $s = c \tan \psi$.

$$\therefore \ \rho = \frac{ds}{d\psi} = c \sec^2 \psi \, .$$

(iii) We have $s = 8a \sin^2 \frac{1}{6}\psi$.

$$\therefore \ \rho = \frac{ds}{d\psi} = 8a.\left(2\sin\frac{1}{6}\psi \cos\frac{1}{6}\psi\right).\frac{1}{6} = \left(\frac{4a}{3}\right)\sin\frac{1}{3}\psi \, .$$

Example 8:

Find the radius of curvature of any point (s, ψ) of the curves

(i) $s = a(e^{m\psi} - 1)$.

(ii) $s = c \log \sec \psi$,

(iii) $s = a \log \cot\left(\frac{\pi}{4} - \frac{\psi}{2}\right) + a\frac{\sin \psi}{\cos^2 \psi}$. **(Lucknow, 1998)**

Solution:

(i) We have, $s = a(e^{m\psi} - 1)$.

$$\therefore \ \rho = \frac{ds}{d\psi} = ame^{m\psi} \, .$$

(ii) We have $s = c \log \sec \psi$.

$$\therefore \ \rho = \frac{ds}{d\psi} = c.\frac{1}{\sec \psi}.\sec \psi \tan \psi = c \tan \psi \, .$$

(iii) We have $s = a \log \cot\left(\frac{\pi}{4} - \frac{\psi}{2}\right) + a \sin \psi \sec^2 \psi$.

$$\therefore\ \rho=\frac{ds}{d\psi}=a\frac{1}{\cot\left(\frac{1}{4}\pi-\frac{1}{2}\psi\right)}.\left\{-\operatorname{cosec}^2\left(\frac{\pi}{4}-\frac{\psi}{2}\right)\right\}.\left(-\frac{1}{2}\right)$$

$$+\ a\sin\psi.2\sec\psi\sec\psi\tan\psi+a\cos\psi\sec^2\psi$$

$$=\frac{a}{2\sin\left(\frac{1}{4}\pi-\frac{1}{2}\psi\right)\cos\left(\frac{1}{4}\pi-\frac{1}{2}\psi\right)}+\frac{2a\sin^2\psi}{\cos^3\psi}+\frac{a}{\cos\psi}$$

$$=\frac{a}{\sin\left(\frac{1}{2}\pi-\psi\right)}+\frac{2a\sin^2\psi}{\cos^3\psi}+\frac{a}{\cos\psi}$$

$$=\frac{2a}{\cos\psi}+\frac{2a\sin^2\psi}{\cos^3\psi}=\frac{2a}{\cos^3\psi}\left(\cos^2\psi+\sin^2\psi\right)=2a\sec^3\psi.\quad \textbf{Ans.}$$

Example 9:

Find the radius of curvature of the curve $y = e^x$ at the point where it crosses the y-axis. **(Kanpur, 1998; Magadh, 97)**

Solution:

We have $y = e^x$.

Therefore $\frac{dy}{dx}=e^x$, and $\frac{d^2y}{dx^2}=e^x$.

$$\therefore\ =\rho \text{ at } (x,y)=\frac{\left[1+\left(\frac{dy}{dx}\right)^2\right]^{3/2}}{\frac{d^2y}{dx^2}}=\frac{\left(1+e^{2x}\right)^{3/2}}{e^x}.$$

The curve $y = e^x$ crosses the y-axis (*i.e.*, the straight line $x = 0$) at the point (0, 1).

$$\therefore\ \rho \text{ at } (0,1)=\frac{\left(1+e^0\right)^{3/2}}{e^0}=\frac{(1+1)^{3/2}}{1}=2\sqrt{2}.$$

Example 10:

If C be the centre of curvature corresponding to any point P(x, y) on the curve $y = a\cosh\left(\frac{x}{a}\right)$ and G is the intersection of the normal at P and x-axis, then show that PC = PG.

Solution:

The given curve is $y = a\cosh\left(\frac{x}{a}\right)$. ...(1)

If ρ be the radius of curvature of the curve (1) at the point P(x, y), then ρ = PC.

Proceeding as Ex. 4 we have $\rho = a\cosh^2\left(\frac{x}{a}\right) = PC$.

Also length of normal at P = PG = y sec

$$y = y\sqrt{(1+\tan^2\psi)} = y\sqrt{\left\{1+\left(\frac{dy}{dx}\right)^2\right\}} = y\sqrt{\left\{1+\sinh^2\left(\frac{x}{a}\right)\right\}}$$

$$= a\cosh\left(\frac{x}{a}\right)\cosh\left(\frac{x}{a}\right) = a\cosh^2\left(\frac{x}{a}\right) = PC.$$ **Hence proved.**

Example 11:

Find the radius of curvature at the point (x, y) on the parabola $y^2 = 4ax$.

Solution:

The given curve is $y^2 = 4ax$. ...(1)

Differentiating (1) w.r.t. x, we get

$$2y\frac{dy}{dx} = 4a \text{ or } \frac{dy}{dx} = \frac{2a}{y}.$$

$$\therefore \frac{d^2y}{dx^2} = \frac{d}{dx}\left(\frac{2a}{y}\right) = -\frac{2a}{y^2}\frac{dy}{dx} = -\frac{2a}{y^2}.\frac{2a}{y} = -\frac{4a^2}{y^3}.$$

Now $$\rho = \frac{\left[1+\left(\frac{dy}{dx}\right)^2\right]^{3/2}}{\frac{d^2y}{dx^2}} = \frac{\left(1+\frac{4a^2}{y^2}\right)^{3/2}}{-\frac{4a^2}{y^3}} = -\frac{\left(y^2+4a^2\right)^{3/2}}{y^3}.\frac{y^3}{4a^2}$$

$$= -\frac{1}{4a^2}\left(4ax+4a^2\right)^{3/2}$$ from (1)

$$= -\frac{1}{4a^2}.a^{3/2}.8.(x+a)^{3/2} = \frac{2}{\sqrt{a}}(x+a)^{3/2}$$

neglecting the – ive sign because ρ is a length. **Ans.**

Example 12:

Find the radius of curvature at any point (x_1, y_1) *of the curve* $\sqrt{x}+\sqrt{y}=\sqrt{a}$. **(Kanpur, 1990; Meerut, 92, 94P)**

Solution:

The given curve is $\sqrt{x}+\sqrt{y}=\sqrt{a}$. ...(1)

Differentiating (1) w.r.t. x, we get

$$\frac{1}{2}x^{-1/2}+\frac{1}{2}y^{-1/2}\left(\frac{dy}{dx}\right)=0$$

or $$\frac{dy}{dx}=-\frac{y^{1/2}}{x^{1/2}}=-y^{1/2}x^{-1/2}.$$

$$\therefore \quad \frac{d^2y}{dx^2}=-\left(\frac{1}{2}y^{-1/2}\frac{dy}{dx}\right)x^{-1/2}-y^{1/2}\left(-\frac{1}{2}x^{-3/2}\right)$$

$$=-\frac{1}{2}y^{-1/2}x^{-1/2}\left(-y^{1/2}x^{1/2}\right)+\frac{1}{2}y^{1/2}x^{-3/2}$$

$$=\frac{1}{2x}+\frac{y^{1/2}}{2x^{3/2}}=\frac{1}{2x}\left(1+\frac{y^{1/2}}{x^{1/2}}\right)=\frac{1}{2x}\frac{x^{1/2}+y^{1/2}}{x^{1/2}}=\frac{\sqrt{a}}{2x^{3/2}}.$$

$$\therefore \quad \rho \text{ at } (x_1, y_1)=\frac{\left[1+\left(\frac{y_1}{x_1}\right)\right]^{3/2}}{\sqrt{\frac{a}{\left(2x_1^{3/2}\right)}}}=\frac{2\left(x_1+y_1\right)^{3/2}}{\sqrt{a}}.$$

Example 13:

Find the radius of curvature at any point (x, y) of the curves

(i) $ay^2 = x^3$,

(ii) $xy = c^2$,

(iii) $y = c \log \sec\left(\frac{x}{c}\right)$, **(Lucknow, 1994)**

(iv) $x^{2/3} + y^{2/3} = a^{2/3}$. **(Meerut, 1991 P; Kanpur, 99)**

Solution:

(i) The equation of the given curve is $ay^2 = x^3$. ...(1)

Differentiating (1) w.r.t. x, we get

$$2ay\frac{dy}{dx}=3x^2 \text{ or } \frac{dy}{dx}=\frac{3x^2}{2ay}=\frac{3x^2}{2a\left(\frac{x^3}{a}\right)^{1/2}}, \quad \text{from (1)}$$

$$=\left(\frac{3}{2\sqrt{a}}\right)x^{1/2}.$$

Differentiating again, $\frac{d^2y}{dx^2} = \frac{3}{2\sqrt{a}}\frac{1}{2}x^{-1/2} = \frac{3}{4\sqrt{a}.x^{1/2}}$.

Now $\rho = \frac{\left[1+\left(\frac{dy}{dx}\right)^2\right]^{3/2}}{\frac{d^2y}{dx^2}} = \frac{\left\{1+\left(\frac{9x}{4a}\right)\right\}^{3/2}}{\frac{3}{4\sqrt{ax^{1/2}}}}$

$$= \frac{(4a+9x)^{3/2}}{8a^{3/2}} \cdot \frac{4\sqrt{ax^{1/2}}}{3} = \frac{x^{1/2}(4a+9x)^{1/2}}{6a}.$$ **Ans.**

(ii) We have xy = c or $y = \frac{c^2}{x}$.

$\therefore \frac{dy}{dx} = -\frac{c^2}{x^2}$ and $\frac{d^2y}{dx^2} = \frac{2c^2}{x^3}$.

$$\therefore \rho = \frac{\left[1+\left(\frac{dy}{dx}\right)^2\right]^{3/2}}{\frac{d^2y}{dx^2}} = \frac{\left[1+\left(\frac{c^4}{x^4}\right)\right]^{3/2}}{\frac{2c^2}{x^3}}$$

$$= \frac{\left(x^4+c^4\right)^{3/2}}{2c^3x^3} = \frac{\left(x^4+x^2y^2\right)^{3/2}}{2c^2x^3},$$ $[\because xy = c^2]$

$$= \frac{\left(x^2+y^2\right)^{3/2}}{2c^2} = \frac{r^3}{2c^2},$$ where $r^2 = x^2 + y^2$. **Ans.**

(iii) We have $y = c \log \sec\left(\frac{x}{c}\right)$.

$$\therefore \quad \frac{dy}{dx} = \frac{c}{\sec\left(\frac{x}{c}\right)}.\sec\left(\frac{x}{c}\right).\tan\left(\frac{x}{c}\right).\frac{1}{c} = \tan\left(\frac{x}{c}\right)$$

Also $\frac{d^2y}{dx^2} = \left(\frac{1}{c}\right)\sec^2\left(\frac{x}{c}\right)$.

$$\therefore \rho = \frac{\left[1+\left(\frac{dy}{dx}\right)^2\right]^{3/2}}{\frac{d^2y}{dx^2}} = \frac{\left[1+\tan^2\left(\frac{x}{c}\right)\right]^{3/2}}{\left(\frac{1}{c}\right)\sec^2\frac{x}{c}} = \frac{c\sec^3\left(\frac{x}{c}\right)}{\sec^2\left(\frac{x}{c}\right)}$$

$$= c\sec\left(\frac{x}{c}\right).$$ **Ans.**

(iv) The given curve is $x^{2/3} + y^{2/3} = a^{2/3}$. ...(1)

The co-ordinates (x, y) of any point on (1) may be taken as $x = a\cos^3 t$, $y = a\sin^3 t$, where t is the parameter.

$$\therefore \frac{dx}{dt} = -3a\cos^2 t\sin t \text{ and } \frac{dy}{dt} = 3a\sin^2 t\cos t.$$

Now $\dfrac{dy}{dx} = \dfrac{\frac{dy}{dt}}{\frac{dx}{dt}} = \dfrac{3a\sin^2 t\cos t}{-3a\cos^2 t\sin t} = -\tan t$.

Also $\dfrac{d^2y}{dx^2} = \dfrac{d}{dx}\left(\dfrac{dy}{dx}\right) = \dfrac{d}{dx}(-\tan t) = \left\{\dfrac{d}{dt}(-\tan t)\right\}.\dfrac{dt}{dx}$

$$= -\sec^2 t.\left(\frac{dt}{dx}\right)$$

$$= -\sec^2 t.\frac{1}{-3a\cos^2 t\sin t}, \qquad \left[\because \frac{dx}{dt} = -3a\cos^2 t\sin t\right]$$

$$= -\left(\frac{1}{3a}\right)\sec^4 t\,\text{cosec}\, t.$$

$$\therefore \rho \text{ at the point 't'} = \frac{\left[1+\left(\frac{dy}{dx}\right)^2\right]^{3/2}}{\frac{d^2y}{dx^2}} = \frac{(1+\tan^2 t)^{3/2}}{\left(\frac{1}{3a}\right)\sec^4 t\,\text{cosec}\, t}$$

$$= \frac{3a\sec^3 t}{\sec^4 t\,\text{cosec}\, t} = 3a\cos t\sin t.$$

But $\cos^3 t = \dfrac{x}{a}$ and $\sin^3 t = \dfrac{y}{a}$. Therefore, $\cos t = \left(\dfrac{x}{a}\right)^{1/3}$ and

$$\sin t = \left(\frac{y}{a}\right)^{1/3}.$$

Hence ρ at the point (x, y)

$$= 3a\left(\frac{x}{a}\right)^{1/3}\left(\frac{y}{a}\right)^{1/3} = 3a^{1/3}x^{1/3}y^{1/3}.$$

Alternative Solution: Differentiating both sides of (1) with respect to x, we get

$$\frac{2}{3}x^{-1/3}+\frac{2}{3}y^{-1/3}.\left(\frac{dy}{dx}\right)=0$$

or $$\frac{dy}{dx}=-\frac{x^{-1/3}}{y^{-1/3}}=-y^{1/3}x^{-1/3}.$$

$\therefore$ $$\frac{d^2y}{dx^2}=-\left(\frac{1}{3}y^{-2/3}\frac{dy}{dx}\right).x^{-1/3}+y^{1/3}.\frac{1}{3}x^{-4/3}$$

$$=-\frac{1}{3}y^{-2/3}x^{-1/3}.\left(y^{1/3}x^{-1/3}\right)+\frac{1}{3}y^{1/3}x^{-4/3}$$

$$=\frac{1}{3}y^{-1/3}x^{-2/3}+\frac{1}{3}y^{1/3}x^{-4/3}$$

$$=\frac{1}{3}x^{-4/3}y^{-1/3}\left(x^{2/3}+y^{2/3}\right)=\frac{1}{3}x^{-4/3}y^{-1/3}a^{2/3}.$$

$\therefore$ r at the point (x, y) $$=\frac{\left[1+\left(\frac{dy}{dx}\right)^2\right]^{3/2}}{\frac{d^2y}{dx^2}}$$

$$=\frac{\left[1+\left(\frac{x^{2/3}}{x^{2/3}}\right)\right]^{3/2}}{\frac{1}{3}x^{-4/3}y^{-1/3}a^{2/3}}=\frac{3\left(x^{2/3}+y^{2/3}\right)^{3/2}}{\left(x^{2/3}\right)^{3/2}x^{-4/3}y^{-1/3}a^{2/3}}$$

$$=\frac{3\left(a^{2/3}\right)^{3/2}}{xx^{-4/3}y^{-1/3}a^{2/3}}=\frac{3a}{x^{-1/3}y^{-1/3}a^{2/3}}=3a^{1/3}x^{1/3}y^{1/3}.$$

Example 14:

Prove that for the curve

$$s = a\ (\sec^3 \psi - 1),\ r = 3a \tan \psi \sec^3 \psi$$

and hence prove that $3a\frac{dy}{dx}.\frac{d^2y}{dx^2}=1$. *Also prove that this differential equation is satisfied by the curve* $27ay^2 = 8x^3$.

Solution:

We have s = $(\sec^3 \psi - 1)$.

$$\therefore\ \rho=\frac{ds}{d\psi}=a\ (3 \sec^2 \psi \sec \psi \tan \psi) = 3a \tan \psi \sec^3 \psi. \qquad ...(1)$$

Also $$\rho=\frac{\left[1+\left(\frac{dy}{dx}\right)^2\right]^{3/2}}{\frac{d^2y}{dx^2}}=\frac{\left(1+\tan^2\psi\right)^{3/2}}{\frac{d^2y}{dx^2}}=\frac{\sec^3\psi}{\frac{d^2y}{dx^2}}.$$

$$\left[\because \frac{dy}{dx}=\tan\psi\right]$$

$$\therefore \frac{d^2y}{dx^2}=\frac{\sec^3\psi}{\rho}=\frac{\sec^3\psi}{3a\tan\psi\sec^3\psi}, \quad \text{from (1)}$$

$$=\frac{1}{3a\left(\frac{dy}{dx}\right)}. \qquad \therefore 3a\frac{dy}{dx}\frac{d^2y}{dx^2}=1 \qquad ...(2)$$

Now, for the curve $27ay^2 = 8x^3$, we have on differentiating w.r.t. x,

$$54ay\left(\frac{dy}{dx}\right)=24x^2$$

or $$\frac{dy}{dx}=\frac{4x^2}{9ay}$$

or $$\left(\frac{dy}{dx}\right)^2=\frac{16x^4}{81a^2y^2}=\frac{16x^4}{81a\left(\frac{8x^3}{27}\right)}. \qquad \left[\because ay^2=\frac{8x^3}{27}\right]$$

Thus, $$\left(\frac{dy}{dx}\right)^2=\frac{\left(16x^4\right).27}{(81a).8x^3}=\frac{2x}{3a} \qquad ...(3)$$

Differentiating w.r.t. x, we have

$$2\frac{dy}{dx}\frac{d^2y}{dx^2}=\frac{2}{3a} \text{ or } 3a\frac{dy}{dx}\frac{d^2y}{dx^2}=1. \qquad ...(4)$$

The differential equations (2) and (4) are the safe. Hence the differential equation (2) is satisfied by the curve $27ay^2 = 8x^3$. **Hence proved.**

Example 15:

Prove that for the ellipse $\frac{x^2}{a^2}+\frac{y^2}{b^2}=1$, $\rho=\frac{a^2b^2}{p^3}$, *p being the perpendicular from the centre upon the tangent at (x, y).*

(Delhi, 1993, 90; Rohilkhand, 98; Agra, 96; Meerut, 91; Kanpur, 92; Allahabad, 92; Lucknow, 93, 90; Indore, 93; Bhopal, 91)

Solution:

Proceeding as in Ex. 2, we get

$$\rho \text{ at the point } (x, y) = \frac{\left(b^4x^2 + a^4y^2\right)^{3/2}}{a^4b^4}. \quad ...(1)$$

Now, the equation of the tangent to the given ellipse at (x, y) is

$$\frac{Xx}{a^2} + \frac{Yy}{b^2} = 1.$$

$\therefore$ p = the length of the perpendicular from the centre (0, 0) to the tangent

$$= \frac{1}{\sqrt{\left(\frac{x^2}{a^4} + \frac{y^2}{b^4}\right)}} = \frac{a^2b^2}{\sqrt{\left(b^4x^2 + a^2y^2\right)}}.$$

$$\therefore \quad p^3 = \frac{a^6b^6}{\left(b^4x^2 + a^4y^2\right)^{3/2}} \text{ or } \left(b^4x^2 + a^4y^2\right)^{3/2} = \frac{a^6b^6}{p^3}.$$

$\therefore$ substituting it in (1), we get $\rho = \dfrac{\frac{a^6b^6}{p^3}}{a^4b^4} = \dfrac{a^2b^2}{p^3}$. **Hence proved.**

Example 16:

If ρ, ρ' be the radii of curvature at the extremities of two conjugate diameters of an ellipse, prove that

$$\{(\rho)^{2/3} + (\rho')^{2/3}\} (ab)^{2/3} = (a^2 + b^2).$$

(Meerut, 1993 S; Gorakhpur, 96; Indore, 94; Kanpur, 97, 92)

Solution:

Let the equation of the ellipse be

$$\frac{x^2}{a^2} + \frac{y^2}{b^2} = 1. \quad ...(1)$$

Let CP and CQ be a pair of conjugate semi-diameters of (1), where C is the centre of the ellipse. Let 't' be the eccentric angle of the point P. Then by co-ordinate geometry,, the eccentric angle of the point Q is $'t + \frac{1}{2}\pi'$.

Now, in terms of the eccentric angle 't the co-ordinates (x, y) of the point P are given by

$$x = a \cos t \text{ and } y = b \sin t.$$

$$\therefore \frac{dx}{dt} = -a \sin t, \frac{dy}{dt} = b \cos t.$$

Hence $\dfrac{dy}{dx} = \dfrac{\frac{dy}{dt}}{\frac{dx}{dt}} = \dfrac{b \cos t}{-a \sin t} = -\dfrac{b}{a}\cot t$.

Also $\dfrac{d^2y}{dx^2} = \dfrac{d}{dx}\left(\dfrac{dy}{dx}\right) = \dfrac{d}{dx}\left(-\dfrac{b}{a}\cot t\right) = \left\{\dfrac{d}{dt}\left(-\dfrac{b}{a}\cot t\right)\right\}.\dfrac{dt}{dx}$

$$= \frac{b}{a}\text{cosec}^2 t.\frac{1}{-a \sin t} = -\frac{b}{a^2}\text{cosec}^3 t.$$

$\therefore$ if ρ be the radius of curvature at the point P, *i.e.*, at the point 't', we have

$$\rho = \frac{\left[1+\left(\frac{dy}{dx}\right)^2\right]^{3/2}}{\frac{d^2y}{dx^2}} = \frac{\left(1+\frac{b^2 \cos^2 t}{a^2 \sin^2 t}\right)^{3/2}}{\left(-\frac{b}{a^2}\right)\text{cosec}^3 t}$$

$$= -\frac{\left(a^2 \sin^2 t + b^2 \cos^2 t\right)^{3/2}}{ab}.$$

Neglecting the negative sign, we get

$$\rho = \frac{\left(a^2 \sin^2 t + b^2 \cos^2 t\right)^{3/2}}{ab} \text{ or } ab.\rho = (a^2 \sin^2 t + b^2 \cos^2 t)^{3/2}$$

or $(ab)^{2/3} \rho^{2/3} = a^2 \sin^2 t + b^2 \cos^2 t$. ...(2)

If ρ' be the radius of curvature at the point Q, *i.e.*, at the point '$t + \frac{1}{2}\pi$', then replacing ρ by ρ' and t by $t + \frac{1}{2}\pi$ in (2), we get

$$(ab)^{2/3} \rho'^{2/3} = a^2 \sin^2\left(\frac{1}{2}\pi + t\right) + b^2 \cos^2\left(\frac{1}{2}\pi + t\right)$$

$$= a^2\cos^2 t + b^2 \sin^2 t. \qquad ...(3)$$

Adding (2) and (3), we get

$$(ab)^{2/3} \rho^{2/3} + (ab)^{2/3} \rho'^{2/3} = a^2 + b^2$$

or $(ab)^{2/3} (\rho^{2/3} + \rho'^{2/3}) = a^2 + b^2$. **Hence proved.**

Example 17:

Find the radius of curvature of $y = 4 \sin x - \sin 2x$ at $x = \frac{1}{2}\pi$.

(Kanpur, 1994)

Solution:

We have y = 4 sin x – sin 2x.

$$\therefore \quad \frac{dy}{dx} = 4\cos x - 2\cos 2x \text{ and } \frac{d^2y}{dx^2} = -4\sin x + 4\sin 2x.$$

$$\therefore \quad \text{at } x = \frac{1}{2}\pi,$$

$$\frac{dy}{dx} = 4\cos\frac{1}{2}\pi - 2\cos\pi = 2,$$

and

$$\frac{d^2y}{dx^2} = -4\sin\frac{1}{2}\pi + 4\sin\pi = -4.$$

$$\therefore \quad \rho \text{ at } \left(x = \frac{\pi}{2}\right) = \frac{(1+4)^{3/2}}{-4}$$

$$= \left(5\sqrt{\frac{5}{4}}\right), \text{ neglecting the – ive sign.}$$

Example 18:

For the curve $y = \frac{ax}{a+x}$*, if* ρ *is the radius of curvature at any point (x, y), show that*

$$\left(\frac{2\rho}{a}\right)^{2/3} = \left(\frac{y}{x}\right)^2 + \left(\frac{x}{y}\right)^2.$$

(Meerut, 1990 P; Kumaun, 93)

Solution:

We have $y = \frac{ax}{a+x}$, ...(1)

Differential w.r.t. x we have

$$\therefore \quad \frac{dy}{dx} = a\frac{(a+x)-x}{(a+x)^2} = \frac{a^2}{(a+x)^2} = a^2(a+x)^{-2}$$

and

$$\frac{d^2y}{dx^2} = \frac{d}{dx}\left(\frac{dy}{dx}\right) = 2a^2(a+x)^{-3} = \frac{-2a^2}{(a+x)^3} = \frac{-2a}{\left(\frac{ax}{y}\right)^3} = -\frac{2y^3}{ax^3}.$$

$$\text{Now } 1+\left(\frac{dy}{dx}\right)^2 = 1+\frac{a^4}{(a+x)^4} = 1+\frac{a^4}{\left(\frac{ax}{y}\right)^4} = 1+\frac{y^4}{x^4} = \frac{x^4+y^4}{x^4}.$$

$$\therefore\ \rho=\frac{\left[1+\left(\dfrac{dy}{dx}\right)^2\right]^{3/2}}{\dfrac{d^2y}{dx^2}}=\frac{\left[\dfrac{\left(x^4+y^4\right)}{x^4}\right]^{3/2}}{\left(\dfrac{2y^3}{ax^3}\right)}$$

(negative sign being neglected)

$$=\frac{a}{2}\frac{\left(x^4+y^4\right)^{3/2}}{x^6\left(\dfrac{y^3}{x^3}\right)}=\frac{a}{2}\frac{\left(x^4+y^4\right)^{3/2}}{x^3y^3}$$

Hence $\left(\dfrac{2\rho}{a}\right)^{2/3}=\dfrac{x^4+y^4}{x^2y^2}=\dfrac{x^4}{x^2y^2}+\dfrac{y^4}{x^2y^2}=\dfrac{x^2}{y^2}+\dfrac{y^2}{x^2}=\left(\dfrac{x}{y}\right)^2+\left(\dfrac{y}{x}\right)^2.$

Hence proved.

Example 19:

Prove that for the curve

$$y = a \log \cot\left(\frac{1}{4}\pi-\frac{1}{2}\psi\right) + a \sin y \sec^2 \psi,\ \rho = 2a \sec^3 \psi$$

and hence prove that $\dfrac{d^2y}{dx^2}=\dfrac{1}{2a}$, *and that this differential equation is satisfied by the parabola* $x^2 = 4ay$. **(Gorakhpur, 1997)**

Solution:

We have $s = a \text{ lo} \cot\left(\dfrac{1}{4}\pi-\dfrac{1}{2}\psi\right) + a \sin \psi \sec^2 \psi$

$= a \log \cot\left(\dfrac{1}{4}\pi-\dfrac{1}{2}\psi\right) + a \sec \psi \tan \psi.$

$$\therefore\ \rho=\frac{ds}{d\psi}$$

$$=\frac{a\left\{-\operatorname{cosec}^2\left(\dfrac{1}{4}\pi-\dfrac{1}{2}\psi\right)\right\}}{\cot\left(\dfrac{1}{4}\pi-\dfrac{1}{2}\psi\right)}.\left(-\frac{1}{2}\right)+a\left(\tan^2\psi\sec\psi+\sec^3\psi\right)$$

$$=\frac{a}{2\sin\left(\dfrac{1}{4}\pi-\dfrac{1}{2}\psi\right)\cos\left(\dfrac{1}{4}\pi-\dfrac{1}{2}\psi\right)}+a\sec\psi\left(\sec^2\psi+\tan^2\psi\right)$$

$$= \frac{a}{\sin\left(\frac{1}{2}\pi - \psi\right)} + a \sec \psi \left(\sec^2 \psi + \tan^2 \psi\right)$$

$= a \sec \psi + a \sec \psi (\sec^2 \psi + \tan^2 \psi)$

$= a \sec \psi (1 + \tan^2 \psi + \sec^2 \psi)$

$= a \sec \psi (\sec^2 \psi + \sec^2 \psi) = 2a \sec^3 \psi.$

But
$$\rho = \frac{\left[1 + \left(\frac{dy}{dx}\right)^2\right]^{3/2}}{\left(\frac{d^2y}{dx^2}\right)} = \frac{\left(1 + \tan^2 \psi\right)^{3/2}}{\left(\frac{d^2y}{dx^2}\right)}$$

$$= \frac{\sec^3 \psi}{\left(\frac{d^2y}{dx^2}\right)}.$$

$$\therefore \quad 2a \sec^3 \psi = \frac{\sec^3 \psi}{\frac{d^2y}{dx^2}}; \quad \text{or} \quad \frac{d^2y}{dx^2} = \frac{1}{2a}. \quad \text{...(1)}$$

Now for the parabola $x^2 = 4ay$,

$$\frac{dy}{dx} = \frac{2x}{4a} = \frac{x}{2a}; \qquad \therefore \frac{d^2y}{dx^2} = \frac{1}{2a} \quad \text{...(2)}$$

We see that the differential equations (1) and (2) are the same. Hence the differential equation (1) is satisfied by the parabola

$$x^2 = 4ay$$

Example 20:

(a) *Show that in the parabola $y^2 = 4ax$, the radius of curvature r at any point P is twice the part of the normal intercepted between the curve and the directrix.*

(b) *Also prove that r^2 varies as $(SP)^3$, where S is the focus.*

(Meerut, 1991S, 94)

Solution:

(a) We have $y^2 = 4ax$.

$$\therefore \quad 2y\frac{dy}{dx} = 4a \text{ or } \frac{dy}{dx} = \frac{2a}{y} = \frac{2a}{(4ax)^{1/2}} = \frac{\sqrt{a}}{\sqrt{x}}.$$

Also
$$\frac{d^2y}{dx^2} = \frac{d}{dx}\left(\frac{dy}{dx}\right) = \sqrt{a}.\left(-\frac{1}{2}\right)x^{-3/2} = \frac{-\sqrt{a}}{2x^{3/2}}.$$

$\therefore$ at any point P(x, y), we have

$$\rho = \frac{\left[1+\left(\frac{dy}{dx}\right)^2\right]^{3/2}}{\frac{d^2y}{dx^2}} = -\frac{\left[1+\left(\frac{a}{x}\right)\right]^{3/2}}{\frac{-\sqrt{a}}{\left(2x^{3/2}\right)}}$$

$$= \left(\frac{2}{\sqrt{a}}\right)(x+a)^{3/2}, \text{ neglecting the } - \text{ ive sign.}$$

Now, the equation of the normal to the given parabola at the point P(x, y) is

$$\frac{dy}{dx}(Y-y)+(X-x)=0 \text{ or } \frac{2a}{y}(Y-y)+(X-x)=0, \quad \left[\because \frac{dy}{dx}=\frac{2a}{y}\right]$$

or $\quad yX + 2aY = xy + 2ay.$ ...(1)

Also directrix of the parabola is $X = -a$. ...(2)

Now the co-ordinates of the point of intersection of (1) and (2) are obtained by solving (1) and (2), we get

$$-ay + 2aY = xy + 2ay, \quad \text{[putting } X = -a \text{ in (1)]}$$

or $\quad Y = \left(\frac{1}{2a}\right)(xy+3ay) = \left(\frac{y}{2a}\right)(x+3a).$

$\therefore$ the point of intersection of the normal and the directrix is

$$\left[-a, \left(\frac{y}{2a}\right)(x+3a)\right].$$

Therefore the length of the normal intercepted between the curve and the directrix

= the distance between (x, y) and $\left\{-a, \left(\frac{y}{2a}\right)(x+3a)\right\}$

$$= \sqrt{\left[(x+a)^2 + \left\{y - \frac{y}{2a}(x+3a)\right\}^2\right]}$$

$$= \frac{1}{2a}\sqrt{\left[4a^2(x+a)^2 + y^2(x+a)^2\right]}$$

$$= \frac{1}{2a}.(x+a).\sqrt{(4a^2+y^2)} = \frac{1}{2a}.(x+a)\sqrt{(4a^2+4ax)}, \quad [\because y^2 = 4ax]$$

$$\frac{(x+a)^{3/2}}{a^{1/2}}.$$

As already proved, ρ at (x, y) $= \dfrac{2(x+a)^{3/2}}{\sqrt{a}}$

= 2 × the part of normal intercepted between the curve and directrix.

(b) The focus S is (a, 0).

$$\therefore\ SP = \sqrt{\left[(x-a)^2+(y-0)^2\right]} = \sqrt{\left\{(x-a)^2+4ax\right\}}, \quad [\because\ y^2 = 4ax]$$

$$= \sqrt{\left\{(x+a)^2\right\}} = x+a.$$

But $\rho = \dfrac{2}{\sqrt{a}}.(x+a)^{3/2} = \dfrac{2}{\sqrt{a}}.(SP)^{3/2}$. $\therefore\ \rho^2 \propto (SP)^3$.

9.5 RADIUS OF CURVATURE FOR PARAMETRIC CURVES

(Kanpur, 1998; Meerut, 97)

Let the curve be defined by means of parametric equations

$$x = f(t) \text{ and } y = \phi(t).$$

Then
$$\frac{dy}{dx} = \frac{\dfrac{dy}{dt}}{\dfrac{dx}{dt}} = \frac{y'}{x'}.$$

(Here dashes or accets denote differentiation w.r.t. 't'.)

Also
$$\frac{d^2y}{dx^2} = \frac{d}{dx}\left(\frac{dy}{dx}\right) = \frac{d}{dx}\left(\frac{y'}{x'}\right) = \left\{\frac{d}{dt}\left(\frac{y'}{x'}\right)\right\}\frac{dt}{dx}$$

$$= \frac{x'y''-y'x''}{(x')^2}\cdot\frac{1}{x'}, \qquad \left[\because\ \frac{dx}{dt} = x' \text{ and } \frac{dt}{dx} = \frac{1}{x'}\right]$$

$$= \frac{x'y''-y'x''}{(x')^3}.$$

$$\therefore \qquad \rho = \frac{\left\{1+\left(\dfrac{dy}{dx}\right)^2\right\}^{3/2}}{\dfrac{d^2y}{dx^2}} = \frac{\left[1+\left(\dfrac{y'}{x'}\right)^2\right]^{3/2}}{\dfrac{(x'y''-y'x'')}{(x')^3}}$$

$$= \frac{\left(x'^2+y'^2\right)^{3/2}}{x'y''-y'x''}, \text{ where } x'y'' - y'x'' \neq 0.$$

Note: Curvature $= \frac{1}{\rho} = \frac{x'y''-y'x''}{\left(x'^2+y'^2\right)^{3/2}}$.

Example 1:

Find the radius of curvature at any point of the cycle x=a(t + sin t), y = a(1 − cos t).

(Meerut, 1991, 93, 98S, Gorakhpur, 90, 98; Delhi, 97; Rohilkhand, 99)

Solution:

We have $x = a(t + \sin t),\ y = a(1 - \cos t)$

$$\therefore \quad \frac{dx}{dt} = a(1+\cos t),\ \frac{dy}{dt} = a \sin t.$$

Now $$\frac{dy}{dx} = \frac{\frac{dy}{dt}}{\frac{dx}{dt}} = \frac{a \sin t}{a(1+\cos t)} = \frac{2\sin\frac{1}{2}t\cos\frac{1}{2}t}{2\cos^2\frac{1}{2}t} = \tan\frac{1}{2}t.$$

Also $$\frac{d^2y}{dx^2} = \frac{d}{dx}\left(\frac{dy}{dx}\right) = \frac{d}{dx}\left(\tan\frac{1}{2}t\right) = \left\{\frac{d}{dt}\left(\tan\frac{1}{2}t\right)\right\}.\frac{dt}{dx}$$

$$= \frac{1}{2}\sec^2\frac{1}{2}t.\frac{1}{a(1+\cos t)} = \frac{1}{2}\sec^2\frac{1}{2}t.\frac{1}{2a\cos^2\frac{1}{2}t} = \frac{1}{4a}\sec^4\frac{1}{2}t$$

Hence $$\rho = \frac{\left[1+\left(\frac{dy}{dx}\right)^2\right]^{3/2}}{\frac{d^2y}{dx^2}} = \frac{\left(1+\tan^2\frac{1}{2}t\right)^{3/2}}{\left(\frac{1}{4a}\right)\sec^4\frac{1}{2}t} = \frac{4a\sec^3\frac{1}{2}t}{\sec^4\frac{1}{2}t}.$$

$$= 4a\cos\frac{1}{2}t.$$

Note: This question can also by denote by using the formula. Thus here

$$x' = \frac{dx}{dt} = a(1+\cos t),\ x'' = \frac{d^2x}{dt^2} = -a\sin t,$$

$$y' = \frac{dy}{dt} = a\sin t,\ y'' = \frac{d^2y}{dt^2} = a\cos t.$$

Now $$\rho = \frac{\left(x'^2+y'^2\right)^{3/2}}{x'y''-y'x''} = \frac{\left[a^2(1+\cos t)^2 + a^2\sin^2 t\right]^{3/2}}{a(1+\cos t)\,a\cos t - a\sin t(-a\sin t)}$$

$$= a\frac{[2(1+\cos t)]^{3/2}}{1+\cos t} = (2\sqrt{2})\, a\,(1+\cos t)^{1/2}$$

$$= (2\sqrt{2})a.\left(2\cos^2\frac{1}{2}t\right)^{1/2} = 4a\cos\frac{1}{2}t.$$

Example 2:

If ρ_1 and ρ_2 be the radii of curvature at the extremities of a focal chord of the parabola $y^2 = 4ax$, prove that

$(\rho_1)^{-2/3} + (\rho_2)^{2/3} = (2a)^{-2/3}.$

Solution:

Let the extremities of a focal chord be the points $P(at_1^2, 2at_1)$ and $Q(at_2^2, 2at_2)$.

Then by co-ordinate geometry, $t_1t_2 = -1$. ...(1)

Also $x = at^2$, $y = 2at$ are the parametric equations of the parabola $y^2 = 4ax$.

We have $x' = \frac{dx}{dt} = 2at$, $y' = \frac{dy}{dt} = 2a$,

$$x'' = \frac{d^2x}{dt^2} = 2a, \quad y'' = \frac{d^2y}{dt^2} = 0.$$

$\therefore$ ρ at the point 't' $= \dfrac{\{(x')^2 + (y')^2\}^{3/2}}{(x'y'' - y'x'')}$,

$$= \frac{[4a^2t^2 + 4a^2]^{3/2}}{2at.0 - 2a.2a} = \frac{8a^3(1+t^2)^{3/2}}{-4a^2}$$

$= 2a(1 + t^2)^{3/2}$, [Negative sign being neglected].

$\therefore$ ρ_1 = radius of curvature at the point $P(at_1^2, 2at_1)$

$$= 2a(1+t_1^2)^{3/2},$$

and ρ_2 = radius of curvature at the point $Q(at_2^2, 2at_2) = 2a(1+t_2^2)^{3/2}$.

Hence $(\rho_1)^{-2/3} + (\rho_2)^{-2/3} = (2a)^{-2/3}\left[\dfrac{1}{1+t_1^2} + \dfrac{1}{1+t_2^2}\right]$

$$= (2a)^{-2/3}\left[\frac{t_1^2+t_2^2+2}{t_1^2+t_2^2+1+1}\right]. \qquad \left[\because t_1^2 t_2^2 = 1\right]$$

$$= (2a)^{-2/3}.$$

Example 3:

Find the radius of curvature at the vertex of the cycloid
$x = a(\theta + \sin\theta)$. $y = a(1 - \cos\theta)$.

Solution:

Proceeding as in Ex. 1, we get $\rho = 4a\cos\frac{1}{2}\theta$.

At the vertex $\theta = 0$ and $\rho = 4a\cos 0 = 4a$.

Example 4:

Prove that in the curve

$4x = \theta + \sin\theta$, $4y = 1 - \cos\theta$, $\rho = \cos\frac{1}{2}\theta$. **(Rohilkhand, 1996)**

Solution:

Proceed exactly as in Ex. 1.

Example 5:

Find the radius of curvature at any point 'θ' of the curve

(i) $x = a\cos^3\theta$, $y = a\sin^3\theta$.

(Meerut, 1995, 97, 95, 96; Rohilkhand, 96)

(ii) $x = a\cos\theta$, $y = b\sin\theta$. **(Gorakhpur, 1992)**

(iii) $x = a(\theta - \sin\theta)$, $y = a(1 - \cos\theta)$.

(iv) $x = a\left(\cos\theta + \log\tan\frac{1}{2}\theta\right)$, $y = a\sin\theta$.

Solution:

(i) Do your self

(ii) Do your self

(iii) We have $x = a(\theta - \sin\theta)$, $y = a(1 - \cos\theta)$.

$$\therefore \qquad \frac{dx}{d\theta} = a(1-\cos\theta) = 2a\sin^2\frac{1}{2}\theta,$$

$$\text{and} \qquad \frac{dy}{d\theta} = a\sin\theta = 2a\sin\frac{1}{2}\theta\cos\frac{1}{2}\theta.$$

$$\text{Now} \qquad \frac{dy}{dx} = \frac{\frac{dy}{d\theta}}{\frac{dx}{d\theta}} = \frac{2a\sin\frac{1}{2}\theta.\cos\frac{1}{2}\theta}{2a\sin^2\frac{1}{2}\theta} = \cot\frac{1}{2}\theta$$

Also $$\frac{d^2y}{dx^2}=\frac{d}{dx}\left(\frac{dy}{dx}\right)=\frac{d}{dx}\left(\cot\frac{1}{2}\theta\right)=\frac{d}{d\theta}\left(\cot\frac{1}{2}\theta\right).\frac{d\theta}{dx}$$

$$=-\frac{1}{2}\text{cosec}^2\frac{1}{2}\theta.\frac{1}{2a\sin^2\frac{1}{2}\theta}=-\frac{1}{4a}\text{cosec}^4\frac{1}{2}\theta.$$

$$\therefore \qquad \rho=\frac{\left[1+\cot^2\frac{1}{2}\theta\right]^{3/2}}{-\left(\frac{1}{4a}\right)\text{cosec}^4\frac{1}{2}\theta}=-\frac{4a}{\text{cosec}\frac{1}{2}\theta}=4a\sin\frac{1}{2}\theta.$$

(Negative sign being neglected).

(iv) We have $x=a\left(\cos\theta+\log\tan\frac{1}{2}\theta\right)$, $y=a\sin\theta$.

$$\therefore \frac{dx}{d\theta}=-a\sin\theta+a.\frac{1}{\tan\frac{1}{2}\theta}.\frac{1}{2}\sec^2\frac{1}{2}\theta$$

$$=-a\sin\theta+\frac{a}{2\sin\frac{1}{2}\theta\cos\frac{1}{2}\theta}=-a\sin\theta+\frac{a}{\sin\theta}=\frac{a\cos^2\theta}{\sin\theta}$$

and $$\frac{dy}{d\theta}=a\cos\theta.$$

Now $$\frac{dy}{dx}=\frac{\frac{dy}{d\theta}}{\frac{dx}{d\theta}}=\frac{a\cos\theta}{\frac{a\cos^2}{\sin\theta}}=\tan\theta.$$

Also $$\frac{d^2y}{dx^2}=\frac{d}{d\theta}(\tan\theta).\frac{d\theta}{dx}=\sec^2\theta.\frac{\sin\theta}{a\cos^2\theta}=\frac{\sin\theta}{a\cos^4\theta}$$

$$\therefore \rho=\frac{\left(1+\tan^2\theta\right)^{3/2}}{\frac{\sin\theta}{\left(a\cos^4\theta\right)}}=\frac{a\cos^4\theta}{\sin\theta}.\sec^3\theta=a\cos\theta.$$

9.6 RADIUS OF CURVATURE FOR PEDAL CURVES

(Meerut, 1993, 90, 91, 96, 96P, 96 BP, 97; Gorakhpur, 97; Rohilkhand, 92, 91; Agra, 91, 99; Kanpur, 97)

We have $p=r\sin\phi$.

Differentiating w.r.t. r, we get

$$\frac{dp}{dr} = \sin\phi + r\cos\phi\left(\frac{d\phi}{dr}\right)$$

$$= r\frac{d\theta}{ds} + r\frac{d\phi}{dr}\frac{dr}{ds}, \qquad \left[\because \sin\phi - r\frac{d\theta}{ds} \text{ and } \cos\phi = \frac{dr}{ds}\right]$$

$$= r\left(\frac{d\theta}{ds} + \frac{d\phi}{ds}\right) = r\frac{d}{ds}(\theta + \phi) = r\frac{d\psi}{ds}, \qquad [\because = \psi = \theta + \phi]$$

$$= r.\left(\frac{1}{\rho}\right), \qquad \left[\because \rho = \frac{ds}{d\psi}\right].$$

Thus, $\frac{dp}{dr} = \frac{r}{\rho}$. Therefore $\rho = r\frac{dr}{dp}$.

Example 1:

Find the radius of curvature of the curve $p = \frac{r^{m+1}}{a^m}$; *show that the radius of curvature varies inversely as the* $(m - 1)^{th}$ *power of the radius vector.*

Solution:

We have $a^m p = r^{m+1}$. ...(1)

Differentiating (1) w.r.t. p, we get

$$a^m = (m + 1)\, r^m \frac{dr}{dp}. \quad \therefore \frac{dr}{dp} = \frac{a^m}{(m+1)\, r^m}.$$

Now $\rho = r\frac{dr}{dp} = \frac{ra^m}{(m+1)r^m} = \frac{a^m}{(m+1)\, r^{m-1}}$.

$\therefore \rho \propto \frac{1}{r^{m-1}}$ *i.e.*, the radius of curvature varies inversely as the $(m - 1)^{th}$ power of the radius vector.

Example 2:

Find the radius of curvature at the point (p, r) on the spiral

$$p^2 = \frac{r^4}{\left(r^2 + a^2\right)}.$$

Solution:

The equation of the curve is $p^2 = \frac{r^4}{\left(r^2 + a^2\right)}$. ...(1)

$$\therefore \quad \frac{1}{p^2} = \frac{r^2 + a^2}{r^4} = \frac{r^2}{r^4} + \frac{a^2}{r^4} = \frac{1}{r^2} + \frac{a^2}{r^4}.$$

Now differentiating both sides w.r.t. r, we have

$$-\frac{2}{p^3}\frac{dp}{dr}=-\frac{2}{r^3}-\frac{4a^2}{r^5}=-2.\frac{r^2+2a^2}{r^5}.$$

$$\therefore \quad \frac{dp}{dr}=p^3\frac{\left(r^2+2a^2\right)}{r^5}=\left(p^2\right)^{3/2}\frac{\left(r^2+2a^2\right)}{r^5} \qquad \textbf{(Note)}$$

$$=\left(\frac{r^4}{r^2+a^2}\right)^{3/2}\left(\frac{r^2+2a^2}{r^5}\right), \qquad \text{from (1)}$$

$$=\frac{r\left(r^2+2a^2\right)}{\left(r^2+a^2\right)^{3/2}}.$$

Hence $\rho=r\dfrac{dr}{dp}=r.\dfrac{\left(r^2+a^2\right)^{3/2}}{r\left(r^2+2a^2\right)}=\dfrac{\left(r^2+a^2\right)^{3/2}}{r^2+2a^2}.$

Example 3:

Find the radius of curvature at any point of the ellipse

$$\frac{1}{p^2}=\frac{1}{a^2}+\frac{1}{b^2}-\frac{r^2}{a^2b^2}. \qquad \textbf{(Delhi, 1992)}$$

Solution:

Differentiating the given equation w.r.t. 'p', we have

$$-\frac{2}{p^3}=-\frac{2r}{a^2b^2}\frac{dr}{dp}; \qquad \therefore \ \rho=r\frac{dr}{dp}=\frac{a^2b^2}{p^3}$$

Example 4:

Find the radius of curvature at the point (p, r) on the following curves:

(i) $p^2 = ar$,

(ii) $r^3 = 2ap^2$,

(iii) $r^3 = a^2p$. **(Meerut, 1997)**

Solution:

(i) We have $p^2 = ar$. ...(1)

Differentiating (1) w.r.t. p, we get

$$2p=a\frac{dr}{dp} \qquad \therefore \ \frac{dr}{dp}=\frac{2p}{a}.$$

Now $\rho=r\dfrac{dr}{dp}=\dfrac{2pr}{a}=\dfrac{2p}{a}\left(\dfrac{p^2}{a}\right)$ from (1)

$$= \frac{2p^3}{a^2}.$$

(ii) We have $r^3 = 3ap^2$. ...(1)

Differentiating (1) w.r.t. p, we have

$$3r^2 \frac{dr}{dp} = 4ap \text{ or } r\frac{dr}{dp} = \frac{4ap}{3r}.$$

$$\therefore \rho = r\frac{dr}{dp} = \frac{4ap}{3r} = \frac{4a}{3r}\sqrt{\left(\frac{r^3}{2a}\right)}, \quad \text{from (1)}$$

$$= \frac{2}{3}\sqrt{(2ar)}.$$

(iii) We have $r^3 = a^2p$. ...(1)

Differentiating (1) w.r.t. p, we get

$$3r^2\left(\frac{dr}{dp}\right) = a^2 \qquad \therefore r = \rho\left(\frac{dr}{dp}\right) = \frac{a^2}{3r}.$$

9.7 RADIUS OF CURVATURE FOR POLAR CURVES

(Gorakhpur, 1991, 99 Gurunanak, 91; Meerut, 90)

We know that $\frac{1}{p^2} = \frac{1}{r^2} + \frac{1}{r^4}\left(\frac{dr}{d\theta}\right)^2$. ...(1)

Differentiating (1) w.r.t. 'r', we get

$$-\frac{2}{p^3}.\frac{dp}{dr} = -\frac{2}{r^3} - \frac{4}{r^5}\left(\frac{dr}{d\theta}\right)^2 + \frac{1}{r^4}2\left(\frac{dr}{d\theta}\right).\frac{d}{dr}\left(\frac{dr}{d\theta}\right)$$

$$= -\frac{2}{r^3} - \frac{4}{r^5}\left(\frac{dr}{d\theta}\right)^2 + \frac{2}{r^4}\left(\frac{dr}{d\theta}\right)\left\{\frac{d}{d\theta}\left(\frac{dr}{d\theta}\right)\right\}.\frac{d\theta}{dr}$$

$$= -\frac{2}{r^3} - \frac{4}{r^5}\left(\frac{dr}{d\theta}\right)^2 + \frac{2}{r^4}\frac{d^2r}{d\theta^2}. \qquad \left[\because \frac{dr}{d\theta}.\frac{d\theta}{dr} = 1\right]$$

$$\therefore \frac{1}{p^3}\frac{dp}{dr} = \frac{1}{r^3} + \frac{2}{r^5}\left(\frac{dr}{d\theta}\right)^2 - \frac{1}{r^4}\frac{d^2r}{d\theta^2}.$$

Now $$\rho = r\frac{dr}{dp} = \frac{r\left(\frac{1}{p^3}\right)}{\left(\frac{1}{r^3}\right) + \left(\frac{2}{r^3}\right)\left(\frac{dr}{d\theta}\right)^2 - \left(\frac{1}{r^4}\right)\left(\frac{d^2r}{d\theta^2}\right)}$$

$$=\frac{r^6\left[\left(\frac{1}{r^2}\right)+\left(\frac{1}{r^4}\right)\left(\frac{dr}{d\theta}\right)^2\right]^{3/2}}{r^2+2\left(\frac{dr}{d\theta}\right)^2-r\left(\frac{d^2r}{d\theta^2}\right)}.$$

$$\left[\because \text{ from (1), } \frac{1}{p^3}=\left\{\left(\frac{1}{r^2}\right)+\left(\frac{1}{r^4}\right)\left(\frac{dr}{d\theta}\right)^2\right\}^{3/2}\right].$$

$$\therefore\quad \rho=\frac{\left[r^2+\left(\frac{dr}{d\theta}\right)^2\right]^{3/2}}{r^2+2\left(\frac{dr}{d\theta}\right)^2-r\left(\frac{d^2r}{d\theta^2}\right)}.$$

Cor. If we put $r=\frac{1}{u}$, we have

$$\frac{dr}{d\theta}=-\frac{1}{u^2}\frac{du}{d\theta} \text{ and } \frac{d^2r}{d\theta^2}=-\frac{1}{u^2}\frac{d^2u}{d\theta^2}+\frac{2}{u^3}\left(\frac{du}{d\theta}\right)^2.$$

$$\therefore\quad \rho=\frac{\left[r^2+\left(\frac{dr}{d\theta}\right)^2\right]^{3/2}}{r^2+2\left(\frac{dr}{d\theta}\right)^2-r\left(\frac{d^2r}{d\theta^2}\right)}$$

$$=\frac{\left[\frac{1}{u^2}+\frac{1}{u^4}\left(\frac{du}{d\theta}\right)^2\right]^{3/2}}{\frac{1}{u^2}+2.\frac{1}{u^4}\left(\frac{du}{d\theta}\right)^2-\frac{1}{u}\left[-\frac{1}{u^2}.\frac{d^2u}{d\theta^2}+\frac{2}{u^3}\left(\frac{du}{d\theta}\right)^2\right]}$$

$$=\frac{1}{u^6}\left[u^2+\left(\frac{du}{d\theta}\right)^2\right]^{3/2}\Big/\left[\frac{1}{u^2}+\frac{1}{u^3}\frac{d^2u}{d\theta^2}\right].$$

Hence $$\rho=\frac{\left[u^2+\left(\frac{du}{d\theta}\right)^2\right]^{3/2}}{u^3\left[u+\frac{d^2u}{d\theta^2}\right]}$$

Important Note: We see that the pedal formula for ρ is simpler than the polar formula. Therefore, if we are given the equation of a curve in polar form, it is often convenient to change it first to pedal equation and then to find ρ with the help of pedal formula.

Example 1:

Find the radius of curvature of the curve $\frac{2a}{r} = 1 + \cos\theta$; *hence show that the square of the radius of curvature varies as the cube of the focal distance.*

(Meerut, 1996)

Solution:

The given curve is $\frac{2a}{r} = 1 + \cos\theta$, ...(1)

which is a parabola with focus as pole.

Proceeding as in the chapter on tangents and normals, we get pedal equation of the curve as $p^2 = ar$. ...(2)

Differentiating (2) w.r.t. p, we get

$$2p = a\left(\frac{dr}{dp}\right); \therefore \frac{dr}{dp} = \frac{2p}{a}.$$

$$\text{Now } \rho = r\left(\frac{dr}{dp}\right) = r\left(\frac{2p}{a}\right) = \left(\frac{2r}{a}\right)p = \left(\frac{2r}{a}\right).\sqrt{(ar)}\text{, from (2) } = \left(\frac{2}{\sqrt{a}}\right)r^{3/2}.$$

$\therefore \rho^2 = \left(\frac{4}{a}\right)^3$. Thus, $r^2 \propto r^3$, where r is the distance of the point from the pole (*i.e.* focus). Hence ρ^2 varies as the cube of the focal distance.

Example 2:

Prove that for any curve $\frac{r}{\rho} = \sin\phi\left(1 + \frac{d\phi}{d\theta}\right)$, *where* ρ *is the radius of curvature and* $\tan\phi = r\frac{d\theta}{dr}$.

Solution:

We know that $\psi = \theta + \phi$. ...(1)

Differentiating (1) w.r.t. s, we get

$$\frac{d\psi}{ds} = \frac{d\theta}{ds} + \frac{d\phi}{ds} = \frac{d\theta}{s} + \frac{d\phi}{d\theta}\frac{d\theta}{ds} = \frac{d\theta}{ds}\left(1 + \frac{d\phi}{d\theta}\right).$$

$$\therefore \frac{1}{\rho} = \frac{\sin\phi}{r}\left(1 + \frac{d\phi}{d\theta}\right), \qquad \left[\because \rho = \frac{ds}{d\psi} \text{ and } \sin\phi = r\frac{d\theta}{ds}\right]$$

$$\text{or } \frac{r}{\rho} = \sin\phi\left(1 + \frac{d\phi}{d\theta}\right).$$

Example 3:

Find the radius of curvature at the point (r, θ) on each of the following curves:

(i) $r = a\ (1 - \cos\theta)$ **(Meerut, 1990, 93, 94, 96, 98; Kanpur, 97)**

(ii) $r = a\ (1 + \cos\theta)$ **(Rohilkhand, 1991; Meerut, 94P)**

(iii) $r = 3\ (1 + \cos\theta)$ **(Agra, 1995)**

(iv) $r^2 \cos 2\theta = a^2$ **(Rohilkhand, 1997; Jhansi, 98; Kanpur, 90)**

(v) $r^n = a^n \cos n\theta$ **(Kanpur, 1999; Meerut, 94, 96P)**

(vi) $r^n = a^n \sin n\theta$ **(Rohilkhand, 1993; Meerut, 98)**

(vii) $r^2 = a^2 \cos 2\theta$

(viii) $\frac{l}{r} = 1 + e\cos\theta.$ **(Meerut, 1990S)**

Solution:

(i) The given curve is $r = a\ (1 - \cos\theta)$. ...(1)

$$\therefore \frac{dr}{d\theta} = a\sin\theta \text{ and } \frac{d^2r}{d\theta^2} = a\cos\theta.$$

$$\text{Hence } \rho = \frac{\left[r^2 + \left(\frac{dr}{d\theta}\right)^2\right]^{3/2}}{r^2 + 2\left(\frac{dr}{d\theta}\right)^2 - r\left(\frac{d^2r}{d\theta^2}\right)}$$

$$= \frac{\left[a^2(1-\cos\theta)^2 + a^2\sin^2\theta\right]^{3/2}}{a^2(1-\cos\theta)^2 + 2a^2\sin^2\theta - a^2\cos\theta\,(1-\cos\theta)}$$

$$= \frac{a^2\left[1+\cos^2\theta - 2\cos\theta + \sin^2\theta\right]^{3/2}}{a^2\left[1+\cos^2\theta - 2\cos\theta + 2\sin^2\theta - \cos\theta + \cos^2\theta\right]}$$

$$= \frac{a\left[2(1-\cos\theta)\right]^{3/2}}{3(1-\cos\theta)} = \frac{2\sqrt{2}}{3}\,a(1-\cos\theta)^{1/2}$$

$$= \frac{2\sqrt{2}}{3} a \sqrt{\left(\frac{r}{a}\right)}, \text{ from (1)}$$

$$= \frac{2}{3}\sqrt{(2ar)}.$$

Note: We could have solved this problem more easily by changing the equation of the curve to pedal form. The method is given in the part (ii) of this example that just follows:

(ii) The given curve is $r = a(1 + \cos\theta)$. ...(1)

Taking log of (1), we get $\log r = \lg a + \log(1 + \cos\theta)$.

Differentiating w.r.t. θ, we get

$$\frac{1}{r}\frac{dr}{d\theta} = \frac{-\sin\theta}{1+\cos\theta} = \frac{-2\sin\frac{1}{2}\theta\cos\frac{1}{2}\theta}{2\cos^2\frac{1}{2}\theta}.$$

$$\therefore \quad \cot\phi = \frac{1}{r}\frac{dr}{d\theta} = -\tan\frac{1}{2}\theta = \cot\left(\frac{1}{2}\pi + \frac{1}{2}\theta\right)$$

so that $\quad \phi = \frac{1}{2}\pi + \frac{1}{2}\theta.$

Now $p = r\sin\phi = r\sin\left(\frac{1}{2}\pi + \frac{1}{2}\theta\right) = r\cos\frac{1}{2}\theta.$

From (1), $r = 2a\cos^2\frac{1}{2}\theta = 2a\left(\frac{p}{r^2}\right)^2$, $\quad \left[\because \frac{p}{r} = \cos\frac{1}{2}\theta\right]$

$\therefore$ the pedal equation of the curve (1) is

$r^3 = 2ap^2$. ...(2)

Differentiating (2) w.r.t. ρ, we get $3r^2\frac{dr}{dp} = 4ap$.

$$\therefore \ \rho = r\frac{dr}{dp} = \frac{4ap}{3r} = \frac{4a}{3r}\left(\frac{r^3}{2a}\right)^{1/2} \qquad \left[\because \text{ from (2), } p^2 = \frac{r^3}{2a}\right]$$

$$= \frac{2}{3}\sqrt{(2ar)}.$$

If we want the value of ρ in terms of θ, then putting $r = a(1+\cos\theta)$ in the value of ρ, we get

$$\rho = \frac{2}{3}\sqrt{\{2a^2(1+\cos\theta)\}} = \frac{2}{3}\sqrt{\left(4a^2\cos^2\frac{1}{2}\theta\right)} = \frac{4a}{3}\cos\frac{1}{2}\theta.$$

(iii) Proceed as in part (ii). The value of ρ is $4\cos\frac{1}{2}\theta$.

(iv) The given curve is $r^2 \cos 2\theta = a^2$. ...(1)

Taking log of (1), we get $2 \log r + \log \cos 2\theta = \log a^2$.

Differentiating w.r.t. θ, we get $\frac{2}{r}\frac{dr}{d\theta}+\frac{-2\sin 2\theta}{\cos 2\theta}=0$.

$\therefore\ \cot\phi=\frac{1}{r}\frac{dr}{d\theta} = \tan 2\theta = \cot\left(\frac{1}{2}\pi-2\theta\right)$; so that $\phi=\frac{1}{2}\pi-2\theta$.

Now $p = r \sin\phi = r\sin\left(\frac{1}{2}\pi-2\theta\right) = r\cos 2\theta = r\left(\frac{a^2}{r^2}\right)$, from (1).

∴ the pedal equation of the curve (1) is $p = \frac{a^2}{r}$.

$$\therefore \qquad \frac{dp}{dr}=-\frac{a^2}{r^2}.$$

Hence $\rho=r\frac{dr}{dp}=r.\left(\frac{-r^2}{a^2}\right)=-\frac{r^3}{a^2}$.

Neglecting the negative sign, we have $\rho=\frac{r^3}{a^2}$.

(v) The given curve is $r^n = a^n \cos n\theta$. ...(1)

Taking logarithm and differentiating w.r.t. θ, we get

$\frac{n}{r}\frac{dr}{d\theta}=\frac{-n\sin n\theta}{\cos n\theta}=-n\tan n\theta$; $\therefore\ \frac{1}{r}\frac{dr}{d\theta}=-\tan n\theta$.

$\therefore\ \cot\phi=\frac{1}{r}\frac{dr}{d\theta}=-\tan n\theta=\cot\left(\frac{1}{2}\pi+n\theta\right)$

so that $\phi=\frac{1}{2}\pi+n\theta$.

Now $p = r\sin\phi = r\sin\left(\frac{1}{2}\pi+n\theta\right) = r\cos n\theta = r\left(\frac{r^n}{a^n}\right)$, from (1).

∴ the pedal equation of (1) is $p = \frac{r^{n+1}}{a^n}$.

$$\therefore \qquad \frac{dp}{dr}=\frac{1}{a^n}(n+1)r^n.$$

Hence $\rho=r\frac{dr}{dp}=r.\frac{a^n}{(n+r)r^n}=\frac{a^n}{(n+1)\,r^{n-1}}$.

(vi) Proceed exactly as in part (v). **Ans.** $\rho = \dfrac{a^n}{[(n+1)r^{n-1}]}$.

(vii) The given curve is $r^2 = a^2 \cos 2\theta$.
Taking logarithm, $2 \log r = \log \cos 2\theta + 2 \log a$. ...(1)

Differentiating w.r.t. θ, we get $\dfrac{2}{r}\dfrac{dr}{d\theta} = -\dfrac{2 \sin 2\theta}{\cos 2\theta}$.

$\therefore\ \cot \phi = \dfrac{1}{r}\dfrac{dr}{d\theta} = -\tan 2\theta = \cot\left(\dfrac{1}{2}\pi + 2\theta\right)$;

so that $\phi = \dfrac{1}{2}\pi + 2\theta$.

Now $p = r \sin \phi = r \sin\left(\dfrac{1}{2}\pi + 2\theta\right) = r \cos 2\theta = \left(\dfrac{r^2}{a^2}\right)$, from (1).

Hence the pedal equation of the curve (1) is $a^2 p = r^3$. ...(2)
Differentiating (2) w.r.t. r, we get

$$a^2\left(\frac{dp}{dr}\right) = 3r^2 \text{ or } \left(\frac{dr}{dp}\right) = \frac{a^2}{3r^2}.$$

Hence $\rho = r\left(\dfrac{dr}{dp}\right) = r.\left(\dfrac{a^2}{3r^2}\right) = \dfrac{a^2}{3r}$.

(viii) The given curve is $\dfrac{l}{r} = 1 + e \cos \theta$. Let $\dfrac{l}{r} = u$. Then the equation of the curve is

$$u = \left(\frac{1}{l}\right)(1 + e \cos \theta).$$

Differentiating, we get

$$\frac{du}{d\theta} = \frac{1}{l}(-e \sin \theta) \text{ and } \frac{d^2u}{d\theta^2} = \frac{1}{l}(-e \cos \theta).$$

$$\therefore \quad \rho = \frac{\left[u^2 + \left(\dfrac{du}{d\theta}\right)^2\right]^{3/2}}{u^3\left(u + \dfrac{d^2u}{d\theta^2}\right)}$$

$$= \frac{\left(\dfrac{1}{l^3}\right)\left[(1 + e \cos \theta)^2 + e^2 \sin^2 \theta\right]^{3/2}}{\left(\dfrac{1}{l}\right)^4 (1 + e \cos \theta)^3 \left[(1 + e \cos \theta) + (-e \cos \theta)\right]}$$

$$= \frac{l\left[1+e^2+2e\cos\theta\right]^{3/2}}{(1+e\cos\theta)^3}.$$

Example 4:

Show that at any point on the equiangular spiral $r = ae^{\theta \cot \alpha}$, $\rho = r$ *cosec* α, *and that it subtends a right angle at the pole.*

(Meerut, 1997, 94; Agra, 97; Kanpur, 90, 98, 91)

Solution:

The equation of the given curve is $r = ae^{\theta \cot \alpha}$. ...(1)

Differentiating (1) w.r.t. θ, we have

$$\frac{dr}{d\theta} = ae^{\theta \cot \alpha}.\cot \alpha = r \cot \alpha.$$

$$\therefore \left(\frac{1}{r}\right)\frac{dr}{d\theta} = \cot \alpha \text{ or } \cot \phi = \cot a \text{ or } \phi = \alpha.$$

Now, $p = r \sin \phi = r \sin \alpha$. Thus the pedal equation of (1) is $p = r \sin \alpha$. Therefore $\frac{dp}{dr} = \sin \alpha$.

Now $\rho = r\frac{dr}{dp} = \frac{r}{\sin \alpha} = r \operatorname{cosec} \alpha$. **(First part proved)**

Second Part: Let $P(r, \theta)$ be any point on the given curve; PT is the tangent and PC is the normal to the curve at P. Let C be the centre of curvature of the point P of the curve. Then PC = the radius of curvature of the curve at P = r cosec α.

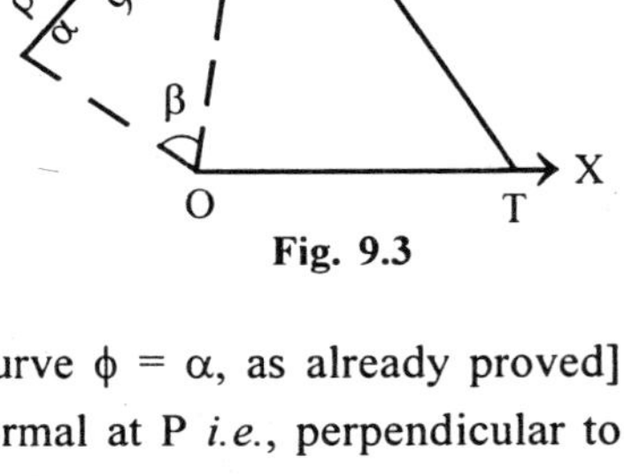

Fig. 9.3

Join OP and OC, where O is the pole. Let $\angle$ POC = β. Then to prove that $\beta = 90°$.

We have $\angle OPT = \phi = \alpha$.

[$\because$ for this curve $\phi = \alpha$, as already proved]

$\therefore \angle OPC = 90° - \alpha$, since PC is normal at P *i.e.*, perpendicular to the tangent PT.

Now in Δ OPC, we have

$$\angle OCP = 180° - \{(90° - \alpha) + \beta\} = (90° + \alpha - \beta).$$

Hence applying the sine theorem for Δ OPC, we get

$$\frac{OP}{\sin\angle OCP}=\frac{PC}{\sin\beta};\ \text{or}\ \frac{r}{\sin(90^\circ+\alpha-\beta)}=\frac{r}{\sin\beta}$$

or $\dfrac{r}{\cos(\alpha-\beta)}=\dfrac{r\ \text{cosec}\ \alpha}{\sin\beta}$, [$\because$ r = cosec α]

or $\sin\alpha\sin\beta=\cos(\alpha-\beta)=\cos\alpha\cos\beta+\sin\alpha\sin\beta$,
[$\because r\neq 0$]

or $\cos\alpha\cos\beta=0$ or $\cos\beta=0$, [$\because \cos\alpha\neq 0$]

$\therefore$ $\beta=90^\circ$.

Example 5:

Show that the curvatures of the curves $r=a\theta$ and $r\theta=a$ at their common point are in the ratio 3: 1. **(Bhopal, 1993; Meerut, 97)**

Solution:

The given curves are

$$r=a\theta \qquad ...(1)$$

and

$$r\theta=a \qquad ...(2)$$

Eliminating r between (1) and (2), we get $a\theta^2=a$ or $\theta^2=1$.

Thus, at the common point of (1) and (2) we have $\theta^2=1$.

For the first curve, $\dfrac{dr}{d\theta}=a$ and $\dfrac{d^2r}{d\theta^2}=0$.

$$\therefore\quad \rho=\frac{\left[r^2+\left(\frac{dr}{d\theta}\right)^2\right]^{3/2}}{r^2+2\left(\frac{dr}{d\theta}\right)^2-r\left(\frac{d^2r}{d\theta^2}\right)}$$

$$=\frac{\left[a^2\theta^2+a^2\right]^{3/2}}{a^2\theta^2+2a^2}=\frac{a\left(\theta^2+1\right)^{3/2}}{\theta^2+2}.$$

$$\therefore\quad \rho(\text{at } \theta^2=1)=\frac{a.(1+1)^{3/2}}{1+2}=\frac{2a\sqrt{2}}{3}.$$

But curvature $=\dfrac{1}{\rho}$. Therefore, the curvature of the first curve at $\theta^2=1$ is $\dfrac{3}{(2a\sqrt{2})}$.

For the second curve, $r = \frac{a}{\theta}; \frac{dr}{d\theta} = -\frac{a}{\theta^2}; \frac{d^2r}{d\theta^2} = \frac{2a}{\theta^3}.$

$$\therefore \rho = \frac{\left[\left(\frac{a^2}{\theta^2}\right)+\left(\frac{a^2}{\theta^4}\right)\right]^{3/2}}{\left(\frac{a^2}{\theta^2}\right)+\left(\frac{2a^2}{\theta^4}\right)-\left(\frac{2a^2}{\theta^4}\right)} = \frac{a\left(\theta^2+1\right)^{3/2}}{\theta^4}.$$

$$\therefore \rho(\text{at } \theta^2 = 1) = \frac{a(1+1)^{3/2}}{1} = 2a\sqrt{2}.$$

$\therefore$ curvature of the second curve at $\theta^2 = 1$ is $\frac{1}{\left(2a\sqrt{2}\right)}$.

Hence at their common point

$$\frac{\text{curvature of first curve}}{\text{curvature of second curve}} = \frac{\frac{3}{\left(2a\sqrt{2}\right)}}{\frac{1}{\left(2a\sqrt{2}\right)}} = \frac{3}{1}.$$

Example 6:

Prove that for any curve $\frac{d^2r}{ds^2} = \frac{\sin^2\phi}{r} - \frac{\sin\phi}{\rho}.$

Solution:

We have $\frac{dr}{ds} = \cos\phi.$

Differentiating w.r.t. 's', we get

$$\frac{d^2r}{ds^2} = -\sin\phi.\frac{d\phi}{ds} = -\sin\phi.\frac{d}{ds}(\psi-\theta), \qquad [\because \phi = \psi - \theta]$$

$$= -\sin\phi\left(\frac{d\psi}{ds} - \frac{d\theta}{ds}\right) = -\sin\phi\left[\frac{1}{\rho} - \frac{1}{r}\sin\phi\right],$$

$$\left[\because \rho = \frac{ds}{d\psi} \text{ and } \sin\phi = r\frac{d\theta}{ds}\right]$$

$$= -\frac{\sin\phi}{\rho} + \frac{1}{r}\sin^2\phi = \frac{\sin^2\phi}{r} - \frac{\sin\phi}{\rho}.$$

Example 7:

If ρ_1, ρ_2 be the radii of curvature at the extremities of any chord of the cardioid $r = a\,(1 + \cos\theta)$ which passes through the pole, then show that

$$\rho_1^2 + \rho_2^2 = \frac{16a^2}{9}.$$ **(Meerut, 1990, 98)**

Solution:

The given curve is $r = a\,(1 + \cos\theta)$ ...(1)

Let PQ be any chord of the curve (1) passing through the pole, and let P be the point (r_1, θ_1) and Q be the point (r_2, θ_2). Then $\theta_2 = \pi + \theta_1$.

[**Note:** To understand this point draw the figure of a chord passing through the pole.]

Since both the points (r_1, θ_1) and (r_2, θ_2) lie on the given cardioid (1), therefore

$$r_1 = a\,(1 + \cos\theta_i) \text{ and } r_2 = a\,(1 + \cos\theta_2) \qquad ...(2)$$

Now, let ρ be the radius of curvature of (1) at the point (r, θ). We get

$$\rho = \frac{2}{3}\sqrt{(2ar)} \text{ or } \rho^2 = \frac{8}{9}ar.$$

If ρ_1 and ρ_2 be radii of curvature at the points P and Q, we have

$$\rho_1^2 + \frac{8}{9}ar_1 \text{ and } \rho_2^2 = \frac{8}{9}ar_2.$$

$$\therefore\ \rho_1^2 + \rho_2^2 = \frac{8}{9}a(r_1 + r_2) = \frac{8}{9}a\,[a(1 + \cos\theta_1) + a\,(1 + \cos\theta_2)],$$

[from (2)]

$$= \left(\frac{8a^2}{9}\right)[1 + \cos\theta_1 + 1 + \cos(\pi + \theta_1)], \qquad [\because\ \theta_2 = \pi + \theta_1]$$

$$= \left(\frac{8a^2}{9}\right)[1 + \cos\theta_1 + 1 - \cos\theta_1] = \left(\frac{16a^2}{9}\right).$$

Example 8:

(a) Find the radius of curvature of the curve

$$r = a \sin n\theta.$$

(b) Find the radius of curvature at (r, θ) on the curve

$$r = 6\left(1 - \sin^2\frac{1}{2}\theta\right).$$

Solution:

(a) Here r = a sin nθ.

Therefore $\frac{dr}{d\theta}$ = an cos nθ; $\frac{d^2r}{d\theta^2}$ = – an² sin nθ = – n²r.

$$\therefore\ \rho = \left[r^2 + \left(\frac{dr}{d\theta}\right)^2\right]^{3/2} \Big/ \left[r^2 + 2\left(\frac{dr}{d\theta}\right)^2 - r\left(\frac{d^2r}{d\theta^2}\right)\right]$$

$$= \frac{\left[r^2 + a^2n^2\cos^2 n\theta\right]^{3/2}}{\left[r^2 + 2a^2n^2\cos^2 n\theta + n^2r^2\right]}.$$

Now $a^2 \cos^2 n\theta = a^2 (1 - \sin^2 n\theta) = a^2 - a^2 \sin^2 n\theta = a^2 - r^2$.

So putting $a^2\cos^2 n\theta = a^2 - r^2$ in the value of ρ, we get

$$\rho = \frac{\left[r^2 + n^2(a^2 - r^2)\right]^{3/2}}{\left[r^2 + 2n^2(a^2 - r^2) + n^2r^2\right]}$$

$$= \frac{\left[r^2 + n^2a^2 - n^2r^2\right]^{3/2}}{\left[r^2 + 2a^2n^2 - n^2r^2\right]}.$$

$$\therefore\ \text{Curvature} = \frac{1}{\rho} = \frac{\left[r^2 + 2a^2n^2 - n^2r^2\right]}{\left[r^2 + n^2a^2 - n^2r^2\right]^{3/2}}.$$

(b) Proceed as in part (a) of this Example. Here $\rho = 4\cos\frac{1}{2}\theta$.

Example 9:

Find the radius of curvature at the point (r, θ) on the curve

$$\theta = a^{-1}(r^2 - a^2)^{1/2} - \cos^{-1}\left(\frac{a}{r}\right).$$ **(Agra, 1992)**

Solution:

We have $$\frac{d\theta}{dr} = \frac{1}{a}\cdot\frac{1}{2}\frac{2r}{\sqrt{(r^2 - a^2)}} + \frac{1}{\sqrt{\left\{1 - \left(\frac{a}{r}\right)^2\right\}}}\left(-\frac{a}{r^2}\right)$$

$$= \frac{r}{a\sqrt{(r^2 - a^2)}} - \frac{a}{r\sqrt{(r^2 - a^2)}} = \frac{r^2 - a^2}{ar\sqrt{(r^2 - a^2)}} = \frac{\sqrt{(r^2 - a^2)}}{ar}.$$

$$\therefore \quad \frac{dr}{d\theta} = \frac{ar}{\sqrt{(r^2 - a^2)}}.$$

Now $\frac{1}{p^2} = \frac{1}{r^2} + \frac{1}{r^4}\left(\frac{dr}{d\theta}\right)^2 = \frac{1}{r^2} + \frac{1}{r^4}.\frac{a^2r^2}{r^2 - a^2} = \frac{1}{r^2} + \frac{a^2}{r^2(r^2 - a^2)}$

$$= \frac{r^2 - a^2 + a^2}{r^2(r^2 - a^2)} = \frac{1}{r^2 - a^2}.$$

$\therefore$ the pedal equation of the given curve is $\frac{1}{p^2} = \frac{1}{r^2 - a^2}$

$$p^2 = r^2 - a. \qquad ...(1)$$

Differentiating (1) w.r.t. p, we get $2p = 2r\left(\frac{dr}{dp}\right)$.

$\therefore \rho = r\left(\frac{dr}{dp}\right) = p = (r^2 - a^2)^{1/2}$, from (1)

Example 10:

If the equation to a curve be given in polar co-ordinates and if $u = \frac{1}{r}$, prove that the curvature is given by $\left(\frac{d^2u}{d\theta^2} + u\right)\sin^3\phi$, where tan $f = r\left(\frac{d\theta}{dr}\right)$.

Solution:

We know that $\frac{1}{p^2} = \frac{1}{r^2} + \frac{1}{r^4}\left(\frac{dr}{d\theta}\right)^2$ ...(1)

Let $u = \frac{1}{r}$. Then $\frac{du}{d\theta} = -\frac{1}{r^2}\frac{dr}{d\theta}$ so that $\left(\frac{du}{d\theta}\right)^2 = \frac{1}{r^4}\left(\frac{dr}{d\theta}\right)^2$.

Therefore equation (1) becomes $\frac{1}{p^2} = u^2 + \left(\frac{du}{d\theta}\right)^2$. ...(2)

Differentiating equation (2) w.r.t. r, we get

$$-\frac{2}{p^3}\frac{dp}{dr} = 2u\frac{du}{dr} + \left(2\frac{du}{d\theta}\frac{d^2u}{d\theta^2}\right)\frac{d\theta}{dr} \quad \left[\because \frac{d}{dr}\left(\frac{du}{d\theta}\right)^2 = \left\{\frac{d}{d\theta}\left(\frac{du}{d\theta}\right)^2\right\}\frac{d\theta}{dr}\right]$$

$$= 2\left[u\frac{du}{dr} + \frac{d^2u}{d\theta^2}\frac{du}{dr}\right] \qquad \left[\because \frac{du}{d\theta}\frac{d\theta}{dr} = \frac{du}{dr}\right]$$

$$= 2\left(u + \frac{d^2u}{d\theta^2}\right)\frac{du}{dr}.$$

$$\therefore \quad -\frac{1}{r^3 \sin^3\phi}\frac{dp}{dr} = \left(u + \frac{d^2u}{d\theta^2}\right)\left(-\frac{1}{r^2}\right)$$

$$\left[\because r = r\sin\phi. \text{ Also } u = \frac{1}{r} \Rightarrow \frac{du}{dr} = -\frac{1}{r^2}\right]$$

or $$\frac{1}{r}\frac{dp}{dr} = \sin^3\phi\left(u + \frac{d^2u}{d\theta^2}\right). \qquad \left[\because \rho = r\frac{dr}{dp}\right].$$

9.8 TANGENTIAL POLAR FORMULA FOR RADIUS OF CURVATURE

(Agra, 1993; Lucknow, 99; Gorakhpur, 90, 97; Allahabad, 94)

A relation between p and ψ, holding for every point of a curve, is called the tangential polar equation of the curve. Thus the tangential polar equation of the curve is of the form $p = f(\psi)$.

To find the radius of curvature of a curve for which the relation between p and ψ is given.

We know that $p = r\sin\phi$. ...(1)

Also $$\frac{dp}{d\psi} = \frac{dp}{dr}\cdot\frac{dr}{ds}\cdot\frac{ds}{d\psi}$$

$$= \left(\frac{dp}{dr}\right).\cos\phi.\rho, \quad \left[\because \frac{ds}{d\psi} = \rho \text{ and } \frac{dr}{ds} = \cos\phi\right]$$

$$= \left(\frac{dp}{dr}\right)\cos\phi.r\left(\frac{dr}{dp}\right) \qquad \left[\because \rho = r\left(\frac{dr}{dp}\right)\right]$$

$$\therefore \quad \frac{dp}{d\psi} = r\cos\phi \qquad ...(2)$$

Squaring and adding (1) and (2), we have

$$r^2 = p^2 + \left(\frac{dp}{d\psi}\right)^2 \qquad ...(3)$$

Differentiating both sides of (3) w.r.t. 'p', we get

$$2r\frac{dr}{dp} = 2p + 2\frac{dp}{d\psi}.\frac{d^2p}{d\psi^2}.\frac{d\psi}{dp}$$

or $$r\frac{dr}{dp} = p + \frac{d^2p}{d\psi^2}.$$

Hence $$\rho = p + \frac{d^2p}{d\psi^2}.$$

Example 1:

Find the radius of curvature of the curve, $p = a \sin b\psi$.

Solution:

We have $p = a \sin b\psi$.

$$\therefore \quad \frac{dp}{d\psi} = ab \cos b\psi \text{ and } \frac{d^2p}{d\psi^2} = -ab^2 \sin b\psi.$$

$$\rho = p + \left(\frac{d^2p}{d\psi^2}\right) = a \sin b\psi - ab^2 \sin b\psi$$

$$= a\left(\frac{p}{a}\right) - ab^2.\left(\frac{p}{a}\right), \qquad \text{from (1)}$$

$= p - b^2p = (1 - b^2)p$, *i.e.*, ρ varies as p.

Example 2:

Find the radius of curvature at any point of the ellipse

$p^2 = a^2 \cos^2 \psi + b^2 \sin^2 \psi$.

Solution:

We have $p^2 = a^2 \cos^2 \psi + b^2 \sin^2 \psi$.

We have

$$p + \frac{d^2p}{d\psi^2} = \frac{a^2b^2}{p^3}. \qquad \text{[Give complete proof here]}$$

$$p = p + \frac{d^2p}{d\psi^2} = \frac{a^2b^2}{p^3}.$$

9.9 MISCELLANEOUS FORMULAE RADIUS OF CURVATURE, WHEN x AND y ARE GIVEN AS FUNCTIONS OF ARC LENGTH

(a) We have $\cos \psi = \frac{dx}{ds}$.

Differentiating w.r.t., 's', we get $(-\sin \psi)\left(\frac{d\psi}{ds}\right) = \frac{d^2x}{ds^2}$

or $$-\sin \psi.\frac{1}{\rho} = \frac{d^2x}{ds^2}, \qquad \left[\because \frac{1}{\rho} = \frac{d\psi}{ds}\right] \qquad ...(1)$$

$$\therefore \qquad \rho = -\frac{\dfrac{dy}{ds}}{\dfrac{d^2x}{ds^2}}. \qquad \left[\because \sin\psi = \frac{dy}{ds}\right].$$

(b) We have $\sin\psi = \dfrac{d\psi}{ds}$

Differentiating w.r.t. 's', we get

$$\cos\psi.\frac{1}{\rho} = \frac{d^2y}{ds^2}. \qquad ...(2)$$

$$\therefore \quad \rho = \frac{\dfrac{dx}{ds}}{\dfrac{d^2y}{ds^2}}.$$

(c) Squaring and adding (1) and (2), we get

$$\frac{1}{\rho^2} = \left(\frac{d^2x}{ds^2}\right)^2 + \left(\frac{d^2y}{ds^2}\right)^2.$$

(d) We know that $\cos\psi = \dfrac{dx}{ds} = \dfrac{dx}{d\psi}.\dfrac{d\psi}{ds} = \dfrac{1}{\rho}.\dfrac{dx}{d\psi}$

and $\sin\psi = \dfrac{dy}{ds} = \dfrac{dy}{d\psi}.\dfrac{d\psi}{ds} = \dfrac{1}{\rho}.\dfrac{dy}{d\psi}.$

Squaring and adding these, we obtain

$$\cos^2\psi + \sin^2\psi = \frac{1}{\rho^2}\left[\left(\frac{dx}{d\psi}\right)^2 + \left(\frac{dy}{d\psi}\right)^2\right]$$

or $$\rho^2 = \left(\frac{dx}{d\psi}\right)^2 + \left(\frac{dy}{d\psi}\right)^2.$$

9.10 RADIUS OF CURVATURE AT THE ORIGIN

(a) Newton's Method of Finding the Radius of Curvature at the Origin

(i) When the curve passes through the origin, and the axis of x is the tangent at the origin, we have

$$y(0) = 0 \text{ and } \left(\frac{dy}{dx}\right)_{x=0} = 0, \textit{ i.e.}, \; y_1(0) = 0. \qquad ...(1)$$

Now, by Maclaurin's theorem, y can be expanded as

$$y = y(0) + xy_1(0) + \left(\frac{x^2}{2!}\right)y_2(0) + \left(\frac{x^3}{3!}\right)y_3(0) + \ldots$$

or $$y = 0 + 0 + \left(\frac{x^2}{2!}\right)y_2(0) + \left(\frac{x^3}{3!}\right)y_3(0) + \cdots, \qquad \text{[by (1)]}.$$

Dividing both sides by x^2, we get

$$\frac{y}{x^2} = \frac{1}{2!}y_2(0) + \frac{x}{3!}y_3(0) + \ldots \qquad \ldots(2)$$

Since the curve passes through the origin, therefore $x \to 0$ $\Rightarrow y \to 0$. Hence taking limit of both sides of (2) when $x \to 0$, $y \to 0$, we get

$$\lim_{x \to 0} \frac{y}{x^2} = \frac{1}{2!}y_2(0). \qquad \text{[Other terms vanish]}.$$

Therefore, ρ (at the origin) $= \dfrac{\left[1 + \{y_1(0)\}^2\right]^{3/2}}{y_2(0)} = \dfrac{1}{y_2(0)},$

$[\because y_1(0) = 0]$

$$= \lim_{x \to 0,\, y \to 0} \frac{x^2}{2y}.$$

(ii) *Similarly are we can prove that if a curve passes through the origin, and the axis of y is the tangent there, then*

$$\rho \text{ (at the origin)} = = \lim_{x \to 0,\, y \to 0} \frac{y^2}{2x}.$$

The above two formulae are known as **Newton's Formulae.**

Note: *If a curve passes through the origin and is given by a rational integral, algebraic equation, then the tangents at the origin are easily obtained by equating to zero the lowest degree terms in the equation of the curve*

(b) Expansion Method for Finding the Radius of Curvature at the Origin: When a curve passes through the origin, but neither of the co-ordinate axes is a tangent at the origin, we cannot apply Newton's formula to find the radius of curvature at the origin. In such cases we can apply the following method known as the **Expansion Method.**

Since the curve passes through the origin, therefore $y(0) = 0$, *i.e.*, the value of y at $x = 0$ is 0.

Let $\left(\frac{dy}{dx}\right)_{(0,0)} = y_1(0) = p$ and $\left(\frac{d^2y}{dx^2}\right)_{(0,0)} = y_2(0) = q$.

Then ρ (at origin) $= \frac{(1+p^2)^{3/2}}{q}$. ...(1)

Now by Maclaurin's theorem, we have

$$y = y(0) + xy_1(0) + \left(\frac{1}{2!}\right)x^2y_2(0) + \ldots$$

$$= px + \frac{1}{2}qx^2 + \ldots$$

because the curve passes through the origin.

Thus, to get the values of p and q, we should obtain from the equation of the curve an expansion for y in ascending powers of x by algebraic or trigonometric methods. The coefficient of x^2 will be equal to $\frac{1}{2}q$, as is obvious from the Maclaurin's expansion for y. Putting these values of p and q in (1), we shall get the ρ at the origin.

(c) Radius of Curvature at the Pole: Suppose the curve passes through the pole and the initial line is the tangent at the pole. The relations between the cartesian and polar co-ordinates are $x = r\cos\theta$, $y = r\sin\theta$. By Newton's Formula, we have

$$\rho \text{ (at the pole)} = \lim_{x\to 0, y\to 0}\frac{x^2}{2y} = \lim_{\theta\to 0}\frac{r^2\cos^2\theta}{2r\sin\theta}$$

$$= \lim_{\theta\to 0}\left(\frac{r}{2\theta}\cdot\frac{\theta}{\sin\theta}.\cos^2\theta\right) \quad \textbf{(Note)}$$

$$= \lim_{\theta\to 0}\left(\frac{r}{2\theta}\right), \qquad \left[\because \text{as } \theta\to 0, \frac{\theta}{\sin\theta}\to 1 \text{ and } \cos\theta\to 1\right].$$

Example 1:

Find the radius of curvature at the origin for the curve

$$3x^2 + 4x^3 - 12y = 0.$$

Solution:

The given curve is $3x^2 + 4x^3 - 12y = 0$. ...(1)

The curve (1) passes through the origin.

To find the tangents at the origin, equating to zero the lowest degree terms in (1), we get y = 0 *i.e.*, x-axis as the tangent at origin.

$\therefore$ by Newton's formula ρ (at the origin) $= \lim_{x\to 0} \frac{x^2}{2y}$. ...(2)

Dividing both sides of (1) by 2y, we get

$$2.\left(\frac{x^2}{2y}\right)+4x\left(\frac{x^2}{2y}\right)-6=0. \qquad ...(3)$$

Taking limits of both sides of (3) as $x \to 0$ and $y \to 0$, we have

$$3\lim_{x\to 0}\frac{x^2}{2y}+4.\lim_{x\to 0} x.\lim_{x\to 0}\frac{x^2}{2y}-6=0$$

or $3\rho + 4.0.\rho - 6 = 0,$ from (2)

$\therefore$ $= \rho = 2.$

Example 2:

Find the radius of curvature for the curve

$$x = c\ log\left\{s+\sqrt{\left(s^2+c^2\right)}\right\},\ y=\sqrt{\left(s^2+c^2\right)}.$$

Solution:

We have $$\frac{dx}{ds}=c\frac{1+\frac{1}{2}\left(s^2+c^2\right)^{-1/2}(2s)}{s+\sqrt{\left(s^2+c^2\right)}}=\frac{c}{\sqrt{\left(s^2+c^2\right)}},$$

and $$\frac{dy}{ds}=\frac{s}{\left(s^2+c^2\right)^{1/2}}.$$

$$\therefore\ \frac{d^2y}{ds^2}=\frac{1.\sqrt{\left(s^2+c^2\right)}-s.\frac{1}{2}\left(s^2+c^2\right)^{-1/2}(2s)}{s^2+c^2}=\frac{c^2}{\left(s^2+c^2\right)^{3/2}},$$

$$\therefore\ \rho=\frac{\frac{dx}{ds}}{\frac{d^2y}{ds^2}}=\frac{c}{\sqrt{\left(s^2+c^2\right)}}.\frac{\left(s^2+c^2\right)^{3/2}}{c^2}=\frac{1}{c}\left(s^2+c^2\right)=\frac{y^2}{c}.$$

Example 3:

Show that for the curve $s^2 = 8ay$,

$$\rho=4a\sqrt{\left[1-\left(\frac{y}{2a}\right)\right]}.$$ **(Lucknow, 1991; Agra, 91; Indore, 91)**

Solution:

We have $s^2 = 8ay$. Differentiating it w.r.t. s, we get

$$2s = 8a\frac{dy}{ds} = 8a \sin\psi . \qquad \left[\because \sin\psi = \frac{dy}{ds}\right]$$

$\therefore \quad s = 4a \sin\psi.$

$$\text{Now } \rho = \frac{ds}{d\psi} = 4a \cos y = 4a\sqrt{\left(1-\sin^2\psi\right)} = 4a\sqrt{\left[1-\frac{s^2}{16a^2}\right]}$$

$$= 4a\sqrt{\left[1-\frac{8ay}{16a^2}\right]}, \qquad [\because s^2 = 8ay]$$

$$= 4a\sqrt{\left[1-\left(\frac{y}{2a}\right)\right]}.$$

Example 4:

Find the radius of curvature at the origin for the curve

$$x = t - \frac{1}{3}t^3,\ y = t^2.$$

Solution:

$$\text{We have } \frac{dy}{dx} = \frac{\frac{dy}{dt}}{\frac{dx}{dt}} = \frac{2t}{1-t^2}.$$

$$\therefore \quad \left(\frac{dy}{dx}\right)_{\text{at origin}} = 0, \qquad [\because \text{ at origin } x = 0,\ y = 0 \Rightarrow t = 0].$$

Thus, the tangent at origin is x-axis.

$$\therefore \ \rho \text{ (at the origin) } = \lim_{x\to 0}\left(\frac{x^2}{2y}\right)$$

$$= \lim_{t\to 0}\left\{\frac{\left(t-\frac{1}{3}t^3\right)^2}{2t^2}\right\} = \lim_{t\to 0}\left(1-\frac{1}{3}t^2\right)^2 = \frac{1}{2}.$$

Example 5:

Apply Newton's method to prove that the radius of curvature at the lowest point of the catenary $y = \cosh\left(\frac{x}{c}\right)$ *is equal to c.*

Solution:

Shifting the origin to the lowest point of the catenary *i.e.*, the point (0, c), we have $y + c = c\cosh\left(\frac{x}{c}\right)$

$$\text{or}\quad y + c = 0\left[1+\frac{x^2}{2!c^2}+\frac{x^4}{4!c^4}+\dots\right]$$

$$\text{or}\quad y = \frac{x^2}{2c}+\frac{x^4}{24c^3}+\dots \qquad \dots(1)$$

Now the tangent to (1) at the new origin is $y = 0$, (*i.e.*, the new x-axis).

Dividing both sides of (1) by 2y, we get

$$\frac{1}{2}=\frac{1}{2c}\cdot\frac{x^2}{2y}+\frac{x^2}{24c^3}\cdot\frac{x^2}{2y}+\dots \qquad \dots(2)$$

Taking the limits of both sides of (2) when $x \to 0$, $y \to 0$, we have

$$\frac{1}{2}=\frac{1}{2c}\cdot\rho, \qquad \text{(Other terms vanish)}$$

or $\rho = c$.

Example 6:

Show that for the curve $s = ae^{x/a}$, $a\rho = s(s^2 - a^2)^{1/2}$.

(Rohilkhand, 1999; Kanpur, 95; Agra, 96)

Solution:

We have $s = ae^{x/a}$.

Therefore $\frac{ds}{dx} = ae^{x/a}\cdot\frac{1}{a} = e^{x/a} = \frac{s}{a}$.

$$\therefore\ s = a\left(\frac{ds}{dx}\right) = a\sec\psi. \qquad \left[\because \cos\psi = \frac{dx}{ds} \Rightarrow \sec\psi = \frac{ds}{dx}\right]$$

$$\text{Now}\ \rho = \frac{ds}{d\psi} = a\sec\psi\tan\psi = s\sqrt{(\sec^2\psi - 1)}$$

$$= s\sqrt{\left(\frac{s^2}{a^2}-1\right)} = \frac{s}{a}\left(s^2 - a^2\right)^{1/2}.$$

Hence $a\rho = s(s^2 - a^2)^{1/2}$.

Example 7:

Find the radius of curvature at the origin of the curve $y = x^3 + 5x^2 + 6x$.

Solution:

The given curve is $y = 6x + 5x^2 + x^3$, ...(1)
which obviously passes through the origin.

Let $\left(\frac{dy}{dx}\right)_{(0,0)} = p$ and $\left(\frac{d^2y}{dx^2}\right)_{(0,0)} = q$.

They by Maclaurin's Expansion, we get for this curve

$$y = px + \frac{1}{2}qx^2 + \ldots \qquad \ldots(2)$$

Comparing (1) and (2), we get $p = 6$, $\frac{q}{2} = 5$, *i.e.*, $q = 10$.

Hence ρ at the origin $= \frac{(1+p^2)^{3/2}}{q} = \frac{(1+36)^{3/2}}{10} + \frac{1}{10}37\sqrt{37}$.

Example 8:

Find the radius of curvature at the origin for the following curves:

(a) $a(y^2 - x^2) = x^3$,

(b) $5x^3 + 7y^3 + 4x^2y + xy^2 + 2x^2 + 3xy + y^2 + 4x = 0$. **(Kanpur, 1999)**

(c) $x^3 - y^3 - 2x^2 + 6y = 0$,

(d) $a_1x + a_2y + b_1x^2 + b_2xy + b_3y^2 + c_1x^3 + \ldots = 0$.

Solution:

(a) The given curve passes through the origin. The tangents at origin $y^2 - x^2 = 0$, *i.e.*, $y = \pm x$. Thus, neither of the co-ordinate axes is a tangent at the origin. So we cannot apply Newton's formula for finding r at origin. From the equation of the curve, we have

$$y^2 = x^2 + \frac{x^3}{a} = x^2\left(1 + \frac{x}{a}\right).$$

$$\therefore \quad y = \pm x\left(1 + \frac{x}{a}\right)^{1/2} = \pm x\left(1 + \frac{1}{2}.\frac{x}{a} + \ldots\right), \qquad \ldots(1)$$

(by binomial theorem).

Let $\left(\frac{dy}{dx}\right)_{(0,0)} = p$ and $\left(\frac{d^2y}{dx^2}\right)_{(0,0)} = q$.

Then by Maclaurin's Expansion, we get for this curve

$$y = px + \frac{1}{2}qx^2 + \ldots \qquad \ldots(2)$$

Comparing equations (1) and (2), we get

$$p = 1,\ q = \frac{1}{a};\ \text{or}\ p = -1,\ q = -\frac{1}{a}.$$

Now ρ (at origin) $= \dfrac{\left(1+p^2\right)^{3/2}}{q}$.

$\therefore$ When $p = 1$, $q = \dfrac{1}{a}$, we have

$$\rho \text{ (at origin)} = \frac{(1+1)^{3/2}}{\frac{1}{a}} = \left(2\sqrt{2}\right)a,$$

and when $p = 1$, $q = -\dfrac{1}{a}$, we have

$$\rho \text{ (at origin)} = \frac{(1+1)^{3/2}}{-\frac{1}{a}} = -\left(2\sqrt{2}\right)a.$$

(b) The given curve passes through the origin (0, 0). Equating to zero the lowest degree terms in the equation of the curve, we get the tangent at origin as $4x = 0$, *i.e.*, $x = 0$, *i.e.*, y-axis.

$\therefore$ by Newton's formula

$$\rho \text{ (at the origin)} = \lim_{x\to 0,\ y\to 0}\left(\frac{y^2}{2x}\right)$$

Now dividing each term in the equation of the curve by 2x, we get

$$\frac{3}{2}x^2 + 7y\left(\frac{y^2}{2x}\right) + 2xy + \frac{1}{2}y^2 + x + \frac{3}{2}y + \frac{y^2}{2x} + 2 = 0. \qquad \text{...(1)}$$

Taking limits of both sides of (1) when $x \to 0$, $y \to 0$ and remembering that $\lim\limits_{x\to 0}\left(\dfrac{y^2}{2x}\right) = \rho$ (at origin), we get

$$0 + 0.\rho + 0 + 0 + 0 + 0 + \rho + 2 = 0$$

i.e., ρ(at origin) $= -2$ or 2 (numerically).

(c) The given curve passes through the origin (0, 0). The tangent at origin obtained by equating to zero the lowest degree terms is $y = 0$, *i.e.*, x-axis. Therefore by Newton's formula

$$\rho\text{(at origin)} = \lim_{x\to 0,\ y\to 0}\left(\frac{x^2}{2y}\right).$$

Now dividing each term in the equation of the curve by 2y, we get

$$x\left(\frac{x^2}{2y}\right)-\frac{1}{2}y^2-2\left(\frac{x^2}{2y}\right)+3=0. \qquad ...(1)$$

Taking limits of both sides of (1) when $x \to 0$, $y \to 0$, we get

$$0.\rho-\frac{1}{2}.0-2.\rho+3=0; \quad \text{or } \rho \text{ (at the origin)} = \frac{3}{2}.$$

(d) The given curve passes through the origin. Since neither of the co-ordinate axes is a tangent at the origin, therefore Newton's method cannot be applied to find ρ at origin. Also we cannot put the equation of the curve in the form

$$y = px + \frac{1}{2}qx^2 + ...$$

So substituting $px+\frac{1}{2}qx^2+...$ for y in the equation of the curve, we get the identity

$$a_1x + a_2\left(px+\frac{1}{2}qx^2+...\right)+b_1x^2+b_2x\left(px+\frac{1}{2}qx^2+...\right)+...=0$$

or $x(a_1+a_2p)+x^2\left(\frac{1}{2}a_2q+b_1+b_2p+b_3p^2\right)+...=0.$

Equating to zero the coefficients of x and x^2, we get

$$a_1 + a_2p = 0 \text{ and } \frac{1}{2}a_2q + b_1 + b_2p + b_3p^2 = 0.$$

Solving these, we get

$$p=-\frac{a_1}{a_2}, \text{ and } q=\left(\frac{2}{a_2^3}\right)\left(a_1b_2a_2-b_1a_2^2-a_1^2b_3\right).$$

$$\therefore \rho \text{ (at the origin)} = \frac{\left(1+p^2\right)^{3/2}}{q} = \frac{\left[1+\left(\frac{a_1^2}{a_2^2}\right)\right]^{3/2}}{\left(\frac{2}{a_2^3}\right)\left(a_1a_2b_2-a_2^2b_1-a_1^2b_3\right)}$$

$$=\frac{\left(a_1^2+a_2^2\right)^{3/2}}{2\left(a_1a_2b_2-a_2^2b_1-a_1^2b_3\right)}.$$

Example 9:

Prove that for any curve

$$\frac{1}{\rho}=\frac{d}{dx}\left(\frac{dy}{ds}\right).$$ **(Meerut, 1999S; Rohilkhand, 95; Kanpur, 97)**

Solution:

We have $\frac{d}{dx}\left(\frac{dy}{ds}\right)=\frac{d}{dx}(\sin\psi)$

$$=\cos y.\frac{d\psi}{dx}=\cos\psi.\frac{d\psi}{ds}.\frac{ds}{dx}=\cos\psi.\frac{1}{\rho}.\sec\psi, \qquad \left[\because \sec\psi=\frac{ds}{dx}\right]$$

$$=\frac{1}{\rho}.$$

Example 9 (a):

Find the radius of curvature at the origin of the curve

$y = x^4 - 4x^3 - 18x^2$.

Solution:

Here $\frac{dy}{dx} = 4x^3 - 12x^2 - 36x$, $\frac{d^2y}{dx^2} = 12x^2 - 24x - 36$.

$\therefore$ At (0, 0), $\frac{dy}{dx}=0$ and $\frac{d^2y}{dx^2}=-36$.

$$\therefore \quad \rho \text{ at } (0,0)=\frac{\left\{1+\left(\frac{dy}{dx}\right)^2\right\}^{3/2}}{\frac{d^2y}{dx^2}} \text{ at } (0,0)$$

$$=\frac{(1+0)^{3/2}}{-36}=\frac{1}{36}, \text{ (numerically).}$$

Example 10:

Apply Newton's formula to find the radius of curvature at the origin of the cycloid

$x = a(\theta + \sin\theta)$, $y = a(1 - \cos\theta)$.

Solution:

Differentiating, $\frac{dx}{d\theta}=a(1+\cos\theta)$ and $\frac{dy}{d\theta}=a\sin\theta$.

$$\therefore \frac{dy}{dx}=\frac{\frac{dy}{d\theta}}{\frac{dx}{d\theta}}=\frac{a\sin\theta}{a(1+\cos\theta)}=\tan\frac{1}{2}\theta.$$

$$\therefore \left(\frac{dy}{dx}\right)_{\text{at origin}} = \tan 0 = 0, \qquad [\because \theta = 0 \text{ at } x = 0,\ y = 0].$$

Thus, the axis of x is tangent at the origin.

$$\therefore \rho(\text{at the origin}) = \lim_{x \to 0}\left(\frac{x^2}{2y}\right)$$

$$= \lim_{\theta \to 0}\left[\frac{a^2(\theta + \sin\theta)^2}{2a(1-\cos\theta)}\right], \qquad \left[\text{form } \frac{0}{0}\right]$$

$$= \lim_{\theta \to 0}\left[\frac{a^2(\theta + \sin\theta)}{2a(1-\cos\theta)}\right], \qquad \left[\text{form } \frac{0}{0}\right]$$

$$= \lim_{\theta \to 0}\left[\frac{a}{2}\,\frac{2(\theta + \sin\theta)(1+\cos\theta)}{\sin\theta}\right], \qquad \text{by L'Hospital's Rule}$$

$$= \lim_{\theta \to 0}\left[a\,\frac{(\theta + \sin\theta)(-\sin\theta) + (1+\cos\theta)^2}{\cos\theta}\right] = 4a.$$

Example 11:

Show that the radii of curvature at the origin on the curve $x^3 + y^3 = 3axy$ is each equal to $\frac{3a}{2}$.

Solution:

The curve is $x^3 + y^3 = 3axy$, ...(1)

which obviously passes through the origin.

Equating to zero the lowest degree terms in (1), the tangents at the origin are given by $3axy = 0$, *i.e.*, are $x = 0$ and $y = 0$.

$$\therefore \qquad \rho \text{ (at the origin)} = \lim_{x \to 0}\left(\frac{y^2}{2x}\right) = \rho_1 \text{ (say)} \qquad ...(2)$$

$$\text{and} \qquad = \lim_{x \to 0}\left(\frac{x^2}{2y}\right) = \rho_2 \text{ (say)}$$

Now dividing both sides of (1) by 2xy, we get

$$\left(\frac{x^2}{2y}\right) + \left(\frac{y^2}{2x}\right) = \frac{3a}{2}. \qquad ...(3)$$

Taking limits of (3) as $x \to 0$ and $y \to 0$, we get

$$\lim_{x\to 0}\frac{x^2}{2y}+\lim_{x\to 0}\left(\frac{y^2}{2x}\right)=\frac{3a}{2}$$

or $$\lim_{x\to 0}\frac{x^2}{2y}+\lim_{x\to 0}\frac{1}{4}xy\frac{2y}{x^2}=\frac{3a}{2}$$

or $$\rho_2=0.\left(\frac{1}{\rho_2}\right)=\frac{3a}{2} \text{ or } \rho_2=\frac{3a}{2}$$

Similarly $$\rho_1=\frac{3a}{2}.$$

Example 12:

Show that the radii of curvature of the curve

$$y^2=x^2\frac{(a+x)}{(a-x)}$$

at the origin are $\pm a\sqrt{2}$. **(Magadh, 1992)**

Solution:

Obviously the given curve passes through the origin. From the equation of the given curve, we have

$$y^2=\frac{x^2(a+x)}{a-x}$$

or $$y=\pm\frac{x(x+a)^{1/2}}{(a-x)^{1/2}}=\pm x\left(1+\frac{x}{a}\right)^{1/2}\left(1-\frac{x}{a}\right)^{-1/2}$$

$$=\pm x\left\{1+\frac{1}{2}\frac{x}{a}+\ldots\right\}\left\{1+\frac{1}{2}\frac{x}{a}+\ldots\right\},$$

expanding by binomial theorem

$$=\pm x\left(1+\frac{x}{a}+\ldots\right). \quad \ldots(1)$$

Let $\left(\frac{dy}{dx}\right)_{(0,\,0)}=p$ and $\left(\frac{d^2y}{dx^2}\right)_{(0,\,0)}=q$.

Then by Maclaurin's expansion, we get for this curve

$$y=px+\frac{1}{2}qx^2+\ldots \quad \ldots(2)$$

Comparing (1) and (2), we get

$$p = 1,\ q = \frac{2}{a};\ \text{or } p = -1,\ q = -\frac{2}{a}.$$

But ρ (at origin) $= \dfrac{\left(1+p^2\right)^{2/3}}{q}$.

$\therefore$ when $p = 1$, $q = \dfrac{2}{a}$, we have ρ (at origin)

$= \dfrac{(1+1)^{3/2}}{\frac{2}{a}} = a\sqrt{2}$ and when $p = -1$, $q = -\dfrac{2}{a}$, we have

$$\rho \text{ (at origin)} = \frac{(1+1)^{3/2}}{-\frac{2}{a}} = -a\sqrt{2}.$$

Example 13:

Find the radius of curvature of the curve $r = a \sin n\theta$ at the origin (pole).

Solution:

As $\theta = 0 \Rightarrow r = 0$, so the pole lies on the given curve. Also $\theta = 0$ is a tangent at pole.

$$\therefore\ \rho \text{ at the pole} = \lim_{\theta \to 0}\left(\frac{r}{2\theta}\right) = \lim_{\theta \to 0} \frac{a \sin n\theta}{2\theta}$$

$$= \lim_{\theta \to 0}\left(\frac{1}{2} na.\frac{\sin n\theta}{n\theta}\right).\frac{1}{2} na.1 = \frac{1}{2} na.$$

Example 14:

Prove that co-ordinates (α, β) of the centre of curvature at any point (x, y) can be expressed in the form

$$\alpha = x - \left(\frac{dy}{d\psi}\right) \text{ and } \beta = y + \left(\frac{dx}{d\psi}\right). \qquad \textbf{(Kanpur, 1996)}$$

Solution:

Let (α, β) be the co-ordinates of the centre of curvature at any point (x, y) of a given curve.

We have $\rho = \dfrac{\left[1+\left(\dfrac{dy}{dx}\right)^2\right]^{3/2}}{\dfrac{d^2y}{dx^2}} = \dfrac{\left(1+\tan^2\psi\right)^{3/2}}{\dfrac{d^2y}{dx^2}}$, $\qquad \left[\because \dfrac{dy}{dx} = \tan\psi\right]$

$$\therefore \qquad \frac{d^2y}{dx^2} = \left(\frac{1}{\rho}\right)\sec^3\psi.$$

$$\text{Now, } \alpha = x - \frac{\left(\frac{dy}{dx}\right)\left[1+\left(\frac{dy}{dx}\right)^2\right]}{\frac{d^2y}{dx^2}} = x - \frac{\tan\psi(1+\tan^2\psi)}{\left(\frac{1}{\rho}\right)\sec^3\psi}$$

$$= x - \rho\sin\psi$$

$$= x - \frac{ds}{d\psi}\cdot\frac{dy}{ds}, \qquad \left[\because \rho = \frac{ds}{d\psi} \text{ and } \sin\psi = \frac{dy}{ds}\right]$$

$$\therefore \qquad \alpha = x - \left(\frac{dy}{d\psi}\right).$$

$$\text{Similarly, } \beta = y + \frac{(1+\tan^2\psi)}{\left(\frac{1}{\rho}\right)\sec^3\psi} = y + \rho\cos\psi$$

$$= y + \frac{ds}{d\psi}\cdot\frac{dx}{ds}, \qquad \left[\because \rho = \frac{ds}{d\psi} \text{ and } \cos\psi = \frac{dx}{ds}\right]$$

$$\therefore \qquad \beta = y + \left(\frac{dx}{d\psi}\right).$$

9.11 CO-ORDINATES OF CENTRE OF CURVATURE

To find the centre of curvature for any point (x, y) of the curve

y = f(x) or f(x, y) = 0.

Let the equation of the curve be y = f(x). Let P be the given point (x, y) on the curve and Q a point (x + δx, y + δy) in the neighbourhood of P. Let N be the point of intersection of the normals at P and Q. As Q → P, suppose N → C. Then C is the centre of curvature for the point P. Let the co-ordinates of C be (α, β).

From the equation of the curve, we have

$$\frac{dy}{dx} = f'(x) = \phi(x), \text{ (say)}.$$

The equation of normal at P(x, y) is

$$(Y - y)\,\phi(x) + (X - x) = 0. \qquad \ldots(1)$$

The equation of the normal at Q(x + δx, y + δy) is

$$\{Y - (y + \delta y)\}\ \phi\ (x + \delta x) + \{X - (x + \delta x)\} = 0. \qquad ...(2)$$

Subtracting (1) from (2), we get

$$(Y - y)\ \{\phi\ (x + \delta x) - \phi(x)\} - \delta y.\phi\ (x + \delta x) - \delta x = 0.$$

Dividing by dx, we get

$$(Y - y)\ \left\{\frac{\phi(x+\delta x)-\phi(x)}{\delta x}\right\} - \phi(x+\delta x)\frac{\delta y}{\delta x} - 1 = 0. \qquad ...(3)$$

The value of Y obtained from this equation gives us the y-co-ordinate of the point of intersection of (1) and (2).

Now as Q → P, δx → 0 and Y obtained from (3) → β.

$$(\beta - y)\lim_{\delta x\to 0}\left\{\frac{\phi(x+\delta x)-\phi(x)}{\delta x}\right\} - \lim_{\delta x\to 0}\phi(x+\delta x).\lim_{\delta x\to 0}\frac{\delta y}{\delta x} - 1 = 0$$

or $$(\beta - y)\frac{d}{dx}\phi(x) - \phi(x).\frac{dy}{dx} - 1 = 0$$

or $$(\beta - y)\frac{d}{dx}\left(\frac{dy}{dx}\right) - \frac{dy}{dx}.\frac{dy}{dx} - 1 = 0, \qquad \left[\because \phi(x) = \frac{dy}{dx}\right]$$

or $$(\beta - y)\frac{d^2y}{dx^2} - \left\{\left(\frac{dy}{dx}\right)^2 + 1\right\} = 0.$$

$$\therefore \qquad \beta = y + \frac{1+\left(\frac{dy}{dx}\right)^2}{\frac{d^2y}{dx^2}}. \qquad ...(4)$$

Also (α, β) lines on (1). Therefore, we get

$$(\beta - y)\left(\frac{dy}{dx}\right) + (\alpha - x) = 0$$

i.e., $$(\alpha - x) = -(\beta - y)\frac{dy}{dx} = -\frac{dy}{dx}.\frac{1+\left(\frac{dy}{dx}\right)^2}{\frac{d^2y}{dx^2}}, \qquad \text{[from (4)]}$$

$$\therefore \qquad \alpha = x - \frac{\left(\frac{dy}{dx}\right)\left\{1+\left(\frac{dy}{dx}\right)^2\right\}}{\frac{d^2y}{dx^2}}. \qquad ...(5)$$

Hence, the co-ordinates (α, β) of the centre of curvature are given by (4) and (5).

Evolute of a Curve

Definition: *The locus of the centres of curvature of all points of a given plane curve is called the evolute of the curve.*

Cor. Equation of the Circle of Curvature at P(x, y)

If (α, β) be the co-ordinates of the centre of curvature and ρ be the radius of curvature at any point (x, y) on a curve then the equation of the circle of curvature at that point is

$$(X - \alpha)^2 + (Y - \beta)^2 = \rho^2.$$

Example 1:

Find the co-ordinates of the centre of curvature for the point (x, y) on the parabola $y^2 = 4ax$. **(Kanpur, 1998; Kurukshetra, 94)**

Also find the equation of the evolute of the parabola.

(Kanpur, 1994, 95; R.U., 99)

Solution:

We have $y^2 = 4ax$. ...(1)

Differentiating, $\dfrac{dy}{dx} = \dfrac{2a}{y} = \dfrac{2a}{\sqrt{(4ax)}} = a^{1/2}x^{-1/2} = \sqrt{\left(\dfrac{a}{x}\right)}$,

and $\dfrac{d^2y}{dx^2} = -\dfrac{1}{2}a^{1/2}x^{-3/2}$.

If (α, β) be the centre of curvature for the point (x, y), then

$$\alpha = x - \frac{\left(\dfrac{dy}{dx}\right)\left\{1+\left(\dfrac{dy}{dx}\right)^2\right\}}{\dfrac{d^2y}{dx^2}} = x - \frac{\left[1+\left(\dfrac{a}{x}\right)\right]\cdot\sqrt{\left(\dfrac{a}{x}\right)}}{-\dfrac{1}{2}a^{1/2}x^{-3/2}}$$

$= x + 2(x + a) = 3x + 2a.$ ...(2)

Also $\beta = y + \dfrac{\left[1+\left(\dfrac{dy}{dx}\right)^2\right]}{\dfrac{d^2y}{dx^2}} = y + \dfrac{\left[1+\left(\dfrac{a}{x}\right)\right]}{-\dfrac{1}{2}a^{1/2}x^{-3/2}}$

$= 2a^{1/2}x^{1/2} - 2a^{-1/2}x^{1/2}(x + a) = 2a^{-1/2}x^{1/2}[a - (x + a)]$

$= -2a^{-1/2}.x^{3/2}.$ ...(3)

Hence the required centre of curvature is the point

$$(3x + 2a, - 2a^{-1/2}x^{3/2}).$$

Eliminating x from (2) and (3), we get

$$\left(\frac{\alpha - 2a}{3}\right)^3 = \frac{a\beta^2}{4} \text{ or } 27a\beta^2 = 4(\alpha - 2a)^3.$$

Therefore, the locus of (α, β) is

$$27\, ay^2 = 4(x - 2a)^3,$$

which is the required evolute of the parabola.

Example 2:

Find the co-ordinates of the centre of curvature of the ellipse $\frac{x^2}{a^2} + \frac{y^2}{b^2} = 1$ *at a given point (x, y).*

(Gorakhpur, 1993)

Solution:

Here $\frac{dy}{dx} = -\frac{b^2x}{(a^2y)}$

and $\frac{d^2y}{dx^2} = -\frac{b^4}{(a^2y^3)}$.

$$\therefore \qquad \alpha = x - \frac{\left(\frac{dy}{dx}\right)\left[1 + \left(\frac{dy}{dx}\right)^2\right]}{\frac{d^2y}{dx^2}} = x - \frac{(b^2x^2 + a^4y^2)}{a^4b^2}$$

and $$\beta = y + \frac{\left[1 + \left(\frac{dy}{dx}\right)^2\right]}{\frac{d^2y}{dx^2}} = y - \frac{(b^4x^2 + a^4y^2)}{a^2b^4}.$$

Example 2 (a):

Find the centre of curvature of the curve

$y = x^3 - 6x^2 + 3x + 1$ at the point (1, – 1). **(Kanpur, 1995)**

Solution:

We have $\frac{dy}{dx} = 3x^2 - 12x + 3$; $\frac{d^2y}{dx^2} = 6x - 12$.

$$\therefore \quad \left(\frac{dy}{dx}\right)_{(1,-1)} = -6; \left(\frac{d^2y}{dx^2}\right)_{(1,-1)} = -6$$

$$\therefore \quad \alpha = 1 - \frac{\left[1+(-6)^2\right](-6)}{-6} = 1 - 37 = -36$$

$$\text{and } \beta = -1 + \frac{\left[1+(-6)^2\right]}{-6} = -1 - \frac{37}{6} = -\frac{43}{6}$$

$\therefore$ the required centre of curvature is $\left(-36, -\frac{43}{6}\right)$

Example 3:

Find the centre of curvature at the point 't' on the ellipse $x = a\cos t$, $y = b\sin t$.

Solution:

Differentiating, $\frac{dx}{dt} = -a\sin t$, $\frac{dy}{dt} = b\cos t$

$$\therefore \quad \frac{dy}{dx} = \frac{\frac{dy}{dt}}{\frac{dx}{dt}} = \frac{b\cos t}{-a\sin t} = -\frac{b}{a}\cot t,$$

$$\text{and } \frac{d^2y}{dx^2} = \frac{d}{dx}\left(\frac{dy}{dx}\right) = \left\{\frac{d}{dt}\left(-\frac{b}{a}\cot t\right)\right\}.\frac{dt}{dx}$$

$$= \frac{b}{a}\operatorname{cosec}^2 t\left(\frac{1}{-a\sin t}\right) = -\frac{b}{a^2}\operatorname{cosec}^3 t.$$

If (α, β) be the centre of curvature for the point 't', then

$$\alpha = x - \frac{\left[1+\left(\frac{dy}{dx}\right)^2\right]\left(\frac{dy}{dx}\right)}{\frac{d^2y}{dx^2}}$$

$$= x - \frac{\left\{1+\left(\frac{b^2}{a^2}\right)\cot^2 t\right\}\left(-\left(\frac{b}{a}\right)\cot t\right)}{-\left(\frac{b}{a^2}\right)\operatorname{cosec}^3 t}$$

$$= a\cos t - \left(\frac{1}{a}\right)\left(a^2\sin^2 t + b^2\cos^2 t\right)\cos t$$

$$= \frac{a^2\left(\cos t - \sin^2 t\cos t\right) - b^2\cos^3 t}{a} = \frac{a^2 - b^2}{a}\cos^3 t.$$

Also $$\beta = y + \frac{\left[1+\left(\frac{dy}{dx}\right)^2\right]}{\frac{d^2y}{dx^2}} = y + \frac{\left\{1+\left(\frac{b^2}{a^2}\right)\cot^2 t\right\}}{-\left(\frac{b}{a^2}\right)\operatorname{cosec}^3 t}$$

$$= b\sin t - \frac{\sin t}{b}.\left(a^2\sin t + b^2\cos^2 t\right)$$

$$= \frac{1}{b}\sin t\left[b^2 - a^2\sin^2 t - b^2\cos^2 t\right]$$

$$= \frac{b^2 - a^2}{b}\sin^3 t = -\frac{a^2 - b^2}{b}\sin^3 t.$$

$\therefore$ the required centre of curvature is the point

$$\left(\frac{a^2 - b^2}{a}\cos^3 t, -\frac{a^2 - t^2}{b}\sin^3 t\right).$$

Example 4:

Prove that the centres of curvature at points of the cycloid x=a(t–sin t), y = a (1 – cos t) lie on an equal cycloid.

Solution:

We have $\frac{dx}{dt} = a(1 - \cos t), \frac{dy}{dt} = a\sin t.$

$$\therefore \quad \frac{dy}{dx} = \frac{\frac{dy}{dt}}{\frac{dx}{dt}} = \frac{a\sin t}{a(1-\cos t)} = \cot\frac{1}{2}t,$$

and $$\frac{d^2y}{dx^2} = \frac{d}{dx}\left(\cot\frac{1}{2}t\right) = \left\{\frac{d}{dt}\left(\cot\frac{1}{2}t\right)\right\}.\frac{dt}{dx}$$

$$= -\frac{1}{2}\operatorname{cosec}^2\frac{1}{2}t.\frac{1}{a(1-\cos t)} = \frac{-1}{4a\sin^4\frac{1}{2}t}.$$

$$\therefore \quad \alpha = x - \frac{\left(\frac{dy}{dx}\right)\left[1+\left(\frac{dy}{dx}\right)^2\right]}{\frac{d^2y}{dx^2}}$$

$$= a(t - \sin t) - \frac{\cot\frac{1}{2}t\left[1 + \cot^2\frac{1}{2}t\right]}{-\left(\frac{1}{4a}\right)\operatorname{cosec}^2\frac{1}{2}t}$$

$$= a(t - \sin t) + 4a\cos\frac{1}{2}t.\sin\frac{1}{2}t$$

$$= at - a\sin t + 2a\sin t = a(t + \sin t) \qquad \text{...(1)}$$

Also $\beta = y + \dfrac{\left[1 + \left(\frac{dy}{dx}\right)^2\right]}{\frac{d^2y}{dx^2}} = y + \dfrac{1 + \cot^2\frac{1}{2}t}{-\left(\frac{1}{4a}\right)\operatorname{cosec}^4\frac{1}{2}t}$

$$= a(1 - \cos t) - 4a\sin^2\frac{1}{2}t$$

$$= a - a\cos t - 2a + 2a\cos t = -a(1 - \cos t). \qquad \text{...(2)}$$

$\therefore$ the centre of curvature at the point 't' of the given cycloid is the point

$$[a(t + \sin t), -a(1 - \cos t)].$$

Now the locus of the centre of curvature is obtained by generalising (α, β) obtained from (1) and (2). The required locus is the curve

$$x = a(t + \sin t),\ y = -a(1 - \cos t),$$

which is an equal cycloid.

Example 5:

Find the co-ordinates of the centre of curvature of the curve $a^2y = x^3$.

Solution:

Here $a^2y = x^3$;

$$\therefore \quad \frac{dy}{dx} = \frac{3x^2}{a^2} \text{ and } \frac{d^2y}{dx^2} = \frac{6x}{a^2}.$$

$$\therefore \quad \alpha = x - \frac{\left[1 + \left(\frac{dy}{dx}\right)^2\right]\left(\frac{dy}{dx}\right)}{\frac{d^2y}{dx^2}} = x - \frac{1}{2}x\left(1 + \frac{9x^4}{a^4}\right)$$

$$= \frac{x}{2}.\left(1 - \frac{9x^4}{a^4}\right).$$

Also $\beta = y + \frac{\left[1+\left(\frac{dy}{dx}\right)^2\right]}{\frac{d^2y}{dx^2}} = \frac{x^3}{a^2} + \frac{\left[1+\left(\frac{9x^4}{a^4}\right)\right]}{\frac{6x}{a^2}} = \frac{5x^3}{2a^2} + \frac{a^2}{6x}$.

$\therefore$ the required centre of curvature is

$$\left[\frac{x}{2}\left(1-\frac{9x^4}{a^4}\right), \left(\frac{5x^3}{2a^2}+\frac{a^2}{6x}\right)\right].$$

Example 6:

Prove that the centre of curvature (α, β) *for the curve*

$x = 3t,\ y = t^2 - 6$

is $\alpha = -\frac{4}{3}t^3,\ \beta = 3t^2 - \frac{3}{2}$. **(Meerut, 1999S)**

Solution:

We have $y = \frac{a^2}{x}$;

$\therefore \quad \frac{dy}{dx} = -a^2x^{-2}$ and $\frac{d^2y}{dx^2} = 2a^2x^{-3}$.

Let (α, β) be the co-ordinates of the centre of curvature at the point (x, y). Then

$$\alpha = x - \frac{\left(\frac{dy}{dx}\right)\left[1+\left(\frac{dy}{dx}\right)^2\right]}{\frac{d^2y}{dx^2}} = x - \frac{\left(-\frac{a^2}{x^2}\right)\left(1+\frac{a^4}{x^4}\right)}{\frac{2a^3}{x^3}}$$

$$= x + \frac{\left(x^4+a^4\right)}{2x^3} = \frac{3x^4+x^2y^2}{2x^3} = \frac{3x}{2} + \frac{y^2}{2x}, \qquad [\because a^4 = x^2y^2]$$

Also $\beta = y + \frac{\left[1+\left(\frac{dy}{dx}\right)^2\right]}{\frac{d^2y}{dx^2}} = y + \frac{\left[1+\left(\frac{a^4}{x^4}\right)\right]}{\frac{2a^2}{x^3}}$

$$y + \frac{x^4+a^4}{2a^2x} = y + \frac{x^4+x^2y^2}{2x^2y}, \qquad [\because a^2 = xy]$$

$$= y + \frac{x^2+y^2}{2y} = \frac{3y^2+x^2}{2y} = \frac{3}{2}y + \frac{x^2}{2y}.$$

Hence the required centre of curvature is the point

$$\left(\frac{3x}{2}+\frac{y^2}{2x}, \frac{3y}{2}+\frac{x^2}{2y}\right).$$

9.12 CHORD OF CURVATURE THROUGH THE ORIGIN (POLE)

Let C be the centre of curvature at the point P on any given curve. Then CP = ρ = radius of curvature at P. The circle whose centre is C and radius CP, is the circle of curvature at P. Any chord of this circle through P is a chord of curvature.

Let O be the pole. Join OP to meet the circle of curvature in E. Then PE is the chord of curvature through the origin.

PD is the diameter of the circle of curvature. We have PD = 2ρ and $\angle$PED = 90°, being an angle in a semi-circle.

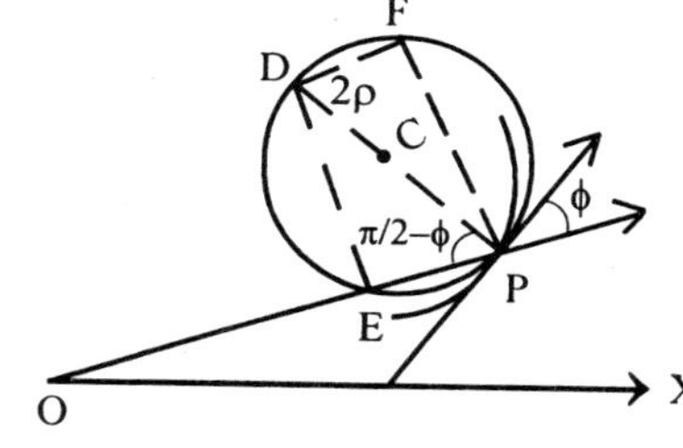

Fig. 9.4

Also PD is normal to the curve at P; therefore $\angle EPD = \frac{1}{2}\pi - \phi$.

Now from the right-angled triangle PED, we have

$$PE = PD \cos\left(\frac{1}{2}\pi - \phi\right) = 2\rho \sin \phi.$$

$\therefore$ **chord of curvature through the pole = $2\rho \sin \phi$.**

Deduction: *To find the chord of curvature through the pole for the curve $p = f(r)$.* **(Lucknow, 1992)**

The required chord of curvature

$$= 2\rho \sin \phi = 2r\frac{dr}{dp}.\frac{p}{r}, \qquad \left[\because \rho = r\frac{dr}{dp} \text{ and } p = r \sin \phi\right]$$

$$= 2p\frac{dr}{dp} = \frac{2f(r)}{f'(r)}, \qquad \left[\because p = f(r) \text{ and } \frac{dp}{dr} = f'(r)\right]$$

9.13 CHORD OF CURVATURE PERPENDICULAR TO THE RADIUS VECTOR

In the figure of this article suppose a line through P, perpendicular to the radius vector OP, meets the circle of curvature in F. Then PF is the chord of curvature perpendicular to the radius vector.

We have $PF = ED = 2\rho \sin\left(\frac{1}{2}\pi - \phi\right) = 2\rho \cos \phi.$

9.14 CHORDS OF CURVATURE PARALLEL TO THE CO-ORDINATE AXES

In the adjoining figure, C is the centre of curvature at any point P on a given curve. We have CP = ρ and PD = 2ρ. PE and PF are the chords of curvature parallel to the x-axis and y-axis respectively.

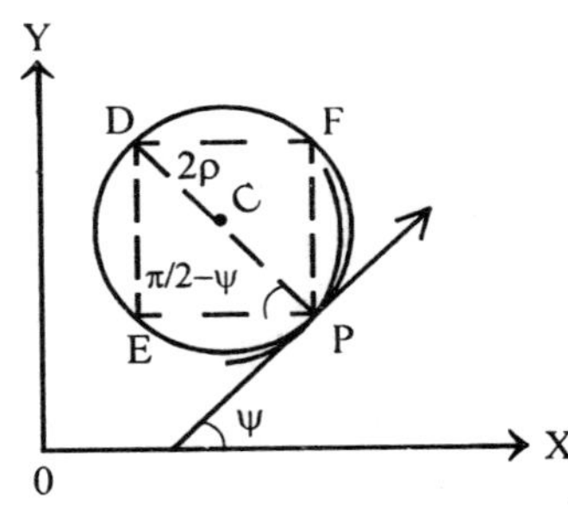

Fig. 9.5

Also $\angle$ PED $= \frac{1}{2}\pi$,

$\angle$ EPT $= \angle$ PTX $= \psi$ and

$\angle$ EPD $= \frac{1}{2}\pi - \psi$.

We have PE $=$ PD $\left(\frac{1}{2}\pi - \psi\right) = 2\rho \sin \psi$.

$\therefore$ **chord of curvature parallel to the axis of x = $2\rho \sin \psi$.**

Again PF $=$ ED $=$ PD $\sin\left(\frac{1}{2}\pi - \psi\right) = 2\rho \cos \psi$.

$\therefore$ **chord of curvature parallel to the axis of y = $2\rho \cos \psi$.**

Example 1:

Show that the length of the chord of curvature, parallel to y-axis at the origin, in the parabola $y = mx + \left(\frac{x^2}{a}\right)$ is $(1 + m^2)\, a$. **(Allahabad, 1991)**

Solution:

We have $y = mx + \left(\frac{x^2}{a}\right)$.

$$\therefore \quad \frac{dy}{dx} = m + \left(\frac{2x}{a}\right), \frac{d^2y}{dx^2} = \frac{2}{a}.$$

$\therefore$ at the origin (0, 0), we have $\frac{dy}{dx} = m$ and $\frac{d^2y}{dx^2} = \frac{2}{a}$.

$$\therefore \quad \rho \text{ (at origin)} = \frac{\left[1 + \left(\frac{dy}{dx}\right)^2\right]^{3/2}}{\frac{d^2y}{dx^2}} \text{ at } (0, 0) = \frac{\left(1 + m^2\right)^{3/2}}{\frac{2}{a}}$$

$$= \frac{a\left(1+m^2\right)^{3/2}}{2}.$$

Also at the origin, we have $\tan \psi = \frac{dy}{dx}$ at (0, 0) = m;

$$\therefore \quad \cos \psi = \frac{1}{\sqrt{\left(1+m^2\right)}}.$$

Now the chord of curvature parallel to the y-axis = 2ρ cos ψ.

∴ at the origin, the chord of curvature parallel to the y-axis

$$= 2.\frac{a\left(1+m^2\right)^{3/2}}{2}.\frac{1}{\sqrt{\left(1+m^2\right)}} = a\left(1+m^2\right).$$

Example 2:

At the point upon the Archimedean spiral $r = a\theta$ at which tangent makes half a right angle with the radius vector, prove that

$$C_r = C_\theta = \frac{4a}{3},$$

where C_r and C_θ denote the chords of curvature through the pole and perpendicular to the radius vector.

Solution:

The curve is r = aθ. ...(1)

$$\therefore \qquad \frac{dr}{d\theta} = a, \ \frac{d^2r}{d\theta^2} = 0.$$

Now $\quad \tan \phi = r\frac{d\theta}{dr} = r.\frac{1}{a} = (a\theta).\frac{1}{a} = \theta.$

∴ at the point where φ = 45°, we have tan φ = 1 and so θ = 1.

$$\text{Now } \rho = \frac{\left[r^2 + \left(\frac{dr}{d\theta}\right)^2\right]^{3/2}}{r^2 + 2\left(\frac{dr}{d\theta}\right)^2 - r\left(\frac{d^2r}{d\theta^2}\right)}.$$

Therefore at the point θ = 1, we have

$$\rho = \frac{\left(a^2+a^2\right)^{3/2}}{a^2+2a^2}, \qquad [\because \text{ on the curve } r = a\theta,\ r = a \text{ when } \theta = 1]$$

$$= \frac{\left(2\sqrt{2a}\right)}{3}.$$

$$\therefore\ C_r = 2\rho \sin\phi = 2.\frac{2\sqrt{2a}}{3}\sin 45^\circ = 2.\frac{2\sqrt{2a}}{3}.\frac{1}{\sqrt{2}} = \frac{4a}{3},$$

and $C_\theta = 2\rho \cos\phi = 2.\frac{2\sqrt{2a}}{2}\cos 45^\circ = 2.\frac{2\sqrt{2a}}{3}.\frac{1}{\sqrt{2}} = \frac{4a}{3}.$

Hence $C_r = C_\theta = \frac{4a}{3}$.

Example 2 (a):

Show that in any curve the chord of curvature perpendicular to the radius vector is $\frac{2\rho\sqrt{\left(r^2 - p^2\right)}}{r}$. **(Kanpur, 1995)**

Solution:

The chord of curvature perpendicular to the radius vector

$$= 2\rho \cos\phi = 2\rho\sqrt{\left(1 - \sin^2\phi\right)}$$

$$= 2\rho\sqrt{\left[1 - \left(\frac{p}{r}\right)^2\right]} \qquad [\because\ p = r\sin\phi]$$

$$= \frac{\left[2\rho\sqrt{\left(r^2 - p^2\right)}\right]}{r}.$$

Example 3:

Show that in the curve $y = a\cosh\left(\frac{x}{a}\right)$ *the chord of curvature parallel to the axis of x is of length* $a\sinh\left(\frac{2x}{a}\right)$.

(Jiwaji, 1992; Gorakhpur, 96)

Solution:

We have $y = a\cosh\left(\frac{x}{a}\right)$,

Therefore $\frac{dy}{dx} = a\left\{\sinh\left(\frac{x}{a}\right)\right\}.\left(\frac{1}{a}\right) = \sinh\left(\frac{x}{a}\right);$

$$\frac{d^2y}{dx^2} = \left(\frac{1}{a}\right)\cosh\left(\frac{x}{a}\right).$$

Now, the chord of curvature parallel to the axis of x = $2\rho \sin\psi$

$$= 2\rho.\frac{1}{\operatorname{cosec}\psi} = 2\rho\frac{1}{\sqrt{\left(1+\cos^2\psi\right)}} = 2\rho\frac{\tan\psi}{\sqrt{\left(1+\tan^2\psi\right)}}$$

$$= 2.\frac{\left[1+\left(\frac{dy}{dx}\right)^2\right]^{3/2}}{\frac{d^2y}{dx^2}}.\frac{\frac{dy}{dx}}{\sqrt{\left[1+\left(\frac{dy}{dx}\right)^2\right]}} = \frac{2\left(\frac{dy}{dx}\right)\left[1+\left(\frac{dy}{dx}\right)^2\right]}{\frac{d^2y}{dx^2}}$$

$$= \frac{2\sinh\left(\frac{x}{a}\right)\left[1+\sinh^2\left(\frac{x}{a}\right)\right]}{\left(\frac{1}{a}\right)\cosh\left(\frac{x}{a}\right)} = \frac{2a\sinh\left(\frac{x}{a}\right)\cosh^2\left(\frac{x}{a}\right)}{\cosh\left(\frac{x}{a}\right)}$$

$$= 2a\sinh\left(\frac{x}{a}\right)\cosh\left(\frac{x}{a}\right) = a\sinh\left(\frac{2x}{a}\right).$$

Example 4:

Show that the chord of curvature through the pole of the curve $r^n = a^n \cos n\theta$ *is* $\frac{2r}{(n+1)}$. **(Meerut, 1993; Lucknow, 97; Gorakhpur, 99)**

Solution:

We have $r^n = a^n \cos n\theta$.

Taking logarithm, $n \log r = n \log a + \log \cos n\theta$.

Differentiating w.r.t. θ, we have $\frac{n}{r}\frac{dr}{d\theta} = -n\tan n\theta$.

Differentiating w.r.t. θ, we have $\frac{n}{r}\frac{dr}{d\theta} = -n\tan n\theta$.

$\therefore \cot\phi = \frac{1}{r}\frac{dr}{d\theta} = -\tan n\theta = \cot\left(\frac{1}{2}\pi + n\theta\right)$; so that $\phi = \frac{1}{2}\pi + n\theta$.

Now $\sin\phi = \sin\left(\frac{1}{2}\pi + n\theta\right) = \cos n\theta = \left(\frac{r}{a}\right)^n$. ...(1)

Also $p = r\sin\phi = r\left(\frac{r}{a}\right)^n = \frac{r^{n+1}}{a^n}$.

Thus, $p = \frac{r^{n+1}}{a^n}$ is the pedal equation of the curve. Differentiating w.r.t. r, we have

$$\frac{dp}{dr} = \frac{(n+1)r^n}{a^n}.$$

$$\therefore \qquad \rho = r\frac{dr}{dp} = r.\frac{a^n}{(n+1)r^n} = \frac{a^n}{(n+1)\,r^{n-1}}. \qquad ...(2)$$

Hence the chord of curvature through the pole

$$= 2\rho \sin\phi = 2\frac{a^n}{(n+1)\,r^{n-1}}\left(\frac{r^n}{a^n}\right), \qquad \text{[from (1) and (2)]}$$

$$= \frac{2r}{(n+1)}.$$

Example 5:

Show that the chord of curvature through the focus of a parabola is four times the focal distance of the point and the chord of curvature parallel to the axis has the same length. **(Lucknow, 1999)**

Solution:

Let the equation of the parabola referred to focus as pole and axis as initial line, (*i.e.*, x-axis) be

$$\frac{2a}{r} = 1 + \cos\theta. \qquad ...(1)$$

Taking log of both sides of (1), we get

$$\log 2a - \log r = \log(1 + \cos\theta).$$

Differentiating w.r.t. θ, we get

$$-\frac{1}{r}\frac{dr}{d\theta} = \frac{-\sin\theta}{1+\cos\theta} = \frac{-2\sin\frac{1}{2}\theta\cos\frac{1}{2}\theta}{2\cos^2\frac{1}{2}\theta} = -\tan\frac{1}{2}\theta.$$

$$\therefore \quad \cot\phi = \frac{1}{r}\frac{dr}{d\theta} = \tan\frac{1}{2}\theta = \cot\left(\frac{1}{2}\pi - \frac{1}{2}\theta\right);$$

so that $\phi = \frac{1}{2}\pi - \frac{1}{2}\theta$.

Now $p = r\sin\phi = r\sin\left(\frac{1}{2}\pi - \frac{1}{2}\theta\right) = r\cos\frac{1}{2}\theta$.

$$\therefore \quad p^2 = r^2\cos^2\frac{1}{2}\theta = \frac{1}{2}r^2\left(2\cos^2\frac{1}{2}\theta\right) = \frac{1}{2}r^2(1+\cos\theta)$$

$$=\frac{1}{2}r^2\left(\frac{2a}{r}\right).$$ from (1)

$\therefore$ the pedal equation of the parabola (1) is $p^2 = ar$.
Differentiating it with respect to p, we get

$$2p = a\frac{dr}{dp}; \therefore \frac{dr}{dp} = \frac{2p}{a}.$$

Now the chord of curvature through the focus
= the chord of curvature through the pole

$$= 2\rho \sin\phi = 2r\frac{dr}{dp}.\frac{p}{r}, \qquad \left[\because p = r\sin\phi \text{ and } \rho = r\frac{dr}{dp}\right]$$

$$= 2p\frac{dr}{dp} = 2p\left(\frac{2p}{a}\right) = \frac{4p^2}{a} = \frac{4ar}{a} = 4r$$

= 4 × the distance of the point from the pole
= 4 × the focal distance of the point. This **proves the first result.**
Again $\psi = \theta + \phi$.

But for this curve $\phi = \frac{1}{2}\pi - \frac{1}{2}\theta$ or $\theta = \pi - 2\phi$.

$\therefore \quad \psi = \pi - 2\phi + \phi = \pi - \phi$.

Now the chord of curvature parallel to the axis of the parabola = the chord of curvature parallel to x-axis

$$= 2\rho \sin\psi = 2\rho\sin(\pi - \phi) = 2\rho\sin\phi = 2r\frac{dr}{dp}\frac{p}{r} = 2p\frac{dr}{dp}$$

$$= 2p\left(\frac{2p}{a}\right) = \frac{4p^2}{a} = \frac{4ar}{a} = 4r$$

= four times the focal distance of the point.

Example 6:

In the curve $y = a\log\sec\left(\frac{x}{a}\right)$ prove that the chord of curvature parallel to the axis of y is of constant length. **(Lucknow, 1990; Kanpur, 96)**

Solution:

We have $y = a\log\sec\left(\frac{x}{a}\right)$. ...(1)

Differentiating, we get

$$\frac{dy}{dx} = a.\frac{1}{\sec\left(\frac{x}{a}\right)}.\sec\left(\frac{x}{a}\right).\tan\left(\frac{x}{a}\right).\frac{1}{a} = \tan\frac{x}{a},$$

and $$\frac{d^2y}{dx^2} = \frac{1}{a}\sec^2\left(\frac{x}{a}\right).$$

Now chord of curvature parallel to y-axis

$$= 2\rho\cos\psi = \frac{2\rho}{\sec\psi} = \frac{2\rho}{\sqrt{\left\{1+\left(\frac{dy}{dx}\right)^2\right\}}}$$

$$= 2\frac{\left\{1+\left(\frac{dy}{dx}\right)^2\right\}^{3/2}}{\frac{d^2y}{dx^2}}\cdot\frac{1}{\left\{1+\left(\frac{dy}{dx}\right)^2\right\}^{1/2}}$$

$$= 2\frac{1+\left(\frac{dy}{dx}\right)^2}{\frac{d^2y}{dx^2}} = 2\frac{1+\tan^2\left(\frac{x}{a}\right)}{\left(\frac{1}{a}\right)\sec^2\left(\frac{x}{a}\right)} = 2a \ = \text{constant}.$$

Example 7:

Find the chord of curvature through the pole of the cardioid, $r = a(1 + \cos\theta)$. **(Gorakhpur, 1998; Jiwaji, 92; Meerut, 94)**

Solution:

We have $r = a(1 + \cos\theta)$.

Differentiating, $\frac{dr}{d\theta} = -a\sin\theta$.

$$\therefore \quad \tan\phi = r\frac{d\theta}{dr} = \frac{a(1+\cos\theta)}{-a\sin\theta} = -\cot\frac{1}{2}\theta = \tan\left(\frac{1}{2}\pi + \frac{1}{2}\theta\right);$$

so that $\phi = \frac{1}{2}\pi + \frac{1}{2}\theta$.

Now $p = r\sin\phi = r\sin\left(\frac{1}{2}\pi + \frac{1}{2}\theta\right) = r\cos\frac{1}{2}\theta$.

$$\therefore \quad 2p^2 = r^2\left(2\cos^2\frac{1}{2}\theta\right) = r^2(1+\cos\theta) = r^2.\left(\frac{r}{a}\right) = \frac{r^3}{a}.$$

Thus, $2p^2 a = r^3$ is the pedal equation of the curve.

Differentiating w.r.t. r, we have $4ap\frac{dp}{dr} = 3r^2$.

$$\therefore \qquad \rho = r\frac{dr}{dp} = r.\frac{4ap}{3r^2} = \frac{4ap}{3r}.$$

Hence the chord of curvature through the pole

$$= 2\rho \sin\phi = 2.\frac{4ap}{3r}.\frac{p}{r}, \qquad [\because p = r \sin\phi]$$

$$= \frac{8ap^2}{3r^2} = \frac{8}{r^2}.\frac{r^3}{2}, \qquad [\because 2ap^2 = r^3]$$

$$= \frac{4r}{3}.$$

Example 8:

Prove that the points on the curve $r = f(\theta)$, the circle of curvature at which passes through the origin are given by the equation

$$f(\theta) + f''(\theta) = 0.$$

Solution:

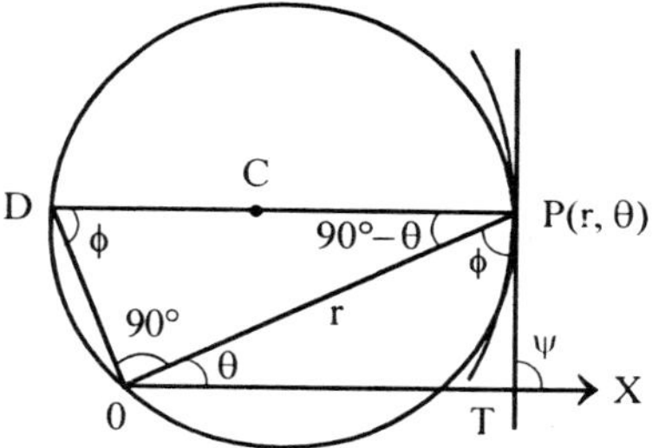

Fig. 9.6

Let $P(r, \theta)$ be any point on the curve $r = f(\theta)$ and let C be the centre of curvature at P. Also let the circle of curvature at pass through the pole O.

Let PC produced meet the circle of curvature in D. Tangent at P, (*i.e.*, PT) makes an angle ϕ with the radius vector OP. Thus in the figure, $OP = r$, $PD = 2\rho$, $\angle OPT = \phi$.

From D POD, we get

$$OP = PD \sin\phi \text{ or } r = 2\rho \sin\phi. \qquad ...(1)$$

$$\text{Now } \tan\phi = r\frac{d\theta}{dr} = \frac{r}{r_1}, \qquad \left(\text{where } r_1 = \frac{dr}{d\theta}\right)$$

$$\therefore \quad \sin\phi = \frac{r}{\sqrt{(r^2 + r_1^2)}}. \qquad ...(2)$$

$$\text{Also } \rho = \frac{(r^2 + r_1^2)^{3/2}}{r^2 + 2r_1^2 - rr_2}, \qquad \left(\text{where } r_2 = \frac{dr^2}{d\theta^2}\right) \qquad ...(3)$$

Now putting the values of $\sin\phi$ and ρ from (2) and (3) in (1), we get

$$\rho = \frac{2(r^2 + r_1^2)^{3/2}}{r^2 + 2r_1^2 - rr_2}.\frac{r}{\sqrt{(r^2 + r_1^2)}} = \frac{2r(r^2 + r_1^2)}{r^2 + 2r_1^2 - rr_2}$$

or $r^2 + 2r_1^2 - rr_2 - 2r^2 + 2r_1^2$, $[\because r \neq 0]$

or $r^2 + rr_2 = 0$ or $r + r_2 = 0$, $[\because r \neq 0]$

or $f(\theta) + f''(\theta) = 0$, $\left[\because r = f(\theta), r_2 = \dfrac{d^2r}{d\theta^2} = f''(\theta)\right]$

Example 9:

Show that the chord of curvature through the pole of the curve

$r = a\, e^{m\theta}$ *is* $2r$.

Solution:

We have $r = a\, e^{m\theta}$. Therefore $\dfrac{dr}{d\theta} = a\, e^{m\theta}.m = mr$.

$\therefore \quad \tan\phi = r\dfrac{d\theta}{dr} = r.\dfrac{1}{mr} = \dfrac{1}{m} = \tan\alpha$. **(Note)**

$\therefore \quad \phi = \alpha$.

Now $p = r\sin\phi = r\sin\alpha$. Therefore, $\dfrac{dp}{dr} = \sin\alpha$.

$\therefore \quad \rho = r.\dfrac{dr}{dp} = r.\dfrac{1}{\sin\alpha} = r\,\text{cosec}\,\alpha$.

Hence the chord of curvature through the pole $= 2\rho\sin\phi$

$= 2r\,\text{cosec}\,\alpha\sin\alpha = 2r$.

Example 10:

If C_x, C_y be the chords of curvature parallel to the axes at any point of the curve $y = ae^{x/a}$, prove that

$$\frac{1}{C_x^2} + \frac{1}{C_y^2} = \frac{1}{2a\,C_x}.$$ **(Lucknow, 1991; Meerut, 99)**

Solution:

We have, C_x = the chord of curvature parallel to the x-axis

$$= 2\rho\sin\psi = 2\rho\frac{1}{\text{cosec}\,\psi} = 2\rho\frac{1}{\sqrt{(1+\cot^2\psi)}} = 2\rho\frac{\tan\psi}{\sqrt{(1+\tan^2\psi)}}$$

$$= 2\frac{\left(1+y_1^2\right)^{3/2}}{y_2}.\frac{y_1}{\sqrt{\left(1+y_1^2\right)}} = \frac{2y_1}{y_2}\left(1+y_1^2\right). \quad ...(1)$$

Here $y = ae^{x/a}$, $y_1 = \dfrac{dy}{dx} = e^{x/a}$; and $y_2 = \dfrac{d^2y}{dx^2} = \left(\dfrac{1}{a}\right)e^{x/a}$.

$\therefore$ from equation (1), we get

$$C_x = \frac{2e^{x/a}}{\left(\frac{1}{a}\right)e^{x/a}}\ [1 + e^{2x/a}] = 2a(1 + e^{2x/a}).$$

Again C_y = the chord of curvature parallel to y-axis

$$= 2\rho \cos\psi = \frac{2\rho}{\sec\psi} = \frac{2\rho}{\sqrt{\left(1+\tan^2\psi\right)}} = 2\frac{\left(1+y_1^2\right)^{3/2}}{y_2}.\frac{1}{\sqrt{\left(1+y_1^2\right)}}$$

$$= \frac{2\left(1+y_1^2\right)}{y_2} = \frac{2}{\left(\frac{1}{a}\right)e^{x/a}}\left[1+e^{2x/a}\right] = \frac{2a}{e^{x/a}}\left(1+e^{2x/a}\right).$$

$$\therefore \quad \frac{1}{C_x^2} + \frac{1}{C_y^2} = \frac{1}{4a^2\left(1+e^{2x/a}\right)^2}\left[1+e^{2x/a}\right]$$

$$= \frac{1}{4a^2\left(1+e^{2x/a}\right)} = \frac{1}{2a\,C}.$$

Example 11:

If C_r and C_p be the chords of curvature of the cardioid, through the pole and perpendicular to the radius vector, then

$$3\left(C_1^2 + C_p^2\right) = 8a.C_r.$$

Solution:

The equation of the cardioid is r = a (1 + cos θ). ...(1)

Taking long of both sides of (1), we get

$$\log r = \log a + \log (1 + \cos\theta).$$

Differentiating w.r.t. θ, we get

$$\frac{1}{r}\frac{dr}{d\theta} = \frac{-\sin\theta}{1+\cos\theta}\frac{-2\sin\frac{1}{2}\theta\cos\frac{1}{2}\theta}{2\cos^2\frac{1}{2}\theta} = -\tan\frac{1}{2}\theta.$$

$$\therefore \quad \cot\phi = -\tan\frac{1}{2}\theta = \cot\left(\frac{1}{2}\pi + \frac{1}{2}\theta\right);\ \text{so that}\ \phi = \frac{1}{2}\pi + \frac{1}{2}\theta.$$

Now $p = r\sin\phi = r\sin\left(\frac{1}{2}\pi + \frac{1}{2}\theta\right) = r\cos\frac{1}{2}\theta.$

$$\therefore \quad p^2 = r^2 \cos^2\frac{1}{2}\theta = \frac{1}{2}r^2\left(2\cos^2\frac{1}{2}\theta\right) = \frac{1}{2}r^2(1+\cos\theta) = \frac{1}{2}r^2\left(\frac{r}{a}\right),$$

from (1).

Thus, the pedal equation of (1) is $p^2 = \frac{r^3}{2a}$.

Differentiating itw.r.t. p, we get

$$2p = \frac{3r^2}{2a}\frac{dr}{dp}; \qquad \therefore \frac{dr}{dp} = \frac{4ap}{3r^2}.$$

Now $\quad \rho = r\frac{dr}{dp} = r\left(\frac{4ap}{3r^2}\right) = \frac{4ap}{3r}$

$\therefore \quad C_r = 2\rho \sin\phi = \frac{8ap}{3r}\frac{p}{r} = \frac{8ap^2}{3r^2}$

and $\quad C_p = 2\rho \cos\phi$.

Now $\quad C_1^2 + C_p^2 = (2\rho \sin\phi)^2 + (2\rho\cos\phi)^2 = 4\rho_2$

$$= 4.\frac{16a^2p^2}{9r^2} = \frac{8a}{3}.\frac{8ap^2}{3r^2} = \frac{8a}{3}C_r.$$

Hence $\quad 3\left(C_1^2 + C_p^2\right) = 8aC_r$.

EXERCISES

1. Prove that the point $x = \frac{\pi}{2}$ of the curve $y = 4\sin x - \sin 2x$, $\rho = \frac{5\sqrt{5}}{4}$.
2. In the cycloid

   $x = a(\theta + \sin\theta)$, $y = a(1 - \cos\theta)$

   Prove that $\rho = 4a\cos\frac{\theta}{2}$.
3. Prove that for the ellipse $\frac{x^2}{a^2} + \frac{y^2}{b^2} = 1$

   $$\rho = \frac{a^2b^2}{p^3}$$

   where ρ being the perpendicular from the centre upon the tangent at (x, y).
4. Show that the radius of curvature at point $(a\cos^3\theta, a\sin^3\theta)$ on the curve

   $x^{2/3} + y^{2/3} = a^{2/3}$ is

   $3a\sin\theta\cos\theta$.

5. Prove that the radius of curvature of the catenary

$$y = \frac{1}{2}a\left(e^{x/a} + e^{-x/a}\right) \text{ is } \frac{y^2}{a}$$

and that of the catenary of uniform strength

$$y = C \log \sec\left(\frac{x}{c}\right) \text{ is } C \sec\left(\frac{x}{c}\right).$$

6. If the coordinates of a point on a curve be given by the equation.

$$x = c \sin 2\theta\ (1 + \cos 2\theta),\ y = c \cos 2\theta\ (1 - \cos 2\theta)$$

show that the radius of curvature at the point is $4c \cos 3\theta$.

7. The co-ordinate of a point on a curve are given by

$$x = a \sin t - b \sin\left(\frac{at}{b}\right)$$

$$y = a \cos t - b \cos\left(\frac{at}{b}\right)$$

show that the radius of curvature at this point is $\frac{4ab}{a+b}\sin\frac{a-b}{2b}t$.

8. Find the radius of curvature at the point (p, r) of the curve $pa^n = r^{n+1}$.

9. Find the radius of curvature at the point (p, r) of the curve $r^3 = a^2p$.

10. Show that the curve for which $y = a\, e^{m\psi}$, the radius of curvature is m times the tangent.

11. Prove that for any curve

$$\frac{r}{\rho} = \sin\phi\left(1 + \frac{d\psi}{d\theta}\right) \text{ where } \rho \text{ is the radius of curvature and } \tan\phi = r.\frac{d\theta}{dr}.$$

12. The curve $r = a\, e^{\theta} \cot\alpha$ cuts any radius vector in the consecutive points $P_1, P_2, \ldots P_n, P_{n+1}$. If ρ_n denotes the radius of curvature at P_n. Prove that

$$\frac{1}{m-n}\log\frac{\rho_m}{\rho_n} \text{ is constant so r all integral values of m and n.}$$

13. Show that in any curve the chord of curvature. Perpendicular to the radius vector is

$$\frac{2p\left(r^2 - p^2\right)^{1/2}}{r}.$$

14. Prove that for any curve

$$\frac{r}{\rho} = \sin\phi\left(1 + \frac{d\phi}{d\theta}\right)$$

where ρ is the radius of curvature and $\tan\phi = r\frac{d\theta}{dr}$.

10

ASYMPTOTES

Asymptote: Some curve are limited in extent. Other extend to infinity, it is interesting to enquire what happens to the tangent when the point at which the tangent is drawn moves further and further away from the origin. There are three possibilities corresponding to the three possibilities in the lose of a series *viz*. That a series may be divergent oscillatory or convergent. Thus, it is possible that the tangent may go further and further away from the origin or it may keep oscillating or it may the straight line to which the tangent tends is called the asymptote.

10.1 DEFINITION

I. *An asymptote is a straight line which cuts a curve in two points at an infinite distance from the origin and yet is not itself wholly at infinity.*

II. *An asymptote is a straight line, at a finite distance from the origin, to which a tangent to a curve tends, as the distance from the origin of the point of contact tends to infinity.*

10.2 DETERMINATION OF ASYMPTOTES

If $y + mx + c$ is an asymptote to the curve $y = f(x)$, to show that

$$m = \lim_{x\to\infty}\left(\frac{y}{x}\right).$$

Let $y = mx + c$ be an asymptote of the curve $y = f(x)$ or $f(x, y) = 0$ so that m and c are both finite.

The equation of the tangent at $P(x, y)$ to the curve $y = f(x)$ is

$$Y - y = \left(\frac{dy}{dx}\right)(X - x)$$

or $$Y = \left(\frac{dy}{dx}\right)X + \left[y - x\left(\frac{dy}{dx}\right)\right]. \qquad ...(1)$$

From (1), $x \to \infty$, $\frac{dy}{dx}$ and $y - x\frac{dy}{dx}$ must both tend to finite limits, say m and c, in order that an asymptote might exist. Thus the tangent (1) tends to the asymptote $y = mx + c$ if

$$\lim_{x\to\infty}\frac{dy}{dx} = m \text{ and } \lim_{x\to\infty}\left(y - x\frac{dy}{dx}\right) = c.$$

Then we have $\lim_{x\to\infty}\dfrac{y - x\left(\frac{dy}{dx}\right)}{x} = 0$, [∵ c is finite]

or $\lim_{x\to\infty}\left(\frac{y}{x} - \frac{dy}{dx}\right) = 0$ or $\lim_{x\to\infty}\frac{y}{x} = \lim_{x\to\infty}\frac{dy}{dx} = m$.

Also $c = \lim_{x\to\infty}\left(y - x\frac{dy}{dx}\right) = \lim_{x\to\infty}(y - mx)$, $\left[\because \lim_{x\to\infty}\frac{dy}{dx} = m\right]$

Therefore, if $y = mx + c$ is an asymptote to the curve $y = f(x)$, then

$$m = \lim_{x\to\infty}\frac{y}{x} \text{ and } c = \lim_{x\to\infty}(y - mx).$$

10.3 THE ASYMPTOTES OF THE GENERAL RATIONAL ALGEBRAIC CURVE

Let $f(x, y) = 0$ be the equation of any rational algebraic curve of the nth degree.

Let the equation of the curve on being arranged in groups of homogeneous terms, be

$$a_0y^n + a_1y^{n-1} + a_2y^{n-2}x^2 + ... + a_{n-1}yx^{n-1} + a_nx^n$$
$$+ b_1y^{n-1} + b_2y^{n-2}x + ... + b_{n-1}yx^{n-2} + b_nx^{n-1}$$
$$+ c_2y^{n-2} + ... + ... = 0. \qquad ...(1)$$

This equation can also be written as

$$x^n\phi_n\left(\frac{y}{x}\right) + x^{n-1}\phi_{n-1}\left(\frac{y}{x}\right) + ... = 0, \qquad ...(2)$$

where $\phi_r\left(\frac{y}{x}\right)$ is a polynomial in $\frac{y}{x}$ of degree r.

Let y = mx + c be a line which is not parallel to the y-axis. It meets (2) in points whose abscissae are given by

$$x_n\phi_n\left(\frac{mx+c}{x}\right)+x^{n-1}\phi_{n-1}\left(\frac{mx+c}{x}\right)+\ldots=0$$

or $$x_n\phi_n\left(m+\frac{c}{x}\right)+x^{n-1}\phi_{n-1}\left(m+\frac{c}{x}\right)+\ldots=0.$$

Expanding each term of the type $\phi_r\left(m+\frac{c}{x}\right)$, by Taylor's theorem, we get

$$x^n\left[\phi_n(m)+\left(\frac{c}{x}\right)\phi_n'(m)+\left(\frac{c^2}{2x^2}\right)\phi_n''(m)+\ldots\right]$$

$$+x^{n-1}\left[\phi_{n-1}(m)+\left(\frac{c}{x}\right)\phi_{n-1}(m)+\ldots\right]$$

$$+x^{n-2}\left[\phi_{n-2}(m)+\left(\frac{c}{x}\right)\phi_{n-2}'(m)+\ldots\right]+\ldots=0.$$

Arranging the terms according to the descending powers of x, we have

$$x^n\phi_n(m)+x^{n-1}[\phi_{n-1}(m)+c\phi_n'(m)]$$

$$+x^{n-2}\left[\phi_{n-2}(m)+c\phi_{n-1}'(m)+\frac{1}{2}c^2\phi_n''(m)\right]+\ldots=0. \quad \ldots(3)$$

If y = mx + c is an asymptote, this equation must have two infinite roots and consequently

$$\phi_n(m)=0, \quad \ldots(4)$$

and $$\phi_{n-1}(m)+c\phi_n'(m)=0, \quad \ldots(5)$$

as can be seen on dividing (3) by x^n.

Solving equations (4) and (5) simultaneously, we get the values for m and c, and hence the asymptotes are determined by substituting the corresponding values of m and c in the equation y = mx + c.

Note: All the imaginary values of m will be rejected.

Working Rule for Finding the Asymptotes

(i) Substitute mx + c for y in the equation of the curve and arrange it in descending powers of x.

(ii) By equating the coefficients of the two highest powers of x to zero find m and c.

(iii) Substitute these values of m and c in y = mx + c.

(iv) If a value of m, makes the coefficient of x^{n-1} identically zero, find c from the equation obtained by equating to zero the coefficient of x^{n-2}; and so on.

A Short Cut Method

(i) Put x = 1 any y = m in the highest *i.e.* nth degree terms and get $\phi_n(m)$. Put $\phi_n(m) = 0$ and solve the resulting equation for m. Let

$$m = m_1 m_2, \ldots, m_n \text{ be its roots.}$$

(ii) Form $\phi_{n-1}(m)$ by putting x = 1 and y = m in the terms of degree n − 1. Then the values of c, say $c_1, c_2, \ldots, c_n$ are found by substituting $m = m_1, m_2 \ldots, m_n$, in turn, in the formula

$$c = -\frac{\phi_{n-1}(m)}{\phi_n'(m)}.$$

The asymptotes then are

$$y = m_1 x + c_1;\ y = m_2 x + c_2;\ \ldots$$

Example 1:

Find all asymptotes of the curve

$$3x^3 + 2x^2y - 7xy^2 + 2y^3 - 14xy + 7y^2 + 4x + 5y = 0.$$

(Meerut, 1994 P; Gorakhpur, 92; Agra, 98)

Solution:

Putting x = 1 and y = m in the third and second degree terms separately, we have

$$\phi_3(m) = 3 + 2m - 7m^2 + 2m^3 \text{ and } \phi_2(m) = 7m^2 - 14m.$$

$$\therefore \quad \phi_3'(m) = 2 - 14m + 6m^2.$$

Putting $\phi_3(m) = 0$, we get $3 + 2m - 7m^2 + 2m^3 = 0$

or $(m - 1)(2m^2 - 5m - 3) = 0$ (or $m - 1)(2m + 1)(m - 3) = 0$.

$$\therefore \quad m = 1, 3, -\frac{1}{2}.$$

Also $$c = -\frac{\phi_2(m)}{\phi_3'(m)} = \frac{14m - 7m^2}{2 - 14m + 6m^2}.$$

Putting m = 1, we get $c = -\frac{7}{6}$;

when m = 3, c = − 1, and when $m = -\frac{1}{2}, c = -\frac{5}{4}$.

Hence asymptotes are

$$y = x = \frac{7}{6};\ y = 3x - 1 \text{ and } y = -\frac{1}{2}x - \frac{5}{6},$$

or $6x - 6y - 7 = 0$, $y = 3x - 1$ and $6y + 3x + 5 = 0$.

Example 2:

Find all the asymptotes of the curve

$$y^3 - 3x^2y + xy^2 - 3x^3 + 2y^2 + 2xy + 4x + 5y + 6 = 0.$$

(Magadh, 1990)

Solution:

Putting x = 1 and y = m in the third and second degree terms separately, we have

$$\phi_3(m) = m^3 - 3m + m^2 - 3 \text{ and } \phi_2(m) = 2(m^2 + m).$$

$$\therefore \quad \phi_3'(m) = 3m^2 + 2m - 3.$$

Equating $\phi_3(m)$ to zero, we get $m^3 + m^2 - 3m - 3 = 0$

or $m^2(m + 1) - 3(m + 1) = 0$ or $(m + 1)(m^2 - 3) = 0$

i.e., $\quad m = -1, +\sqrt{3} - \sqrt{3}$.

Also we know that $c = -\dfrac{\phi_2(m)}{\phi_3'(m)} = -\dfrac{2(m^2 + m)}{3m^2 + 2m - 3}$.

$\therefore$ when $m = 1$, $c = -2$; when $m = \sqrt{3}$, $c = -1$;

and when $m = -\sqrt{3}$, $c = -1$.

Hence the asymptotes are

$$y = x - 2;\ y = \pm x\sqrt{3} - 1.$$

Example 3:

Find the asymptotes of the curve

$$y^3 - x^2y - 2xy^2 + 2x^3 - 7xy + 3y^2 + 2x^2 + 2x + 2y + 1 = 0.$$

(Lucknow, 1992; Meerut, 97)

Solution:

Putting y = mx + c, we have

$$(mx + c)^3 - x^2(mx + c) - 2x(mx + c)^2 + 2x^3 - 7x(mx + x)$$
$$+ 3(mx + c)^2 + 2x^2 + 2x + 2(mx + c) + 1 = 0$$

or $\quad x^3(m^3 - m - 2m^2 + 2) + x^2(3m^2c - c - 4mc - 7m + 3m^2 + 2)$

$$+ \ldots = 0.$$

Equating the coefficient of the highest degree term (*i.e.*, x^3) to zero, we get

$$m^3 - m - 2m^2 + 2 = 0$$

or $(m - 1)(m + 1)(m - 2) = 0.$

Therefore, $m = 1, -1, 2.$

Now equating the coefficient of x^2 to zero, we get

$$c(3m^2 - 4m - 1) + 3m^2 - 7m + 2 = 0$$

or $$c = -\frac{(3m^2 - 7m + 2)}{(3m^2 - 4m - 1)}.$$

Therefore, when $m = 1$, $c = -1$; when $m = -1$, $c = -2$; and when $m = 2$, $c = 0$.

Substituting these pairs of values of m and c in $y = mx + c$, the required asymptotes are

$$y = x - 1,\ y = -x - 2 \text{ and } y = 2x.$$

Example 4:

Find the asymptotes of

$$x^3 - y^3 - x^2 - 5y^2 - 11x - 5y + 7 = 0.$$

Solution:

Putting $x = 1$, $y = m$ in the third degree and second degree terms separately, we have

$$\phi_3(m) = 1 - m^3 \text{ and } \phi_2(m) = -1 - 5m^2.$$

$\therefore$ $$\phi_3'(m) = -3m^2.$$

Now $\phi_3(m) = 1 - m^3 = (1 - m)(1 + m + m^2) = 0$ gives $m = 1$ as the only real root.

Also $$c = -\frac{\phi_2(m)}{\phi_3'(m)} = -\frac{-1-5m^2}{-3m^2}.$$

Putting $m = 1$, we have $c = -2$.

Hence the only asymptote of the curve is $y = x - 2$.

Example 5:

Find the asymptotes of the curve

$$x^3 - 2y^3 + 2x^2y - xy^2 + xy - y^2 + 1 = 0.$$

(Agra, 1997; Meerut, 97, 94 S, 92)

Solution:

Putting $y = mx + c$, we have

$$x^3 - 2(mx + c)^3 + 2x^2(mx + c) - x(mx + c)^2 + x(mx + c) - (x + c)^2 + 1 = 0$$

or $x^3(-2m^3 - m^2 + 2m + 1) + x^2(-6m^2c - 2mc + 2c + m - m^2) + \dots = 0.$

Equating to zero the coefficient of x^3, we have

$$2m^3 + m^2 - 2m - 1 = 0, \ \textit{i.e.}, \ (m^2 - 1)(2m + 1) = 0.$$

$\therefore \quad m = 1, -1, -\frac{1}{2}.$

Equating to zero the coefficient of x^2, we have

$$-2(3m^2 + m - 1)c - (m^2 - m) = 0,$$

i.e., $c = \dfrac{(m - m^2)}{2(3m + m - 1)}.$

Therefore, when $m = 1$, $c = 0$; when $m = -1$, $y = -1$; and when

$$m = -\frac{1}{2}, c = \frac{1}{2}.$$

$\therefore$ the asymptotes are $y = x$, $y = -x - 1$ and $y = -\frac{1}{2}x + \frac{1}{2}$;

or $\quad x - y = 0$, $x + y + 1 = 0$ and $x + 2y - 1 = 0$.

Example 6:

Find the asymptotes of the curve

$$2x^3 + 3x^2y - 3xy^2 - 2y^3 + 3x^2 - 3y^2 + y - 3 = 0.$$

Solution:

Here $\phi_3(m) = 2 + 3m - 3m^2 - 2m^3$,

$\phi_3'(m) = 3 - 6m - 6m^2$ and $\phi_2(m) = 3 - 3m^2$; $\phi_3(m) = 0$, gives

$$m = 1, -\frac{1}{2}, -2.$$

Also $c = -\dfrac{\phi_2(m)}{\phi_3'(m)} = \dfrac{3m^2 - 3}{3 - 6m - 6m^2}.$

When $m = 1$, $c = 0$; when $m = -\frac{1}{2}$, $c = -\frac{1}{2}$;

and when $m = -2$, $c = -1$.

$\therefore$ $y = x$, $y = -\frac{1}{2}x - \frac{1}{2}$ and $y = -2x - 1$ are the required asymptotes.

Example 7:

Find the asymptotes of

$$4x^3 - x^2y - 4xy^2 + y^3 + 3x^2 + 2xy - y^2 - 7 = 0.$$

Solution:

Putting x = 1 and y = m in the third degree and second degree terms separately, we have

$$\phi_3(m) = 4 - m - 4m^2 + m^3;\ \phi_2(m) = 3 + 2m - m^2.$$

Now solving the equation $\phi_3(m) = 4 - m - 4m^2 + m^3 = 0$

i.e., $(4 - m)(1 - m^2) = 0$, we get $m = 1, -1, 4$.

Also $\phi_3'(m) = -1 - 8m + 3m^2$.

$$\therefore\ c = -\frac{\phi_2(m)}{\phi_3'(m)} = \frac{3+2m-m^2}{1+8m-3m^2}.$$

Putting m = 1, we get $c = \frac{2}{3}$;

when m = – 1, c = 0 and when m = 4, $c = \frac{1}{3}$.

Hence the asymptotes are

$$y = x + \frac{2}{3};\ y = -x \text{ and } y = 4x + \frac{1}{3}.$$

Example 8:

Find the asymptotes of

$$x^3 - 2y^3 + xy(2x - y) + y(x - 1) + 1 = 0.$$

Solution:

Putting x = 1, y = m in the third and second degree terms separately, we have

$$\phi_3(m) = 1 - 2m^3 + 2m - m^2 \text{ and } \phi_2(m) = m.$$

$$\therefore\ \phi_3'(m) = -6m^2 + 2 - 2m.$$

Now the slopes of the asymptotes are given by the equation $\phi_3(m) = 0$ or $(1 - 2m^3 + 2m - m^2) = 0$ or $(1 - m^2)(1 + 2m) = 0$.

$$\therefore\ m = 1, -1, -\frac{1}{2}.$$

Also we know that $c = -\frac{\phi_2(m)}{\phi_3'(m)} = \frac{m}{6m^2+2m-2}$.

$\therefore$ when m = 1, $c = \frac{1}{6}$; when m = – 1, $c = -\frac{1}{2}$

and when $m = -\frac{1}{2}$, $c = \frac{1}{3}$.

Hence the asymptotes are

$$y = x + \frac{1}{6};\ y = -x - \frac{1}{2} \text{ and } y = -\frac{1}{2}x + \frac{1}{3}$$

i.e., 6x – 6y + 1 = 0; 2x + 2y + 1 = 0 and 6y + 3x – 2 = 0.

Example 9:

Find the asymptotes of the curve $x^3 + y^3 - 3axy = 0$.

(Rohilkhand, 1992; Delhi, 92; Magadh, 96; Meerut, 98, 97; Gorakhpur, 97)

Solution:

Putting x = 1 and y = m in the third and second degree terms separately, we get

$\phi_3(m) = 1 = m^3$, and $\phi_2(m) = -3am$.

Now we solve the equation $\phi_3(m) = (1 + m)(1 - m + m^2) = 0$.

Obviously m = – 1 is the only real root of this equation.

To determine c, we have the formula

$$c = -\frac{\phi_2(m)}{\phi_3'(m)} = -\frac{3am}{3m^2} = \frac{a}{m}.$$

Putting m = 1, we get c = – a.

Hence the only asymptote of the curve is

y = – x – a or x + y + a = 0.

Example 10:

Find the asymptotes of the curve $y^3 = x^3 + ax$.

(Kumaun, 1993; Meerut, 99)

Solution:

The given curve is $y^3 - x^3 - ax^2 = 0$.

Putting y = m and x = 1 in the highest *i.e.*, third degree terms in the equation of the curve, we get $\phi_3(m) = m^3 - 1$.

Solving the equation $\phi_3(m) = 0$.

i.e., $m^3 - 1 = 0$, *i.e.*, $(m - 1)(m^2 + m + 1) = 0$, we get m = 1 as the only real root.

The other two roots are imaginary.

Again putting $y = m$ and $x = 1$ in the second degree terms in the equation of the curve, we get $\phi_2(m) = -a$.

Now c is given by $c\phi_3'(m) + \phi_2(m) = 0$, *i.e.*, $c(3m^2) - a = 0$.

Putting $m = 1$, we get $3c - a = 0$ or $c = \frac{a}{3}$.

Hence the only real asymptote of the curve is

$$y = x + \left(\frac{a}{3}\right).$$

10.4 NON-EXISTENCE OF ASYMPTOTES

If $\phi_n(m) = 0$, gives one or more values of m, such that they make $\phi_n'(m) = 0$, whereas $\phi_{n-1}(m) \neq 0$, then from the equation $c = -\frac{\phi_{n-1}(m)}{\phi_n'(m)}$, we get $c = +\infty$ or $-\infty$ and this corresponds to the case when the tangent goes farther and farther away from the origin as $x \to \infty$. Therefore, for such values of m, we shall get no asymptotes.

Example 1:

Find the asymptotes of the curve $y^3 = x^2 + 3x$.

Solution:

Putting $x = 1$, $y = m$ in the third degree and second degree terms separately, we have

$$\phi_3(m) = m^3 \text{ and } \phi_2(m) = -1.$$

$$\therefore \quad \phi_3'(m) = 3m^2.$$

Now $\phi_3(m) = 0$ *i.e.*, $m^3 = 0$ gives $m = 0, 0, 0$.

Also $c = -\frac{\phi_2(m)}{\phi_3'(m)} = -\frac{-1}{3m^2}$ which is ∞ for $m = 0$.

Hence $y^3 = x^2 + 3x$ has no asymptotes.

10.5 CASE OF PARALLEL ASYMPTOTES

(i) Two Parallel Asymptotes: Suppose the equation $\phi_n(m) = 0$ gives us two equal values of m. This repeated value of m makes $\phi_n'(m) = 0$. In case it does not make $\phi_{n-1}(m)$ equal to zero, the value of c determined by the equation $c = -\frac{\phi_{n-1}(m)}{\phi'_n(m)}$ comes out to be infinite, and hence the asymptotes corresponding to this value of m do not exist. Thus, for the existence of the asymptotes corresponding to this value of m, it is necessary that it must make $\phi_{n-1}(m)$ equal to zero. But then the equation $c\phi'_n(m) + \phi_{n-1}(m) = 0$ from which c is usually determined reduces to the identity

$$0.c + 0 = 0,$$

and we cannot find the value of c in this way. To determine c in this case, we equate to zero the coefficient of x^{n-2} in the equation (3) of 3, and we get the equation

$$\frac{c^2}{2!}\phi''_n(m) + \frac{c}{1!}\phi'_{n-1}(m) + \phi_{n-2}(m) = 0.$$

This equation is a quadratic in c. It gives us two values of c, say c_1 and c_2, corresponding to that repeated value of m. The corresponding asymptotes are $y = mx + c_1$ and $y = mx + c_2$, which are obviously parallel.

(ii) Three Parallel Asymptotes: If the equation $\phi_n(m) = 0$ gives us three equal values of m, then this repeated value of m makes $\phi'_n(m)$ and $\phi''_n(m)$ equal to zero. For the existence of the corresponding asymptotes it must $\phi_{n-1}(m)$ equal to zero. If it also makes $\phi'_{n-1}(m)$ and $\phi_{n-2}(m)$ equal to zero, then the equation to determine c reduces to the identity

$$0.c^2 + 0.c + 0 = 0,$$

and we cannot find the values of c in this way. To determine c in this case, we equate to zero the coefficient of x^{n-3} in the equation (3) of 3, and we get the equation

$$\frac{c^3}{3!}\phi'''_n(m) + \frac{c^2}{2!}\phi''_{n-1}(m) + \frac{c}{1!}\phi'_{n-2}(m) + \phi_{n-3}(m) = 0.$$

This equation gives us three values of c corresponding to that repeated value of m and accordingly we get three parallel asymptotes. In a similar way we can discuss the case of more than three parallel asymptotes.

Example 1:

Find the asymptotes of $x^3 + 3x^2y - 4y^3 - x + y + 3 = 0$.

(Rohilkhand, 1991; Gorakhpur, 98; Agra, 96; Meerut, 91S, 93, 96)

Solution:

Putting y = m and x = 1 in the third degree and second degree terms separately, we get

$$\phi_3(m) = 1 + 3m - 4m^3 \text{ and } \phi_2(m) = 0.$$

(Not that there are no second degree terms).

$$\therefore \quad \phi_3'(m) = 3 - 12m^2.$$

Now the slopes of the asymptotes are given by the equation

$$\phi_3(m) = 0 \textit{ i.e., } 1 + 3m - 4m^3 = 0 \textit{ i.e., } (1 - m)(1 + 2m)^2 = 0.$$

$$\therefore \quad m = 1, -\frac{1}{2}, -\frac{1}{2}.$$

Again, c, is given by the equation

$$c = -\frac{\phi_2(m)}{\phi_3'(m)} = -\frac{0}{3 - 12m^2}. \qquad ...(1)$$

When m = 1, we have c = 0 and the corresponding asymptote is

$$y = x + 0 \textit{ i.e., } y = x.$$

When $m = -\frac{1}{2}$, the equation (1) fails to give c. In this case c is to be determined from the equation

$$\frac{c^2}{2!}\phi_3''(m) + \frac{c}{1!}\phi_2'(m) + \phi_1(m) = 0.$$

Putting y = m and x = 1 in the first degree terms of the equation of the curve, we get $\phi_1(m) = -1 + m$. Also $\phi_3''(m) = -24m$ and $\phi_2'(m) = 0$. Hence for $m = -\frac{1}{2}$, c is to be given by

$$\frac{1}{2}c^2(-24m) + 0 + m - 1 = 0.$$

Putting $m = -\frac{1}{2}$ in this equation, we get

$$6c^2 - \frac{3}{2} = 0 \text{ or } c^2 = \frac{1}{4} \text{ or } c = \pm\frac{1}{2}.$$

Hence $y = -\frac{1}{2}x + \frac{1}{2}$ and $y = -\frac{1}{2}x - \frac{1}{2}$ are two parallel asymptotes corresponding to the slope $m = -\frac{1}{2}$.

Therefore the required asymptotes are

$$y = x \text{ and } x + 2y = \pm 1.$$

Example 2:

Find the asymptotes of $y^3 + x^2y + 2xy^2 - y + 1 = 0.$

(Mysore, 1991; U.P.P.C.S., 92)

Solution:

Putting y = m and x = 1 in the third degree and second degree terms separately, we get

$$\phi_3(m) = m^3 + m + 2m^2, \text{ and } \phi_2(m) = 0.$$

$$\therefore \qquad \phi_3'(m) = 3m^2 + 4m + 1.$$

Now the slopes of the asymptotes are given by the equation $\phi_3(m) = 0$

i.e., $\quad m^3 + 2m^2 + m = 0$ *i.e.,* $m(m^2 + 2m + 1) = 0$

i.e., $\quad m(m + 1)^2 = 0.$

$\therefore \quad m = 0, -1, -1.$

Again, c, is given by the equation

$$c = -\frac{\phi_2(m)}{\phi_3'(m)} = -\frac{0}{3m^2 + 4m + 1} \qquad \text{...(1)}$$

When m = 0, we have $c = -\frac{0}{1} = 0$, and the corresponding asymptote is y = 0.x + 0 *i.e.,* y = 0.

When m = – 1, the equation (1) gives $c = -\frac{0}{0}$ which is indeterminate form. So the equation (1) fails to give c when m = – 1. In this case c is to be determined from the equation

$$\frac{c^2}{2!}\phi_3''(m) = \frac{c}{1!}\phi_2'(m) + \phi_1(m) = 0.$$

Putting y = m and x = 1 in the first degree terms of the equation of the curve, we get $\phi_1(m) = -m$. Also $\phi_3''(m) = 6m + 4$, $\phi_2'(m) = 0$.

Therefore for m = – 1, c is given by the equation

$$\frac{1}{2}c^2(6m + 4) + c.0 - m = 0 \text{ i.e., } (3m + 2)c^2 - m = 0.$$

Putting m = – 1 in this equation, we get

$$-c^2 + 1 = 0 \text{ or } c^2 = 1 \text{ or } c = \pm 1.$$

Hence y = – x + 1 and y = – x – 1 are two parallel asymptotes corresponding to the slope m = – 1.

∴ all the three asymptotes of the curve are

$$y = 0,\ y + x - 1 = 0 \text{ and } y + x + 1 = 0.$$

Example 3:

Find all the asymptotes of the curve

$$(x + y)^2 (x + 2y + 2) = x + 9y + 2.$$

(Kanpur, 1999; Rohilkhand, 90; Meerut, 93S)

Solution:

The equation of the given curve can be written as

$$(x + y)^2 (x + 2y) + 2(x + y)^2 - (x + 9y) - 2 = 0.$$

Putting y = m and x = 1 in the third degree and second degree terms separately, we have

$$\phi_3(m) = (1 + m)^2 (1 + 2m) \text{ and } \phi_2(m) = 2(1 + m)^2.$$

$$\therefore \quad \phi_3'(m) = 2(1 + m)(1 + 2m) + 2(1 + m)^2.$$

The slopes of the asymptotes are given by the equation

$$\phi_3(m) = 0 \text{ i.e., } (1 + m)^2 (1 + 2m) = 0.$$

The determine c, we have the equation

$$c = -\frac{\phi_2(m)}{\phi_3'(m)} = -\frac{2(1+m)^2}{2(1+m)(1+2m)+2(1+m)^2} \qquad ...(1)$$

When $m = -\frac{1}{2}$, we have

$$c = -\frac{2\left(1-\frac{1}{2}\right)^2}{2\left(1-\frac{1}{2}\right)(1-1)+2\left(1-\frac{1}{2}\right)^2} = -1,$$

and the corresponding asymptote is $y = -\frac{1}{2}x - 1$ *i.e.*, 2y + x + 2 = 0.

When m = – 1, the equation (1) fails to give c, because then the value of c takes the indeterminate form $\frac{0}{0}$. In this case c is to be determined from the equation

$$\frac{c^2}{2!}\phi_3''(m) + \frac{c}{1!}\phi_2'(m) + \phi_1(m) = 0.$$

Now $\phi_3''(m) = 2(1 + 2m) + 4(1 + m) + 4(1 + m)$,

$\phi_2'(m) = 4(1 + m)$. Also putting $y = m$ and $x = 1$ in the first degree terms of the equation of the curve, we get $\phi_1(m) = -(1 - 9m)$. Hence for $m = -1$, c is given by

$$\frac{1}{2}c^2\{2(1 + 2m) + 8(1 + m)\} + c\{4(1 + m)\} - (1 + 9m) = 0.$$

Putting $m = -1$ in this equation, we get

$$-c^2 + 8 = 0 \text{ or } c = \pm 2\sqrt{2}.$$

Hence $y = -x + 2\sqrt{2}$ and $y = -x - 2\sqrt{2}$ are two parallel asymptotes corresponding to the slope $m = -1$.

$\therefore$ the required asymptotes are

$$2y + x + 2 = 0 \text{ and } y + x = \pm 2\sqrt{2}.$$

10.6 ASYMPTOTES PARALLEL TO THE CO-ORDINATE AXES

(i) Asymptotes Parallel to y-axis of a Rational Algebraic Curve: Asymptotes parallel to y-axis cannot be determined by the method given in 3 because the equation to a straight line parallel to y-axis cannot be put in the form $y = mx + c$. Here we shall discuss the method of finding asymptotes parallel to y-axis.

Let the equation of the curve, when arranged in descending powers of y, be

$$y^m\phi(x) + y^{m-1}\phi_1(x) + y^{m-2}\phi_2(x) + \ldots = 0, \quad \ldots(1)$$

where $\phi(x)$, $\phi_1(x)$, $\phi_2(x)$ etc., are polynomials in x.

Dividing the equation (1) by y^m, we get

$$\phi(y) + \frac{1}{y}\phi_1(x) + \frac{1}{y^2}\phi_2(x) + \ldots = 0. \quad \ldots(2)$$

If $x = k$ is an asymptote parallel to y-axis of (1), then obviously $k = \lim_{y \to \infty} x$, where (x, y) lies on the curve (1). Therefore taking limit of (2) as $x \to \infty$ and remembering that $x \to k$ as $y \to \infty$, we get $\phi(k) = 0$.

Therefore, k is a root of the equation $\phi(x) = 0$.

If k_1, k_2 etc., be the roots of $\phi(x) = 0$, then the asymptotes of (1) parallel to y-axis are $x = k_1$, $x = k_2$, etc.

From algebra, we know that if k_1 is a root of $\phi(x) = 0$, then $x - k_1$ must be a factor of $\phi(x)$. Also $\phi(x)$ is the coefficient of the highest power of y *i.e.*, y^m in the equation of the curve. Hence we have the following simple rule:

The *asymptotes parallel to the axis of y are obtained by equating to zero the coefficient of the highest power of y in the equation of the curve.* In case the coefficient of highest power of y, is a constant or if its linear factors are all imaginary, there will be no asymptotes parallel to y-axis.

(ii) Asymptotes Parallel to x-axis: Proceeding as above, we have the following rule for finding the asymptotes parallel to x-axis of a rational algebraic curve:

The asymptotes parallel to the axis of x are obtained be equating to zero the coefficient of the highest power of x, in the equation of the curve. In case the coefficient of the highest power of x, is a constant or its linear factors are all imaginary, there will be no asymptotes parallel to x-axis.

10.7 TOTAL NUMBER OF ASYMPTOTES OF A CURVE

The number of asymptotes, real or imaginary, of an algebraic curve of the nth degree cannot exceed n.

The slopes of the asymptotes which are not parallel to y-axis are given as the roots of the equation $\phi_n(m) = 0$ which is of degree n at the most. If the equation of the curve possesses some asymptotes parallel to y-axis, then we can easily see that the degree of $\phi_n(m) = 0$ will be smaller than n by atleast the same number. In general, one value of m gives only one value of c. In case the equation for determining c is a quadratic, the equation $\phi_n(m) = 0$ has two equal roots. Similarly if the equation for determining c is a cubic, the equation $\phi_n(m) = 0$ has three equal roots.

Hence *a curve of degree n cannot have more than n asymptotes.* But the number of real asymptotes can be less than n. Some roots of the equation $\phi_n(m) = 0$ may come out to be imaginary or even corresponding to a real value of m the value of c may come out to be infinite.

10.8 COMPLETE WORKING RULE FOR FINDING THE ASYMPTOTES OF RATIONAL ALGEBRAIC CURVES

(i) A curve of degree n cannot have more than n asymptotes real or imaginary.

(ii) Equating to zero the coefficient of the highest power of y in the equation of the curve, we get asymptotes parallel to y-axis. Similarly equation to zero the coefficient of the highest power of x in the equation of the curve, we get asymptotes parallel to x-axis.

If y = mx = c is an asymptote not parallel to y-axis, then the values of m and c are found as follows:

(iii) Putting y = m and x = 1 in the highest *i.e.*, nth degree terms in the equation of the curve, we get $\phi_n(m)$. Solving the equation $\phi_n(m) = 0$, we get the slopes of the asymptotes. If some values of m are imaginary, we reject them.

(iv) Corresponding to a value of m, the value of c is given by the equation

$$c\phi'_n(m) + \phi_{n-1}(m) = 0,$$

where $\phi_{m-1}(m)$ is obtained by putting y = m and x = 1 in the (n – 1)th degree terms in the equation of the curve. The asymptotes corresponding to m = 0 are already found in (ii). So we need not find the value of c corresponding to m = 0.

(v) If corresponding to two equal values of m, the equation for determining c, given in (iv) reduces to the identity 0.c + 0 = 0, then the values of c are given by

$$\frac{c^2}{2!}\phi''_n(m) + \frac{c}{1!}\phi'_{n-1}(m) + \phi_{n-2}(m) = 0.$$

(vi) Similarly, if three values of m are equal and the equation for determining c, given in (v), reduces to the identity

$0.c^2 + 0.c + 0 = 0,$

then the corresponding values of c are given by

$$\frac{c^3}{3!}\phi'''_n(m) + \frac{c^2}{2!}\phi''_{n-1}(m) + \frac{c}{1!}\phi'_{n-2}(m) + \phi_{n-3}(m) = 0.$$

Example 1:

Find the asymptotes of the curve

$y^2(x^2 - a^2) = x^2(x^2 - 4a^2).$ **(Meerut, 1992; 95, 96, 98S)**

Solution:

The given curve is

$$y^2(x^2 - a^2) = x^2(x^2 - 4a^2)$$

or $$y^2x^2 - x^4 - a^2y^2 + 4a^2x^2 = 0. \qquad ...(1)$$

Since the curve (1) is of degree 4, therefore it cannot have more than four asymptotes.

Equating to zero the coefficient of the highest power of y(*i.e.*, of y^2) the asymptotes parallel to y-axis are given by

$x^2 - a^2 = 0$ *i.e.*, $x = \pm a$. (Two asymptotes)

The coefficient of the highest power of x *i.e.*, of x is simply a constant and so there is no asymptote parallel to x-axis.

To find the remaining oblique asymptotes, we have put y = m and x = 1 in the highest *i.e.*, fourth degree terms in the equation (1) and we get

$$\phi_4(m) = m^2 - 1.$$

The slopes of the asymptotes are given by the equation

$$\phi_4(m) = 0 \text{ i.e., } m^2 - 1 = 0.$$

$\therefore$ $m = \pm 1.$

Again putting y = m and x = 1 in the next highest *i.e.*, third degree terms, we get

$\phi_3(m) = 0$. [Note that there are no terms of degree 3 in the equation of the curve and so $\phi_3(m) = 0$.]

Now c is given by the equation $c\phi_4'(m) + \phi_3(m) = 0$

i.e., $c(2m) + 0 = 0$ *i.e.*, $2cm = 0$.

When $m = 1$, $c = 0$; and when $m = -1$, $c = 0$.

$\therefore$ the asymptotes are $y = 1x + 0$ and $y = (-1)x + 0$

i.e., the asymptotes $y = x$ and $y = -x$.

Hence all the your asymptotes of the curve are $x = \pm a$, $y = \pm x$.

Example 2:

Find the asymptotes parallel to the coordinate axes of the curve

$$x^2(x - y)^2 + a^2(x^2 - y^2) - a^2xy = 0.$$

Solution:

Equating to zero the coefficient of the highest power of y (*i.e.*, of y^2) the asymptotes parallel to y-axis are given by

$$x^2 - a^2 = 0 \text{ i.e., } x = \pm a.$$

The coefficient of the highest power of (*i.e.*, of x^4) is merely a constant. Hence there is no asymptote parallel to x-axis.

Example 3:

Show that the asymptotes of the curve

$$x^2y^2 - a^2(x^2 + y^2) - a^3(x + y) + a^4 = 0$$

form a square through two of whose angular points the curve passes.

(Gorakhpur, 1996; Rohilkhand, 90; Meerut, 90, 93; U.P.P.C.S., 94)

Solution:

Since the curve is of degree 4, therefore it cannot have more than four asymptotes.

Equating to zero the coefficient of the highest power of x (*i.e.*, of x^2), we get the asymptotes parallel to x-axis as $y^2 - a^2 = 0$ *i.e.*, $x = \pm a$.

Also equating to zero the coefficient of the highest power of x (*i.e.*, of x^2), we get the asymptotes parallel to x-axis as $y^2 - a^2 = 0$ *i.e.*, $y = \pm a$.

Thus all the four asymptotes of the curve have been found and they are $x = \pm a$, $y = \pm a$. These lines form a square whose sides are parallel to the axes each of length 2a.

The angular points of the square are (a, a), (a, – a), (– a, a) and (– a, – a). It is evident that the curve passes through two angular points (a, – a) and (– a, a).

Example 4:

Find all the asymptotes of the curve

$x^3 - x^2y - xy^2 + y^3 + 2x^2 - 4y^2 + 2xy + x + y + 1 = 0.$

(Gorakhpur, 1993)

Solution:

Putting y = m and x = 1 in the third and second degree terms separately, we get

$$\phi_3(m) = 1 - m - m^2 + m^3 \text{ and } \phi_2(m) = 2 - 4m^2 + 2m.$$

$$\therefore \quad \phi_3'(m) = -1 - 2m + 3m^2.$$

Now the slopes of the asymptotes are given by the equation

$$\phi_3(m) = 0 \text{ } i.e., \text{ } 1 - m - m^2 + m^3 = 0 \text{ } i.e., \text{ } (1 - m)^2 (1 + m) = 0.$$

$$\therefore \quad m = 1, 1, -1.$$

Again c is given by the equation

$$c = -\frac{\phi_2(m)}{\phi_3'(m)} = -\frac{2-4m^2+2m}{-1-2m+3m^2} = \frac{2-4m^2+2m}{1+2m-3m^2}. \qquad \text{...(1)}$$

When m = – 1, we have from (1), c = 1. The corresponding asymptotes is y = – x *i.e.*, x + y = 1.

When m = 1, the equation (1) gives $c = \frac{0}{0}$ and thus the equation (1) fails to give c for m = 1.

In this case c is to be determined from the equation

$$\left(\frac{c^2}{2!}\right)\phi_3''(m) = \left(\frac{c}{1!}\right)\phi_2'(m) + \phi_1(m) = 0.$$

Putting y = m and x = 1 in the first degree terms of the equation of the curve, we get

$\phi_1(m) = 1 + m$. Also $\phi_3''(m) = -2 + 6m$, and $\phi_2'(m) = -8m + 2$.

Hence for m = 1, c is to be given by

$$\frac{1}{2}c^2(-2 + 6m) + c(-8m + 2) + 1 + m = 0.$$

Putting m = 1 in this equation, we get

$$2c^2 - 6c + 2 = 0 \text{ i.e., } c^2 - 3c + 1 = 0.$$

$$\therefore\ c = \frac{3 \pm \sqrt{(9-4)}}{2} = \frac{3 \pm \sqrt{5}}{2}.$$

Hence the two parallel asymptotes corresponding to the slope m = 1 are

$$y = x + \frac{1}{2}\left(3 \pm \sqrt{5}\right).$$

∴ the required asymptotes are

$$x + y = 1,\ 2x - 2y + 3 \pm \sqrt{5} = 0.$$

Example 5:

Find the asymptotes of the curve

$y^2(x - 2) = x^2(y - 1)$. **(Meerut, 1991, 96P)**

Solution:

The given curve is

$$y^2(x - 2) - x^2(y - 1) = 0$$

or $$y^2x - x^2y - 2y^2 + x^2 = 0. \quad \text{...(1)}$$

Since the curve (1) is of degree 3, it cannot have more than three asymptotes.

Equating to zero the coefficient of the highest power of y(*i.e.*, of y^2) the asymptote parallel to y-axis is given by

$$x - 2 = 0 \text{ i.e., } x = 2.$$

Again equating to zero the coefficient of the highest power of x(*i.e.*, of x^2) the asymptote parallel to x-axis is given by

$$y - 1 = 0 \text{ i.e., } y = 1.$$

To find the remaining oblique asymptotes, we put y = m and x = 1 in the highest *i.e.*, third degree terms in the equation (1) and we get

$$\phi_3(m) = m^2 - m.$$

The slopes of the asymptotes are given by the equation

$$\phi_3(m) = 0 \text{ i.e., } m^2 - m = 0 \text{ i.e., } m(m - 1) = 0.$$

$$\therefore \quad m = 0, 1.$$

Again putting y = m and x = 1 in the second degree terms, we get

$$\phi_2(m) = -2m^2 + 1.$$

Now c is given by the equation

$$c\phi_3'(m) + \phi_2(m) = 0$$

i.e., $$c(2m - 1) - 2m^2 + 1 = 0.$$

The asymptote corresponding to m = 0 is parallel to x-axis and has already been found. So no need of finding c for m = 0.

When m = 1, we have c – 1 = 0 *i.e.*, c = 1 and so the corresponding asymptote is y = 1x + 1 *i.e.*, y = x + 1. Hence all the three asymptotes of the given curve are x = 2, y = 1 and y = x + 1.

Example 6:

Find the asymptotes of the curve

$$\left(\frac{a^2}{x^2}\right) - \left(\frac{b^2}{y^2}\right) = 1.$$ **(Meerut, 1990)**

Solution:

The equation of the given curve can be written as

$$a^2y^2 - b^2x^2 = x^2y^2 \text{ or } x^2y^2 - a^2y^2 + b^2x^2 = 0.$$

Since the curve is of degree 4, therefore it cannot have more than four asymptotes.

Equating to zero the coefficient of the highest power of y(*i.e.*, of y^2) the asymptotes parallel to y-axis are given by

$$x^2 - a^2 = 0 \text{ i.e., } x = \pm a.$$

Also equating to zero the coefficient of the highest power of x (*i.e.*, of x^2) the asymptotes parallel to x-axis are given by

$$y^2 + b^2 = 0,$$

which gives two imaginary asymptotes.

Thus, all the four possible asymptotes of the curve have been found and the only real asymptotes are x = ± a.

Example 7:

Find the asymptotes of the curve

$$x^3 + x^2y - xy^2 - y^3 - 3x - y - 1 = 0.$$

(Meerut, 1991, 95, 96 P. 98)

Solution:

The given curve is of degree 3, so it cannot have more than three asymptotes.

There are no asymptotes parallel to y-axis because the coefficient of the highest power of y *i.e.*, of y^3 is merely a constant. Similarly, there are no asymptotes parallel to x-axis.

Now putting y = m and x = 1 in the highest *i.e.*, third degree terms in the equation of the curve, we get

$$\phi_3(m) = 1 + m - m^2 - m^3.$$

The slopes of the asymptotes are given by the equation

$$\phi_3(m) = 0 \text{ i.e., } 1 + m - m^2 - m^3 = 0$$

or $\quad (1 + m) - m^2(1 + m) = 0$

or $\quad (1 + m)(1 - m^2) = 0$

or $\quad (1 + m)^2 (1 - m) = 0.$

$\therefore \quad m = 1, -1, -1.$

Again putting y = m and x = 1 in the second degree terms in the equation of the curve, we get

$$\phi_2(m) = 0, \text{ since there are no second degree terms.}$$

To determine c we have the equation

$$c\phi'_3(m) + \phi_2(m) = 0 \text{ i.e., } (1 - 2m - 3m^2) + 0 = 0. \quad ...(1)$$

For m = 1, the equation (1) gives $-4c = 0$ or $c = 0$ and the corresponding asymptote is $y = 1.x + 0$ *i.e.*, $y = x$.

For m = − 1, the equation (1) reduces to the identity c. 0 = 0 and thus it fails to give c. In this case c is to be determined by the equation

$$\frac{c^2}{2!}\phi''_3(m) + \frac{c}{1!}\phi'_2(m) + \phi_1(m) = 0.$$

Now putting y = m and x = 1 in the first degree terms in the equation of the curve, we get

$$\phi_1(m) = -3 - m.$$

So for m = – 1, c is to be given by the equation

$$\frac{1}{2}c^2.(-2-6m)+0.c-3-m=0$$

or $$c^2(1+3m)+3+m=0.$$

Putting m = – 1 in this equation, we get

$$-2c^2+2=0 \text{ or } c^2=1 \text{ or } c=\pm 1.$$

∴ the asymptotes corresponding to m = – 1 are

$$y=-1x+1 \text{ and } y=-1x-1.$$

Hence all the three asymptotes of the given curve are

$$y=x,\ y=-x+1 \text{ and } y=-x-1.$$

Example 8:

Find the asymptotes of the curve

$y^2(x^2-a^2)=x.$ **(Meerut, 1991, 95 S, 96P)**

Solution:

Since the given curve is of degree 4, therefore it cannot have more than four asymptotes.

Equating to zero the coefficient of the highest power of y(*i.e.*, of y^2) the asymptotes parallel to y-axis are given by

$$x^2-a^2=0 \text{ i.e., } x=\pm a.$$

Again equating to zero the coefficient of the highest power of x(*i.e.*, of x^2) the asymptotes parallel to x-axis are given by

$$y^2=0 \text{ i.e., } y=0,\ y=0 \quad \text{(two coincident asymptotes).}$$

Thus all the four asymptotes have been found.

Example 8 (a):

Find the asymptotes of the curve

$x^2y^2=a^2(x^2+y^2).$ **(Meerut, 1990, 94, 96 BP)**

Solution:

The given curve is of degree 4, so it cannot have more than four asymptotes.

Equating to zero the coefficient of the highest power of y(*i.e.*, of y^2) the asymptotes parallel to y-axis are given by

$$x^2-a^2=0 \text{ i.e., } x=\pm a.$$

Again equating to zero the coefficient of the highest power of x (*i.e.*, of x^2) the asymptotes parallel to x-axis are given by

$$y^2 - a^2 = 0 \text{ i.e., } y = \pm a.$$

Thus all the four asymptotes have been found and they are

$$x = \pm a. y = \pm a.$$

Example 9:

Find the asymptotes of the curve

$$x^2(x^2 + y^2) = a^2(y^2 - x^2).$$

Solution:

The equation of the curve is $x^2 (x^2 + y^2) - a^2(y^2 - x^2) = 0$. Since the curve is of degree 4, therefore it cannot have more than four asymptotes.

Equating to zero the coefficient of the highest power of y(*i.e.*, of y^2) the asymptotes parallel to y-axis are given by

$$x^2 - a^2 = 0 \text{ i.e., } x = \pm a.$$

The coefficient of the highest power of x, *i.e.*, (of x^4) is merely a constant. Hence there is no asymptote parallel to x-axis.

To find the remaining oblique asymptotes, we put y = m and x = 1 in the highest *i.e.*, fourth degree terms of and we get

$$\phi_4(m) = 1 + m^2.$$

The roots of the equation $\phi_4(m) = 0$, *i.e.*, $1 + m^2 = 0$ are imaginary and consequently the corresponding asymptotes are imaginary.

Hence the only real asymptotes of the curve are $x = \pm a$.

Example 10:

Find the asymptotes of the curve

$$x^2y^3 + x^3y^2 = x^3 + y^3.$$ **(Rohilkhand, 1999)**

Solution:

The given curve is

$$x^2y^3 + x^3y^2 - x^3 - y^3 = 0 \quad ...(1)$$

The curve (1) is of degree 5, so it cannot have more than five asymptotes.

The asymptotes parallel to y-axis are given by

$$x^2 - 1 = 0 \text{ i.e., } x = \pm 1.$$

To find the remaining oblique asymptotes, we put y = m and x = 1 in the highest *i.e.*, fifth degree terms in the equation (1) and we get

$$\phi_5(m) = m^3 + m^2.$$

The slopes of the asymptotes are given by the equation

$\phi_5(m) = 0$ *i.e.*, $m^3 + m^2 = 0$ *i.e.*, $m^2 (m + 1) = 0$.

$\therefore$ $m = 0, 9, - 1.$

Again putting y = m and x = 1 in the next highest *i.e.*, fourth degree terms in the equation (1), we get

$\phi_4(m) = 0.$ [Note that there are no fourth degree terms in the equation (1) and so $f_4(m) = 0$.]

Now c is given by the equation

$$c\phi_5'(m)+\phi_4(m)=0$$

i.e., $c(3m^2 + 2m) + 0 = 0.$ ...(2)

The asymptotes corresponding to m = 0 are parallel to x-axis and have already been found and so no need of finding c for m = 0.

When m = – 1, we have from (2), c = 0 and so the corresponding asymptotes is y = – 1x + 0 *i.e.*, y = – x.

Hence all the five asymptotes of the given curve are

$$x = \pm 1,\ y = \pm 1,\ y = - x.$$

Example 11:

Find the asymptotes of the following curves:

(a) $x^2y - xy^2 + 3x^2 - 2y^2 = 0$;

(b) $x^3 - xy^2 + 6y^2 = 0$.

Solution:

(a) The given curve is of degree 3. So it cannot have more than 3 asymptotes.

Equating to zero the coefficient of the highest power of y(*i.e.*, of y^2), we get the asymptote parallel to y-axis as – x – 2 = 0 *i.e.*, x = – 2. Also equating to zero the coefficient of the highest power of x(*i.e.*, of x^2) the asymptote parallel to x-axis is given by y + 3 = 0, *i.e.*, y = – 3.

Now we proceed to find the remaining oblique asymptotes.

Putting y = m and x = 1 in the third degree and second degree terms separately, we get

$\phi_3(m) = m - m^2$ and $\phi_2(m) = 3 - 2m^2$.

The slopes of the asymptotes are given by the equation

$\phi_3(m) = 0$ *i.e.*, $m - m^2 = 0$ *i.e.*, $m(1 - m) = 0$. $\therefore$ $m = 0, 1$.

To determine c, we have the equation

$$c\phi_3'(m)+\phi_2(m)=0 \text{ i.e. } c(1-2m)+3-2m^2=0. \quad ...(1)$$

The asymptote corresponding to m = 0 (*i.e.*, parallel to x-axis) has already been found. So putting m = 1 in (1), we get

c(– 1) + 1 = 0 *i.e.*, c = 1 and the corresponding asymptote is y = x + 1.

∴ the required asymptotes are x = – 2.y = – 3, and y = x + 1.

(b) The given curve is of degree 3 and so it cannot have more than three asymptotes. The coefficients of the highest powers of y and x are (– x + 6) and 1 respectively. The asymptote parallel to y-axis is given by – x + 6 = 0, *i.e.*, x = 6. Also there is no asymptote parallel to x-axis.

For oblique asymptotes, we have

$$\phi_3(m) = 1 - m^2 \text{ and } \phi_2(m) = 6m^2.$$

The equation $\phi_3(m) = 0$ gives m = 1, – 1.

The determine c, we have the equation

$$c=-\frac{\phi_2(m)}{\phi_3'(m)}=-\frac{6m^2}{-2m}=\frac{3m^2}{m}=3m.$$

When m = 1, c = 3 and when m = – 1, c = – 3.

Hence the oblique asymptotes are y = x + 3 and y = – x – 3.

∴ the required asymptotes are x = 6, y = x + 3 and y = – x – 3.

Example 12:

Find all the asymptotes of the curve

$$y^3 - xy^2 - x^2y + x^3 + x^2 - y^2 - 1 = 0.$$

(Gorakhpur, 1999; Meerut, 91P, 97)

Solution:

Putting y = m and x = 1 in the highest *i.e.*, third degree terms of the equation of the curve, we get

$$\phi_3(m) = m^3 - m^2 - m + 1.$$

The slopes of the asymptotes are given by

$$\phi_3(m) = 0 \text{ i.e., } m^3 - m^2 - m + 1 = 0 \text{ i.e., } (m-1)^2(m+1) = 0.$$

∴ m = 1, 1, – 1.

Now putting y = m and x = 1 in the next highest *i.e.*, second degree terms, we get $\phi_2(m) = 1 - m^2$.

To determine c, we have $c\phi_3'(m)+\phi_2(m)=0$

i.e., $c(3m^2 - 2m - 1) + (1 - m^2) = 0.$ ...(1)

When $m = -1$, we have $c = 0$ and the corresponding asymptote is

$$y = -x + 0 \text{ i.e., } y + x = 0.$$

When $m = 1$, the equation (1) reduces to the identity

$$c.0 + 0 = 0$$

and we cannot determine c from it. In this case c is to be determined from the equation

$$\frac{c^2}{2!}\phi_3''(m)+\frac{c}{1!}\phi_2'(m)+\phi_1(m)=0.$$

Putting $y = m$ and $x = 1$ in the first degree terms in the equation of the curve, we get $\phi_1(m) = 0$, since there are no first degree terms.

Hence for $m = 1$, c is to be given by

$$\left(\frac{c^2}{2}\right)(6m - 2) + c(-2m) = 0$$

i.e., $(3m - 1)c^2 - 2mc = 0.$

For $m = 1$, this becomes $2c^2 - 2c = 0$ *i.e.*, $c(c - 1) = 0.$

$\therefore$ $c = 0$ and 1.

Hence $y = x + 1$ and $y = x + 0$ are two parallel asymptotes corresponding to the slope $m = 1$.

$\therefore$ the required asymptotes are

$$y + x = 0,\ y - x = 0,\ y - x - 1 = 0.$$

Example 13:

Find the asymptotes of the curve

$$x^3 - 2x^2y + xy^2 + x^2 - xy + 2 = 0.$$

(Lucknow, 1990; Gorakhpur, 99; Meerut, 93, 98)

Solution:

The given curve is of degree 3. So it cannot have more than three asymptotes.

Equating to zero the coefficient of the highest power of y (*i.e.*, of y^2), we get the asymptote parallel to y-axis as $x = 0$. Also there is no asymptote parallel to x-axis because the coefficient of x^2 is merely a constant.

Now, we proceed to find the remaining oblique asymptotes.

Putting y = m and x = 1 in the third degree and second degree terms separately, we get

$$\phi_3(m) = 1 - 2m + m^2 \text{ and } \phi_2(m) = 1 - m.$$

The slopes of the asymptotes are given by the equation

$$\phi_3(m) = 0 \text{ i.e., } 1 - 2m + m^2 = 0 \text{ i.e., } (1 - m)^2 = 0. \therefore m = 1, 1.$$

To determine c, we have the equation

$$c\phi_3'(m) + \phi_2(m) = 0 \text{ i.e., } c(-2 + 2m) + (1 - m) = 0. \quad ...(1)$$

For m = 1, the equation (1) reduces to the identity c.0 + 0 = 0 and thus it fails to give c. In this case c is to be determined by the equation

$$\frac{c^2}{2!}\phi_3''m + \frac{c}{1!}\phi_2'(m) + \phi_1(m) = 0.$$

Now $\phi_3''(m) = 2$, $\phi_2'(m) = -1$ and $\phi_1(m) = 0$ because there are no first degree terms in the equation of the curve. So for m = 1, c is to be given by

$$\frac{1}{2}c^2.(2) + c.(-1) + 0 = 0 \text{ i.e., } c^2 - c = 0 \text{ i.e., } c(c - 1) = 0.$$

$\therefore$ c = 0, 1.

Hence y = x + 0 and y = x + 1 are two parallel asymptotes corresponding to the slope m = 1.

$\therefore$ the required asymptotes are x = 0, y = x and y = x + 1.

Example 14:

Find the asymptotes of the curve

$$x^2(x - y)^2 + a(x^2 - y^2) - a^2xy = 0.$$

Solution:

Since the given curve is of degree 4, therefore it cannot have more than four asymptotes.

Equating to zero the coefficient of the highest power of y (*i.e.*, of y^2), we get the asymptotes parallel to y-axis as $x^2 - a^2 = 0$ *i.e.*, $x = \pm a$. But there is no asymptote parallel to x-axis because the coefficient of the highest power of x (*i.e.*, of x^4) is merely a constant.

To find the remaining oblique asymptotes, we put y = m and x = 1 in the highest *i.e.*, fourth degree terms of the equation of the curve, and we get $\phi_4(m) = (1 - m)^2$.

The slopes of the asymptotes are given by the equation

$$\phi_4(m) = 0 \text{ i.e., } (1 - m)^2 = 0. \therefore m = 1, 1.$$

Now $\phi_3(m) = 0$ because there are no third degree terms in the equation of the curve. To determine c we have the equation

$$c\phi_3'(m) + \phi_3(m) = 0$$

i.e., $$c\{-2(1 - m)\} + 0 = 0. \quad ...(1)$$

For m = 1, the equation (1) reduces to the identity c.0 + 0 = 0 and thus it fails to give c. In this case c is to be determined by the equation

$$\frac{c^2}{2!}\phi_4''(m) + \frac{c}{1!}\phi_3'(m) + \phi_2(m) = 0. \quad ...(2)$$

Now $\phi_4''(m) = 2, \phi_3'(m) = 0$ and $\phi_2(m) = -a^2m$.

Putting these values in (2), we get $c^2 - a^2m = 0$.

Putting m = 1 in this equation, we get $c^2 - a^2 = 0$ *i.e.,* $c = \pm a$.

Hence y = x + a and y = x – a are two parallel asymptotes corresponding to the slope m = 1.

∴ the required asymptotes are

$x = \pm a$, $y = x + a$ and $y = x - a$.

19.9. ASYMPTOTES BY EXPANSION

To show that $y = mx + c$, is an asymptote of the curve

$$y = mx + c + \left(\frac{A}{x}\right) + \left(\frac{B}{x^2}\right) + \left(\frac{C}{x^3}\right) + \ldots, \quad ...(1)$$

where the series $\left(\frac{A}{x}\right) + \left(\frac{B}{x^2}\right) + \left(\frac{C}{x^3}\right) + \ldots$ *is convergent for sufficiently large values of x.*

Differentiating (1), we get

$$\frac{dy}{dx} = m - \frac{A}{x^2} - \frac{2B}{x^3} - \ldots$$

∴ the equation of the tangent to (1) at (x, y) is

$$Y - y = \left(m - \frac{A}{x^2} - \frac{2B}{x^3} - \ldots\right)(X - x)$$

or $$Y = \left(m - \frac{A}{x^2} - \frac{2B}{x^3} - \ldots\right)X + y - \left(m - \frac{A}{x^2} - \frac{2B}{x^3} - \ldots\right)x$$

or $$Y=\left(m-\frac{A}{x^2}-\frac{2B}{x^3}-...\right)X+c+\frac{2A}{x}+\frac{3B}{x^2}+... \quad ...(2)$$

substituting the value of y from (1).

Now when $x \to \infty$, equation (2) tends to the equation $Y = mX + c$.

Hence $y = mx + c$ is the asymptote of the curve

$$y = mx + c + \frac{A}{x} + \frac{B}{x^2} + \frac{C}{x^3} + ...$$

Example 1:

Find the asymptotes of the hyperbola

$$\left(\frac{x^2}{a^2}\right)-\left(\frac{y^2}{b^2}\right)=1.$$

Solution:

The given curve may be written as $y^2 = \frac{b^2}{a^2}\left(x^2 - a^2\right)$.

$$\therefore \quad y = \pm\frac{b}{a}x\left(1-\frac{a^2}{x^2}\right)^{1/2}$$

$$=\pm\frac{b}{a}x\left[1-\frac{a^2}{2x^2}-\frac{a^4}{8x^4}...\right], \quad \text{(by binomial theorem)}$$

$$=\pm\frac{b}{a}\left[x-\frac{a^2}{2x}-\frac{a^4}{8x^3}-...\right].$$

Hence by 9, the asymptotes of the curve are $y=\pm\left(\frac{b}{a}\right)x$.

Example 2:

Find the asymptotes of the curve, $y^3 = x^2(x - a)$.

Solution:

The given curve may be written as

$$y = x\left(1-\frac{a}{x}\right)^{1/3} = x\left[1-\frac{a}{3x}-\frac{a^2}{9x^2}-...\right], \quad \text{(by binomial theorem)}$$

or $$y = x - \frac{a}{3} - \frac{a^2}{9x} - ...$$

Hence by 9, the asymptote of the curve is $y = x - \frac{1}{3}a$.

10.10 ALTERNATIVE METHODS OF FINDING ASYMPTOTES OF ALGEBRAIC CURVES

The methods of finding asymptotes of algebraic curves already discussed are sufficient to determine the asymptotes. But the following methods of finding asymptotes are sometimes found to be more convenient than the previous methods.

Let the equation of the curve be of degree n in x and y. Obviously the asymptotes of the curve are parallel to the lines obtained by equating to zero the linear factors of the highest degree terms in its equation.

Case I: Let $(y - m_1x)$ be a non-repeated linear factor of the nth degree terms in the given equation of the curve.

Then the equation of the curve can be put into the form

$$(y - m_1x)\, F_{n-1} + P_{n-1} = 0, \quad ...(1)$$

where F_{n-1} contains only terms of degree $(n - 1)$, and P_{n-1} contains terms of various degrees, not higher than $(n - 1)$. Obviously, one of the values of m (slope of an asymptote) in this case is m_1 and so there is an asymptote $y = m_1x = c_1$, provided we can find a finite value of c_1.

Also $c_1 = \lim_{x\to\infty,\, y/x\to m_1} (y - m_1x)$, where (x, y) lies on (1).

But when (x, y) lies on (1),

$$y - m_1x = -\frac{P_{n-1}}{F_{n-1}}. \quad \therefore \; c_1 = \lim_{x\to\infty,\, y/x\to m_1}\left(-\frac{P_{n-1}}{F_{n-1}}\right).$$

Hence $y - m_1x + \lim_{x\to\infty,\, y/x\to m_1}\left(-\frac{P_{n-1}}{F_{n-1}}\right) = 0$, is an asymptote of the curve (1).

Thus, dividing (1) by F_{n-1} and taking limit as $x \to \infty$, $\frac{y}{x} \to m_1$, we get an asymptote of (1). Similarly, we can find asymptotes corresponding to the other non-repeated linear factors.

Case II: Let $(y - m_1x)^2$ be a factor of the nth degree terms of the equation of the curve and $(y - m_1x)$ be not a factor of the $(n - 1)^{th}$ degree terms.

Then, proceeding as in Case I, we find that

$$\lim_{x\to\infty,\ y/x\to m_1} (y - m_1x)^2$$ is either $+\infty$ or $-\infty$.

Hence there is no asymptote corresponc .g to the factor $(y - m_1x)^2$ in this case.

Case III: Let the equation to the curve be of the form $(y - m_1x)^2 F_{n-2} + (y - m_1x) G_{n-2} + P_{n-2} = 0$, where F_{n-2} and G_{n-2} contain only terms of $(n - 2)^{th}$ degree and P_{n-2} contains terms none of which is of a degree higher than $(n - 2)$.

Dividing the equation of the curve by F_{n-2}, we have

$$(y - m_1x)^2 + (y - m_1x).\frac{G_{n-2}}{F_{n-2}} + \frac{P_{n-2}}{F_{n-2}} = 0.$$

Taking limits as $x \to \infty$ and $\frac{y}{x} \to m_1$, we get an equation of the form $(y - m_1x)^2 + \lambda(y - m_1x) + \mu = 0$, which on being solved for $(y - m_1x)$ gives us two parallel asymptotes of the form $y = m_1x + c_1$ and $y = m_1x + c_2$.

Case IV: Let the nth degree terms contain $(y - m_1x)^3$ or a higher power of $(y - m_1x)$, as a factor.

In this case we can proceed exactly in the same way as in case III, to find the asymptotes.

Remark: If the equation to the curve is of the form,

$$(ax + by + c)\ F_{n-1} + P_{n-1} = 0,$$

where F_{n-1} and P_{n-1} contain terms none of which is of degree higher than $n - 1$, and F_{n-1} contains at least one term of degree $n - 1$, a little consideration will show that the asymptote corresponding to the factor $ax + by + c$ is

$$(ax + by + c) + \lim_{x\to\infty,\ y/x\to -a/b} \left(\frac{P_{n-1}}{F_{n-1}}\right) = 0.$$

A similar modification can be made in cases III and IV.

Working Rule: In order to find the asymptotes (nor parallel to y-axis) of any given curve.

1. Factorise the highest degree terms or (the terms including all the highest degree terms) into real linear factors.
2. Proceed as in Case I or Cases II, III and IV according as a factor $(y - m_1x)$ or $(ax + by + c)$ is a non-repeated one or a repeated one, and get asymptotes.

Example 1:

Find the asymptotes of the curve

$(x + y)^2 (x + 2y + 2) = x + 9y - 2.$ **(Meerut, 1993 S)**

Solution:

The given curve is of degree 3. So it cannot have more than three asymptotes. Obviously the asymptotes of the curve are parallel to the lines x + y = 0 and x + 2y + 2 = 0.

Asymptotes parallel to the line x + y = 0 are given by

$$(x+y)^2 = \lim_{x\to\infty,\, y/x\to -1} \frac{x+9y-2}{x+2y+2}$$

$$= \lim_{x\to\infty,\, y/x\to -1} \frac{1+9\left(\frac{y}{x}\right)-\frac{2}{x}}{1+2\left(\frac{y}{x}\right)+\frac{2}{x}} = \frac{1-9}{1-2} = 8.$$

Thus, $x + y = \pm\sqrt{8} = \pm 2\sqrt{2}$ are two asymptotes of the curve.

The asymptote parallel to the line x + 2y + 2 = 0 is given by

$$x + 2y + 2 = \lim_{x\to\infty,\, y/x\to -\frac{1}{2}} \frac{x+9y-2}{(x+y)^2}$$

$$= \lim_{x\to\infty,\, y/x\to -\frac{1}{2}} \frac{\frac{1}{x}+9\left(\frac{y}{x}\right)\left(\frac{1}{x}\right)-\frac{2}{x^2}}{\left(1+\frac{y}{x}\right)^2} = 0.$$

Hence the required asymptotes are

$x + x+y=\pm\sqrt{2}$ and x + 2y + 2 = 0.

Example 2:

Find the all the asymptotes of $y^2(x^2 - a^2) = x^2(x^2 - 4a^2)$.

(Agra, 1999; Meerut, 92; 95, 98S; Jiwaji, 90)

Solution:

Factorising the highest degree terms, the given equation can be re-written as

$$x^2(y - x)(y + x) + 4a^2x^2 - y^2a^2 = 0.$$

Equating the coefficient of the highest power of y (*i.e.*, of y^2) to zero, we get $x^2 - a^2 = 0$ or $x = \pm a$ as the asymptotes parallel to y-axis.

$$y - x = \lim_{x\to\infty,\ y/x\to 1}\left[\frac{a^2y^2 - 4a^2x^2}{x^2(y+x)}\right]$$

$$= \lim_{x\to\infty,\ y/x\to 1}\left[\frac{a^2\left(\dfrac{y^2}{x^2}\right)\left(\dfrac{1}{x}\right) - 4a^2\left(\dfrac{1}{x}\right)}{\left(\dfrac{y}{x+1}\right)}\right] = 0, \text{ (on taking limit)}.$$

$\therefore$ $y - x = 0$ is an asymptote of the curve.

Again the asymptote corresponding to the factor $(y + x)$ is

$$y + x = \lim_{x\to\infty,\ y/x\to 1}\left[\frac{a^2y^2 - 4a^2x^2}{x^2(y-x)}\right] = 0, \qquad \text{(as above).}$$

Hence the required asymptotes are

$x = \pm a$, $y = \pm x$.

Example 3:

Find all the asymptotes of the curve

$(x^2 - y^2)(x + 2y + 1) + x + y + 1 = 0.$ **(Meerut, 1999)**

Solution:

Factorising the highest degree terms into real linear factors, the given equation of the curve may be written as

$$(x + y)(x - y)(x + 2y) + (x^2 - y^2) + (x + y + 1) = 0.$$

The slope m of the asymptote corresponding to the factor $x + y$ is -1.

Hence the asymptote corresponding to the factor $(x + y)$ is

$$x + y = \lim_{x\to\infty,\ y/x\to -1}\frac{-(x^2 - y^2) - (x - y + 1)}{(x+2y)(x-y)}, \qquad [\because m = -1]$$

$$= \lim_{x\to\infty,\ y/x\to -1}\frac{(y^2 - x^2)}{(x-y)(x+2y)} - \lim_{x\to\infty,\ y/x\to -1}\frac{x+y+1}{(x-y)(x+2y)}$$

$$= \lim_{x\to\infty,\ y/x\to -1} \frac{\left\{\left(\frac{y^2}{x^2}\right)-1\right\}}{\left\{1-\left(\frac{y}{x}\right)\right\}\left\{1+2\left(\frac{y}{x}\right)\right\}}$$

$$- \lim_{x\to\infty,\ y/x\to -1} \frac{\frac{1}{x}+\frac{y}{x}.\frac{1}{x}+\frac{1}{x^2}}{\left\{1-\left(\frac{y}{x}\right)\right\}\left\{1+2\left(\frac{y}{x}\right)\right\}},$$

on dividing the numerator and denominator by x^2

$$= \frac{1-1}{(1+1)(1-2)} - 0 = 0.$$

Thus, $x + y = 0$ is one asymptote of the curve. Again the asymptote corresponding to the factor $(x - y)$ is

$$x - y = \lim_{x\to\infty,\ y/x\to -1} \frac{(y^2 - x^2)}{(x+y)(x+2y)} - \lim_{x\to\infty,\ y/x\to -1} \frac{x+y+1}{(x+y)(x+2y)},$$

$[\because m = 1]$

$$= \lim_{x\to\infty,\ y/x\to -1} \frac{\left(\frac{y^2}{x^2}\right)-1}{\left\{1+\left(\frac{y}{x}\right)\right\}\left\{1+2\left(\frac{y}{x}\right)\right\}}$$

$$- \lim_{x\to\infty,\ y/x\to -1} \frac{\frac{1}{x}+\frac{y}{x}.\frac{1}{x}+\frac{1}{x^2}}{\left\{1+\left(\frac{y}{x}\right)\right\}\left\{1+2\left(\frac{y}{x}\right)\right\}}$$

$$= \frac{(1-1)}{(1+1)(1+2)} - 0 = 0.$$

Thus, $x - y = 0$ is another asymptote of the curve. The third asymptote of the curve is

$$x + 2y = \lim_{x\to\infty,\ y/x\to -1/2} \frac{(y^2 - x^2) + \text{terms of degree lower than 2}}{(x-y)(x+y)}$$

$$= \lim_{x\to\infty,\ y/x\to -1/2} \frac{\left\{\left(\dfrac{y^2}{x^2}\right)-1\right\} + \text{terms which} \to 0}{\left\{1-\left(\dfrac{y}{x}\right)\right\}\left\{1+\left(\dfrac{y}{x}\right)\right\}}$$

$$= \frac{\left(\dfrac{1}{4}-1\right)+0}{\left(1+\dfrac{1}{2}\right)\left(1-\dfrac{1}{2}\right)} = \frac{-\dfrac{3}{4}}{\dfrac{3}{4}} = -1.$$

Therefore, $x + 2y + 1 = 0$ is the third asymptote of the curve.

Important Remark: It should be noted that while taking limit we should reject all the terms in the numerator whose degree is lower than the degree denominator. All such terms $\to 0$ as $x \to \infty$.

Example 4:

Find all the asymptotes of the curve

$(a + x)^2 (b^2 + x^2) = x^2y^2.$ **(Raj., 1997)**

Solution:

The given curve may be written as

$$x^2(y - x)(y + x) = 2ax^3 + (a^2 + b^2)x^2 + 2ab^2x + a^2b^2.$$

∴ the asymptotes are

$x^2 = 0$ *i.e.*, $x = 0$, $x = 0$ (the asymptotes parallel to y-axis)

$$\text{and } y - x = \lim_{x\to\infty,\ y/x\to -1}\left[\frac{2ax^3 + (a^2 + b^2)x^2 + 2ab^2x + a^2b^2}{x^2(y + x)}\right]$$

$= - a.$

∴ the required asymptotes are $x = 0$, $y = x + a$, $y = - x - a$.

Example 4 (a):

Find the asymptotes of the curve

$y^2(x - 2) = x^2(y - 1).$ **(Meerut, 91; Gorakhpur, 93)**

Solution:

The given curve may be written as

$$xy(y - x) - 2y + x^2 = 0.$$

Here y − 1 = 0 is the asymptote parallel to x-axis,
and x − 2 = 0 is the asymptote parallel to y-axis.
The other asymptote is given by

$$y - x = \lim_{x\to\infty,\, y/x\to 1}\left[\frac{2y^2 - x^2}{xy}\right]$$

$= 1,$ [dividing the numerator and the denominator by x^2 and taking the limits].

∴ the required asymptotes are x = 2, y = 1 and y = x + 1.

Example 5:

Find the asymptotes of the curve

$x^2y - xy^2 + xy + y^2 + x - y = 0.$ **(Meerut, 1990)**

Solution:

Factorising the highest degree terms, the given equation can be re-written as $xy\,(y - x) = xy + y^2 + x - y$.

Equating to zero the coefficient of the highest powers of x and y we get y = 0 and x = 0 as asymptotes parallel to the co-ordinate axes.

Now the asymptote corresponding to the factor (y − x) is

$$y - x = \lim_{x\to\infty,\, y/x\to 1}\left[\frac{xy + y^2 + x - y}{xy}\right]$$

$$= \lim_{x\to\infty,\, y/x\to 1}\left[\frac{\left(\frac{y}{x}\right) + \left(\frac{y^2}{x^2}\right) + \left(\frac{1}{x}\right) - \left(\frac{y}{x}\right)\left(\frac{1}{x}\right)}{\left(\frac{y}{x}\right)}\right].$$

Taking limits, we have y − x = 2 or y = x + 2.

Hence the required asymptotes are x = 0, y = 0 and y = x + 2.

Example 6:

Find the asymptotes of the curve

$(x^2 - y^2)(y^2 - 4x^2) - 6x^3 + 5x^2y + 3xy^2 - 2y^3 - x^2 + 3xy = 1.$

(Meerut, 1993; Agra, 90; Lucknow, 98, 91; Gorakhpur, 91)

Solution:

Factorising the highest degree terms, the given equation can be re-written as

$$(x + y)(x - y)(y + 2x)(y - 2x) = 6x^3 - 5x^2y - 3xy^2 + 2y^3 + x^2 - 3xy + 1.$$

$\therefore$ the asymptotes are parallel to the lines $x + y = 0$, $x - y = 0$, $y + 2x = 0$ and $y - 2x = 0$.

Asymptote parallel to $x + y = 0$ is

$$x + y = \lim_{x\to\infty,\, y/x\to -1}\left[\frac{6x^3 - 5x^2y - 3xy^2 + 2y^3 + x^2 - 3xy + 1}{(x-y)(y+2x)(y-2x)}\right]$$

$[\because m = -1]$

$$= \lim\left[\frac{6 - 5\left(\frac{y}{x}\right) - 3\left(\frac{y^2}{x^2}\right) + 2\left(\frac{y^3}{x^3}\right) + \left(\frac{1}{x}\right) - 3\left(\frac{y}{x^2}\right) + \left(\frac{1}{x^3}\right)}{\left(1 - \frac{y}{x}\right)\left(2 + \frac{y}{x}\right)\left(-2 + \frac{y}{x}\right)}\right],$$

on dividing the numerator and denominator by x^3

$$= \frac{6 + 5 - 3 - 2}{2.1.(-3)} = \frac{6}{-6} = -1.$$

Hence $x + y + 1 = 0$ is one asymptote of the curve.

The second asymptote parallel to $x - y = 0$ is

$$x - y = \lim_{x\to\infty,\, y/x\to 1}\left[\frac{6x^3 - 5x^2 - 3xy^2 + 2y^3 + x^2 - 3xy + 1}{(x+y)(y+2x)(y-2x)}\right].$$

Taking limits as above, we get

$x - y = 0$ as the second asymptote of the curve.

The third asymptote parallel to $y + 2x = 0$ is

$$x - y = \lim_{x\to\infty,\, y/x\to -2}\left[\frac{6x^3 + 5x^2y - 3xy^2 + 2y^3 + x^2 - 3xy + 1}{(x-y)(x+y)(y-2x)}\right].$$

Taking limits as above, we get

$y + 2x = -1$, as the third asymptote of the curve.

The fourth asymptote parallel to $(y - 2x)$ is

$$y - 2x = \lim_{x\to\infty,\, y/x\to 2}\left[\frac{6x^3 - 5x^2y - 2y^3 + x^2 - 3xy + 1}{(x-y)(x+y)(y+2x)}\right].$$

Taking limits as above, we get

$y - 2x = 0$, as the fourth asymptote of the curve.

Hence the required asymptotes are

$$x + y + 1 = 0,\ x - y = 0,\ 2x + y + 1 = 0 \text{ and } 2x - y = 0.$$

Example 7:

Find the asymptotes of the curve

$(x - y)^2 (x^2 + y^2) - 10(x - y)x^2 + 12y^2 + 2x + y = 0.$ **(Agra, 1993)**

Solution:

The given curve is

$$(x - y)^2 (x^2 + y^2) - 10(x - y)x^2 + 12y^2 + 2x + y = 0.$$

The equation of the curve is of fourth degree and so there are at most four asymptotes. Dividing each term by the coefficient $x^2 + y^2$ of $(x - y)^2$ and taking limits, the asymptotes. Dividing each term by the coefficient $x^2 + y^2$ of $(x - y)^2$ and taking limits, the asymptotes corresponding to the factor $(x - y)^2$ are given by

$$(x - y)^2 (x^2 + y^2) - 10(x - y)\, x^2 + 12y^2 + 2x + y = 0.$$

$$(x - y)^2 - 10(x - y) \lim_{x\to\infty,\ y/x\to 1} \frac{x^2}{x^2 + y^2} + \lim_{x\to\infty,\ y/x\to 1} \frac{12y^2 + 2x + y}{x^2 + y^2} = 0$$

or
$$(x - y)^2 - 10\,(x - y) \lim_{x\to\infty,\ y/x\to 1} \frac{1}{1 + \left(\frac{y}{x}\right)^2}$$

$$+ \lim_{x\to\infty,\ y/x\to 1} \frac{12\left(\frac{y}{x}\right)^2 + \left(\frac{2}{x}\right) + \left(\frac{y}{x}\right)\left(\frac{1}{x}\right)}{1 + \left(\frac{y}{x}\right)^2} = 0,$$

or
$$(x - y)^2 - 10(x - y).\left(\frac{1}{2}\right) + \left(\frac{12}{2}\right) = 0$$

or
$$(x - y)^2 - 5\,(x - y) + 6 = 0$$

i.e., $x - y = \dfrac{5 \pm \sqrt{(25 - 24)}}{2} = \dfrac{5 \pm 1}{2} = 2 \text{ or } 3$

i.e., $x - y = 2$ and $x - y = 3$.

Again $x^2 + y^2 = (x + iy)(x - iy)$ implies that the remaining linear factors of the fourth degree terms in the equation of the curve are imaginary and so the other two asymptotes are imaginary.

Example 8:

Find the asymptotes of the curve

$(x^2 - y^2)(x + 2y) = y^2 - y + 1.$ **(Utkal, 1990)**

Solution:

Factorising the highest degree terms, the given equation can be re-written as $(x - y)(x + y)(x + 2y) = y^2 - y + 1$.

$\therefore$ the asymptotes are parallel to the lines $x - y = 0$, $x + y = 0$ and $x + 2y = 0$.

The asymptote parallel to $x - y = 0$ is

$$x - y = \lim_{x\to\infty,\ y/x\to 1}\left[\frac{y^2 - y + 1}{(x+y)(x+2y)}\right]$$

$$= \lim_{x\to\infty,\ y/x\to 1}\left[\frac{\left(\frac{y}{x}\right)^2 - \left(\frac{y}{x}\right)\left(\frac{1}{x}\right) + \left(\frac{1}{x^2}\right)}{\left\{1+\left(\frac{y}{x}\right)\right\}\left\{1+2\left(\frac{y}{x}\right)\right\}}\right],$$

on dividing the numerator and the denominator by x^2

$$= \frac{1-0+0}{2.(1+2)} = \frac{1}{6}.$$

Hence $x - y = \frac{1}{6}$ is an asymptote of the curve.

The second asymptote parallel to $x + y = 0$ is

$$x + y = \lim_{x\to\infty,\ y/x\to -1}\left[\frac{y^2 - y + 1}{(x-y)(x+2y)}\right]$$

$$= \lim_{x\to\infty,\ y/x\to -1}\left[\frac{\left(\frac{y^2}{x^2}\right) - \left(\frac{y}{x}\right)\left(\frac{1}{x}\right) + \left(\frac{1}{x^2}\right)}{\left\{1-\left(\frac{y}{x}\right)\right\}\left\{1+2\left(\frac{y}{x}\right)\right\}}\right] = -\frac{1}{2}.$$

Thus, $x + y = -\frac{1}{2}$ is the second asymptote of the curve.

The third asymptote parallel to $x + 2y = 0$ is

$$x + 2y = \lim_{x\to\infty,\ y/x\to -1/2}\left[\frac{y^2 - y + 1}{(x+y)(x-y)}\right]$$

$$= \lim_{x\to\infty,\ y/x\to -1/2}\left[\frac{\left(\frac{y}{x}\right)^2 - \left(\frac{y}{x}\right)\left(\frac{1}{x}\right) + \left(\frac{1}{x}\right)^2}{\left\{1+\left(\frac{y}{x}\right)\right\}\left\{1-\left(\frac{y}{x}\right)\right\}}\right]$$

$$= \frac{\left(-\frac{1}{2}\right)^2 - \left(-\frac{1}{2}\right).0+0}{\left\{1+\left(-\frac{1}{2}\right)\right\}\left\{1-\left(-\frac{1}{2}\right)\right\}} = \frac{\frac{1}{4}}{\frac{1}{2}.\frac{3}{2}} = \frac{1}{3}.$$

Thus, $x + 2y = \frac{1}{3}$ is the third asymptote of the curve.

Hence the required asymptotes are

$$x - y = \frac{1}{6},\ x + y = -\frac{1}{2},\ x + 2y = \frac{1}{2}.$$

Example 9:

Find all the asymptotes of the curve $y^2(x - b) = x^3 + a^3$.

Solution:

The given curve is $x(y - x)(y + x) = a^3 + by^2$.

∴ the required asymptotes are

$$y - x = \lim_{x\to\infty,\ y/x\to 1}\left[\frac{a^3 + by^2}{x(y+x)}\right]$$

$$= \lim_{x\to\infty,\ y/x\to 1}\left[\frac{\left(\frac{a^3}{x^2}\right) + b\left(\frac{y^2}{x^2}\right)}{\left(\frac{y}{x}+1\right)}\right] = \frac{1}{2}b,$$

and $y + x = \lim_{x\to\infty,\ y/x\to -1}\left[\frac{a^3 + by^2}{x(y-x)}\right] = -\frac{1}{2}b.$

Also equating the coefficient of y^2 to zero, we get $x - b = 0$ as an asymptote parallel to y-axis.

∴ $\pm y = x \pm \frac{1}{2}b$ and $x = b$ are the required asymptotes.

Example 10:

Find the asymptotes of the curve

$$(y - x)(y - 2x)^2 + (y + 3x)(y - 2x) + 2x + 2y - 1 = 0.$$

(Meerut 1994)

Solution:

The asymptotes corresponding to the factor $(y - 2x)^2$ are given by

$$(y-2x)^2 + (y-2x)\lim_{x\to\infty,\, y/x\to 2}\left[\frac{y+3x}{y-x}\right] + \lim_{x\to\infty,\, y/x\to 2}\left[\frac{2x+2y-1}{y-x}\right] = 0$$

$$\text{or } (y-2x)^2 + (y-2x)\lim_{x\to\infty,\, y/x\to 2}\left[\frac{\left(\frac{y}{x}\right)+3}{\left(\frac{y}{x}\right)-1}\right]$$

$$+\lim_{x\to\infty,\, y/x\to 2}\left[\frac{2+2\left(\frac{y}{x}\right)-\left(\frac{1}{x}\right)}{\left(\frac{y}{x}\right)-1}\right] = 0$$

or $(y - 2x)^2 + 5(y - 2x) + 6 = 0$

i.e., $y - 2x = \frac{1}{2}\left[-5 \pm \sqrt{(25-24)}\right] = \frac{1}{2}(-5 \pm 1)$

i.e., $y - 2x = \frac{1}{2}(-5+1)$ and $y - 2x = \frac{1}{2}(-5-1)$.

$\therefore$ $y = 2x - 2$ and $y = 2x - 3$ are the asymptotes corresponding to the factor $(y - 2x)^2$.

Also the asymptote corresponding to the factor $(y - x)$ is

$$y - x = \lim_{x\to\infty,\, y/x\to 1}\left[\frac{-(y+3x)(y-2x)-2x-2y+1}{(y-2x)^2}\right]$$

$$= \lim_{x\to\infty,\, y/x\to 1}\left[\frac{-\left\{\left(\frac{y}{x}\right)+3\right\}\left\{\left(\frac{y}{x}\right)-2\right\}-2\left(\frac{1}{x}\right)-2\left(\frac{y}{x^2}\right)+\left(\frac{1}{x^2}\right)}{\left\{\left(\frac{y}{x}\right)-2\right\}^2}\right] = 4,$$

on taking limits.

Hence $y = x + 4$ is another asymptote of the given curve.

Example 11:

Find the asymptotes of the curve

$$(y - x)^2 x - 3y(y - x) + 2x = 0.$$

Solution:

The given equation is of third degree. So there are at most three asymptotes.

Dividing each term by x and taking limits, the asymptotes corresponding to the factor $(y-x)^2$ are given by

$$(y-x)^2 - 3(y-x) = \lim_{x\to\infty,\, y/x\to 1}\left(\frac{y}{x}\right) + \lim_{x\to\infty,\, y/x\to 1}\left(\frac{2x}{x}\right) = 0$$

or $$(y-x)^2 - 3(y-x) + 2 = 0.$$

Hence $y - x = \frac{1}{2}\left[3 \pm \sqrt{(9-8)}\right]$ *i.e.*, $= 2$, or 1.

Therefore, $y - x = 2$ and $y - x = 1$ are the two asymptotes.

Also equating to zero the coefficient of the highest power of y, we get the asymptote parallel to y-axis as $x = 3$.,

Hence the required asymptotes are

$$x = 3,\ y - x = 2 \text{ and } y - x = 1.$$

Example 12:

Find all the asymptotes of the curve

$$x^2(x+y)(x-y)^2 + ax^2(x-y) - a^2y^3 = 0.$$

(Lucknow, 1998; Meerut, 91)

Solution:

The given equation is of fifth degree. So there are at most five asymptotes. Dividing each term by the coefficient of $(x-y)^2$ and taking the limits, the asymptotes corresponding to the factor $(x-y)^2$ are given by

$$(x-y)^2 + (x-y)\lim_{x\to\infty,\, y/x\to 1}\left[\frac{a}{x+y}\right] - \lim_{x\to\infty,\, y/x\to 1}\left[\frac{a^3y^3}{x^2(x+y)}\right] = 0$$

or $$(x-y)^2 + (x-y)\lim\left[\frac{\frac{a}{x}}{1+\frac{y}{x}}\right] - \lim\left[\frac{a^2\left(\frac{y^3}{x^2}\right)}{1+\left(\frac{y}{x}\right)}\right] = 0.$$

Taking limits, we get $(x-y)^2 + (x-y).0 - \frac{1}{2}a^2 = 0$

or $x - y = \pm\frac{a}{\sqrt{2}}$ as the two asymptotes.

Also the asymptote corresponding to the factor $(x+y)$ is

$$x+y=\lim_{x\to\infty,\ y/x\to-1}\left[\frac{ax^2(y-x)+a^2y^3}{x^2(x-y)^2}\right]$$

$$=\lim_{x\to\infty,\ y/x\to-1}\left[\frac{a\left\{\left(\frac{y}{x}\right)-1\right\}\left(\frac{1}{x}\right)+a^2\left(\frac{y^3}{x^3}\right)\left(\frac{1}{x}\right)}{\left\{1-\left(\frac{y}{x}\right)\right\}^2}\right].$$

Taking limits we have x + y = 0, as an asymptote.

Also equating to zero the coefficient of the highest power of y, we get the asymptotes parallel to y-axis as $x^2 - a^2 = 0$ *i.e.*, $x = \pm a$.

Hence the required asymptotes are

$$x = \pm a,\ x - y = \frac{a}{\sqrt{2}} \text{ and } x + y = 0.$$

Example 13:

Find the asymptotes of the curve

$$y(x + y)^2 - y + 1 = 0.$$

Solution:

Here y = 0 is an asymptote parallel to x-axis.

The asymptotes corresponding to the factor $(y + x)^2$ are given by

$$(y+x)^2=\lim_{x\to\infty,\ y/x\to-1}\left[\frac{y-1}{y}\right]$$

$$=\lim_{x\to\infty,\ y/x\to-1}\left[\frac{\left(\frac{y}{x}\right)-\left(\frac{1}{x}\right)}{\left(\frac{y}{x}\right)}\right]=1$$

∴ the required asymptotes are $y + x = \pm 1$, and y = 0.

Example 13 (a):

Find all the asymptotes of the curve

$(y - a)^2 (x^2 - a^2) = x^4 + a^4$. **(Rohilkhand, 1997; Kanpur 99)**

Solution:

The given curve is

$$y^2(x^2 - a^2) - 2ay(x^2 - a^2) + a^2(x^2 - a^2) = x^4 + a^4$$

or
$$x^2(y - x)(y + x) - 2ax^2y + a^2(x^2 - y^2) + 2a^3y - 2a^4 = 0.$$

Equating to zero the coefficient of the highest power of y (*i.e.*, of y^2), we get $x^2 - a^2 = 0$ or $x = \pm a$ as the asymptotes parallel to y-axis.

The other asymptotes are

$$y - x = \lim_{x\to\infty,\ y/x\to 1}\left[\frac{2ax^2y - a^2(x^2 - y^2) - 2a^3y + 2a^4}{x^2(y + x)}\right]$$

$$= \lim_{x\to\infty,\ y/x\to 1}\left[\frac{2a\left(\frac{y}{x}\right) - a^2\left\{\left(\frac{1}{x}\right) - \left(\frac{y^2}{x^3}\right)\right\} - 2a^3\left(\frac{y}{x^3}\right) + 2a^4\left(\frac{1}{x^3}\right)}{\left[1 + \left(\frac{y}{x}\right)\right]}\right]$$

$$= a,$$

and
$$= y + x = \lim_{x\to\infty,\ y/x\to -1}\left[\frac{2ax^2y - a^2(x^2 - y^2) - 2a^3y + 2a^4}{x^2(y - x)}\right]$$

$$= \lim_{x\to\infty,\ y/x\to -1}\left[\frac{2a\left(\frac{y}{x}\right) - a^2\left\{\left(\frac{1}{x}\right) - \left(\frac{y^2}{x^3}\right)\right\} - 2a^3\left(\frac{y}{x^3}\right) + 2a^4\left(\frac{1}{x^3}\right)}{\left(\frac{y}{x}\right) - 1}\right]$$

$$= a.$$

$\therefore$ the required asymptotes are $x = \pm a$ and $y = \pm a$ and $y = \pm x + a$.

Example 14:

Find the asymptotes of the curve

$$x^2(x - y)^2 + 9(x^2 - y^2) = 9xy. \qquad \textbf{(Agra, 1990)}$$

Solution:

The given equation can be written can b re-written as

$$x^2(x - y)^2 + 9(x - y)(x + y) = 9xy.$$

Dividing each term by the coefficient of $(x - y)_2$ (*i.e.*, x^2) and taking limits, the asymptotes corresponding to the factor $(x - y)^2$ are given by

$$(x - y)^2 + 9(x - y)\lim_{x\to\infty,\ y/x\to 1}\left[\frac{(x + y)}{x^2}\right] - \lim_{x\to\infty,\ y/x\to 1}\left[\frac{9xy}{x^2}\right] = 0$$

or $\qquad (x - y)^2 = 9$. Then $x - y = \pm 3$ are two asymptotes.

Also equating zero the coefficient of highest power of y, we get the asymptotes parallel to y-axis as $x^2 - 9 = 0$ *i.e.*, $x = \pm 3$.

Hence the required asymptotes are

$$x = \pm 3 \text{ and } x - y = \pm 3.$$

Example 15:

Find the asymptotes of the curve

$$x(y - 3)^3 = 4y(x - 1)^2. \qquad \textbf{(Agra, 1992)}$$

Solution:

The equation of the given curve can be written as

$$xy(y - 2x)(y + 2x) = 9xy^2 - 12yx^2 - 15xy + 27x - 4y.$$

Here the coefficient of the highest power of $y = x$.

Therefore $x = 0$ is an asymptote parallel to y-axis.

Again the coefficient of the highest power of $x = 4y$.

Therefore $y = 0$ is an asymptote parallel to x-axis.

The other asymptotes are given by

$$y - 2x = \lim_{x\to\infty,\ y/x\to 2}\left[\frac{9xy^2 - 12yx^2 - 15xy + 27x - 4y}{xy(y+2x)}\right]$$

$$= \frac{3}{2};\ \textit{i.e.},\ 2y - 4x = 3;$$

and $$y + 2x = \lim_{x\to\infty,\ y/x\to -2}\left[\frac{9xy^2 - 12yx^2 - 15xy + 27x - 4y}{xy(y-2x)}\right]$$

$$= \frac{15}{2}\ \textit{i.e.},\ 2y + 4x = 15.$$

Example 16:

Find the asymptotes of the curve

$$x(y^2 - x^2) = 2ay^2 - a^3.$$

Solution:

The given curve is $(x - 2a)\,y^2 - x^3 + a^3 = 0$. ...(1)

The asymptote parallel to y-axis is $x - 2a = 0$ or $x = 2a$.

The equation (1) can be written as

$$x(y - x)(y + x) = 2ay^2 - a^3.$$

Hence the other asymptotes are

$$y-x=\lim_{x\to\infty,\ y/x\to 1}\left[\frac{2ay^2-a^3}{x(y+x)}\right]=\frac{2a}{2}=a,$$

and $$y+x=\lim_{x\to\infty,\ y/x\to -1}\left[\frac{2ay^2-a^3}{x(y-x)}\right]=\frac{2a}{-2}=-a.$$

$\therefore$ the required asymptotes are $x = 2a$, $y = x + a$, $y = -x - a$.

Example 17:

Find the asymptotes of the curve

$$(y-x)(y-2x)^2+(y+3x)(y-2x)+2x+2y-1=0.$$

(Meerut, 1993, 94)

Solution:

The given equation of third degree and so there are almost three asymptotes. Dividing each term by the coefficient of $(y-2x)^2$ and taking limits, the asymptotes corresponding to the factor $(y-2x)^2$ are given by

$$(y-2x)^2+(y-2x)\lim_{x\to\infty,\ y/x\to 2}\left[\frac{(y+3x)}{(y-x)}\right]+\lim_{x\to\infty,\ y/x\to 2}\left[\frac{2x+2y-1}{y-x}\right]=0$$

or $$(y-2x)^2+(y-2x)\lim\left[\frac{\left(\frac{y}{x}\right)+3}{\left(\frac{y}{x}\right)-1}\right]+\lim\left[\frac{2+2\left(\frac{y}{x}\right)-\frac{1}{x}}{\left(\frac{y}{x}\right)-1}\right]=0$$

Taking limits, we have

$$(y-2x)^2+5(y-2x)+6=0$$

giving $$y-2x=\frac{1}{2}[-5\pm 1]=-2, \text{ or } -3.$$

Thus $y-2x+2=0$ and $y-2x+3=0$ are the two asymptotes.

Also the asymptote corresponding to the factor $(y-x)$ is

$$y-x=\lim_{x\to\infty,\ y/x\to 1}\left[\frac{-(y+3x)(y-2x)-2x-2y+1}{(y-2x)^2}\right]$$

$$=\lim_{x\to\infty,\ y/x\to 1}\left[\frac{-\left(\frac{y}{x}+3\right)\left(\frac{y}{x}-2\right)-2\left(\frac{1}{x}\right)-2\left(\frac{y}{x^2}\right)+\left(\frac{1}{x^2}\right)}{\left\{\left(\frac{y}{x}\right)-2\right\}^2}\right]$$

Taking limits, we have y − x = 4 as another asymptote.

Hence the required asymptotes of the curve are

$$x - y + 4 = 0,\ y - 2x + 2 = 0 \text{ and } y - 2x + 3 = 0.$$

Example 18:

Find all the asymptotes of the curve

$$(x - y + 1)(x - y + 2)(x + y) = 8x - 1. \qquad \textbf{(Meerut, 1991)}$$

Solution:

The given curve may be written as

$$(x + y)(x - y)^2 - (x^2 - y^2) - 10x - 2y + 1 = 0.$$

The asymptote corresponding to the factor (x + y) is

$$x + y = \lim_{x\to\infty,\ y/x\to -1}\left[\frac{8x-1}{(x-y+1)(x-y-2)}\right]$$

= 0, [dividing the numerator and the denominator by x^2 and taking the limits].

Also the asymptotes corresponding to the factor $(x - y)^2$ are given by

$$(x - y)^2 - (x - y) = \lim_{x\to\infty,\ y/x\to 1}\left[\frac{10x+2y-1}{x+y}\right] = 6$$

or $$x - y = \frac{1}{3}\left[1 \pm \sqrt{(1+24)}\right] = 3 \text{ or } -2$$

i.e., $$y = x - 3 \text{ and } y = x + 2.$$

∴ the required asymptotes are

$$y + x = 0,\ y = x - 3 \text{ and } y = x + 2.$$

Example 19:

Find the asymptotes of the curve

$$x^2(x - y)^2 + a^2(x^2 - y^2) = a^2xy.$$

Solution:

The equation of the curve is

$$x^2(x - y)^2 + a^2(x - y)(x + y) = a^2xy.$$

The asymptotes corresponding to the factor $(x - y)^2$ are given by

$$(x - y)^2 + a^2(x - y) \lim_{x\to\infty,\ y/x\to 1}\left[\frac{x+y}{x^2}\right] = \lim_{x\to\infty,\ y/x\to 1} \frac{a^2xy}{x^2}$$

or $\quad (x - y)^2 = a^2$ or $x - y = \pm a$.

Also $x = \pm a$ are asymptotes parallel to y-axis.

$\therefore$ the required asymptotes are $x = \pm a$ and $x - y = \pm a$.

Example 20:

Find all the asymptotes of the curve

$y^2(x - 2) = x^2(y - 1)$ *or* $xy(y - x) = 2y^2 - x^2$. **(Gorakhpur, 1990)**

Solution:

Equating the coefficients of the highest powers of x and y to zero, we get

$y - 1 = 0$ as an asymptote parallel to x-axis

and $x - 2 = 0$ as an asymptote parallel to y-axis.

The other asymptote corresponding to the factor $(y - x)$ is

$$y - x = \lim_{x\to\infty,\, y/x\to 1}\left[\frac{2y^2 - x^2}{xy}\right] = 1.$$

$\therefore$ the required asymptotes are $y = 1$, $x = 2$ and $y - x = 1$.

Example 21:

Find the asymptotes of the curve

$x^3 + 3x^2y - 4y^3 - x + y + 3 = 0.$

Solution:

The equation of the curve may be written as

$$(x - y)(x + 2y)^2 = x - y - 3.$$

The asymptote corresponding to the factor $(x - y)$ is

$$x - y = \lim_{x\to\infty,\, y/x\to 1}\left[\frac{x - y - 3}{(x + 2y)^2}\right] = 0.$$

Also the asymptotes corresponding to the factor $(x + 2y)^2$ are given by

$$(x + 2y)^2 = \lim_{x\to\infty,\, y/x\to -1/2}\left[\frac{x - y - 3}{x - y}\right] = 1$$

or $\quad x + 2y = \pm 1.$

$\therefore$ the required asymptotes are $x - y = 0$ and $x + 2y = \pm 1$.

Example 22:

Find all the asymptotes of the curve

$$(x^2 - y^2)^2 = 4y^2 - y.$$

The asymptotes corresponding to the factors $(x - y)^2$ and $(x + y)^2$ are

$$(x - y)^2 = \lim_{x\to\infty,\ y/x\to 1}\left[\frac{4y^2 - y}{(x+y)^2}\right] = 1, \text{ i.e., } x - y = \pm 1,$$

and $$(x + y)^2 = \lim_{x\to\infty,\ y/x\to -1}\left[\frac{4y^2 - y}{(x-y)^2}\right] = 1, \text{ i.e., } x + y = \pm 1.$$

∴ the required asymptotes are $x - y = \pm 1$ and $x + y = \pm 1$.

Example 23:

Find the asymptotes of the curve

$x^3 + 2x^2y + xy^2 - x^2 - xy + 2 = 0.$ **(Meerut, 1990 P, 98 P)**

Solution:

The given curve may be written as

$$x(x + y)^2 - x(x + y) + 2 = 0.$$

Here an asymptote parallel to y-axis is $x = 0$.

The asymptotes corresponding to the factor $(x + y)^2$ are given by

$$(x + y)^2 - (x + y) + \lim_{x\to\infty,\ y/x\to -1}\left[\frac{2}{x}\right] = 0$$

or $$(x + y)^2 - (x + y) + 0 = 0$$

or $$(x + y)(x + y - 1) = 0.$$

∴ the required asymptotes are

$$x = 0,\ x + y = 0 \text{ and } x + y - 1 = 0.$$

10.11. ASYMPTOTES BY INSPECTION

If the equation of a curve is of the form $F_n + F_{n-2} = 0$, where F_n is of degree n (i.e., contains terms of degree n and may also contain terms of lower degree) and F_{n-2} is of degree $(n - 2)$ at the most, and if $F_n = 0$ can be broken up into n linear factors so as to represent n straight lines no two of which are parallel or coincident, then $F_n = 0$ gives all the asymptotes of the curve.

Example 1:

Find the asymptotes of the curve

$$xy(x^2 - y^2)(x^2 - 4y^2) + 3xy(x^2 - y^2) + x^2 + y^2 - 7 = 0.$$

(Meerut, 1993)

Solution:

The given equation can be put in the form

$$[xy(x^2 - y^2)(x^2 - 4y^2)] + [3xy(x^2 - y^2) + x^2 + y^2 - 7] = 0.$$

This equation is of the form $F_n + F_{n-2} = 0$, where F_n can be broken up into n linear factors so as to represent n straight lines no two of which are parallel or coincident.

$\therefore$ all the asymptotes are given by $F_n = 0$

or $$xy(x - y)(x + y)(x - 2y)(x + 2y) = 0.$$

Then by inspection, the asymptotes are

$$x = 0,\ y = 0,\ x - y = 0,\ x + y = 0,\ x - 2y = 0$$

and $$x + 2y = 0.$$

Example 2:

Find the asymptotes of the curve

$$x^3 - 4xy^2 + x^2y - 4y^3 - 2x^3 + 8y^2 + 3x + 4y - 5 = 0.$$

Solution:

The given equation can be put in the form

$$(x - 2y)(x + 2y)(x + y - 2) + (3x + 4y - 5) = 0.$$

By inspection, the asymptotes are

$$x - 2y = 0,\ x + 2y = 0,\ x + y - 2 = 0.$$

10.12. INTERSECTION OF A CURVE AND ITS ASYMPTOTES

To prove that any asymptote of an algebraic curve of the nth degree cuts the curve in (n – 2) points.

A straight line $y = mx + c$ ...(1)

cuts the curve of the nth degree

$$x^n\phi_n\left(\frac{y}{x}\right) + x^{n-1}\phi_{n-1}\left(\frac{y}{x}\right) + x^{n-2}\phi_{n-2}\left(\frac{y}{x}\right) + \ldots = 0 \qquad \text{...(2)}$$

in n points real or imaginary.

Eliminating y from (1) and (2), we get

$$x^n\phi_n\left(m+\frac{c}{x}\right)+x^{n-1}\phi_{n-1}\left(m+\frac{c}{x}\right)+x^{n-2}\phi_{n-2}\left(m+\frac{c}{x}\right)+\ldots=0.$$

Expanding each term by Taylor's theorem and arranging in descending powers of x, we get

$$x^n\phi_n(m)|\left[c\phi_n'(m)+\phi_{n-1}(m)\right]x^{n-1}$$

$$+\left[\frac{c^2}{2!}\phi_n''(m)+\frac{c}{1!}\phi_{n-1}'(m)+\phi_{n-2}(m)\right]x^{n-2}+\ldots=0.$$

....(3)

The equation (3) gives the abscissae of the points of intersection of (1) and (2).

If y = mx + c is an asymptote of (2), then $\phi_n(m) = 0$

and $$c\phi_n'(m)+\phi_{n-1}(m)=0.$$

Consequently (3) reduces to an equation of (n – 1 degree in x and therefore the asymptote (1) cuts the curve (2) in (n – 2) points.

Cor. *The, n asymptotes of a curve of the nth degree cut it in n(n – 2) points. In general, a curve of the degree n – 2, or less, can be made to pass through these n (n – 2) points.*

Example 1:

Find the equation of the cubic curve whose asymptotes are x + a = 0, y – a = 0 and x + y + a = 0 and which touches the axis of x at the origin and passes through the point (– 2a, – 2a).

Solution:

The asymptotes are

$$x + a = 0,\ y - a = 0 \text{ and } x + y + a = 0.$$

∴ the combined equation of the asymptotes is

$$(x + a)(y - a)(x + y + a) = 0.$$

Now the equation of any curve, having these asymptotes, is of the form

$$(x + a)(y - a)(x + y + a) + F_1 = 0,$$

where F_1 is of 1st degree in x and y (say F_1 = bx + cy + d).

So let the equation of the required curve be

$$(x + a)(y - a)(x + y + a) + (bx + cy + d) = 0.$$

But it is given that the curve passes through the origin (0, 0), therefore $\qquad d = a^3.$

Thus, the equation of the curve reduces to

$$(x + a)(y - a)(x + y + a) + bx + cy + a^3 = 0$$

or $\qquad xy(x + y) + a(y^2 - x^2 + xy) - 2a^2x + bx + cy = 0.$

Now equating the lowest degree terms to zero the equation of the tangent at origin is $(-2a^2 + b)x + cy = 0$.

Also given that the x-axis (*i.e.*, $y = 0$) is tangent at the origin.

$\therefore \qquad -2a^2 + b = 0$ or $b = 2a^2$.

$\therefore$ the equation of the curve is

$$xy(x + y) + a(y^2 - x^2 + xy) + cy = 0.$$

This equation passes through $(-2a, -2a)$. (given)

$\therefore \qquad 4a^2(-4a) + a(4a^2 - 4a^2 + 4a^2) + c(-2a) = 0$ or $c = -6a^2$.

Therefore the required equation is

$$xy(x + y) + a(y^2 - x^2 + xy) - 6a^2y = 0.$$

Example 2:

Form the equation of a curve which has $x = 0$, $y = 0$, $y = x$ and $y = -x$ four asymptotes and which passes through the point (a, b) and cuts its asymptotes again in eight points lying upon the circle $x^2 + y^2 = a^2$.

Solution:

The combined equation of the asymptotes is

$$xy(y^2 - x^2) = 0. \qquad ...(1)$$

The required curve passes through the points of intersection of the asymptotes and the given circle $x^2 + y^2 - a^2 = 0$.

Let the equation of the curve be

$$xy(y^2 - x^2) + \lambda(x^2 + y^2 - a^2) = 0. \qquad ...(2)$$

Now this curve also passes through the point (a, b).

$\therefore \qquad ab(b^2 - a^2) + 1(a^2 + b^2 - a^2) = 0,$

i.e., $\qquad \lambda = \left(\frac{a}{b}\right)(a^2 - b^2).$

Substituting this value of λ in (2), we have

$$bxy(y^2 - x^2) + a(a^2 - b^2)(x^2 + y^2 - a^2) = 0$$

as the required equation of the curve.

Example 3:

Show that the four asymptotes of the curve

$(x^2 - y^2)(y^2 - 4x^2) + 6x^3 - 5x^2y - 5x^2y - 3xy^2 + 2y^3 - x^2 + 3xy - 1 = 0$

cut the curve in eight points which lie on the circle $x^2 + y^2 = 1$.

(Gorakhpur, 1999; Meerut, 93 S; Agra, 90; I.C.S. 97)

Solution:

The given curve is

$$P \equiv (x^2 - y^2)(y^2 - 4x^2) + 6x^3 - 5x^2y - 3xy^2 + 2y^3 - x^2 + 3xy - 1 = 0. \quad ....(1)$$

Here $\phi_4(m) = (1 - m^2)(m^2 - 4) = -4 + 5m^2 - m^4$.

The slopes of the asymptotes are given by the equation

$$\phi_4(m) = (1 - m^2)(m^2 - 4) = 0.$$

$$\therefore \quad m = \pm 1, \pm 2.$$

Also $\quad \phi_3(m) = 6 - 5m - 3m^2 - 3m^2 + 2m^3$,

and $\phi'(m) = 10m - 4m^3$.

Now c is given by the equation

$$c\phi_4'(m) + \phi_3(m) = 0$$

i.e., $c(10m - 4m^3) + 6 - 5m - 3m^2 + 2m^3 = 0$

i.e., $c = \dfrac{6 - 5m - 3m^2 + 2m^3}{4m^3 - 10m}$.

When $m = 1$, $c = 0$; when $m = -1$, $c = 1$;

when $m = 2$, $c = 0$; and when $m = -2$, $c = 1$.

Thus, the asymptotes of the curve (1) are $y = x$, $y = -x + 1$, $y = 2x$ and $y = -2x + 1$

The combined equation of the asymptotes is

$$(y - x)(y + x - 1)(y - 2x)(y + 2x - 1) = 0$$

or $\quad [(y^2 - x^2) - y + x][(y^2 - 4x^2) - y + 2x] = 0$

or $\quad (y^2 - x^2)(y^2 - 4x^2) - y^3 + 2xy^2 + x^2y - 2x^3 - y^3 + 4x^2y + xy^2 - 4x^3 + y^2 - 2xy - xy + 2x^2 = 0$

or $\quad Q \equiv (y^2 - x^2)(y^2 - 4x^2) - 6x^3 + 5x^2y + 3xy^2 - 2y^3 + y^2 - 3xy + 2x^2 = 0 \quad ...(2)$

Now each asymptote of (1) will cut it in 4 – 2 *i.e.*, 2 points. Therefore, the four asymptotes will cut in 4 × 2 *i.e.*, 8 points.

Now taking $\lambda = 1$, $P + Q = 0$ gives $x^2 + y^2 - 1 = 0$ *i.e.*, $x^2 + y^2 = 1$, which is the equation of a circle.

Hence the eight points of intersection of (1) and (2) lie on the circle $x^2 + y^2 = 1$.

Example 4:

Find the equation of the cubic which has the same asymptotes as the curve $x^3 - 6x^2y + 11xy^2 - 6y^3 + 4x + 5y + 7 = 0$ *and which passes through the points (0, 0), (2, 0) and (0, 2).*

Solution:

The given curve is

$$(x - y)(x - 2y)(x - 3y) + (4x + 5y + 7) = 0. \quad ...(1)$$

By intersection,

$(x - y)(x - 2y)(x - 3y) = 0$, is the combined equation of the asymptotes of (1).

Now the most general equation of the curve, having these asymptotes, is

$$(x - y)(x - 2y)(x - 3y) + ax + by + c = 0. \quad ...(2)$$

If (2) passes through the points (0, 0), (2, 0) and (0, 2), then

$$c = 0;\ 8 + 2a = 0 \text{ or } a = -4;\ -48 + 2b = 0 \text{ or } b = 24.$$

$\therefore$ from (2), the required equation of the curve is

$$x^3 - 6x^2y + 11xy - 6y^3 - 4x + 24y = 0.$$

10.13. ASYMPTOTES OF THE POLAR CURVES

If α *be a root of the equation* $f(\theta) = 0$, *then*

$$r \sin(\theta - \alpha) = \frac{1}{f'(\alpha)}$$

is an asymptote of the curve

$$\frac{1}{r} = f(\theta) \text{ or } v = f(\theta), \text{ where } v = \frac{1}{r}.$$

Let P be any point (r, θ) on the curve $\frac{1}{r} = f(\theta)$ ...(1)

As $P \to \infty$, $r \to \infty$ and consequently $f(\theta) \to 0$.

Let OT be the perpendicular to the radius OP. Then OT (*i.e.*, the polar subtangent of the curve at P) is given by

$$OT = r^2 \frac{d\theta}{dr} = \frac{1}{f'(\theta)}. \qquad \left[\because \text{ from (1)}, -\frac{1}{r^2}\frac{dr}{d\theta} = f'(\theta)\right]$$

Now let $\theta \to \alpha$. Then $f(\theta) \to 0$. [$\because f(\alpha) = 0$]

Therefore $r \to \infty$, PT $\to$ to an asymptote, and

$$OT \to \left(r^2 \frac{d\theta}{dr}\right)_{\theta=\alpha}$$

or $$OT \to -\frac{1}{f'(\alpha)} \text{ if } f'(a) \neq 0.$$

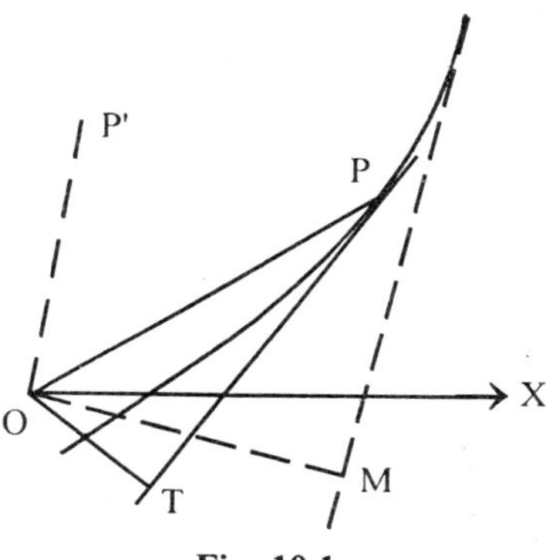

Fig. 10.1

Also OP and PT will tend to become parallel, and the angle OTP will tend to a right angle and OT will tend to OM where OM is perpendicular to the asymptote. Hence, the asymptote is the straight line parallel to the radius vector $\theta = \alpha$ and situated at a distance $\left(r^2 \frac{d\theta}{dr}\right)_{\theta=\alpha}$ from O.

Now the polar equation of the straight line, the perpendicular p on which from the origin makes an angle β with the initial line, is $r \cos(\theta - \beta) = p$.

[Remember]

Here, for the asymptote $p = OM = \left(r^2 \frac{d\theta}{dr}\right)_{\theta=a} = -\frac{1}{f'(\alpha)}$,

and $\beta = -\left(\frac{1}{2}\pi - \alpha\right) = \alpha - \frac{1}{2}\pi$.

Hence the equation of the asymptote is

$$r \cos\left\{\theta - \left(\alpha - \frac{1}{2}\pi\right)\right\} = -\frac{1}{f'(\alpha)}$$

i.e., $$r \sin(\theta - \alpha) = \frac{1}{f'(\alpha)}.$$

Example 5:

Find the circular asymptote of the curve

$r(e^\theta - 1) = a(e^\theta + 1)$. **(Agra, 1991)**

Solution:

The given equation is

$$r = \frac{a(e^\theta + 1)}{(e^\theta - 1)} = f(\theta), \text{ say.}$$

$\therefore$ the circular asymptote is given by

$$r = a \lim_{\theta \to \infty} \frac{e^{\theta}+1}{e^{\theta}-1}, \qquad \left[\text{form } \frac{\infty}{\infty}\right]$$

$$= a \lim_{\theta \to \infty} \frac{1+e^{-\theta}}{1-e^{-\theta}} = a.$$

Hence r = a is the circular asymptote.

Example 5 (a):

Find the circular asymptote of the curve

$$r = \frac{3\theta^2 + 2\theta + 1}{2\theta^2 + \theta + 1}.$$

Solution:

The given equation is

$$r = \frac{3 + \left(\frac{2}{\theta}\right) + \left(\frac{1}{\theta^2}\right)}{2 + \left(\frac{1}{\theta}\right) + \left(\frac{1}{\theta^2}\right)},$$ dividing numerator and denominator by θ^2.

Taking limits when $\theta \to \infty$, we see that the circular asymptote is $r = \frac{3}{2}$.

Example 6:

Find the asymptotes of the curve r sin nθ = a.

(Gorakhpur, 1991; Meerut, 95)

Solution:

The equation of the curve can be written as

$$\frac{1}{r} = \frac{1}{a} \sin n\theta = f(\theta), \text{ say}.$$

Now $f(\theta) = 0$ if $\sin n\theta = 0$ *i.e.*, $n\theta = m\pi$ (where m is any integer).

$$\therefore \qquad \theta = \left(\frac{m\pi}{n}\right) = \alpha, \text{ say}.$$

Also $$f'(\theta) = \left(\frac{1}{a}\right).n \cos n\theta.$$

$$\therefore \qquad f'(\alpha) = \left(\frac{1}{a}\right).n \cos n\alpha = \left(\frac{1}{a}\right).n \cos\left(n.\frac{m\pi}{n}\right)$$

$$= \left(\frac{1}{a}\right).n.\cos m\pi.$$

$\therefore$ the asymptotes are $r \sin\left(\theta - \frac{m\pi}{n}\right) = \frac{a}{n \cos m\pi}$, where m is any integer.

Example 7:

Show that the curve $r = a \sec m\theta + b \tan m\theta$ has two sets of asymptotes; members of the first set touching one fixed circle, and those of the other, another fixed circle.

Solution:

The equation of the curve can be written as

$$\frac{1}{r} = \frac{1}{a \sec m\theta + b \tan m\theta} = \frac{\cos m\theta}{a + b \sin m\theta} = f(\theta), \text{ say.}$$

$$\therefore \quad f'(\theta) = \frac{(a + b \sin m\theta) - (-m \sin m\theta) - \cos m\theta\,(bm \cos m\theta)}{(a + b \sin m\theta)^2}.$$

Now $f'(\theta) = 0 \Rightarrow \cos m\theta = 0$, or $m\theta = (2k + 1).\frac{1}{2}\pi$, (k is may integer).

or $$\theta = (2k+1).\frac{\pi}{2m} = \alpha, \text{ say.}$$

Obviously $\cos m\alpha = 0$.

Also $\sin ma = \sin\left\{(2k+1)\frac{1}{2}\pi\right\} = \sin\left(k\pi + \frac{1}{2}\pi\right) = \cos k\pi = (-1)^k$.

$$\therefore \quad f'(\alpha) = \frac{\{a + b(-1)^k\}\{-m(-1)^k\}}{\{a + b(-1)^k\}^2} = \frac{(-1)^{k+1}.m}{\{a + b(-1)^k\}}.$$

$\therefore$ the asymptotes are given by

$$r \sin\left[\theta - \left\{(2k+1).\frac{\pi}{2m}\right\}\right] = \frac{a + b(-1)^k}{(-1)^{k+1}.m}. \qquad ...(1)$$

Case I: When k is odd, the first set of asymptotes are

$$r \sin\left\{\theta - \frac{(2k+1)\pi}{2m}\right\} = \frac{a-b}{m}, \qquad [\because (-1)^k = -1]$$

or $$r\left[\sin\theta \cos\frac{(2k+1)\pi}{2m} - \cos\theta \sin\frac{(2k+1)\pi}{2m}\right] = \frac{a-b}{m},$$

or $$x\sin\frac{(2k+1)\pi}{2m} - y\cos\frac{(2k+1)\pi}{2m} + \frac{a-b}{m} = 0,$$

[since r cos θ = x and r sin θ = y]

and these asymptotes the circle $x^2 + y^2 = \left(\frac{a-b}{m}\right)^2$, which is fixed.

Case II: When k is even, the second set of asymptotes are

$$r\sin\left(\theta - \frac{2k+1}{2m}.\pi\right) = -\frac{a+b}{m}, \qquad [\because (-1)^k = 1]$$

or $$y\cos\left(\frac{2k+1}{2m}.\pi\right) - x\sin\left(\frac{2k+1}{2m}.\pi\right) = -\frac{a+b}{m}$$

or $$x\sin\left(\frac{2k+1}{2m}.\pi\right) - y\cos\left(\frac{2k+1}{2m}.\pi\right) - \left(\frac{a+b}{m}\right) = 0$$

and these touch another fixed circle

$$x^2 + y^2 = \left(\frac{a+b}{m}\right)^2.$$

Example 8:

Find the asymptotes of the curve

$$r\theta\cos\theta = a\cos 2\theta. \qquad \textbf{(Agra, 1995)}$$

Solution:

The equation of the given curve can be written as

$$\frac{1}{r} = \frac{\theta\cos\theta}{a\cos 2\theta} = f(\theta), \text{ say.}$$

$$\therefore \quad f'(\theta) = \frac{1}{a}\left[\frac{\cos 2\theta.\{\cos\theta - \theta.\sin\theta\} + 2\theta\cos\theta.\sin 2\theta}{\cos^2 2\theta}\right].$$

Now $f(\theta) = 0 \Rightarrow \theta = 0$ or $\cos\theta = 0$ *i.e.*, $\theta = (2k+1).\frac{1}{2}\pi$.

If $\theta = 0 = \alpha$ (say), then $f'(\alpha) = \frac{1}{a}$.

$\therefore$ $r\sin(\theta - 0) = a$

or $r\sin\theta = a$ is the corresponding asymptote.

When $\theta = (2k+1).\frac{1}{2}\pi = \alpha$ (say), we have

$\cos\alpha = 0$, $\sin 2\alpha = 0$, $\sin\alpha = (-1)^k$

and $\cos 2\alpha = \cos(2k\pi + \pi) = \cos\pi = -1$.

$$\therefore \quad f'(\alpha) = \frac{1}{a}\left[\frac{-\left\{(-1)^k (2k+1).\frac{1}{2}\pi\right\}}{1}\right]$$

$$= (-1)^{k+1}(2k+1).\frac{\pi}{2a}.$$

The corresponding asymptote is

$$r\sin\left\{\theta - (2k+1).\frac{\pi}{2}\right\} = \frac{2a}{(-1)^{k+1}.(2k+1)\pi}$$

or $\qquad (-1)^k\, r(\cos(k\pi - \theta))\}\ (2k+1)\,\pi = 2a,$

or $\qquad (-1)^{2k}\, r\cos\theta = \dfrac{2a}{\{(2k+1)\pi\}}.$

$\therefore$ the required asymptotes are $r\sin\theta = a$

and $\qquad r\cos\theta = \dfrac{2a}{\{(2k+1)\pi\}}.$

Example 9:

Find the asymptotes of the curve

$r\theta = a.$ **(Garhwal, 1993; Meerut, 1990, 96, 97)**

Solution:

The equation of the given curve can be written as

$$\frac{1}{r} = \frac{\theta}{a} = f(\theta), \text{ say.}$$

Now $\qquad f(\theta) = 0$ if $\dfrac{\theta}{a} = 0$ *i.e.*, $\theta = 0 = a$, say.

Also $\qquad f'(\theta) = \dfrac{1}{a}$, so that

$$f'(\alpha) = f'(0) = \frac{1}{a},\ \frac{1}{f'}(\alpha) = a.$$

Now the asymptote corresponding to $\theta = \alpha$ is given by

$$r\sin(\theta - \alpha) = \frac{1}{f'}(\alpha)$$

i.e., $\qquad r\sin(\theta - 0) = a,$ $\qquad [\because$ here $\alpha = 0]$

i.e., $\qquad r\sin\theta = a.$

Example 10:

Find the asymptotes of the curve $r = \frac{a}{(1-\cos\theta)}$.

Solution:

The equation to the curve can be written as

$$\frac{1}{r} = \frac{1}{a}(1-\cos\theta) = f(\theta), \text{ say.}$$

Now $f(\theta) = 0 \Rightarrow 1 - \cos\theta = 0$ or $\cos\theta = 1$

i.e., $\theta = 2k\pi$, where k is any integer.

Also $\quad f'(\theta) = \frac{1}{a}\sin\theta$.

$\therefore \quad f'(2k\pi) = \frac{1}{a}\sin(2k\pi) = 0$.

Hence $\quad \frac{1}{f'(2k\pi)} = \infty$

and consequently there is no asymptote of the given curve.

Example 11:

Find the asymptotes of the curve

$$r = \frac{2a}{(1-2\cos\theta)}.$$

Solution:

The equation of the curve can be written as

$$\frac{1}{r} = \frac{1}{2a}(1 - 2\cos\theta) = f(\theta), \text{ say.}$$

Now $f(\theta) = 0$ if $1 - 2\cos\theta = 0$ *i.e.*, $2\cos\theta = 1$ or $\cos\theta = \frac{1}{2}$ *i.e.*,

$\theta = 2n\pi \pm \frac{1}{3}\pi$, where n is any integer.

Also $\quad f'(\theta) = \frac{1}{2a}(2\sin\theta) = \frac{1}{a}\sin\theta$.

$\therefore \quad f'\left(2n\pi \pm \frac{\pi}{3}\right) = \frac{1}{a}\sin\left(2n\pi \pm \frac{\pi}{3}\right) = \pm\frac{1}{a}\sin\frac{\pi}{3} = \pm\frac{\sqrt{3}}{2a}$.

$\therefore$ the asymptotes are given by

$$r\sin\left\{\theta - \left(2n\pi \pm \frac{\pi}{3}\right)\right\} = \pm\frac{2a}{\sqrt{3}}$$

or $$r\sin\theta\cos\left(2n\pi \pm \frac{\pi}{3}\right) - r\cos\theta.\sin\left(2n\pi \pm \frac{\pi}{3}\right) = \pm\frac{2a}{\sqrt{3}}$$

i.e., $$(r\sin\theta)\cos\frac{\pi}{3} \mp r\cos\theta\sin\frac{\pi}{3} = \frac{2a}{\sqrt{3}}$$

i.e., $$r\sin\theta\cos\frac{\pi}{3} - r\cos\theta\sin\frac{\pi}{3} = \frac{2a}{\sqrt{3}}$$

and $$r\sin\theta\cos\frac{\pi}{3} + r\cos\theta\sin\frac{\pi}{3} = -\frac{2a}{\sqrt{3}}$$

i.e., $$r\sin\left(\theta - \frac{\pi}{3}\right) = \frac{2a}{\sqrt{3}} \text{ and } r\sin\left(\theta + \frac{\pi}{3}\right) = -\frac{2a}{\sqrt{3}}.$$

Example 12:

Find the asymptotes of the curve

$r \sin \theta = 2 \cos 2\theta$. **(Meerut, 1992, 94, 96 BP, 99)**

Solution:

The equation to the curve can be written as

$$\frac{1}{r} = \frac{\sin\theta}{2\cos 2\theta} = f(\theta), \text{ say.}$$

Now $f(\theta) = 0 \Rightarrow \sin\theta = 0$ *i.e.,* $\theta = k\pi = \alpha$ (say).

Here k is any integer.

Also $$f'(\theta) = \frac{1}{2}\,\frac{\cos 2\theta\cos\theta - \sin\theta.(-2\sin 2\theta)}{\cos^2 2\theta}.$$

$$\therefore \quad \frac{1}{f'(\alpha)} = \frac{2\cos^2(2k\pi)}{\cos 2k\pi.\cos k\pi + 2\sin k\pi.\sin 2k\pi}$$

$$= \frac{2}{\cos k\pi}, \qquad [\because \cos 2k\pi = 1]$$

$$= \frac{2}{(-1)^k}. \qquad [\because \cos k\pi = (-1)^k]$$

$\therefore$ the required asymptotes are given by

$$r\sin(\theta - k\pi) = \frac{2}{(-1)^k}$$

or $$-r\sin(k\pi - \theta) = \frac{2}{(-1)^k}$$

or $$-r\left\{(-1)^{k-1}\sin\theta\right\} = \frac{2}{(-1)^k}$$

or $r \sin \theta = 2.$

Example 13:

Find the asymptotes of the curve $r = a \operatorname{cosec} \theta + b$.

Solution:

The equation of the curve can be written as

$$\frac{1}{r} = \frac{1}{a \operatorname{cosec}\theta + b} = \frac{\sin\theta}{a + b\sin\theta} = f(\theta), \text{ say.}$$

Now $f(\theta) = 0$, if $\sin \theta = 0$ *i.e.*, $\theta = n\pi$, (where n is any integer).

Also $$f'(\theta) = \frac{\cos\theta(a + b\sin\theta) - b\sin\theta\cos\theta}{(a + b\sin\theta)^2}.$$

$\therefore$ $$f'(n\pi) = \frac{\cos n\pi(a + b\sin n\pi) - b\sin n\pi\cos n\pi}{(a + b\sin n\pi)^2} \cdot \frac{\cos n\pi}{a}.$$

$\therefore$ the asymptotes are given by $r \sin(\theta - n\pi) = \dfrac{a}{\cos n\pi}$.

or $$r(\sin\theta\cos n\pi - \cos\theta\sin n\pi) = \frac{a}{\cos n\pi}$$

or $r \sin\theta \cos^2 n\pi - 0 = a$ or $r \sin\theta = a$.

10.14 CIRCULAR ASYMPTOTES

Definition: *Let the equation of a curve be $r = f(\theta)$.*

If $\lim_{\theta\to\infty} f(\theta) = l$, then the circle $r = l$ is called the circular asymptote of the curve $r = f(\theta)$.

Example 1:

Find the circular asymptote of the curve

$$r = a.\frac{\theta}{\theta - 1}.$$ **(Meerut, 1991)**

Solution:

The circular asymptote is given by

$r = a \lim_{\theta\to\infty} \dfrac{\theta}{\theta - 1} = a$. Thus, $r = a$ is the circular asymptote.

Example 2:

Find the asymptotes of the curve

$$4(x^4 + y^4) - 17x^2y^2 - 4x(4y^2 - x^2) + 2(x^2 - 2) = 0$$

and show that they pass through the points of intersection of the curve with the ellipse $x^2 + 4y^2 = 4$. **(I.C.S., 1996)**

Solution:

The given curve may be written as

$$(4x^4 + 4y^4 - 17x^2y^2) - 4(4xy^2 - x^3) + 2x^2 - 4 = 0. \quad ...(1)$$

Here $\phi_4(m) = 4m^4 - 17m^2 + 4 = 0$ gives $m = \pm \frac{1}{2}, \pm 2$.

Also $c = -\frac{\phi_3(m)}{\phi_4'(m)} = \frac{4(4m^2 - 1)}{16m^3 - 34m}$.

When $m = \frac{1}{2}$, $c = 0$, when $m = -\frac{1}{2}$, $c = 0$,

when $m = 2$, $c = 1$ and when $m = -2$, $c = -1$.

Thus the asymptotes of (1) are

$$y = \frac{1}{2}x,\ y = -\frac{1}{2}x,\ y = 2x + 1 \text{ and } y = -2x - 1.$$

∴ the combined equation of the asymptotes is

$$(x - 2y)(x + 2y)(2x - y + 1)(2x + y + 1) = 0$$

or $$(4x^4 + 4y^4 - 17x^2y^2) - 4(4xy^2 - x^3) + (x^2 - 4y^2) = 0. \quad ...(2)$$

Subtracting (2) from (1), we get

$$x^2 + 4y^2 - 4 = 0. \quad ...(3)$$

Also the four asymptotes cut the curve into $n(n - 2)$ *i.e.*, $4(4 - 2) = 8$ points. These 8 points of intersection lie on the ellipse (3).

Example 3:

Find the equation of the cubic which has the same asymptotes as the curve $x^3 - 6x^2y + 11xy^2 - 6y^3 + x + y + 1 = 0$ *and which touches the axis of y at the origin and passes through the point (3, 2).*

(Rohilkhand, 1996; Lucknow, 90; Kanpur, 98; Meerut, 99)

Solution:

The given curve is

$$(x^3 - 6x^2y + 11xy^2 - 6y^3) + (x + y) + 1 = 0.$$

Here $\qquad \phi_3(m) = 1 - 6m + 11m^2 - 6m^3$

$$= (1 - m)(1 - 2m)(1 - 3m).$$

Therefore, $\phi_3(m) = 0$ gives $m = 1, \frac{1}{2}, \frac{1}{3}$. Also $\phi_2(m) = 0$.

$\therefore \qquad c = -\frac{\phi_2(m)}{\phi_3'(m)} = 0$, for all the three values of m.

$\therefore$ the asymptotes of the given curve are

$$y = x;\ y = \frac{1}{2}x \text{ and } y = \frac{1}{3}x.$$

Hence the combined equation of the asymptotes is

$$(x - y)(x - 2y)(x - 3y) = 0. \qquad ...(1)$$

Now, the most general equation of any curve, having these asymptotes, is of the form

$$(x - y)(x - 2y)(x - 3y) + F_1 = 0,$$

where F_1 is of the first degree in x and y (say $F_1 = ax + by + c$).

If the curve $(x - y)(x - 2y)(x - 3y) + ax + by + c = 0$, passes through the origin (0, 0), then $c = 0$ and the equation of the curve becomes

$$(x - y)(x - 2y)(x - 3y) + ax + by = 0. \qquad ...(2)$$

Equating to zero, the lowest degree terms in (2), we get $ax + by = 0$, as the equation of the tangent at origin.

But $x = 0$ *i.e.*, y-axis is given to be the tangent at origin.

Therefore, $b = 0$ and hence the equation of the curve is

$$(x - y)(x - 2y)(x - 3y) + ax = 0 \qquad ...(3)$$

It passes through the point (3, 2), (given).

$\therefore \qquad (3 - 2)(3 - 4)(3 - 6) + 3a = 0$; or $a = -1$.

Therefore, the required equation of the curve is

$$(x - y)(x - 2y)(x - 3y) - x = 0, \text{ [putting } a = -1 \text{ in (3)]}$$

or $\qquad x^3 - 6x^2y + 11xy^2 - 6y^3 - x = 0.$

Example 4:

Show that the eight points of the curve

$$x^4 - 5x^2y^2 + 4y^4 + x^2 - y^2 + x + y + 1 = 0$$

and its asymptotes lie on a rectangular hyperbola. **(Meerut, 1992)**

Solution:

The equation of the given curve is

$$x^4 - 5x^2y^2 + 4y^4 + x^2 - y^2 + x + y + 1 = 0 \qquad ...(1)$$

or $(x^2 - y^2)(x^2 - 4y^2) + x^2 - y^2 + x + y + 1 = 0$

or $(x - y)(x + y)(x - 2y)(x + 2y) + x^2 - y^2 + x + y + 1 = 0.$

$\therefore$ by inspection, the combined equation of the asymptotes of (1) is

$$(x - y)(x + y)(x - 2y)(x + 2y) = 0$$

or $$x^4 - 6x^2y^2 + 4y^4 = 0. \qquad ...(2)$$

Now each asymptote of (1) will cut it in 4 – 2 *i.e.*, 2 points. Therefore, the four asymptotes will cut it in 4 × 2 *i.e.*, 8 points.

Subtracting (2) from (1), we get

$$x^2 - y^2 + x + y + 1 = 0. \qquad ...(3)$$

The curve (3) passes through the eight points of intersection of (1) and (2). Also the conic (3) is a rectangular hyperbola because in its equation the sum of the coefficients of x^2 and y^2 is zero.

Hence the eight points of intersection of (1) and (2) lie on a rectangular hyperbola.

Example 5:

Show that asymptotes of the cubic

$$x^3 - 2y^3 + xy(2x - y) + y(x - y) + 1 = 0$$

cut the curve again in three points which lie on the straight line

$$x - y + 1 = 0. \qquad \textbf{(Meerut, 1992 P, 98P)}$$

Solution:

The given curve is

$$x^3 - 2y^3 + xy(2x - y) + y(x - y) + 1 = 0. \qquad ...(1)$$

$\therefore$ $\phi_3(m) = 1 - 2m^3 + 2m - m^2 = 0$ gives $m = 1, -1, -\frac{1}{2}$.

Also $\phi_3'(m) - 6m^2 + 2 - 2m$ and $\phi_2(m) = m - m^2$.

$$\therefore \qquad c = -\frac{\phi_2(m)}{\phi_3'} = -\frac{(m - m^2)}{(2 - 2m - 6m^2)}.$$

When $m = 1$, $c = 0$; when $m = -1$, $c = -1$;

and when $m = -\frac{1}{2}$, $c = \frac{1}{2}$.

$\therefore$ the asymptotes of (1) are $y = x$, $y = -x - 1$

and $y = -\frac{1}{2}x + \frac{1}{2}$.

Hence the combined equation of the asymptotes of (1) is

$$(x - y)(x + y + 1)(x + 2y - 1) = 0$$

or $$x^3 - 2y^3 + 2x^2y - xy^2 + xy - y^2 - x + y = 0.$$

Subtracting (2) form (1), we get

$$x - y + 1 = 0,$$

which shows that the points of intersection of the curve and its asymptotes lie on the straight line $x - y + 1 = 0$.

Also the three asymptotes cut the cubic in $n(n - 2)$, *i.e.*, $3(3 - 2) = 3$ points. As shown above, these, three points lie on the straight line

$$x - y + 1 = 0.$$

Example 6:

A tree trunk l feet long is in the shape of frustum of a cone, the radii of its ends being a and b feet (a > b). It is required to cut from it a beam of uniform square section. Prove that the beam of the greatest volume that can be cut is al/{3(a – b)} feet long.

Solution:

Let x be a side of the square base and y e the length BC of the beam to be cut. Then

$$OB = \frac{1}{2}(\text{diagonal of the base}) = x/\sqrt{2}.$$

In the figure AB and CD are the parallel diagonals of the ends of the beam to be cut. Now from the figure we have

$$\frac{MQ}{MR} = \frac{BQ}{BC}$$

or $$\frac{OQ - OM}{l} = \frac{OQ - OB}{y}$$

or $$\frac{a - b}{l} = \frac{a - (x/\sqrt{2})}{y}$$

$\therefore$ $$y = \frac{l}{a - b}\left(a - \frac{x}{\sqrt{2}}\right). \quad ...(1)$$

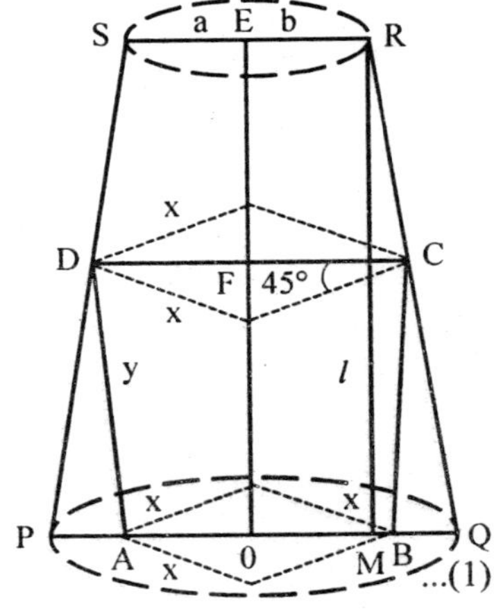

Fig. 10.2

Now, the volume V of the beam is given by

$$V = x^2y = x^2.\frac{1}{a-b}\left(a-\frac{x}{\sqrt{2}}\right), \quad \text{from (1).}$$

For a maximum or a minimum of V, we have

$$\frac{dV}{dx}=\frac{1}{a-b}\left[x^2\left(-\frac{1}{\sqrt{2}}\right)+\left(a-\frac{x}{\sqrt{2}}\right)2x\right]=0$$

or $x\left(2a-\sqrt{2}x-x\sqrt{2}\right)=0$. Therefore, x = 0 or $x=2\sqrt{2}a\sqrt{3}\cdot$.

Obviously x = 0 is not admissible.

Also $$\frac{d^2V}{dx^2}=\frac{1}{a-b}\left[2a-2\sqrt{2}x-\sqrt{2x}\right]=\frac{1}{a-b}\left(2a-3\sqrt{2x}\right)$$

$$= -\text{ ive at } x = \frac{2\sqrt{2}}{3}a.$$

Hence V is maximum when $x=\left(2\sqrt{2a}\right)/3$. The $x=\left(2\sqrt{2a}\right)/3$. Then from (1),

$$y=\frac{l}{a-b}\left[a-\frac{2a}{3}\right]=\frac{al}{3(a-b)} \text{ feet.}$$

Example 7:

A thin closed rectangular box is to have one rectangular edge n times the length of another edge and the volume of the box is given to be S. Prove that the least surface S is given by

$$nS^3 = 54(n+1)^2 V^2.$$

Solution:

Let the lengths of the edges be x, nx and y. Then V = the volume of the rectangular box = x.nx.y = nx^2y = constant. ...(1)

Differentiating (1) w.r.t. x, we get

$$0=nx^2\left(\frac{dy}{dx}\right)+2nxy \text{ of } \left(\frac{dy}{dx}\right)=-\frac{2y}{x}. \quad ...(2)$$

Now S = the surface of the box = $2nx^2 + 2(1 + n)xy$. ...(3)

Here S is a function of x and y. But y is connected with x by (1). So we can regard S as a function of x only. For a maximum or a minimum of S, we have $\frac{dS}{dx} = 0$.

Now $$\frac{dS}{dx}=4nx+2(1+n)\left(x\frac{dy}{dx}+y\right)=4nx-2(1+n)y, \text{ from (2).}$$

$\therefore \qquad \frac{dS}{dx} = 0$ gives $2nx = (1 + n)y$.

Also $\quad \frac{d^2S}{dx^2} = 4n - 2(1+n)\frac{dy}{dx} = 4n + \{4(1 + n)y\}/x$, from (2) = + ive.

Therefore, S is least when $2nx = (1 + n)y$. For this least S, on putting $y = 2nx/(1 + n)$ in (1) and (3), we get

$$V = \frac{2n^2x^3}{1+n} \text{ and } S = 2nx^2 + 2(1 + n)x.\frac{2nx}{1+n} = 6nx^2.$$

$\therefore \qquad S^3 = 6^3n^3x^6 = 6^3n^3[(1 + n)^2V^2/(4n^4)]$.

Hence $\qquad nS^3 = 54(n + 1)^2V^2$, when S is least.

Example 8:

One corner of a long rectangular sheet of paper of width 1 foot is folded so as to reach the opposite edge of the sheet. Find the minimum length of the crease.

Solution:

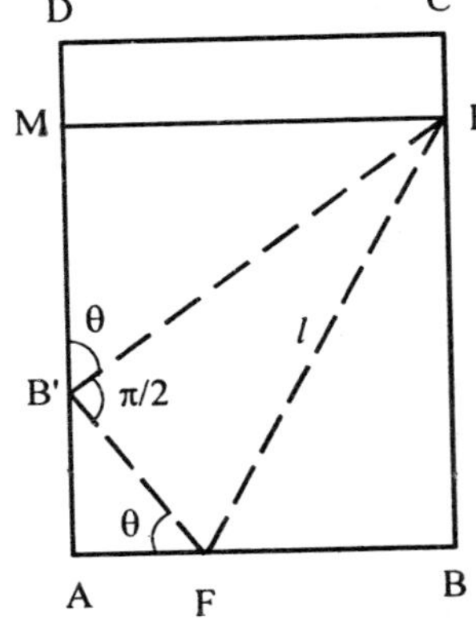

Fig. 10.3

Let the corner B of the rectangular sheet ABCD of width 1 foot be folded along EF so as to reach the opposite edge AD at B'. Let l feet the length of the crease EF.

Also, let BF = x. Then B'F = BF = x; AF = AB − BF = 1 − x.

Let $\angle AFB' = \theta$; then $\angle EB'M = \theta$, where EM is perpendicular to AD

Now $\cos\theta = \frac{AF}{B'F} = \frac{1-x}{x}$. $\qquad$ ...(1)

From $\Delta B'ME.B'E = ME \operatorname{cosec}\theta$

$$= \operatorname{cosec}\theta = 1\sqrt{(1-\cos^2\theta)} = x/\sqrt{(2x-1)}, \quad \text{from (1).}$$

From the right angled triangle EB'F, we have

$$EF^2 = EB'^2 + B'F^2 = \frac{x^2}{(2x-1)} + x^2 = \frac{2x^3}{2x-1}$$

or $\quad l^2 = \frac{2x^3}{2x-1}$. $\qquad$ ...(2)

Now l is maximum or minimum when l^2 is maximum or minimum.

Let l^2 = u. Then $u = \frac{2x^3}{2x-1}$. Therefore

$$\frac{du}{dx} = \frac{6x^2(2x-1)-4x^3}{(2x-1)^2} = \frac{2x(4x-3)}{(2x-1)} = \frac{8x^2}{(2x-1)^2}\left(x-\frac{3}{4}\right).$$

For a maximum of a minimum of u.

$$\frac{du}{dx} = 0 \text{ } i.e., \frac{8x^2}{(2x-1)}\left(x-\frac{3}{4}\right) = 0,$$

i.e.,; $x = \frac{3}{4}$, since x = 0 is not possible if there is to be a folding.

Now $$\frac{d^2u}{dx^2} = \left(x-\frac{3}{4}\right)\frac{d}{dx}\left\{\frac{8x^2}{(2x-1)^2}\right\} + \frac{8x^2}{(2x-1)^2}.1$$

= + ive, when $x = \frac{3}{4}$. Therefore u is minimum, and hence l is minimum, when $x = \frac{3}{4}$. Putting $x = \frac{3}{4}$ in (2), we get

$$l^2 = \frac{2.\left(\frac{27}{64}\right)}{\frac{1}{2}} = \frac{27}{16}.$$

Therefore, $l = \left(3\sqrt{3}\right)/4$ feet is the minimum length of the crease.

Example 9:

Show that the cone of greatest volume which can be inscribed in a given sphere is such that three its altitude is twice the diameter of the sphere.

(Agra, 1998; Gorakhpur, 96; Lucknow, 95)

Solution:

Let x be the radius of the base and y be the height of a cone inscribed in a given sphere of radius a.

Let V be the volume of the cone.

Then $$V = \frac{1}{3}\pi x^2 y. \qquad ...(1)$$

From figure, we have

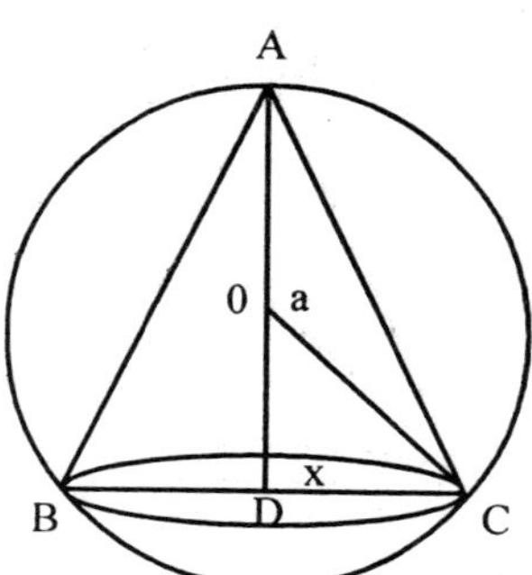

Fig. 10.4

$$OD = AD - AO = y - a.$$

$\therefore$ $c^2 = (y - a)^2 + x^2$

or $x^2 = a^2 - (y - a)^2$

$$= a^2 - y^2 + 2ay - a^2$$

$$= y(2a - y).$$

Putting the value of x^2 in (1), we get

$$V = \frac{1}{3}\pi y^2 (2a - y).$$

$$\frac{dV}{dy} = \frac{1}{3}\pi\left[2y(2a - y) - y^2\right]$$

$$= \frac{1}{3}\pi y(4a - 3y).$$

For a maximum or minimum of V, $\frac{dV}{dy} = 0$

i.e., $y(4a - 3y) = 0$, *i.e.*, $y = \left(\frac{2}{3}\right) = (2a)$, since $y = 0$ is inadmissible.

Also $\frac{d^2V}{dy^2} = \frac{1}{3}\pi(4a - 3y) + \frac{1}{3}\pi y(-3)$

$$= - \text{ ive, when } y = \frac{4a}{3}.$$

Therefore V is maximum when $y = \frac{2}{3}(2a)$,

i.e., $3y = 2(2a)$.

Example 10:

Show that the maximum rectangle inscribed in a circle is a square.

(Kanpur, 1999; Vikram, 94; Ranchi, 96; Meerut, 97; Agra, 93; Gorakhpur, 99)

Solution:

Let x and y be the sides of the rectangle inscribed in a circle of diameter a.

Then $x + y^2 = a^2.$ ...(1)

Let A be the area of the rectangle.

Then $A = xy = x(a^2 - x^2)^{1/2}.$

Since A is positive, therefore, A is maximum or minimum according as A^2 is maximum or minimum.

Let $u = A^2 = x^2(a^2 - x^2) = x^2a^2 - x^4.$

Then $\quad \frac{du}{dx} = 2xa^2 - 4x^3.$

For a maximum or a minimum of u, $\frac{du}{dx} = 0$

i.e., $\quad 2xa^2 - 4x^3 = 0$ *i.e.*, $2x(a^2 - 2x^2) = 0$

i.e., $\quad x = a/\sqrt{2}$, since $x \neq 0$ and x cannot be – ive.

Now $\quad \frac{d^2u}{dx^2} = 2a^2 - 12x^2.$

When $\quad x = a/\sqrt{2}, \frac{d^2u}{dx^2} = 2a^2 - \frac{1}{2} \times 12a^2 = -4a^2 = -\text{ive}.$

Hence u is maximum when $x = a/\sqrt{2}$.

When $x = a/\sqrt{2}$, we have from (1), $y = a/\sqrt{2}$ showing that the maximum rectangle inscribed in a circle is a square.

Example 11:

Find the coordinates of the point on the parabola $y = x^2$ which is nearest to the point (3, 0).

Solution:

Let (x, y) be any point on the parabola $y = x^2$. ...(1)

Let s be the distance of (x, y) from (3, 0).

Then $s^2 = (x - 3)^2 + (y - 0)^2 = (x - 3)^2 + y^2 = (x - 3)^2 + x^4$, [from (1)].

No s is maximum or minimum according as s^2 is maximum or minimum.

Let $u = s^2 = (x - 3)^2 + x^4$. Then $\frac{du}{dx} = 2(x - 3) + 4x^3$. For a maximum or a minimum of u, $\frac{du}{dx} = 0$ *i.e.*,

$$2(x - 3) + 4x^3 = 0 \text{ i.e., } 2x^3 + x - 3 = 0 \text{ giving } x = 1.$$

Now $\frac{d^2u}{dx^2} = 2 + 12x^2 = +\text{ive}$, when $x = 1$. Hence u is minimum when $x = 1$. When $x = 1$, we have from (1), $y = 1$. Thus the required point is (1, 1).

Example 12:

Show that the semi-vertical angle of the cone of maximum volume of given slant height is $\tan^{-1}\sqrt{2}$.

(Meerut, 1997, 93, 98; Agra, 96; Allahabad, 97)

Solution:

Let α be the semi-vertical angle of the cone and OA = l be slant height.

Then V = volume of the cone

$$= \frac{1}{3}\pi(l \sin \alpha)^2 (l \cos \alpha)$$

$$= \frac{1}{3}\pi l^3 \sin^2 \alpha \cos \alpha.$$

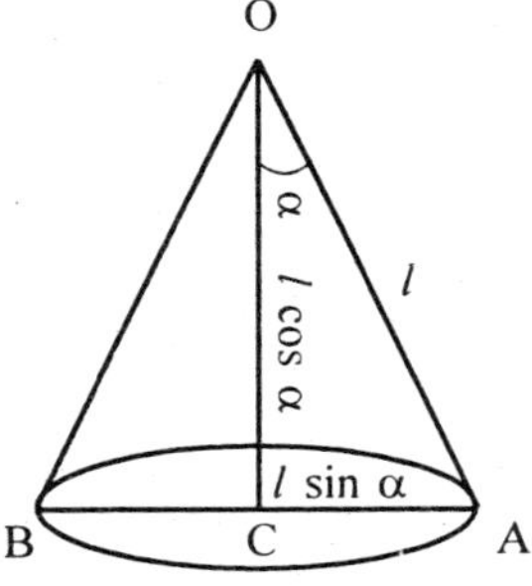

Fig. 10.5

Since l is given, therefore V is a function of α. We have

$$\frac{dV}{d\alpha} = \frac{1}{3}\pi l^3 (2 \sin \alpha \cos^2 \alpha - \sin^3 \alpha)$$

$$= \frac{1}{3} \pi l^3 \sin \alpha (2 \cos^2 \alpha - \sin^2 \alpha).$$

For a maximum or minimum of V,

$$\frac{dV}{d\alpha} = 0$$

i.e., $\sin \alpha (2 \cos^2 \alpha - \sin^2 \alpha) = 0$ *i.e.*, $\sin \alpha = 0$ or $\tan^2 \alpha = 2$

i.e., $\alpha = \tan^{-1} \sqrt{2}$, because the values $\alpha = 0$ and= $\alpha = -\sqrt{2}$ are inadmissible.

Now $$\frac{d^2V}{d\alpha^2} = \frac{1}{3} \pi l^3 \cos \alpha (2 \cos^2 \alpha - \sin^2 \alpha)$$

$$+ \frac{1}{3} \pi l^3 \sin \alpha (- 4 \cos \alpha \sin \alpha - 2 \sin a \cos \alpha)$$

$$< 0, \text{ when } \tan \alpha = \sqrt{2} \text{ i.e., when } 2 \cos^2 \alpha = \sin^2 \alpha.$$

Hence V is maximum when $\alpha = \tan^{-1} \sqrt{2}$.

Example 13:

Show that the volume of the greatest cylinder which can be inscribed in a cone of height h and semi-vertical angle a is (4/27) p1h³ tan² α.

(Meerut, 1992, 97; Lucknow, 91, 99; Agra, 96; Allahabad, 94; Gorakhpur, 94, 92; Andhra, 91; Magadh, 98; Garhwal, 97)

Solution:

Let PQRS be a cylinder of radius x inscribed in the cone ABC of height h and semi-vertical angle α.

From the figure, AM = $x \cot \alpha$.

$\therefore$ height of the cylinder = $h - x \cot \alpha$ and V = volume of the cylinder = $\pi x^2(h - x \cot \alpha)$.

For a maximum or a minimum value of V, $\frac{dV}{dx} = 0$.

Now $\frac{dV}{dx} = \pi\{x^2(-\cot\alpha)+(h-x\cot\alpha)2x\}=0$

gives $x(-3x\cot\alpha + 2h) = 0$

i.e., $x = 0$ or $\left(\frac{2h}{3}\right)\tan\alpha$.

Also $\frac{d^2V}{dx^2} = \pi(2h - 6x\cot\alpha)$,

which is –ive at $x = \frac{2h}{3}\tan\alpha$.

Hence V is maximum when

$$x = \left(\frac{2h}{3}\right) = \tan\alpha.$$

Also from (1), the maximum value of V

$$= \pi.\frac{4h^2}{9}.(\tan^2\alpha)\left(h - \frac{2h}{3}\right) = \frac{4\pi h^3}{27}\tan^2\alpha.$$

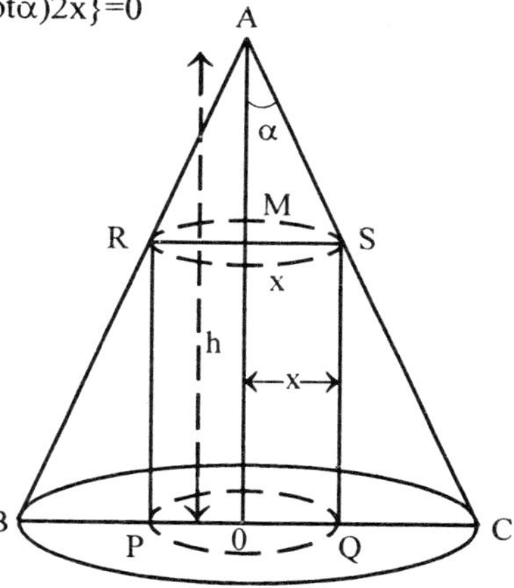

Fig. 10.6

Example 14:

A piece of wire of length l is cut into two parts, one of which is bent in the shape of a circle and the other into the shape of a square. How should the wire be cut so that the sum of the areas of the circle and the square is minimum? **(Meerut, 1993)**

Solution:

Let the lengths of the two parts of the wire be x and y.

Then $x = y = l$. ...(1)

Let the part of length x be bent in the shape of a circle, say of radius r. Then x = the circumference of the circle = $2\pi r$, so that $r = \frac{x}{(2\pi)}$.

The area of the circle = $\pi r^2 = \pi\left(\frac{x^2}{4\pi^2}\right) = \frac{x^2}{4\pi}$.

Further suppose of the areas of the circle and the square.

Then $A = \frac{x^2}{4\pi} + \frac{y^2}{16} = \frac{x^2}{4\pi} + \frac{(l-x)^2}{16}$, [from (1)].

We have $\frac{dA}{dx} = \frac{2x}{4\pi} - \frac{2(l-x)}{16} = \frac{x}{2\pi} - \frac{l}{8} + \frac{x}{8}$.

For a maximum or minimum of A, $\frac{dA}{dx} = 0$ *i.e.*,

$$\frac{x}{8}\left(1+\frac{4}{\pi}\right) - \frac{l}{8} = 0,\ x = \frac{\pi l}{\pi+4}.$$

Now $\frac{d^2A}{dx^2} = \frac{1}{2\pi} + \frac{1}{8} = +\text{ive for } x = \frac{\pi l}{x+4}$.

Hence A is minimum when $x = \frac{\pi l}{(\pi+4)}$.

Putting $x = \frac{\pi l}{\pi+4}$ in (1), we get $y = l - \frac{\pi l}{\pi+4} = \frac{4l}{\pi+4}$.

Hence the wire should be cut in length $\frac{\pi l}{(\pi+4)}$ to be bent in the shape of a circle and the length $\frac{4l}{(\pi+4)}$ to be bent in the shape of a square.

Example 15:

Show that the radius of the right circular cylinder of greatest curved surface which can be inscribed in a given cone is half that of the cone.

Solution:

Here let r be the radius and α the semi-vertical angle of the given cone. Then the height AO of the cone = r cot α.

Let PQRS be a cylinder of radius x inscribed in the cone ABC

From the figure, AM = x cot α.

$\therefore$ the height MO of the cylinder PQRS = r cot α – x cot α.

If S be the curved surface of the cylinder PQRS, then

$$S = 2\pi x(r \cot \alpha - x \cot \alpha) = 2\pi \cot \alpha\ (xr - x^2).$$

We have $\frac{dS}{dx} = 2\pi \cot \alpha\ (r - 2x)$.

For a maximum or a minimum value of S, we must have

$$\frac{dS}{dx} = 0 \text{ i.e., } 2\pi \cot \alpha\ (r - 2x) = 0$$

i.e., r – 2x = 0 *i.e.*, $x = \frac{r}{2}$.

Also $\dfrac{d^2S}{dx^2} = 2\pi \cot \alpha . (-2)$, which is negative.

$\therefore$ S is maximum when $x = \dfrac{r}{2}$ *i.e.*, when the radius of the cylinder is half that of the cone.

Example 16:

The three sides of a trapezium are equal, each being 6 centimetres long. Find the area of the trapezium when it is maximum. **(Agra, 1991)**

Solution:

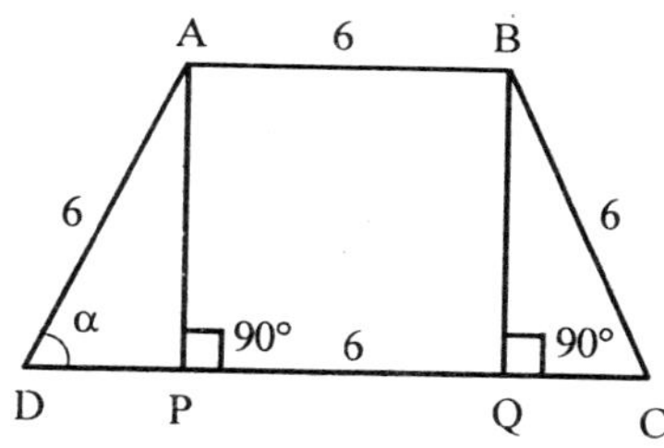

Fig. 10.7

From figure,

$AP = B = 6 \sin \alpha$ and

$DP = QC = 6 \cos \alpha$.

$\therefore \quad CD = 6 + 12 \cos \alpha$.

If S is the area of the trapezium in square centimetres, then

$$S = \frac{1}{2}(AB + CD)\, BQ$$

$$= \frac{1}{2}(6 + 6 + 12 \cos \alpha)\, 6 \sin \alpha$$

$$= (36\, \alpha + \sin \alpha \cos \alpha)$$

$$= 36\left(\sin \alpha + \frac{1}{2}\sin 2\alpha\right).$$

We have $\left(\dfrac{dS}{d\alpha}\right) = 36(\cos \alpha + \cos 2\alpha = 72 \cos\left(\dfrac{3\alpha}{2}\right) \cos\left(\dfrac{\alpha}{2}\right).$

For a maximum or a minimum of S,

$$\frac{ds}{d\alpha} = 0 \text{ i.e., } \cos\left(\frac{3\alpha}{2}\right)\cos\left(\frac{\alpha}{2}\right) = 0$$

i.e., $\dfrac{3\alpha}{2} = \dfrac{\pi}{2}$ or $\dfrac{\alpha}{2} = \dfrac{\pi}{2}$ *i.e.*, $\alpha = \dfrac{\pi}{3}$ or $\alpha = \pi$.

But $\alpha = \pi$ is inadmissible.

Now $\dfrac{d^2S}{d\alpha^2} = 36(-\sin \alpha - 2 \sin 2\alpha) = -$ ive when $\alpha = \dfrac{\pi}{3}$.

$\therefore$ S is maximum when $\alpha = \dfrac{\pi}{3}$. Also the maximum value of S is $36\left(\sin\dfrac{\pi}{3} + \dfrac{1}{2}\sin\dfrac{2\pi}{3}\right) = 27\sqrt{3}$ square centimeters.

Example 17:

A rectangular sheet of metal has four equal square portions removed at the corners and the sides are then turned up so as to form an open rectangular box. Show when the volume contained in the box is a maximum, the depth will be

$$\frac{1}{6}\left\{(a+b)-\sqrt{\left(a^2-ab+b^2\right)}\right\},$$

where a, b are sides of the original rectangle.

Solution:

Let x be the length of the side of each of the square portions removed. Then the length of the box = a – 2x, the breadth = b – 2x, and the depth = x. If V is the volume of the box, then

$$V = x(a-2x)(b-2x),\ (a > b)$$

or $$V = 4x^3 - 2x^2(a+b) + abx.$$

For a maximum of V, we have

$$\frac{dV}{dx} = 12x^2 - 4(a+b)x + ab = 0$$

or $$x = \frac{\left[4(a+b) \pm \sqrt{\left\{16(a+b)^2 - 48ab\right\}}\right]}{24}$$

$$= \frac{1}{6}\left\{(a+b) \pm \sqrt{\left(a^2-ab+b^2\right)}\right\}.$$

Now the +ive sign before the radical is not admissible, because then the value of x becomes $> \frac{1}{2}b$. Obviously the other value of x obtained by taking the –ive sign before the radical is less than $\frac{b}{2}$.

Now $$\frac{d^2V}{dx^2} = 24x - 4(a+b)$$

$$= 4\left\{(a+b) - \sqrt{\left(a^2-ab+b^2\right)}\right\} - 4(a+b),$$

when $$x = \frac{1}{6}\left\{(a+b) - \sqrt{\left(a^2-ab+b^2\right)}\right\}$$

$$= -4\sqrt{\left(a^2-ab+b^2\right)},$$ which is –ive.

Hence Vis maximum for this value of x.

Example 18:

Show that the triangle of maximum area that can be inscribed in a circle of radius a is an equilateral triangle. **(Meerut, 1992 S)**

Solution:

First we observe that among all the inscribed triangles having AB as base, the area of that triangle is greatest for which the altitude of C w.r.t. AB is greatest. Evidently such a triangle is an isosceles triangle. Let θ be the semi-vertical angle of such a triangle ABC inscribed in a given circle of radius a.

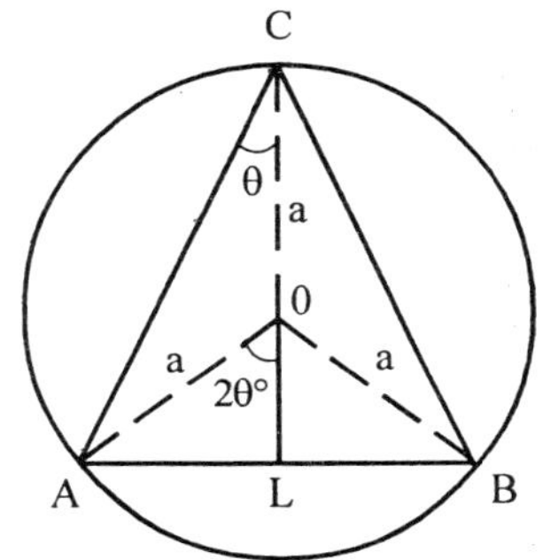

Fig. 10.8

From fig., S = area of the triangle ABC = 2ΔAOC + ΔAOB

$$= 2.\frac{1}{2}a^2 \sin(\pi - 2\theta) + \frac{1}{2}a^2 \sin 4\theta$$

$$= a^2 \sin 2\theta + \frac{1}{2}a^2 \sin 4\theta.$$

$$\therefore \qquad \frac{dS}{d\theta} = 2a^2 \cos 2\theta + 2a^2 \cos 4\theta.$$

For a maximum or minimum of S, we have $\frac{dS}{d\theta} = 0$

i.e., $\cos 2\theta + \cos 4\theta = 0$, *i.e.*, $2 \cos 3\theta \cos \theta = 0$

i.e., $3\theta = \frac{\pi}{2}$ or $\theta = \frac{\pi}{2}$, *i.e.*, $\theta = \frac{\pi}{6}$ or $\theta = \frac{\pi}{2}$.

But $\theta = \frac{\pi}{2}$ is not admissible. Hence $\theta = \frac{\pi}{6}$.

Also, $\frac{d^2S}{d\theta^2} = -\, 4a^2 \sin 2\theta - 8a^2 \sin 4\theta = -$ ive for $\theta = \frac{\pi}{6}$.

Hence area is maximum when $\theta = \frac{\pi}{6}$ or $2\theta = \frac{\pi}{3}$, *i.e.*, when the triangle is equilateral.

Example 19:

Prove that the least perimeter of an isosceles triangle in which a circle of radius r can be inscribed is $6r\sqrt{3}$.

Solution:

Let θ be the semi-vertical angle of an isosceles triangle in which a circle of radius r can be inscribed.

Then $AQ = r \cot\theta$

and $OA = r \operatorname{cosec}\theta$.

$\therefore$ $AP = r + r\operatorname{cosec}\theta = r(1 + \operatorname{cosec}\theta)$, and $BP = AP\tan\theta$

$= r(1 + \operatorname{cosec}\theta)\tan\theta = r(\tan\theta + \sec\theta)$.

$s = 2AQ + 4PB$ [Note that $AQ = AR$, $BQ = BP = CP = CR$]

$= 2r\cot\theta + 4r(\tan\theta + \sec\theta) = 2r(\cot\theta + 2\tan\theta + 2\sec\theta)$.

$$\therefore \quad \frac{ds}{d\theta} = 2r(-\operatorname{cosec}^2\theta + 2\sec^2\theta + 2\sec\theta\tan\theta).$$

For s to be a maximum or a minimum, we have $\frac{ds}{d\theta} = 0$

i.e., $-\operatorname{cosec}^2\theta + 2\sec^2\theta + 2\sec\theta\tan\theta = 0$

$$i.e., \quad -\frac{1}{\sin^2\theta} + \frac{2}{\cos^2\theta} + \frac{2\sin\theta}{\cos^2\theta} = 0$$

or $2\sin^3\theta + 2\sin^2\theta - \cos^2\theta = 0$

or $2\sin^3\theta + 3\sin^2\theta - 1 = 0$.

Solving this equation, we get

$$\sin\theta = -1 \text{ or } \sin\theta = \frac{1}{2}.$$

But $\sin\theta = -1$ is inadmissible.

Therefore, $\sin\theta = \frac{1}{2}$ or $\theta = 30°$.

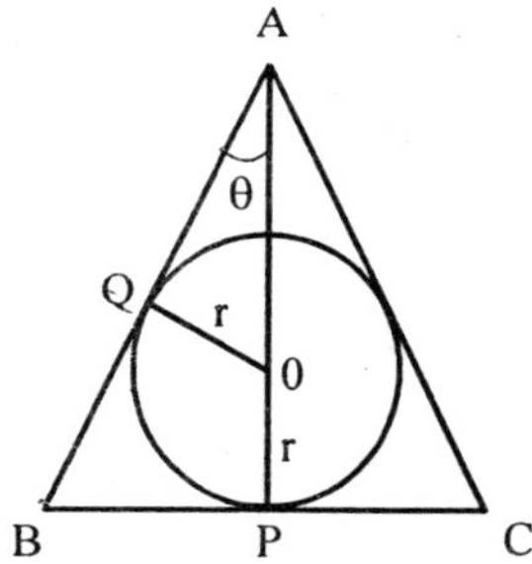

Fig. 10.9

$$\text{Also} \quad \frac{d^2s}{d\theta^2} = 2r[2\operatorname{cosec}^2\theta\cot\theta + 4\sec^2\theta\tan\theta + 2\sec^3\theta + 2\sec\theta\tan^2\theta]$$

$= +$ ive for $\theta = 30°$.

Hence $\theta = 30°$ gives least perimeter, which is

$$= 2r\left[\sqrt{3} + \frac{2}{\sqrt{3}} + 2.\frac{2}{\sqrt{3}}\right] = \frac{18r}{\sqrt{3}} = 6r\sqrt{3}.$$

Example 20:

N is the foot of the perpendicular drawn from the centre O onto the tangent at a variable point P on the ellipse

$$\frac{x^2}{a^2}+\frac{y^2}{b^2}=1,\ (a>b).$$

Find the maximum length of PN and also the maximum area of the triangle OPN.

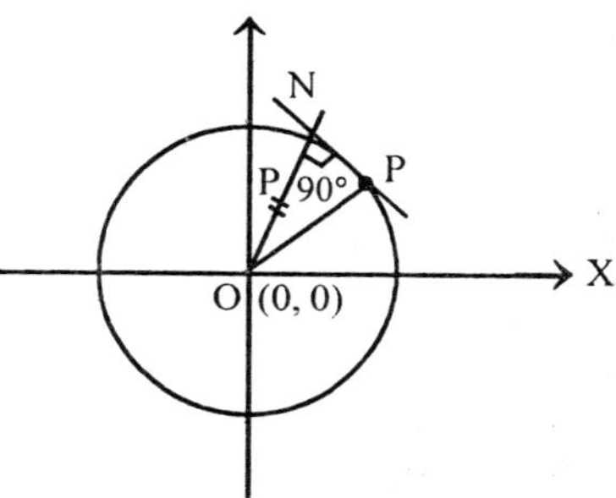

Fig. 10.10

Solution:

Let P be the point (a cos t, b sin t). Then the equation of the tangent at P to the given ellipse is

$$\frac{x}{a}\cos t+\frac{y}{b}\sin t=1ab$$

or bx cos t + ay sin t = ab.

Let ON = p.

Then $$p=\frac{ab}{\sqrt{\left(b^2\cos^2 t+a^2\sin^2 t\right)}}.$$

$\therefore$ $\dfrac{a^2b^2}{p^2}$ = $b^2\cos^2 t + a^2\sin^2 t = b^2(1-\sin^2 t)+a^2(1-\cos^2 t)$

$= a^2 + b^2 - (a^2\cos^2 t + b^2\sin^2 t) = a^2 + b^2 - OP^2$, because $OP^2 = a^2\cos^2 t + b^2\sin^2 t$.

$$\therefore \qquad OP^2 = a^2 + b^2 - \left(\frac{a^2b^2}{p^2}\right).$$

We have

$$PN^2 = OP^2 - ON^2 = a^2 + b^2 - \left(\frac{a^2b^2}{p^2}\right)-p^2=z, \text{ say} \qquad ...(1)$$

Now, PN is maximum or minimum according as PN^2 is maximum or minimum. But PN^2 *i.e.*, z is a function of p. For a maximum or a minimum of z, we have

$$\frac{dz}{dp}\equiv\frac{2a^2b^2}{p^3}-2p=0 \text{ i.e., } p^4 = a^2b^2 \text{ i.e., } p^2 = ab.$$

Also $\dfrac{d^2z}{dp^2}=-\dfrac{6a^2b^2}{p^4}-2=-\text{ive}$, when $p^2 = ab$. Therefore z is maximum when $p^2 = ab$. Putting $p^2 = ab$ in (1), we get

$$PN^2 = a^2 + b^2 - \frac{a^2b^2}{ab}-ab=(a-b)^2; \qquad \therefore\ PN = (a-b).$$

Hence the maximum length of PN = a – b.

Now let S be the area of the triangle OPN.

Then $S = \frac{1}{2}ON.PN = \frac{1}{2}p.PN$. Therefore

$$S^2 = \frac{1}{4}p^2.PN^2 = \frac{1}{4}p^2\left(a^2 + b^2 - \frac{a^2b^2}{p^2} - p^2\right), \text{ from (1).}$$

Let $u = S^2 = \frac{1}{4}\left\{p^2\left(a^2 + b^2\right) - a^2b^2 - p^4\right\}$.

Now S is maximum or minimum according as S^2, *i.e.*, u is maximum or minimum. For a maximum or minimum of u, we have

$$\frac{du}{dp} \equiv \frac{1}{4}\left\{2p\left(a^2 + b^2\right) - 4p^3\right\} = 0, \text{ i.e., } p^2 = \frac{a^2 + b^2}{2}, \text{ since } p \neq 0.$$

Also $\frac{d^2u}{dp^2} = \frac{1}{4}\left\{2\left(a^2 + b^2\right) - 12p^2\right\} = -\text{ive}$, when $p^2 = \frac{a^2 + b^2}{2}$.

Hence u is maximum when $p^2 = \frac{\left(a^2 + b^2\right)}{2}$. Putting $p^2 = \frac{\left(a^2 + b^2\right)}{2}$ in the value of S^2, we get

$$S^2 = \frac{1}{4}\left\{\frac{1}{2}\left(a^2 + b^2\right)^2 - a^2b^2 - \frac{1}{4}\left(a^2 + b^2\right)^2\right\}$$

$$= \frac{1}{16}\left\{\left(a^2 + b^2\right)^2 - 4a^2b^2\right\} = \frac{1}{16}\left(a^2 - b^2\right)^2.$$

Hence the maximum area of the triangle $OPN = \frac{1}{4}\left(a^2 - b^2\right)$.

Example 21:

Investigate the maxima and minima of ax + by when xy = c^2.

(Meerut, 1990; Kanpur, 95)

Solution:

Let z = ax + by ...(1)

Given xy = c^2 or $y = \frac{c^2}{x}$. Substituting this value of y in (1), we get

$$z = \frac{ax + bc^2}{x}.$$

Now for Max. or Min. values of z, we must have $\frac{dz}{dx} = 0$.

$$\therefore \qquad \frac{dz}{dx} = \frac{a - bc^2}{x^2} = 0$$

or $$a = \frac{bc^2}{x^2}$$

or $$x^2 = \frac{bc^2}{a}$$

or $$x = \pm c\sqrt{\left(\frac{b}{a}\right)}.$$

Again, $$\frac{d^2z}{dx^2} = \frac{2bc^2}{x^3} = +\text{ive for } x = c\sqrt{\left(\frac{b}{a}\right)}$$

and $$\frac{d^2z}{dx^2} = -\text{ive for } x = -c\sqrt{\left(\frac{b}{a}\right)}.$$

Hence z is maximum when $x = -c\sqrt{\left(\frac{b}{a}\right)}$ and minimum when $x = c\sqrt{\left(\frac{b}{a}\right)}$

$\therefore$ maximum value of $z = -ac\sqrt{\left(\frac{b}{as}\right)} - \frac{bc^2}{\left\{c\sqrt{\left(\frac{b}{a}\right)}\right\}}$

$$= -c\sqrt{(ab)} - c\sqrt{(ab)} = -2c\sqrt{(ab)}.$$

And minimum value of $z = ac\sqrt{\left(\frac{b}{a}\right)} + \frac{bc^2}{\left\{c\sqrt{\left(\frac{b}{a}\right)}\right\}}$

$$= c\sqrt{(ab)} + c\sqrt{(ab)} = 2c\sqrt{(ab)}.$$

Example 22:

Assuming that the petrol burnt in driving a motor boat varies as the cube of its velocity, show that the most economical speed when going against a current of c kilometers per hour is $\left(\frac{3}{2}\right)$ c kilometers per hour.

(Allahabad, 1999; Meerut, S-90; Agra, 98)

Solution:

Let the speed of the boat be v kilometers per hour and the distance travelled in 't' hours be a kilometers. When the boat is going against the current of c kilometers per hour, its resultant speed is (v – c) kilometers per hour.

$$\therefore \qquad t = \frac{\text{distance}}{\text{speed}} = \frac{a}{(v-c)} \text{ hours.}$$

Now the petrol burnt per hour is kv^3, where k is a constant. If x is the petrol burnt in t hours, then

$$x = \left\{\frac{a}{(v-c)}\right\}kv^3 = \frac{akv^3}{(v-c)}.$$

For the most economical speed, x should be minimum.

Now $$\frac{dx}{dv} = \frac{ak\left\{3v^2(v-c) - v^3\right\}}{(v-c)^2}$$

$$= \frac{ak\left(2v^3 - 3v^2c\right)}{(v-c)^2}$$

$$= ak.v^2(2v-3c).\frac{1}{(v-c)^2}.$$

For max. or min. value of x, $\frac{dx}{dv} = 0$

or $v^2(2v - 3c) = 0$

or $2v - 3c = 0,$ $[\because v \neq 0]$

or $v = \frac{3c}{2}.$

Also $$\frac{d^2x}{dv^2} = ak\left[-\frac{2}{(v-c)^3}.v^2(2v-3c) + \frac{1}{(v-c)^2}.6v(v-c)\right].$$

When $$v = \frac{3c}{2}, \frac{d^2x}{dv^2} = ak.\frac{b.\left(\frac{3c}{2}\right)}{\left(\frac{3c}{2}\right) - c} = 18ak \text{ i.e., +ive.}$$

Hence x is minimum when $v = \frac{3c}{2}$ *i.e.*, the most economical speed is $\frac{c}{2}$ kilometers per hour.

Example 23:

Prove that a conical tent of given capacity will require the least amount of canvas when the height is $\sqrt{2}$ times the radius of the base.

(Meerut, 1993, 99P; Jhansi, 98; Agra, 92; Bundelkhand, 98; Gorakhpur, 97, 95)

Solution:

Let r be the radius and h be the height of the conical tent. Since the capacity of the tent is given, therefore its volume

$$V = \frac{1}{3}\pi r^2 h = \text{constant.} \qquad \ldots(1)$$

The amount of canvas required for the tent = the curved surface of the cone = S, say.

We have, $S = \pi r\sqrt{(r^2 + h^2)}$. Therefore $S^2 = \pi^2 r^2(r^2 + h^2)$

or $$S^2 = \pi^2 r^2 \left\{ r^2 + \frac{9V^2}{(\pi^2 r^4)} \right\}, \qquad \left[\because \text{ from (1), } h = \frac{3V}{(\pi r^2)}\right]$$

$$= \pi^2 r^4 + \frac{9V^2}{r^2} = z, \text{ say.}$$

Now S is maximum or minimum according as S^2 *i.e.*, z is maximum or minimum. We have $\frac{dz}{dr} = 4\pi^2 r^3 - \frac{18V^2}{r3}$. For a maximum or minimum of z, we have $\frac{dz}{dr} = 0$,

$$i.e., \; 4\pi^2 r^3 - \frac{18V^2}{r^3} = 0 \; i.e., \; r^6 = \frac{9V^2}{2\pi^2}.$$

Now $\frac{d^2z}{dr^2} = 12\pi^2 r^2 + \frac{54V^2}{r^4}$ which is +ive when $r^6 = \frac{9V^2}{2\pi^2}$.

Hence z or S is minimum when $r^6 = \frac{9V^2}{2\pi^2}$ or $r^6 = \frac{\pi^2 r^4 h^2}{2\pi^2}$,

$[\because$ from (1), $3V = \pi r^2 h]$

or $$h^2 = 2r^2 \text{ or } h = r\sqrt{2}.$$

Example 24:

A gas-holder is a cylindrical vessel closed at the top and open at the bottom (which dips into water). What should be the ratio of the height to the diameter in order that for a given volume its construction may require the least amount of material?

Solution:

Let h be the height and 2r be the diameter of the cylindrical vessel. Since the capacity of the vessel is given, therefore its volume

$$V = \pi r^2 h = \text{constant}. \qquad \ldots(1)$$

The amount of material required for the construction of the vessel is least when its whole surface, say S, is least. We have

S = area of the top + area of the curved surface

$$= \pi r^2 + 2\pi rh = \pi r^2 + 2\pi r.\frac{V}{(\pi r^2)}, \text{ from (1)}$$

$$= \pi r^2 + \frac{2V}{r}.$$

For a maximum or minimum of S, we must have

$$\frac{dS}{dr} = 2\pi r - \frac{2V}{r^2} = 0 \quad i.e., \ \pi r^3 = V.$$

Now $$\frac{d^2S}{dr^2} = 2\pi + \frac{4V}{r^3} = +\text{ ive when } \pi r^3 = V.$$

Hence S is minimum when $\pi r^3 = V$

or $\pi r^2 h = \pi r^3$, [$\because$ from (1), $V = \pi r^2 h$]

or $h = r$ or $\dfrac{h}{r} = \dfrac{1}{2}$.

Example 25:

An ellipse is inscribed in an isoscetes triangle of height h and base 2k, and having one axis lying along the perpendicular from the vertex of the triangle to the base. Show that the maximum are a of the ellipse is $\sqrt{3}.\dfrac{\pi hk}{9}$.

Solution:

Let BCD be the given isosceles triangle whose base is BC, vertex D and height DA'. Then BC = 2k and DA' = h.

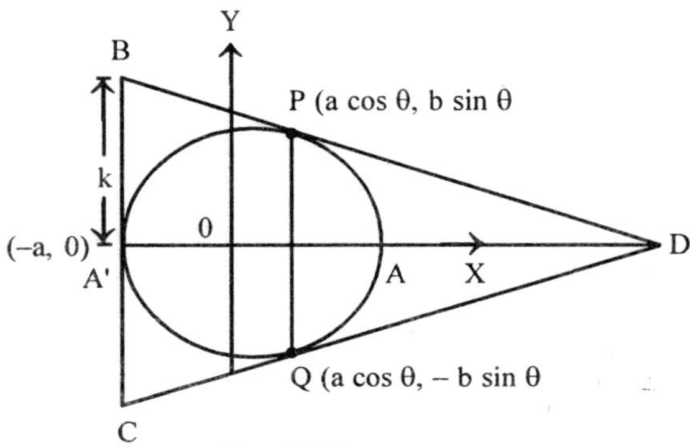

Fig. 10.11

An ellipse is inscribed in this triangle so that one axis of the ellipse is along A'A. Let A'A = 2a and let the length of the other axis of the ellipse be 2b. Let the equation of the ellipse be

$$\frac{x^2}{a^2}+\frac{y^2}{b^2}=1,$$

where the x-axis is along OA and the y-axis is perpendicular to OA.

Here O is the middle point of A'A.

The sides BD, CD and BC of the triangle touch the ellipse at the points P, Q and A' respectively.

Here PQ is a double ordinate.

Let P be the point (a cos θ, b sin θ).

Then the equation of the tangent to the ellipse at the point (a cos θ, b sin θ) is

$$\left(\frac{x}{a}\right)\cos\theta+\left(\frac{y}{b}\right)\sin\theta=1. \qquad ...(1)$$

The coordinates of the point B are (– a, k).

Since the point B lies on (1),

therefore $-\frac{a}{a}\cos\theta+\frac{k}{b}\sin\theta=1$ or $\frac{k}{b}\sin\theta=1+\cos\theta$

or $$b=\frac{k\sin\theta}{(1+\cos\theta)}. \qquad ...(2)$$

The coordinates of the point D are (h – a, 0).

[Note that DA' = h and A'O = a]

Since the point D also lies on (1),

therefore $\left\{\frac{(h-a)}{a}\right\}\cos\theta=1$ or $h\cos\theta-a\cos\theta=a$

or $$a=\frac{h\cos\theta}{(1+\cos\theta)} \qquad ...(3)$$

Let S be the area of the ellipse.

Then $S=\pi ab=\frac{\pi hk\sin\theta\cos\theta}{(1+\cos\theta)^2}$, from (2) and (3).

We have $\frac{dS}{d\theta}$

$$= \pi hk \frac{(\cos^2\theta - \sin^2\theta)(1+\cos\theta)^2 + 2\sin^2\theta\cos\theta(1+\cos\theta)}{(1+\cos\theta)^4}$$

$$= \pi hk \frac{(\cos^2\theta - 1)(1+\cos\theta)^2 + 2(1-\cos^2\theta)\cos\theta(1+\cos\theta)}{(1+\cos\theta)^4}$$

$$= \pi hk \frac{(1+\cos\theta)^2\{(2\cos^2\theta - 1) + 2\cos\theta(1-\cos\theta)\}}{(1+\cos\theta)^4}$$

$$= \pi hk \frac{2\cos\theta - 1}{(1+\cos\theta)^2}, \text{ since } \cos\theta \neq -1.$$

For a maximum or minimum of S, we have

$$\frac{dS}{d\theta} = 0 \text{ i.e., } 2\cos\theta - 1 = 0 \text{ i.e., } \cos\theta = \frac{1}{2} \text{ i.e., } \theta = \frac{\pi}{3}.$$

Now $$\frac{d^2S}{d\theta^2} = \pi hk(2\cos\theta - 1)\frac{d}{d\theta}\left\{\frac{1}{(1+\cos\theta)^2}\right\} + \frac{\pi hk}{(1+\cos\theta)^2}(-2\sin\theta),$$

which is negative when $\cos\theta = \frac{1}{2}$ or $\theta = \frac{1}{3}\pi$.

Hence S is maximum when $\theta = \frac{1}{3}\pi$.

The maximum area of the ellipse

$$= \pi hk \frac{\sin\left(\frac{\pi}{3}\right)\cos\left(\frac{\pi}{3}\right)}{\left\{1+\cos\left(\frac{\pi}{3}\right)\right\}^2}$$

$$= \pi hk \frac{\left(\frac{\sqrt{3}}{4}\right)}{\left(1+\frac{1}{2}\right)^2} = \frac{\sqrt{3}.\pi hk}{9}.$$

Example 26:

The velocity of waves of wavelength λ on deep water is proportional to $\sqrt{\left(\frac{\lambda}{a} + \frac{a}{\lambda}\right)}$, where a is a certain linear magnitude, prove that the velocity is a minimum when $\lambda = a$.

Solution:

Let $v=\mu\sqrt{\left(\frac{\lambda}{a}+\frac{a}{\lambda}\right)}$, where μ is some constant.

Then $v^2=\mu^2\left(\frac{\lambda}{a}+\frac{a}{\lambda}\right)=z$, say.

Now v is maximum or minimum according as v^2 *i.e.*, z is maximum or minimum.

Now $\frac{dz}{d\lambda}=\mu^2\left(\frac{1}{a}-\frac{a}{\lambda^2}\right)$. For a maximum or a minimum of z, we have $\frac{dz}{d\lambda}=0$ *i.e.*, $\frac{1}{a}-\frac{a}{\lambda^2}=0$ *i.e.*, $\lambda^2=a^2$ *i.e.*, $\lambda=a$, because $\lambda=-a$ is inadmissible.

Now $\frac{d^2a}{d\lambda^2}=\mu^2\left(\frac{2a}{\lambda^3}\right)$ = + ive when λ = a. Hence z or v is minimum when λ = a.

Example 27:

A can in the form of a closed right circular cylinder is to be made of sheet metal to have capacity V. Find the height of the can and the diameter of its base so that the metal used may be minimum. **(Meerut, 1998 S)**

Solution:

Let h be the height and 2r be the diameter of the cylindrical can. Since the capacity of the can is given, therefore its volume

$V=\pi r^2h=$ constant. ...(1)

The amount of metal used for the construction of the can is least when the whole surface, say S, is least. We have

S = area of the two circular faces + area of the curved surface

$=2\pi r^2+2\pi rh=2\pi r^2+2\pi r.\left(\frac{V}{\pi r^2}\right)$ from (1)

$=2\pi r^2+\left(\frac{2V}{r}\right)$.

For a maximum or minimum of S, we must have

$$\frac{dS}{dr}=\left(-\frac{2V}{r^2}\right)+4\pi r=0 \text{ i.e., } \pi r^3=V$$

i.e., $$r=\left(\frac{V}{2\pi}\right)^{1/3}.$$

Now $\frac{d^2S}{dr^2} = \left(\frac{4V}{r^3}\right) + 4\pi$ which is > 0 when $2\pi r^3 = V$.

Hence S is minimum when $2\pi r^3 = V$.

When $2\pi r^3 = V$, we have from (1) $2\pi r^3 = \pi r^2 h$ or $h = 2r$.

$\therefore$ the required height of the can $= 2\left(\frac{V}{2\pi}\right)^{1/3}$ = the diameter of the base of the can.

Example 28:

A person being in a boat a kilometers from the nearest point of the beach, wishes to reach as quickly as possible a point b kms. from that point along the shore. The ratio of his rate of walking to his rate of rowing is sec α. Prove that he should land at a distance b – a cot α from the place to be reached. **(Allahabad, 1990)**

Solution:

Let the initial position of the boat be P and the point of the beach nearest to P be A. Then PA = a. Suppose, the point to be reached is B. Then AB = b. Let the rate of rowing be u kms. per hour. Then the rate of rowing be u kms. per hour. Then the rate of walking = u sec α kms. per hour.

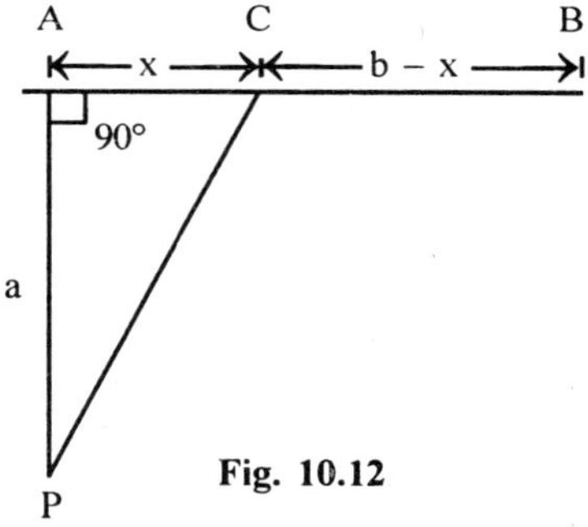

Suppose the person lands at C, where AC = x, and then walks to B. Thus, he covers the distance PC by boat and the distance CB by walking. If the total time taken to reach B is t hours, then

A C B
x
b – x
90°
a
P

Fig. 10.12

$$t = \frac{PC}{u} + \frac{CB}{u \sec \alpha} = \frac{\sqrt{(a^2 + x^2)}}{u} + \frac{(b - x)}{u \sec \alpha}.$$

We have $\frac{dt}{dx} = \frac{x}{u\sqrt{(a^2 + x^2)}} - \frac{1}{u \sec \alpha}$.

For a maximum or minimum of t, we have $\frac{dt}{dx} = 0$

i.e., $$\frac{x}{\sqrt{(a^2 + x^2)}} - \frac{1}{\sec \alpha} = 0$$

i.e., $$\frac{x^2}{a^2+x^2} = \cos^2 \alpha$$

i.e., $x^2 = \alpha^2 \cos^2 \alpha + x^2 \cos^2 \alpha$ *i.e.,* $x^2 \sin^2 \alpha = a^2 \cos^2 \alpha$

i.e., $x = a \cot \alpha.$

Now $$\frac{d^2t}{dx^2} = \frac{1}{u\sqrt{(a^2+x^2)}} - \frac{x^2}{u(a^2+x^2)^{3/2}}.$$

When $x = a \cot \alpha.$

$$\frac{d^2t}{dx^2} = \frac{1}{ua \operatorname{cosec} \alpha} - \frac{d^2 \cot^2 \alpha}{ua^3 \operatorname{cosec}^3 \alpha}$$

$$= \frac{1}{ua}\left[\sin \alpha - \cos^2 \alpha \sin \alpha\right] = \frac{1}{ua} \sin^3 \alpha = +\text{ive}.$$

Therefore, t is minimum when $x = a \cot \alpha$.

Thus, to reach B in the shortest possible time, the person should land at C where $AC = a \cot \alpha$ or $BC = b - a \cot \alpha$.

Example 29:

A figure consists of a semi-circle with a rectangle on its diameter. Given that the perimeter of the figure is 20 feet, find its dimensions in order that its area may be maximum.

Solution:

Let x be the breadth and y be the height of the rectangle. Then the diameter of the semi-circle is x. Therefore, the perimeter of the figure

$$= x + 2y + \frac{\pi x}{2} = 20.$$

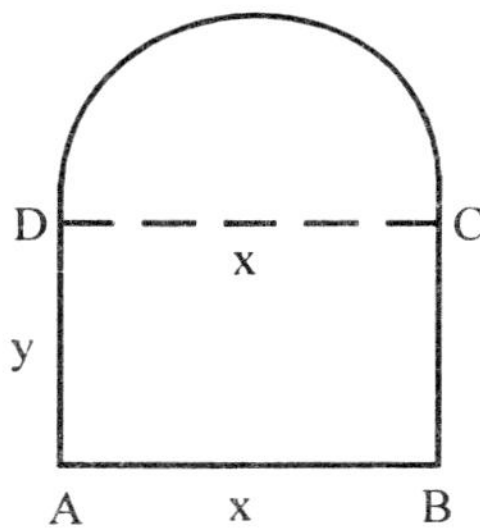

Fig. 10.13

Let A be the area of the figure.

Then $$A = xy + \frac{1}{2}\pi\left(\frac{x}{2}\right)^2$$

$$= x\left(10 - \frac{x}{2} - \frac{\pi x}{4}\right) + \frac{\pi x^2}{8}, \qquad \left[\because \text{from (1)}, y = 10 - \frac{x}{2} - \frac{\pi x}{4}\right]$$

$$= 10x - \frac{x^2}{2} - \frac{\pi x^2}{8}.$$

For a maximum or a minimum of A, we have

$$\frac{dA}{dx} = 10 - x - \frac{\pi x}{4} = 0$$

i.e., $$x\left(1 + \frac{\pi}{4}\right) = 10$$

i.e., $$x = \frac{40}{(\pi + 4)}.$$

Now $\frac{d^2A}{dx^2} = -1 - \frac{1}{4}\pi$, which is negative when $x = \frac{40}{(\pi + 4)}$.

Hence A is maximum when $x = \frac{40}{(\pi + 4)}$.

When $x = \frac{40}{(\pi + 4)}$, we get from (1),

$$y = 10 - \frac{20}{(\pi + 4)} - \frac{10\pi}{(\pi + 4)} = \frac{20}{(\pi + 4)}.$$

Hence the area of the figure is maximum when the radius of the semi-circle = the height of the rectangle $= \frac{20}{(\pi + 4)}$ feet.

Example 30:

In a submarine telegraph cable the speed of signalling varies as $x^2 \log \left(\frac{1}{x}\right)$, where x is the ratio of the radius of the core to that of the covering. Show that the greatest speed is attained when this ratio is $1 : \sqrt{e}$.

Solution:

Let S be the speed of signalling.

Then $S = \mu x^2 \log\left(\frac{1}{x}\right) = -\mu x^2 \log x$, where μ is a constant.

We have $\frac{dS}{dx} = -\mu\left\{2x \log x + x^2\left(\frac{1}{x}\right)\right\} = -\mu x\,(2 \log x + 1)$.

For a max. or a min. of S, we have $\frac{dS}{dx} = 0$

i.e., $x(2 \log x + 1) = 0$ *i.e.,* $x = 0$

or $$\log x = -\frac{1}{2}.$$